MW00717400

RISK AND INSURANCE IN CONSTRUCTION
SECOND EDITION

Those involved in construction have to cope with so much learning in their own discipline that they shun further involvement in subjects such as insurance and law which in themselves are so deeply and intensely complex. However, insurance and law are interwoven in the basic procedures used in the construction industry to undertake work, be it design or construction or supervision or operation or any combination of the foregoing. Furthermore, they both interact with the theory of Risk and the application of Risk Management. From the legal point, this interaction stems from the essence of the construction contract and its purpose, which is to allocate the risks, to which the project is exposed, between the parties. From an insurance aspect, risk forms the basis of insurability and premium calculation.

Insurance costs have escalated to become a major cost factor in any branch of the construction industry. Such escalation makes it essential for decision-makers within the industry to have a thorough understanding of the risks, liabilities and indemnities which play an important role in forming the appropriate relationship between those involved.

This thoroughly revised edition of Nael Bunni's successful *Insurance in Construction* with its new title 'Risk and Insurance in Construction' provides information on risk, construction law and construction insurance for those involved with all aspects of construction. The chapters on risk have been expanded to include recent developments in that area and provides further examples of events which could occur on what can be termed as the most risky human work activity: construction. New chapters are also added to deal with the insurance clauses of the many new standard forms of contract published in the recent years, including FIDIC's new suite of contracts published in September 1999, ICE's seventh edition of the civil engineering standard form of contract; and ICE's second edition of the design/build form.

Nael G. Bunni is a Chartered Engineer, Registered Chartered Arbitrator and Conciliator. He is Past President of the Association of Consulting Engineers of Ireland, Past Chairman of the Irish Branch of the Chartered Institute of Arbitrators, and Past President of the Chartered Institute of Arbitrators, London. At present, he is a member of the Board of Directors of the London Court of International Arbitration and member of various Standing Committees of the Commission on International Arbitration of the ICC, Paris, including the forum on ADR, the Construction Arbitration Section of the Forum on 'Arbitration and New Fields' and the working Party on Turnkey Transactions of the Commission on International Commercial Practice. He is a member of 'ICCA' and a Fellow of the Irish Academy of Engineering.

RISK AND INSURANCE IN CONSTRUCTION

Second edition

Nael G. Bunni

Spon Press
Taylor & Francis Group

LONDON AND NEW YORK

First published 1986
under the title
Construction Insurance
Second edition publised 2003
by Spon Press
2 Park Square, Milton Park Abingdon, Oxon OX14 4RN

Simultaneously published in the USA and Canada
by Spon Press
711 Third Ave, New York NY 10017

Spon Press is an imprint of the Taylor & Francis Group, an informa business

First issued in paperback 2011

© 2003 Nael G. Bunni

The right of Nael G. Bunni to be identified as the Author of this Work has been
asserted by him in accordance with the Copyright, Design and Patents Act 1988

Typeset in Sabon by Exe Valley Dataset Ltd, Exeter

All rights reserved. No part of this book may be reprinted or reproduced or
utilized in any form or by any electronic, mechanical, or other means, now
known or hereafter invented, including photocopying and recording, or in
any information storage or retrieval system, without permission in writing
from the publishers.

The publisher makes no representation, express or implied, with regard to the accuracy
of the information contained in this book and cannot accept any legal responsibility
or liability for any errors or omissions that may be made.

British Library Cataloguing in Publication Data
A catalogue record for this book is available from the British Library

Library of Congress Cataloging in Publication Data
Bunni, Nael G.
Risk and insurance in construction/Nael G. Bunni. – 2nd ed.
p. cm.
Includes bibliographical references and index.
1. Construction industry–Insurance–Law and legislation–Great Britain. 2. Risk
management–Great Britain. I. Title.

KD1641 .B86 2003
346.41´0865–dc21   2002011487

ISBN 13: 978-0-419-21380-2 (hbk)
ISBN 13: 978-0-415-51442-2 (pbk)

TO ANNE,
NADIA, LAYTH, SIOBHAIN,
LARA, LAYLA AND LYDIA

CONTENTS

List of figures   xii
List of tables   xiii
Biographical note   xiv
Foreword to the first edition   xv
Acknowledgements to the first edition   xvi
Acknowledgements to the second edition   xvii
Preface to the first edition   xix
Preface to the second edition   xxi

1   Interaction between construction, insurance and law   1

    A glimpse through history 1
    Interaction between insurance and law 7
    Construction contracts 7
    Construction and the law 12
    Development of insurance clauses in standard forms of
        contract 15
    Construction and society 17

2   Hazards and risks   26

    Instincts v. reasoned decisions 26
    Modern society and awareness of hazards 27
    A hazard or a risk? 28
    Risks in construction 32
    Analysis of hazards 34
    Risk assessment 34
    Risk management 36
    Hazards in construction 39
    Identification and classification of risks in construction 41
    Allocation of risks 47
    Risks inherent in major projects 48
    Assessment of risks in construction 49

3   The spectrum of hazard and risks in construction   52

Classification based on chronology 53
The spectrum of risks in construction 53
E.1.1 Risks associated with the feasibility stage 53
E.1.2 Risks associated with the design stage 61
E.2.1 Risks during construction associated with the site of the
*   project and its location 70*
E.2.2 Risks during construction associated with the technical
*   aspects of the project 81*
E.2.3 Risks during construction associated with Acts of Man 97
E.3.1 Risks Associated with the post-construction stage 114

4   The risks as classified in standard forms of construction contracts   130
Summary 138

5   Responsibility and liability in construction   139

Logic 139
Allocation of risk and responsibility 141
Definitions 142
Responsibility in construction 145
Liability emanating through construction 147
Developments in the laws of liability 148
Levels of liability 149
Liable or not? Developments in contract law 150
Developments in contract law: contractor's duties 164
Liable or not? Developments in the law of torts 166
Future developments 179

6   Indemnity and insurance   181

Indemnity 181
Indemnity through law 181
Indemnity through contract 184
Insurance as a contract of indemnity 184
Construction insurance 186
Characteristics of the construction contract 187
Definition of the word 'accident' 192
The insurability of risks 194
Uninsurable risks 195
Insurance policies required in construction 196
Property insurance 196
Liability insurance 197
Non-negligence insurance 198
Decennial insurance 201
Overlaps and gaps 204
Bonds and guarantees 205

7   The insurance scene   207

Structure of the insurance market 207
The broker 208
The insurer 213
The reinsurer 214
Premium calculations 215
Documentation 218
The insurers' departmentation 220

8   The insurance clauses in standard forms of contract   221

Liability and insurance clauses of the client/consultant Model
    Services Agreement (The White Book) 3rd edition 1998 222
The responsibility, liability, indemnity and insurance clauses of
    the FIDIC Form of Contract between Owner and Contractor
    (The Red Book) 4th edition 1992 224
Alternative one: the FIDIC and ICE insurance clauses modified 227
Clause 20 – Care of the Works and Employer's Risks 230
The FIDIC Red Book, 4th Edition, 1992 Reprint, Clause 20 233
ICE Conditions, 7th Edition – Measurement Version – Clause 20 234
Clause 21 – Insurance of the Works, etc. 245
The FIDIC Red Book, 4th Edition, 1992 Reprint, Sub-Clauses
    21.1 to 21.4 247
ICE Conditions, 7th Edition – Measurement Version – Clause 21 248
Clause 22 – Damage to Persons and Property – Indemnity 253
The FIDIC Red Book, 4th Edition, 1992 Reprint, Sub-Clauses
    22.1 to 22.3 255
ICE Conditions, 7th Edition – Measurement Version – Clause 22 256
Clauses 23 and 24 – Third Party Insurance – Injury to Workmen 257
The FIDIC Red Book, 4th Edition, 1992 Reprint, Sub-Clauses
    23.1 to 23.3 258
ICE Conditions, 7th Edition – Measurement Version – Clause 23 259
ICE Conditions, 7th Edition – Measurement Version – Clause 24 259
Clause 25 – General insurance requirements 261
The FIDIC Red Book, 4th Edition, 1992 Reprint, Sub-Clauses
    25.1 to 25.4 262
ICE Conditions, 7th Edition – Measurement Version – Clause 25 262
Clause 65 – Special risks 265
The FIDIC Red Book, 4th Edition, 1992 Reprint, Special Risks –
    Clause 65. Sub-Clauses 65.1 to 65.8 265
ICE Conditions, 7th Edition – Measurement Version – Clause 63 268

9   The insurance clauses in FIDIC's traditional forms of contract:
a proposed redraft   279

Proposed Model Clauses for a Contract for Works of Civil
    Engineering Construction 280

Notes on the proposed Model Clauses 287
Amendments required for an electrical and mechanical contract 295

10 The insurance clauses of the new 1999 FIDIC Forms of Contract   298

The New Green Book 299
Clauses 13, 14 and 6 of the Green Book 300
Analysis of Clauses 13 and 14 of the Green Book 303
The new major books: Red, Yellow and Silver 305
New provisions that might affect the risks which lead to loss and/or
*   damage, and insurance 306*
Differences between the three new FIDIC books in Clauses 17 to 19 310
Analysis of Clauses 17 to 19 of the New Red Book 312
Clauses 17 to 19 of the New Red Book – 1999 Edition:
*   17 Risk and responsibility 314*
Clauses 17 to 19 of the New Red Book – 1999 Edition:
*   18 Insurance 319*
Clauses 17 to 19 of the New Red Book – 1999 Edition:
*   19 Force Majeure 323*

11 The insurance clauses of the new 1999 FIDIC Forms of Contract:
    a proposed redraft   329

The proposed replacement of Clauses 17 to 19 of the New
*   Red Book 329*
The proposed replacement of Clauses 17 to 19 of the New
*   Yellow and Silver Books 338*

12 Insurances required under the FIDIC agreements   339

Contractors' All Risks insurance policy 340
Policy wording 341
Settlement of claims 354
Insured perils 355
Public liability insurance policy 356
Employer's liability insurance policy 357

13 Professional indemnity insurance   358

Indemnity through insurance 359
Policy wording 360
The Insuring Clause 361
The Schedule 362
The exceptions/exclusions 367
Conditions 368
Memoranda 369
Signature Clause 372
Other aspects of the professional indemnity insurance Cover 372

Professional indemnity insurance group schemes 374
Group schemes with retention of control 'Funded Group Scheme' 376

14 Alternative methods of insuring   379

The conventional method 379
Overlaps and gaps 380
Recent developments in construction, insurance and law 386
Alternative methods of insuring 387

Appendices
A   Questionnaire and proposal for Contractors' All Risks insurance   398
B   Contractors' All Risks Policy   400
C   Notification of Loss or Damage for Contractors' All Risks Insurance 406
D   Employer's Liability Insurance   408
E   Proposal for Professional Indemnity Insurance   413
F   Professional Indemnity Policy   418
G   Do you know your C.I.I.Q?   424
H   The Insurance Clauses of the FIDIC Conditions of Contract for
      Works of Civil Engineering Construction, Part I – General
      Conditions, 4th Edition 1987, reprinted 1992   425
I   The Insurance Clauses of the FIDIC Conditions of Contract for
      Electrical and Mechanical Works, Part I – General Conditions,
      3rd Edition 1987, reprinted 1988   428
J   The Insurance Clauses of the ICE Conditions of Contract,
      Measurement Version, 7th Edition, 1999   433
K   The Insurance Clauses of the ICE Conditions of Contract, Design
      and Construct, 2nd Edition, 2001   436
L   The Insurance Clauses of the FIDIC Conditions of Contract for
      Construction, 1st Edition, 1999   439
M   The Insurance Clauses of the IEI Conditions of Contract,
      3rd Edition, 1980, reprinted in 1990   445
N   The Insurance Clauses of the IEI Conditions of Contract,
      4th Edition, 1995, reprinted in 1998   450
O   Liability and Insurance Clauses of the Client/Consultant Model
      Services Agreement: (The White Book) 3rd Edition, 1998   455

List of cases   457
Index   461

FIGURES

1.1	Rules 230 and 231 of Hammurabi's Code	3
1.2	FIDIC Conditions of Contract for Works of Civil Engineering Construction	10
1.3	FIDIC Conditions of Contract for Electrical and Mechanical Works	11
1.4	Areas of the law affecting construction in legal systems based on common law	13
2.1	Hazards exist	35
2.2	Differing criteria for the classification of risks	45
2.3	Spectrum of risks in construction	46
2.4	Allocation of risks	47
3.1	Risks associated with the feasibility stage	54
3.2	Risks associated with the design stage	62
3.3	Risks during construction associated with the site of the project and its location	71
3.4	Building defects occurring or detected in the first ten years	72
3.5	Flotation of a pipeline due to rainfall	72
3.6	Risks during construction associated with the technical aspects of the project	82
3.7	Brittle fracture in a steel chimney due to oscillation	94
3.8	Defective design of a temporary girder	94
3.9	Risks during construction associated with Acts of Man	98
3.10	General arrangement of framing of the walkway and floor beam detail, Hyatt-Regency Hotel, Kansas City, USA	112
3.11	Risks associated with the post-construction stage	115
4.1	Risks of injury and/or damage	134
4.2	Risks resulting in economic and/or time loss	135
4.3	Risks as specified in FIDIC's Red Book and the ICE Form	136
5.1	Flow of risk into responsibility, liability and insurance indemnity	140
5.2	Logic in the interaction between law, insurance and construction	143
6.1	Classification of risks on the basis of insurability	196
6.2	Insurances which may be required on a construction project	197
6.3	Retained and insured risks	204
7.1	Distribution of risk between insurers and reinsurers	215
7.2	Distribution of risk and premium between insurers and reinsurers	216
8.1	Responsibilities and liabilities	228
8.2	Insurance requirements under FIDIC's Red Book, fourth edition 1992	229
8.3	The Insurance Scheme as in the 4th edition of the Old Red Book	270/271
8.4	Consequences of risks eventuating	274/275
9.1	Risk analysis chart	289
13.1	Details of the funded group scheme	378

TABLES

2.1 Percentage of fatal injuries to workers by kind of accident 1996/97 to 2000/0l p   31

2.2 Percentage of major injuries to employees by kind of accident 1996/97 to 2000/01 p   32

2.3 Percentage of over-3-day injuries to employees by kind of accident 1996/97 to 2000/0l p   32

2.4 Incidence rates of all injuries sustained by employees per 100,000 employees for each of the years 1981 and 1982, in the United Kingdom   40

2.5 Incidence rates of fatal and major injury sustained by employees per 100,000 employees, for 1981 and 1982, in the United Kingdom   41

2.6 Industries with the highest rates of fatal injuries to workers, 1998/99–2000/01 provisional combined, in the United Kingdom   41

2.7 Industries with the highest rates of major injuries to employees, 1998/99–2000/01 provisional combined, in the United Kingdom   42

2.8 An example of grading for probability of occurrence   43

2.9 An example of grading for severity of consequences of events   44

3.1 Distribution of failures of diesel and natural gas engines by fields of application   89

3.2 The principal failure areas in diesel and natural gas engines   89

3.3 Primary causes of damage in diesel and natural gas engines   90

3.4 List of loss to property from natural hazards within the period 1970 to 1980   124

3.5 List of insured losses from natural hazards after 1980 of US$ 1 billion and above   125

9.1 Definitions of the Special Risks   291

13.1 Professional indemnity insurance cost as a percentage of gross fees   363

13.2 Distribution of limit of indemnity by size of firm   365

13.3 Combined loss development, Canadian architects and engineers   374

13.4 Distribution of claims and alerts as analysed by the New Zealand Architects Co-operative Society Ltd. and the Consulting Engineers Advancement Society Inc.   378

13.5 With whom does communication fail?   378

BIOGRAPHICAL NOTE

Nael G. Bunni is a Chartered Engineer, Registered Chartered Arbitrator and Conciliator. He is Past Chairman of FIDIC's Standing Committee on Professional Liability and FIDIC's Task Committee on Construction, Insurance and Law. He is also Past President of the Association of Consulting Engineers of Ireland; Past President of the Chartered Institute of Arbitrators, London; and Past Chairman of its Irish Branch. At present, he is a member of the Board of Directors of the London Court of International Arbitration; and member of various Standing Committees of the Commission on International Arbitration of the ICC, Paris, including the Forum on ADR; the Construction Arbitration Section of the Forum on 'Arbitration and New Fields'; and the Working Party on Turnkey Transactions of the Commission on International Commercial Practice. He is also member or chairman of a number of technical committees in Ireland and internationally.

In 1996, Nael G. Bunni was appointed Visiting Professor at Trinity College, Dublin University, and he continues to hold that position. In 1999, he was elected a Fellow of the Irish Academy of Engineering and in March 2000, elected member of the International Council for Commercial Arbitration, ICCA. He has been involved in many civil and structural engineering projects in Ireland and abroad. His involvement in construction insurance and risk analysis goes back to 1965 when he received special training at the Munich Reinsurance Company, the Swiss Reinsurance Company and the Mercantile General and actively participated in the formation of the Iraq Reinsurance Company's pool for construction insurance. He has won a number of professional awards for his work in engineering design and in dispute resolution.

Nael G. Bunni has acted as conciliator, arbitrator or chairman of arbitral tribunals in numerous domestic and international disputes involving parties from over forty-five jurisdictions. These appointments involved disputes in many areas of construction including construction insurance and risk allocation. He is the author of a large number of technical papers and two other books: *Construction Insurance and the Irish Conditions of Contract*; and *The FIDIC Form of Contract*, now in its second edition. He has lectured extensively and has been invited to speak in many countries in Europe, Asia, Africa, North and South America and in New Zealand.

July 2002

FOREWORD
TO THE FIRST EDITION

The construction industry is currently bedevilled by a savage trinity of forces which are essentially misunderstandings; clients of the industry misunderstand its ability to deliver problem-free products; society as a whole misunderstands the role of insurance; legal tribunals misunderstand the special nature of the construction milieu. In the increasingly litigious consumer-led environment which has been evolving over the last two decades these three misunderstandings are diminishing the professional, commercial and physical resources available to pursue the construction process.

This book makes a significant contribution towards enlightening us on the nature of the misunderstandings. Those of us in the industry, and amongst its clients, who are neither lawyers nor insurance professionals must welcome a treatise on construction insurance by a practising engineer who not only works in the construction trinity as a structural designer of high calibre but also has a long experience in, and a deep understanding of, the insurance world.

What he has done is to demonstrate how heavily we depend on the precision of language when disputes require legal resolution, and at the same time he has demonstrated that language is savagely imprecise when it is ultimately tested. This is familiar ground in the area of contracts for the execution of construction works, where the relationship between the parties has degenerated from 'fundamentally trusting' to 'fundamentally adversarial'. Somehow we did not realise that the same problems lay in wait for insurance contracts, and their manifestation over the last two years has been a shock to the system.

Dr Bunni argues that it is time to sort out the relationships. I agree. This will require us to address seriously the misunderstandings and the excessive expectations that have become inherent in the relationships between the construction industry, its clients and the community at large. A large part of our work must be to place the imprecision of language into perspective, and to show that language has a special use in the construction industry such that it is irresponsible for legal tribunals to determine matters on the basis of 'strict construction'.

This book is a valuable contribution to our understanding of a most important area, and I am grateful for it.

PETER MILLER
Past President of F.I.D.I.C.

ACKNOWLEDGEMENTS TO THE
FIRST EDITION

The material incorporated in this book is a distillation of knowledge acquired since 1964 through work experience with a great number of people in the fields of construction, insurance and law. Of these people, too many to mention by name have helped in one way or another and to them I owe a great deal.

My special thanks, however, are due to Mr Charles O'Farrell, Mr Timothy Sullivan and Miss Josephine Murphy who provided the comments and criticisms to the many drafts produced.

I am also indebted to Dr Peter Miller who has kindly provided the Foreword and gave permission to quote from some of his papers. To all those who gave permission to quote from their own publications, I am also grateful. Special mention must be made to the following:

The Munich Reinsurance Company, Federal Republic of Germany;
Victor O. Schinnerer & Co. Inc., USA;
Fédération Internationale des Ingénieurs–Conseils, Switzerland;
The Institution of Civil Engineers, UK.

I am also grateful to Mr Peter Lanagan on behalf of the Publishers for his continuous encouragement, patience and also for the standard achieved in publication.

Finally, to my daughter Nadia for her perpetual and limitless devotion in typing and word processing the successive drafts of the text and illustrations, I owe a special gratitude.

NAEL G. BUNNI

ACKNOWLEDGEMENTS TO THE SECOND EDITION

Acknowledgement is given to the following for their kind permission to reproduce or use material:

Appendices H, I, L and O
>Clauses 20–25 of the FIDIC Conditions of Contract for Works of Civil Engineering Construction, Part I – General Conditions, 4th Edition 1987, reprinted 1992.
>Clauses 37–44 of the FIDIC Conditions of Contract for Electrical and Mechanical Works, Part I – General Conditions, 3rd Edition 1987, reprinted 1988.
>Clauses 17–19 of the FIDIC Conditions of Contract for Construction, 1st Edition, 1999
>Clauses 16–20 of the FIDIC Client/Consultant Model Services Agreement, 3rd Edition 1998.
>© FIDIC (International Federation of Consulting Engineers). Reproduced by kind permission.
>Copies of these conditions of contract may be obtained from FIDIC, PO Box 311, CH-1215 Geneva 15, Switzerland; www.fidic.org/bookshop

Appendices J and K
>Clauses 20–25 of the ICE Conditions of Contract, Measurement Version, 7th Edition, 1999
>Clauses 20–25 of the ICE Conditions of Contract, Design & Construct, 2nd Edition, 2001
>© Institution of Civil Engineers. Reproduced by kind permission.

Appendices M and N
>Clauses 20–25 and 65 of the IEI Conditions of Contract for Civil Engineering Works, 3rd Edition, 1980
>Clauses 20–25 and 65 of the IEI Conditions of Contract for Civil Engineering Works, 4th Edition, 1995
>© Institution of Engineers of Ireland, Dublin. Reproduced by kind permission.

Extracts have been taken from, and extensive references have been made to, *Schaden Spiegel* and other publications of Munich Reinsurance Company, as noted in the text.

ACKNOWLEDGEMENTS TO THE SECOND EDITION

I would also like to express my gratitude to those who helped in producing this edition of the book. In particular, to Anne my wife for her continued support, understanding and for being a sounding board for ideas and composition. I gratefully acknowledge the help extended by my daughter Siobhain DipDIT, MIDI, MCIArb, in producing the computerised art material for the charts. To my daughters Layla, BA, LL.B., and Lydia, LL.B., I owe a special tribute for their valued research and editorial assistance. To Mary Farrell, my secretary, I owe my thanks for her patiently executed secretarial work. To Mr Tony Harkness, Barrister at Law at the King's Inns, Dublin, I owe a debt of gratitude for his review of some parts of the book.

Finally, I wish to add a special word of thanks to Mr Tony Moore and Ms Sarah Kramer of Taylor & Francis Ltd, for their continual encouragement, patience and support prior and during the production of this book.

NAEL G. BUNNI
October 2002

PREFACE TO THE FIRST EDITION

Those involved in construction have to cope with so much learning in their own discipline that they shun further involvement in subjects such as insurance and law which in themselves are so deeply and intensely complex. However, insurance and law are interwoven in the basic procedures used in the construction industry to undertake work, be it design or construction or supervision or operation or any combinaion of the foregoing.

Insurance costs have recently escalated to become now a major cost factor in any branch of the construction industry. Such escalation makes it essential for decision-makers within the industry to have a thorough understanding of the risks and liabilities which play an important role in the division of responsibility between those involved.

The need to know more about construction law and construction insurance have prompted some lawyers and insurers to become specialists in these topics. But the pursuit of knowledge is hampered by a large gap in published material dealing with construction insurance for people whose discipline is not insurance. I hope that this work will make a contribution towards closing part of that gap in that it has the following features:

—It deals with the subject of construction insurance from its rationale to its day-to-day practice. It also deals with the important interaction with construction law and describes the present problems felt by the Construction Trinity of owner, design professional and contractor through recent developments in the liability issues.
—An analysis of the risks associated with construction is made and a spectrum of these risks is displayed through case histories and legal cases from all over the world.
—The many facets of construction insurance are dealt with separately and in detail. Thus, the contractors' all risks, public liability, employer's liability and professional indemnity insurances are explained. The manner in which they have been traditionally spliced in standard forms of contract to achieve a cover for the construction activities is described and the gaps and overlaps which inevitably form in such an arrangement are discussed. A proposal for possible modification is outlined to serve as an alternative logical and sequential method of describing the risks and liabilities and how they should be shared.
—Having always felt that professional indemnity has served in recent times as a contiguous medium through which vibrations from one sector of society are felt

xix

within other sectors, I decided that it deserves a special section. This section describes in detail the peculiar aspects of this type of insurance.

—Where insurance has become a failing expectation or a promise unfulfilled, there are usually some lessons which must be learned. These have prompted various thoughts on the matter of how insurance should be transacted. The thoughts are outlined.

I have deviated from convention by always using a lower case for the five words 'owner, employer, design professional, engineer and contractor', whatever the form and function they take. I have, however, bowed to convention in using 'he' and not 'he or she' wherever a reference is needed. I beg understanding from the purists.

NAEL G. BUNNI

PREFACE TO THE SECOND EDITION

At the time of publication of the first edition of this book, 1986, the topic of 'Risk' was not as fashionable as it later became or as it is nowadays. One reviewer of the first edition even questioned the logic of dealing with risk in a book on construction insurance. Others understood the link between risk and insurance and the interaction between these two topics and the contractual arrangements between the parties. They also understood that the purpose of a contract is to allocate the risks between the contracting parties and that from such allocation flows their responsibility and liability towards each other and towards third parties. Hence, if the liability is too great to be born by the contracting parties, then it could be shifted through an indemnity provision to an insurer, if it is insurable. However, some questioned the wisdom of not having the two words 'risk' and 'insurance' in the title of the book 'Construction Insurance'. This was a valid criticism and hence the change to the new title 'Risk and Insurance in Construction' adopted in this second edition.

Since 1986, there have been a number of important developments relating to various aspects of both risk and insurance in construction contracts. Amongst these are:

- the amendments published in 1991 to British Standard No. 4778 in relation to the topics of risk and risk management;
- the evolution of the definition of risk in the Australian/New Zealand Standards;
- the introduction of the 4th edition of FIDIC's Red Book;
- the introduction of the 7th edition of the ICE Form of Contract; and
- the publication of the new suite of contracts by FIDIC in 1999.

However, despite the importance of these developments, there has been very little published material on their effect on construction insurance. In fact, what has been published is largely by way of articles in insurance and legal periodicals and journals, which are mainly read by the specialist and rarely by the practising engineer.

It was therefore necessary to update the first edition of this book which now includes three new chapters dealing with:

(a) The risk as classified in the standard forms of construction contracts;
(b) The insurance clauses of the New 1999 FIDIC Forms of Contract; and
(c) The insurance clauses of the New 1999 FIDIC Forms of Contract – A proposed redraft.

Also new in this second edition, is the focus of discussion on the insurance clauses of the 4th edition, instead of the 3rd edition, of FIDIC's Red Book and on the 7th edition of the ICE civil engineering form of contract, instead of its 5th edition.

Chapter 2 'Hazards and Risks' was reconsidered in view of the amendments to British Standard No. 4778 referred to above.

Chapter 5 (Chapter 4 in the first edition) continues to deal with the topic of 'Responsibility and Liability in Construction'. However, it should be noted that statements and comments made in this book on the law have, of necessity, been of a generalised nature referring to general rules and principles, which must not be taken to mean that they apply without exception or qualification. In particular, these rules and principles may differ from one jurisdiction to another and from time to time. Therefore, specialist knowledge and advice must be sought in every specific situation.

Finally, new appendices have been added to reflect the new material that has been included in this edition.

NAEL G. BUNNI
October 2002

1

INTERACTION BETWEEN CONSTRUCTION, INSURANCE AND LAW

A glimpse through history

As man organised himself in settlements around the world, law and order became a necessity to achieve a proper balance between the freedom of choice of the individual and the control of this freedom for the protection of others. Hence, order and ultimately law prevailed and must endure if people are to be enabled to interact within a society devoid of conflict, struggle and friction.

It is significant to the engineer that this idea of the need for law is referred to by some as social engineering, thus expanding the horizon of engineering from a restrictive scene, involving applied science, to a much wider sphere encompassing the analysis and design of the society in which one wishes to live. This reference also brings science, with its powerful means of analysis, design and solution, to bear upon the concept of law.

While the concept of identifying law with a scientific process and applying scientific principles to the analysis of social and legal problems owes a lot to the French philosopher Comte (who in 1837 invented the term 'sociology' for such social studies), the genesis of sociology can be traced to the earliest records of human thought in the ancient civilisations of Assyria, Babylon, China, Egypt, India and Persia.

In most of these civilisations, as the concept of law became acceptable, it was found necessary to ensure that laws, when enacted, were not only enforceable but also enforced. The idea of a supreme power behind that concept was born and the law was attributed to the gods. Thus in Mesopotamia around 2000 BC it was believed that there existed three gods: Anu, the god of sky who issued decrees which commanded obedience as they emanated from supreme divinity; Enlil, the god of earth who executed the sentence of the gods on those who did not obey; and Ea, the god of water and wisdom. The law in Mesopotamia was therefore believed to have been handed down from the gods and was codified for the use of ordinary people as early as the year 2100 BC, by the Sumerian King Ur Nammu of Ur. The most famous of that era is Hammurabi's Code of 1760 BC.

Hammurabi was the sixth and best-known king of Babylon's first dynasty and his code is of special interest here because it contains the earliest available recorded rules of codified construction law. In all, there were 282 rules found inscribed on an imposing stone stele in cuneiform script.[1] The rules were divided into three sections: property law, family law and laws relating to retaliation and restitution. Part of the latter section, entitled 'On the Construction of Houses and of Ships', dealt with construction law and contained thirteen rules pertaining to remuneration and failures.[2] Five of these rules specified the standard required to be achieved in a building contract and prescribed penalties for those who had the misfortune not to comply with it. They were:

§229   If a builder builds a house for a man and does not make its construction firm and the house which he has built collapses and causes the death of the owner of the house that builder shall be put to death.

§230   If it causes the death of the son of the owner of the house they shall put to death a son of that builder.

§231   If it causes the death of a slave of the owner of the house he shall give to the owner of the house a slave of equal value.

§232   If it destroys property, he shall restore whatever it destroyed, and because he did not make the house which he built firm and it collapsed, he shall rebuild the house which collapsed at his own expense.

§233   If a builder builds a house for a man and does not make its construction meet the requirements and a wall falls in, that builder shall strengthen the wall at his own expense.

Figure 1.1 shows rules 230 and 231, two of the five rules mentioned above, written in the original cuneiform script.

The severe penalty imposed by these rules ensured that building work achieved the required standards of construction and safety and helped to ensure that houses were free from the defects resulting from bad design, materials or workmanship. The assurance that this would be so was based on the principle of 'an eye for an eye' in accordance with the law of that time, a principle that still exists today in some legal systems. However, there was little provision for restitution and a lot more retaliation in the rough justice of that era.

Although the current concept of construction insurance was unknown then, the notion of 'risk, responsibility, liability and indemnity'[3] was embodied in the spirit of those rules. It could therefore be said that the first systematic risk management process for the problem of defects in construction was devised at that time and although it was simple in its concept, it was nevertheless to the point.[4]

It is interesting, however, to note that the general principle of insurance of loss-sharing must have been realised even at that early stage of the development of social

---

1 Cuneiform script was one of the earliest writing systems to emerge. It was written on clay tablets and emerged in Sumer in Mesopotamia around 3000 BC, independently of the writing systems that emerged in Egypt and China.

2 *Lessons from Failures of Concrete Structures*, by Jacob Feld, American Concrete Institute and the Iowa State University Press (1964), page 9.

3 See Chapters 2, 3 and 4 below.

4 'Overview – Prudence, Principles and Practice', by David Elms, part of a book entitled *Owning the Future*, edited by Mr Elms and published by the Centre for Advanced Engineering, University of Canterbury, Christchurch, New Zealand, 1998.

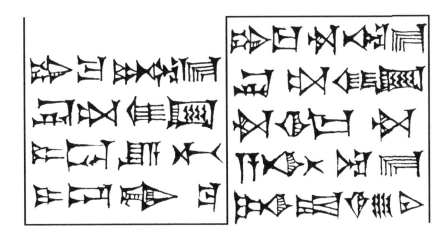

Figure 1.1 Rules 230 and 231 of Hammurabi's Code.

needs. Under section 1 of Hammurabi's Code, which dealt with property law, the principle was embodied in the following text of rule 23:[5]

§22   The man who is taken in the act of stealing will be condemned to death.
§23   But if the thief is not caught, the person that was the victim of the robbery will declare in the sight of god, under oath, what was stolen from him; afterwards the city and the governor of the same, in whose territory and boundary the robbery was committed, will return to him in full all that was stolen from him.

Furthermore, the Code contained rules relating to commercial risk, which entailed the provision for an investor taking on the risk of a voyage in exchange for a share of the profit.[6] Thus, the Babylonians knew at that time the loss-sharing principle through trading customs in what is referred to today as 'bottomry' in the English language. It is a term given to describe a maritime contract, under which a vessel owner borrowed money for a specific period of time to finance a voyage, on the security of the vessel. The loan was not repayable in case of total loss or destruction of the vessel. The interest rate which was applied to these transactions at that time was around 30%, a figure much higher than usually applied then to other loans, thus reflecting the higher risks involved.

Therefore, it could be said that with the above two sets of codified rules, Hammurabi started the whole process of risk management. The risks of injury, loss and damage in construction contracts were identified and allocated between the contracting parties. The commercial risks and the risks of building and owning property were also identified and allocated on a logical basis that remains with us till this day.

---

5 *Codigo de Hammurabi*, Edicion preparada por Federico Lara Peinado, Editora Nacional, Madrid, 1982.
6 Rules 100 and 101 of the Code.

Bottomry was also practised, but much later, by the Hindus in the sixth century BC and understood by the Greeks in the fourth century BC. It developed afterward into marine insurance in different parts of the world as early as the twelfth century AD. A clear definition of the principles of insurance is crystallised in the three-fold aims of the 1601 English Insurance Act 1601 of the English Parliament, viz.:

- to distribute the loss of few onto many others;
- to encourage those who are willing to take risks to do so with the promise of compensation; and
- to attract the young toward venture.

The relevant text from the preamble to the English Insurance Act of 1601 is reproduced here:[7]

> It hath been ... an usage amongst merchants both of the realm and of foreign nations when they make any great adventure (specially into remote parts) to give some consideration of money to other persons (which commonly are in no small number) to have from them assurance made of their goods ... it comes to pass that upon the loss or perishing of any ship there follows not the undoing of any man, but the loss alights rather easily upon many men than heavily upon few, and rather on those that adventure not, than on those that do adventure, whereby all merchants, especially the younger sort, are allured to venture more willingly and freely.

By that time, legal concepts had developed from the law of the gods to a three-tier hierarchy. At the top, the law of the gods had changed to the laws of God due to the evolution of religion. The second tier represented natural law or the law of reason and common sense and the third tier represented man-made law. The latter was subject to evolution from time to time and from place to place as long as it respected the boundaries laid down by the divine and natural laws.

The interaction between legal theories and the development of religion is beyond the scope of this book but the following three paragraphs will illumine the most relevant aspects of this trend as it became established.

As religion evolved, its influence on the social fabric of society brought about a change in attitude towards the principles of restitution and compensation. Thus, for example, Greek, Roman, Christian and Arabic civilisations produced Roman Law, Canon Law and Islamic Law, respectively, which developed and spread to different parts of the world, producing and influencing changes in the prevailing legal systems. Therefore, under Islamic Law, the penalty for certain offences was through payment of 'blood monies'. In the Anglo-Saxon legal system, the principle of compensation was applied in the eleventh century to all types of offencses to control and mitigate personal revenge and to provide instead compensation through payment of money, wherever possible.

---

7 *A Handbook to Marine Insurance*, by Victor Dover Publishers, in association with D. Farrow, 3rd edition, H.F. & G. Witherby, London, 1929.

In the western world, the second and third tiers of law became more established and developed in such a way that in 1610 the English jurist, Lord Coke, stated in *Bonham's Case*:[8]

> When an Act of Parliament is against right and reason or repugnant, or impossible to be performed, the common law will control it, and adjudge that Act to be void.

In another part of the world, Islamic Law, which is sometimes referred to as 'Shari'ah', is not merely a system of law, but a comprehensive code of behaviour embracing both private and public activities. Because, to its followers, it is an expression of divine inspiration, which ceased with the death of Prophet Mohammed in the year 632, Shari'ah Law is static and, subject to the elaboration of Muslim jurists, it does not accommodate any change in society. It is for this reason that society's needs for insurance did not flourish in jurisdictions where strict interpretation of that system of law applies.

Insurance, as we know it today, may be defined as the equitable financial contribution of many for the benefit of an individual who has suffered loss. This concept is more or less universally accepted and, as mentioned earlier, it did not develop until sometime in the fifteenth century with the advent of marine insurance as a result of the expanding world trade through sea-faring people. Marine insurance was followed by life insurance in 1583 and later by fire insurance after the Great Fire of London in 1666. Accident insurance, to which branch construction insurance belongs, did not evolve until the nineteenth century after the industrial revolution and the consequent expanding use of machinery. Construction insurance, as distinct from machinery insurance, did not come into being until the 1930s. It is, however, generally agreed that the period immediately after the Second World War marks an imaginary line after which this type of insurance thrived. The concentrated rebuilding programmes in the devastated areas of the world, accompanied by the rapid technological advances which took place in new materials and methods of construction, consolidated the principles of risk, responsibility, liability and indemnity in that area which gave rise, in one way or another, to a strong need for construction insurance.

With respect to law, however, two major legal systems with contrasting ideologies evolved in more recent history: the common law and the civil law. On the one hand, we have the common law system of England; Wales; jurisdictions within the Commonwealth; Ireland; and the United States of America, which has developed from a distillation of experience gained through accumulated judgments in these countries. On the other hand, we have the Civil Code, the legal system of most of the countries of Continental Europe, from whence it spread, either by choice or by imposition, to South America, parts of Africa, Asia and the Middle East. The Civil Code evolved from the law of ancient Rome, originally codified by Justinian and later in France, in 1804, by Napoleon. It is based on the judicial application of a certain legal code to a particular case by learned jurists and theorists, in conformity with logical and

---

8 *Bonham's Case* (1610) 8 Co. Rep. 114a.

systematic deduction. In this legal system, the law of evidence has certain rules of an inquisitorial type and the presiding judge plays a major role in finding the truth. Changes in the law occur usually when a law commission is formed by the authorities to examine and report on particular aspects of the system.

The common law, by contrast, is based on a piecemeal process of judicial decisions from precedent to precedent. It continues to develop in accordance with decisions given by judges which are based on interpretation of and reference to earlier case decisions. When it is felt necessary, new principles are introduced through judgments which, unless reversed by a higher court, become binding on judges of lower courts. In general terms, a very strict attitude prevails in common law jurisdictions towards the rule of precedent and only the highest court in the jurisdiction has the authority to review its earlier decisions and to depart from them if they are considered to be inappropriate for a new situation.

The law of evidence under the common law is based on the adversarial system and the judge has to decide, on the balance of probabilities, which of the competing versions of the evidence presented to him is the truth. Under the common law, a judgment given in one jurisdiction, while not forming a precedent in another, may be 'persuasive' in analysing the facts and the law and reaching a decision and influencing judgment. Thus, changes in the law may occur overnight without warning and the loser could find himself in the unfortunate situation that, had his case been heard a few days later, he would have been the winner. An example of this is the recurring changes that took place in the 1970s and 1980s in connection with the definition of the Period of Limitation and the extent of professional liability in relation to construction failures.

Under both legal systems, the principle of imposing a severe penalty to ensure the proper execution of a construction contract, as envisaged by Hammurabi, has been replaced by the principles of restitution and financial compensation, although the nature and extent of these principles differ from one system to the other. However, it is worthy of note that in recent years there has been a shift from civil liability in construction towards criminal liability where matters of health and safety are involved.[9]

In countries where the common law system prevails, the development of the law of negligence and the law of tort has been influenced to a large extent by what has taken place in the United Kingdom where a considerable development in these laws has occurred in recent years. It is interesting to note that Hammurabi's Code and his five rules were referred to in 1932 in the dissenting judgment of Lord Buckmaster in *Donoghue v. Stevenson*, which is considered to be one of the most important judgments, in the United Kingdom, where the law of negligence is concerned.[10] The reference is quoted later in Chapter 4, due to its relevance and also because this case is recognised as the turning point in the development of the law of negligence and in the definition of liability in countries where common law is used.

---

9 EU Health and Safety directive: (Design and Management) Regulations 1994, 'CDM', Health and Safety at Work Regulations. The directive has been introduced in various jurisdictions within the European Union.

10 *Donoghue v. Stevenson* [1932] AC 562, see page 168 below.

Interaction between insurance and law

Insurance developed and spread as a result of society's needs and demands. Thus, as mentioned above, marine insurance was followed by life insurance and shortly afterwards in the seventeenth century by fire insurance. Since then, human progress has been marked by developments in the insurance field and a variety of branches in the following classes of insurance sprang up, each forming a subject of its own: property insurance, machinery, loss of profits, engineering, motor, liability, aviation, credit, electronic equipment, off-shore structures and, most recently, space equipment. Each of these branches of insurance represents a milestone in the history of mankind.

However, the fact that insurance was itself available has influenced developments in other facets of society, forming dialogue between insurance and, for example, law or finance. This can be seen very clearly in the development of the law of negligence. The following extract, concluding a chapter on negligence, from *The Discipline of Law* by the great jurist and writer of the twentieth century Lord Denning, illustrates this point:[11]

> During this discussion I have tried to show you how much the law of negligence has been extended; especially in regard to the negligence of professional men. This extension would have been intolerable for all concerned – had it not been for insurance. The only way in which professional men can safeguard themselves – against ruinous liability – is by insurance. . . . The policy behind it all is that, when severe loss is suffered by any one singly, it should be borne, not by him alone, but be spread throughout the community at large. Nevertheless, the moral element does come in. The sufferer will not recover any damages from anyone except when it is that person's fault. It is only by retaining that moral element that society can be kept solvent.

It is doubtful if developments in the laws of contract and negligence would have occurred in this complicated and intensely commercial world of ours without the help of insurance, which has truly shaped some of the relationships in society.

In contrast, it is important to note that there is the view that insurance against tortious liability should be considered unacceptable because it permits the individual to escape from the financial responsibility of negligent acts.

Construction contracts

The simplest definition of a contract is 'A promise enforceable by law'. A slightly more elaborate definition is 'An agreement between two or more parties in which each party binds himself to do or forbear to do some act and each acquires the right to what the other promises'. Under common law, however, the promise has to be accompanied by 'consideration' which, in simple terms, means financial reward but it could also be any legally acceptable act agreed upon. Since it is between one party and at least one other

---

11 *The Discipline of Law* by the Rt. Hon. Lord Denning, Butterworths, (1979), London, page 280.

and since the contract is made with the mutual agreement of the parties, it is necessary to have 'an offer' and 'an acceptance'. If the contract or the promise is not performed, the remedy can be either the specific or actual performance of what was actually promised, or a financial compensation of one sort or another.[12] The enforcement of a contract is one of the most important sections of the legal system.

In any democratic society, the freedom of the individual to contract has been deemed the supreme facet of freedom since the beginning of social intercourse. The extensive growth of commercial activities in the nineteenth and twentieth centuries produced some abuse of this freedom, necessitating intervention by the State in the form of legislation to prevent monopoly and its harmful effects on society. This intervention, however, has not always been by way of legislation. In some cases, it has been initiated by specific groups of people interested in preserving the concept of fair play in a certain commercial activity. Others have done the same to prevent one-sided agreement in which the strong might impose their will on the weak. The result was the Standard Form of Contract consisting of a standardised set of conditions presented in an already printed form best suited to the particular use for which it was envisaged. In construction contracts, where the obligations and responsibilities of the contracting parties can be extremely complex but to a large extent remain unchanged from one project to another, the Standard Form was developed by the relevant professional institutions in order to help make the contracts fair, just and equitable. This development was extremely suitable for the tendering system usually adopted in construction contracts as it ensured a common basis for the comparison and evaluation of tenders.

In Europe, and more particularly in the United Kingdom and in Ireland, such forms were produced as early as the nineteenth century. The RIBA Form, which is used for building work contracts, was issued under the aegis of the Royal Institute of British Architects some time towards the end of the nineteenth century and that was followed by the RIAI Articles of Agreement and Schedule of Conditions of Building Contract, issued by the Royal Institute of the Architects of Ireland. In civil engineering works, the ICE form was first issued by the Institution of Civil Engineers in the United Kingdom in 1945. In civil engineering, various forms which were in use in the English language prior to the Second World War by different employers were fused, in England, into an agreed standard document. This was achieved in December 1945 by the Institution of Civil Engineers and the Federation of Civil Engineering Contractors. The document thereafter was known as the ICE Conditions of Contract. In January 1950, it was revised and issued with the added agreement of the Association of Consulting Engineers, UK. Five further revisions were made, the last of these in September 1999: the document which is in use at present is the seventh edition.

To the credit of those responsible for drafting the ICE document, many professional institutions all over the world based their conditions of contract on its text and made only minor amendments to accommodate differences in matters of law and nomenclature. Amongst these forms are two which will be referred to later in detail due to the relevance of their insurance clauses. These are the IEI Form and

---

12 *Law of Contract*, Cheshire and Fifoot's, 10th edition, Butterworths, London, 1981.

FIDIC's Red Book. The first is issued jointly by the Institution of Engineers of Ireland, the Association of Consulting Engineers of Ireland and the Civil Engineering Contractors Association and is in its 4th edition since 1995. The second document, dating back to 1987, is also in its 4th edition and is prepared by the International Federation of Consulting Engineers (FIDIC). Revisions were implemented in the ICE, IEI and FIDIC Conditions of Contract as a result of demand from one or more of the constituent organisations or from the construction industry. This demand was in response either to a need or to a legal decision given by a court of law in deciding a case based on one of the conditions of the document in question.

Originally, these documents were drafted in precise, legal language, which would be expected to remain unequivocal even when subjected to detailed and hostile scrutiny by astute legal minds. However, as revisions were incorporated, the language became more and more complicated and inscrutable. In certain cases, the number of words in each sentence grew to a level beyond the understanding of the average reader. As can be seen from Figure 1.2, drawn for the 3rd edition of FIDIC's Red Book, Conditions of Contract (International) for Works of Civil Engineering Construction, published in 1977, the number of words was at a level in excess of what a reasonably intelligent person is expected to readily understand, especially in cases where English is not the reader's mother-tongue. It can be seen that only eight sentences fall within zone A, where the number of words in each sentence is less than eighteen. Only twenty-seven sentences fall within zone B, where the number of words is between eighteen and twenty-eight. The majority of sentences, 221 (86%), fall beyond the twenty-eight-word zone. Theoretically, these sentences can only be easily and readily understood by 4% of the population (equivalent to an I.Q. of 130 and over). Figure 1.3 shows the number of words in sentences in the 2nd edition of FIDIC's Yellow Book, the conditions of contract for electrical and mechanical works, published in 1977. A similar observation can be drawn from these clauses. This is particularly important for those clauses that deal with the insurance aspects of the construction contract in view of the fact that insurance is a foreign subject to the basic training of most of those who deal with these clauses.

In the 1st edition of this book, it was suggested that if these documents are to be revised again, it would be important to give serious consideration to a complete change in the language used. It was thought that the time has come for all concerned to present the construction agreement in simple engineering language based on the realities of the industry and for the construction professions to clarify and maintain correct industry practices. It is gratifying to note that a change has taken place in the wording of the recent ICE and FIDIC standard forms of contract in the direction of simpler and clearer language. However, this change is still insufficient to produce simplicity and clarity in the insurance-related clauses in standard forms of contract and could be supplanted by further efforts in that direction. Similar graphs for the number of sentences in the insurance clauses of FIDIC's 4th edition of the Red Book and 3rd edition of the Yellow Book could be superimposed in Figures 1.2 and 1.3, respectively, showing the improvement in the language.

Since the publication of the 1st edition of this book, there has been further research into the psychology of language and it is now understood that ease of comprehension of a sentence depends not only on the number of words contained in that sentence, but also on other factors, including: the number of ideas presented;

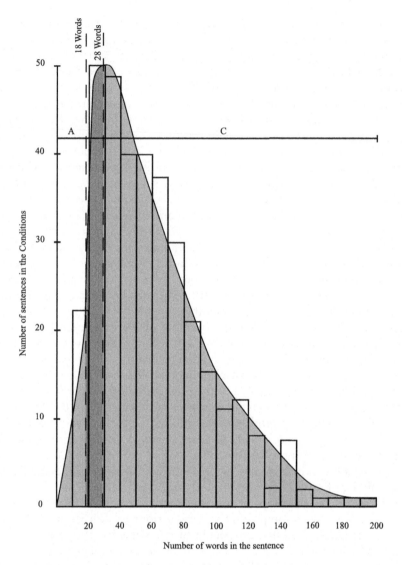

Figure 1.2 FIDIC Conditions of Contract for Works of Civil Engineering Construction.

the number of words which have more than one meaning; and structure and coherence of the text.[13]

Leaving aside the complexity of the language in these documents, the various revisions that have been made reflect the changes sought in matters of principle and philosophy from previous editions. In particular, they depart from previous editions in the allocation of risk to each of the parties involved in the construction contract.

13 *The Psychology of Language – from date to theory*, by Trevor Harley, 2nd edition, Psychology Press Ltd., UK, 2001, Chapters 6 and 11.

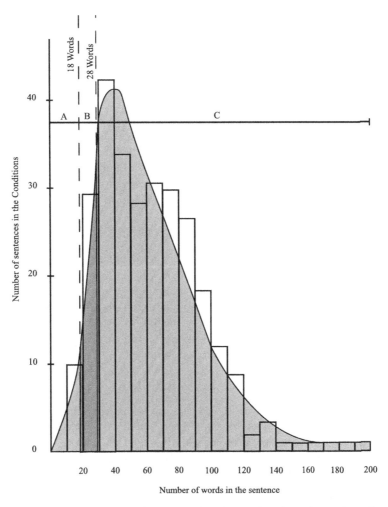

Figure 1.3 FIDIC Conditions of Contract for Electrical and Mechanical Works.

An example of the allocation of responsibilities and therefore of risks in the standard forms of contract can be best stated by reference to the reply to criticisms made of the 5th edition of the ICE conditions of contract after its publication in 1973. The reply was given by Sir William Harris, Chairman of the Joint Contracts Committee, and David Gardam QC, Legal adviser.[14] They explained that the principles of risk allocation they sought to apply in that form of contract were as follows:[15]

> It is a function of a contract to define upon whom the various risks of an enterprise shall fall, and it was decided that the Contractor should only

14 Sir William Harris, Chairman of the Joint Contracts Committee (JCC) responsible for the revision of the ICE documents to the 5th edition.
15 'Clearing the Critics' Confusion', *New Civil Engineer*, 20 December 1973, London, page 33.

price for those risks which an experienced contractor could reasonably be expected to foresee at the time of tender. . . . It is the right and the duty of the Employer to decide, and by his engineer to design and specify that which is to be done, and it is the Employer's duty to allow the Contractor to do that which is to be done without hindrance. It is the duty of the Contractor to do what the Contract requires to be done, as designed and specified by the Engineer, but, subject to any specific requirement in the contract, it is his right and duty to decide the manner in which he will do it. If there are to be exceptional cases where the Contractor is to decide what to do, or to design what is to be done, or where the Employer or the Engineer is to decide how the work is to be done, the contract must expressly provide for this and for the necessary financial consequences for the protection of the Contractor.

Unfortunately, the risk allocation applied above does not deal with all the possible risks and therefore does not go far enough to provide a comprehensive guide for allocation of risks that are (i) foreseeable in general but not in specific terms, (ii) unforeseeable but (at least partly) preventable, or (iii) unforeseeable but insurable.[16]

In this connection, it is worth noting that in respect of the philosophy of the construction contract, it is important to appreciate that besides setting out the scope and cost of the project to be constructed, the purpose of a construction contract is to allocate the risks to which that project is exposed; and to provide a clear statement as to how these risks are to be dealt with and managed.

Construction and the law

The interaction between construction and the law stems from the activity generated by the construction process. It involves matters related to legal concepts that reach far beyond the law of contract. Professionals involved in construction must realise that ignorance of the law is not only a handicap, but is also no defence. Therefore, a certain minimum basic knowledge of the laws governing the areas of their professional involvement is necessary.

To start with, however, it is important to set out two propositions that are vital to this discussion. The first is that, in general terms, we are only dealing with civil wrong acts as distinct from those that are likely to be followed by criminal legal proceedings. The second proposition is that the principles of the law that apply to construction are similar in almost all jurisdictions irrespective of the legal system that applies in a particular jurisdiction. Some authors and commentators even refer to the body of law that applies to construction as 'construction law'.[17] In jurisdictions within the common law group, besides common law, legislation and equity form two integral parts of the whole legal system. Figure 1.4 diagrammatically shows these areas.

In a construction contract one may encounter all of these three areas of the law. For example, a person may be in breach of a statutory duty if he either does not follow or incorrectly follows the legislation of the jurisdiction.

16 See also note 14 to Chapter 2 (the Grove Report, paragraph 6.1)
17 'Moving Toward a Construction Lex Mercatoria: A Lex Constructionis', by Charles Molineaux, *Journal of International Arbitration*, 1997.

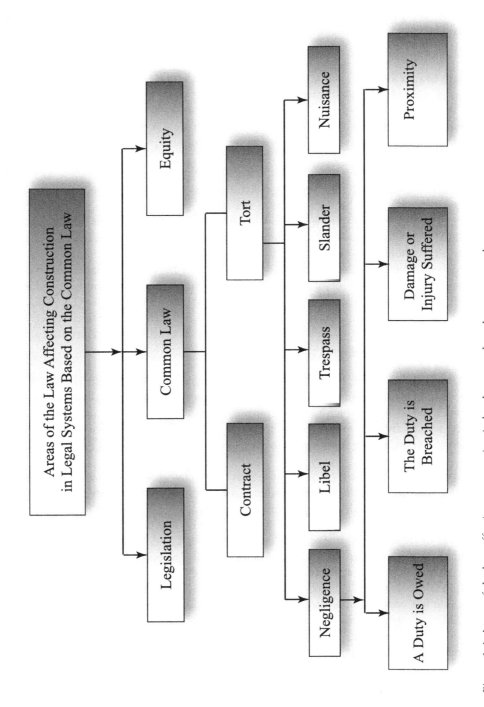

Figure 1.4 Areas of the law affecting construction in legal systems based on common law.

Under common law, he can be in breach of either the law of contract or the law of torts. Under the former, if the contract or the promise given is broken, he will find himself in breach of contract. Under the latter, he will find himself in breach of the law of torts and therefore subject to tortious liability if the four following elements are established:

- duty of care is owed, requiring conformity to a certain standard of conduct for the protection of others against exposure to risk;
- a breach of that duty has been committed;
- damage or injury is suffered as a result of that breach; and
- a proximate connection exists between the conduct in question and the resulting damage.

Thus, torts are essentially civil wrongs dictating no fixed measure of response. A civil wrong can simply be defined as: 'a breach of a legal duty which affects the interests of an individual to a degree which the law regards as sufficient to allow that individual to complain on his or her own account rather than as a representative of society as a whole.'[18] To distinguish tort from contract, one should focus on two main aspects: First, tortious duties are primarily fixed by law, whereas contractual duties are based on consent of the contracting parties. Therefore, contract is concerned with voluntary obligations while tort is concerned with involuntary obligations. Second, tortious duties are owed to persons generally, whereas contractual duties are undertaken towards specific person(s). The following definition is a simple, but encompassing one:[19]

> A tort may therefore be defined as: An unlawful act arising primarily from operation of law and not from breach of agreement between parties, the typical remedy for which is an action for unliquidated damages; and which is not exclusively a breach of contract, or exclusively a breach of trust or other equitable obligation, or exclusively a crime.

A voluntary assumption of risk by the party suffering injury, or damage, or contributory negligence on his behalf, would either negate or lessen liability. There is also a duty imposed on that party to mitigate the loss.

In construction, it is important to note that there is an interaction between the principles of contract and tort. Two situations must be considered. The first is where a claimant is a party to a contract but wishes to utilise a longer period of limitation than that available under the contract but permitted under the law of tort. In this situation, the courts might be willing to accept concurrent liability depending on the applicable law and provided there is nothing in the contract which excludes liability in tort. The second situation arises where there is no contractual relationship between claimant and defendant, but the damage results from the defendant's wrongful performance of a contract to which the claimant is not a party.

---

18 *Clerk & Lindsell on Torts*, 18th edition, section 1–01, Sweet & Maxwell, 2000.
19 *Law of Torts* by J.G.M. Tyas, 4th edition, an M & E Handbook published by Pitman Publishing Ltd, London, 1982.

The third area of the law, in common law jurisdictions, that one may encounter is equity, and so one must consider the rules of equity to find out if one is within the boundaries of justice. The idea of equity stems from the concept that justice is not always achieved by simply being within the bounds of legal acceptability. It evolved to correct the rigour of law when applied to individual cases, which may not conform to the generally applicable legal set of rules and regulations. Thus, in a matter concerning construction, one of the equity rules demands that 'Regard must be paid to the intent and not the form', which means that where the parties to a contract include a term to the effect that on a breach of contract a fixed sum will be payable, and where equity is satisfied that the fixed sum is meant as a penalty and not a reasonable measure of the loss which the breach would entail, then equity will not allow the penalty to be enforced. Also falling under the rules of equity in the context of construction law are specific performance, injunctions and their modes of operation.

Sometimes, the rules of equity are referred to as the rules of natural justice. It is however, arguable whether this reference is accurate or metaphoric insofar as it is debatable whether nature can always be termed as just.

Development of insurance clauses in standard forms of contract

As distinct from engineering insurance, which began in England with the industrial revolution around the middle of the nineteenth century, the necessity to insure various aspects of a construction project during its period of construction was recognised when the standard forms of contract were developed. The earliest contractors' all risks insurance requirement appeared in individual civil engineering contracts as early as 1929 for the construction of the Lambeth Bridge over the Thames in London. In Germany, this type of insurance was introduced in 1934 using terms and conditions derived from erection all risks insurance developed for erection and testing of industrial facilities, which had been launched in 1924.[20]

A standard form used in 1935 by the Electricity Supply Board of Ireland reads as follows:[21]

Clause 26-Insurance
The Contractor shall insure with a Company previously approved by the Board in writing such plant and materials as may for the time being be upon the site and shall keep them insured against destruction or damage for the whole value of such plant and materials until the completion of the works. And he shall, from time to time, when so required by the Engineer, produce the policy and the receipts for the premium for inspection. All monies received under such policies shall be applied in or towards the reconstruction or replacement of the plant and materials destroyed or damaged, but this provision shall not affect the Contractor's liabilities under the Contract.

---

20 *100 Years of Engineering Insurances at Munich Re*, a publication of the Munich Reinsurance Company, Central Division: Corporate Communications, Germany, May 2000.

21 General Conditions of Contract for Works of Civil Engineering Construction issued by the Electricity Supply Board, 1935.

The insurance requirement as set out in the aforementioned clause is limited to material and plant and it is not clear whether it was intended that the insurance should cease upon incorporation of the material in the works.

After the Second World War, the responsibilities and liabilities of the contracting parties in construction contracts increased in extent and in value. Clients, who in many cases were banks and financial institutions, found it imperative to cover their liabilities through insurance. Hence, the 1st edition of the ICE form of contract, issued in 1945, highlighted the importance of insurance by incorporating clauses which remained in force until 1973 when the 5th edition was issued incorporating a revision of the insurance clauses.[22] This revision of the clauses took place to allow for the developments which had occurred in the insurance markets of the world during the intervening period since 1945 and to cater for the technological advancement and the appearance of new construction materials and methods which emerged during this period. These developments created two effects: the first was that the new materials, methods and technology created new sets of risks and remedies which had to be recognised and allocated to one or more of the contracting parties; the second was that the insurers, on their part, varied their insurance policies in accordance with these developments, thus creating a significant difference between the requirements of the conditions of contract and what the insurance market was prepared to insure.

Thus, the 5th edition of the ICE standard form of contract included revised insurance clauses which achieve a certain harmony between the need to insure and the perils that the market was prepared to cover in its standard policies. Some revisions to these clauses, with significant improvements, were made when the 6th and 7th editions were published in 1991 and 1999, respectively. However, there still exist a number of gaps in the insurance cover as sought by the document and a number of anomalies have been left without solution. The 3rd edition of FIDIC's Red Book followed a number of the changes made in the ICE 5th edition, but not all of them. However, extensive revisions were subsequently made in the 4th edition, which was published in 1987 (see Chapter 8 of this book).

In September 1999, FIDIC produced a new suite of contract forms, three for large contracts and one for a small contract of around US$500,000 in value. The three main contracts, the New Red Book, the New Yellow Book and the Silver Book, included similar insurance clauses which differed considerably from those of the 4th edition of the Red Book, 1987. An analysis of these new clauses is given in Chapter 10.

The ICE family of contracts also includes the Conditions of Contract for Design and Construct, 2nd edition, September 2001; the Conditions of Contract for Minor Works, April 2001, where the value of the Works does not exceed £500,000; and the Engineering and Construction Contract, 2nd edition, 1995.

In Ireland, the insurance clauses of the 3rd edition of the General Conditions of Contract of the Institution of Engineers of Ireland, 1980, unlike the other clauses, deviated extensively from those of the 5th edition of the ICE. Further changes

---

22 The Institution of Civil Engineers Form of Contract issued in 1945, 1st edition, Great George Street, Westminster, London.

were made in the 1990–reprint of the 3rd edition resulting in a considerable deviation from its previous editions and from the 5th edition of the ICE form. These changes were intended to rectify certain omissions and anomalies referred to in Chapter 8.

For building work, the Standard Forms of Contract existed in the United Kingdom and in Ireland since the last quarter of the 19th century. In the United Kingdom, the Standard Form has been known since 1903 as the RIBA Contract and, in Ireland, as the RIAI Schedule of Agreement. However, the former is more commonly referred to as the Joint Contracts Tribunal, 'The JCT Contract', and its latest edition in the traditional form is the 1980 edition. A Design & Build version of this form was published in 1981 and a Management form in 1987.[23] The earlier format of the insurance clauses in these documents required the cover provided by the contractor to be 'Fire Extended Cover'. This type of insurance cover gives a very limited protection for a construction project and includes only fire, lightning, explosion, riot and civil commotion, earthquake, storm, flood and bursting or overflowing of water tanks, apparatus and pipes. Later editions were extensively revised to take account of the need for a more comprehensive insurance cover similar to that required by the civil engineering forms of contract.

Construction and society

Construction has existed since time beyond trace. As one of man's basic needs, protection from inclement weather and enemies necessitated the construction of shelter. Construction materials and skills developed very slowly and some of the earlier methods of construction using interwoven palm tree branches, stems, reeds, rushes and mud have survived to this day for over 6,000 years. Modern reed dwellings can still be seen in the marshes of southern Iraq.

As construction materials, timber, masonry and bitumen followed closely behind but it was not until relatively recently, just over 200 years, that metal and concrete were used.

The need to apply science in the construction design process became apparent after the collapse of many iron structures and it is perhaps that which led to the definition of engineering as the art of applying science to the optimum conversion of the resources of nature to benefit mankind. It may have also been the beginning of the divergence in construction between architecture and engineering as crystallised in Le Corbusier's statement:

> The engineer, inspired by the law of economy and led by mathematical calculation, puts us in accord with the laws of the universe. He achieves harmony. The architect, by his arrangement of forms, achieves an order which is a pure creation of his spirit . . . it is there that we experience beauty.

However, the addition of the term 'civil' to engineering owes its origin to 1716 in France with the formation of the Corps des Ponts et Chaussées, 'Bridges and

---

23 There are amendments to these forms of contract issued from time to time up to April 1998.

Highways Corps', out of which grew the Ecole des Ponts et Chaussées in 1747. The word 'civil' was added to distinguish military from civil contrivance.[24]

As distinct from building construction, civil engineering construction such as in canals, bridges and other public works was initiated by engineers with financial backing from private sources. The engineer acted originally not only as a designer but also as an organiser of the construction process, and purchased materials, employed labour and supervised the construction. He assumed part of the risk in the construction process, but those who financed the project assumed the financial risk. In the United Kingdom, the Smeatonian Society was founded in 1771 to bring together experienced engineers, entrepreneurs and lawyers to promote the building of large public works and to secure the parliamentary powers necessary to execute their schemes. Engineers and lawyers worked together for the good of society.

Social and technological changes in the past two centuries brought public works within the scope of government, the civil engineer into two distinct and separate roles (that of the consulting engineer with design and supervision duties and that of the contractor with supply of materials and construction duties), the lawyer into the role of the adviser and many other disciplines into the realm of construction. Banking and financial institutions became involved in construction as the cost of projects increased beyond the capacity of individuals and even governments.

In many aspects of everyday tasks, those involved in construction carry out their duties faced with conflicting requirements. Thus, quality, safety, cost, aesthetics, performance, durability, speed and available time have all to be considered and a balance must be struck between what is to be considered as priority matter and what is not. Striking a balance is the function of the decision-maker.

Problems begin to arise when matters go wrong, and society's attitude towards the design professional is one of double standards. When design professionals are briefed by an owner, they are asked to apply 'appropriate technology': a technology which is tempered to be practical, resulting in action which can be afforded by the owner in fiscal and/or moral terms and thus achieve maximum economy within the strict constraints which normally apply.[25] Yet, when matters go wrong with that appropriate technology, the professional designer is loaded with infinite responsibilities and with the accusation that he should have applied 'available technology' and should have tendered advice encompassing the maximum available wisdom, regardless of practicability and cost.

The past sixty years have witnessed tremendous achievements for mankind in science and particularly in the field of computers. The advent of computers opened a door into what could only have been fantasy to previous generations. Man went to the moon and landed exactly on the spot picked. He came back and is now planning to build cities in space. Society has now become accustomed to precise answers to problems through the input of the manufacturing industry. In the various fields of construction, computers have helped in providing a very powerful tool not only in

---

24 In the English language, the words 'engineer' and 'ingenious' were derived from same Latin root, *ingenerare* meaning 'to create', and the early English verb 'engine' meant 'to contrive'. In other languages such as Arabic, architecture is expressed as part of engineering.

25 'The Future', by Peter O. Miller, Presidential Address, FIDIC Annual Conference, Florence, Italy, June 1983.

day-to-day tasks but also in overall control and management. In particular, engineering problems which could not be solved before the advent of computers, and could only be solved by experienced and highly skilled professionals, have now become simple and easy to tackle using computers.[26]

Engineering moved more and more towards science, and its basic understanding and definition have become obscure. Civil engineering, which was described in the 1828 charter of the Institution of Civil Engineers as

> the art of directing the great sources of power in nature for the use and convenience of man, . . .,

has indulged in the science of analysis. Even its students have become more orientated towards how to write a computer program to analyse a simple structural problem than towards the feeling of how the structure behaves under load. In the drive towards mastering this tool, basic and essential aspects of engineering knowledge have been curtailed. The concept that engineering is an art where experience counts very highly and where for a single problem there are many solutions subject to circumstances has disappeared into the background. Unfortunately, society has followed this trend by expecting from engineering results similar to those expected from science.

Society has in the past twenty years moved very rapidly in the direction of an expectation that the cost of professional services can be based on competition in price.[27] Usually, intelligent administrations have been forced or persuaded to accept this concept and to abandon any thoughts that if the price is cut, then something must have been sacrificed. Too often it is the quality of the service that is sacrificed. In many instances, where such price competition was allowed in certain parts of the world, the result was defensive design at the expense of the owner and society.[28]

What is meant by defensive design can be explained probably by citing the events in one particular project when, at the request of the owner, the structural steelwork had to be designed in the very short period of two weeks and what should have been 900 tonnes of steelwork ended up weighing nearly 2,000 tonnes. There was no available time to consider any refinement of analysis or design. The design costs were much smaller than normal. One chief engineer, two engineers and six draughtsmen worked on the project and the owner saved £25,000 in fees. However, the additional costs incurred by the owner in the end, in terms of construction costs, were nearly £1 million.

In a paper read at the 1983 FIDIC Conference in Florence, Mr Norman Westerberg, a Finnish consulting engineer, quoted a statement by Rear-Admiral D.C. Iselin, the Commander of the US Navy's Naval Facilities Engineering Command, to the US House Sub-committee on Military Construction in 1978. The Commander stated:

---

26 'The Irish Scene in Structural Steelwork', by Nael G. Bunni, Series of lectures on structural steelwork, University College Dublin in association with British Steel Corporation, February, 1985.

27 'FIDIC's View of Design Liability', by Nael G. Bunni, IABSE Colloquium on Design Liability, Cambridge, UK, July 1984.

28 'Is Insurance the Answer?', by Nael G. Bunni, Conference on Liability for Development, Design and Construction of Buildings, the Royal Institute of the Architects of Ireland, October, 1983.

In a typical project, we find that outfitting, operating, maintenance, and repair costs represent 56% of the life cycle cost, construction costs represent 42%, and the design cost represents approximately 2%. This relatively modest cost, notwithstanding the A-E design effort, has critical influence on both the 42% for construction costs, and especially the 56% for operations, maintenance, and repair. It is vitally important that we get the highest possible technical quality in the design effort. In my professional opinion, any proposal which seeks to reap a near term saving by reduction in design costs, but which increases the risk of diminished technical quality of the design effort, is shortsighted in the extreme. We will live with the cost impacts of that diminished technical quality for the full economic life of the facility: this concern is a cornerstone on my opposition to price competition.

This compulsion towards price competition seems to have stemmed from a number of factors, the most important of which are the following:

1 Erosion of the position of a professional and a breakdown in the relationship of trust

There has been a marked erosion of the position of a professional within the community resultant from a breakdown in the traditional relationship of trust between him and his client. In construction-related professions, this breakdown is happening due to financial considerations, litigious tendencies and insensitive legal pressures. The breakdown is also extending to the relationship with the contractor.

In 1983, Dr Peter Miller, then President of the International Federation of Consulting Engineers, FIDIC, wrote:[29]

> The traditional relationship between the advising professions and communities was one of great trust – not necessarily absolute trust, and not necessarily a perfect relationship (one has only to read Dickens' satirical treatment of the legal professional to realise that there were some problems in the 19th century) – but one of great trust nevertheless. Without realising it precisely, communities and their advising professionals developed a subtle social contract in which the professionals traded acceptance of ethical constraints which placed concern for community paramount in exchange for the right of self-regulation. In comparatively recent times there is some evidence that this contract is deemed to be broken by both sides. Communities appear to be developing a view that the professions are dominated by self-interest, rather than community interest. Professionals believe the community is intruding somewhat recklessly into the self-regulation concept. There is a degree of adversarial behaviour intruding into the trust.

---

29 'The Future of the Professions', by Peter O. Miller, paper read at the University of Sydney, October, 1983.

2 Inability to appreciate the difference in quality

Construction artefacts are different from those manufactured to which society has become accustomed. One of the main differences focuses on the difficulty in construction projects of assessing the quality of the completed article when the decision to 'purchase' is made, see pages 189 and 190. When the project is completed and the quality is assumed, again with difficulty and time, it is usually too late to carry out any large-scale corrective measures. Once again, Dr Peter Miller is quoted here, but from an earlier article in 1974, when as a guest he wrote in an editorial for the *Journal of the Institution of Engineers of Australia*:[30]

> The attack on the professions always centers around money. It has been a tradition of the professions that members compete not on the basis of money but on professional excellence. They maintain that there is a clear distinction between a professional activity where there is a free interchange of ideas between practitioners in the interests of the services that can be given to the society as a whole and the protectionism of the business world where new ideas are jealously guarded. It is this tradition which has enabled them to retain sufficient control over their working environment to pursue professional excellence. The tradition has led to scales of professional fees fixed by the professions at levels which will enable existing skills to be maintained and new skills developed. The professions have come to regard it as their right to control their working environment in this way. The price the professionals pay to the community for this right is contained in the concept of adherence to Codes of Ethics. Some of these codes have very ancient roots. The legal code is derived from the Law of Hammurabi which originated some 38 centuries ago. The medical Hippocratic Oath goes back about 24 centuries. Engineers have had Codes of Ethics for much shorter times. The common theme which runs through these codes is the simple but extraordinarily demanding charge on the professional that he must put the interests of his client and the public before his own . . .

And in 1983 he wrote, commenting on his earlier article, as follows:[31]

> I have postulated since then that the thoughts I expressed in that short treatise were an expression of the terms and conditions of a subtle social contract, developed over hundreds of years, between society and what I have called 'the advising professions'. This contract, which is largely unrecognised, committed advising professionals to a strongly disciplined dedication to the interests of the community in exchange for the right to a high degree of self-regulation. Over the last 50 years or so the attitude of both parties suggests that the contract is being broken by both sides. The community sees the professions as succumbing to self-interest, and the professions see the community as interfering in appropriate self-regulation.

30 Editorial, by Peter O. Miller, *Journal of the Institution of Engineers of Australia*.
31 'Professionalism and Competition', by Peter O. Miller, a paper presented at the Australian Professional Consultants Council Workshop, November 1983.

He continued later to give his thoughts on price competition, which had just began to take hold at the time, and is now a serious problem:

Price competition, on the other hand, is relatively new and its popularity is still rising.

. . .

There is a concept in the newly established discipline of marketing which is called 'the life cycle of the product or service'. It is claimed to have been verified in qualitative terms, and is postulated to have four phases:

- Establishment: where the initial service is created by one or a few entrepreneurs who put in a lot of hard work to establish the type of service but gain little return.
- Growth: where the need for the service is recognised and the demand grows rapidly, but there are various levels of quality being provided.
- Maturity: marked by the entry of many competitors to the market with quality generally converging and a basic level of quality being clearly defined in the market place so that differentiation by price becomes a major criterion in the eyes of the purchaser, and because of these factors profitability begins to decline.
- Decline: when the availability of the service is diminishing.

This process has an appealing 'naturalness' about it, a 'birth-life-death' connection. The maturity phase of the cycle, however, bears clear evidence of willful destruction. The convergence of quality to a basic level clearly defined in the market place is effectively acceptance of the 'lowest common de-nominator'. This is 'unnatural'-quite the opposite of human experience. Progress clearly stems from refusal to accept the lowest common denominator. When-ever we accept the lowest common denominator for something we kill it, or at least disable it substantially.

. . .

Clearly, also the debilitating and ultimately destructive maturity phase described is characterised by the phenomenon 'price competition'. The validity of price competition depends upon the comparison of goods or services offered in the market as being of equal quality. The fundamental reason for the destructive nature of this phase lies in the fact that the market can no longer discern quality differences except insofar as the most basic criteria are concerned. Take the example of a refrigerator – the market makes basic judgments about the size, colour and configuration when comparing offered products but can make no judgments about technical quality. Refrigerators now sit on the lowest common denominator. We are struggling to maintain a production facility, despite heavy 'rationalization'. However, the worst feature of all is that refrigerators are no longer seen as a challenge, and no longer attract good minds and the curiosity of research and inventiveness. I submit they will only do so again when price competi-tion is no longer part of the scene.

. . .

It may be valid for a society seduced by the short term ease of 'disposable' living to take this view about some products, and to risk waking up one morning to find that some product producers have finally sickened of the mindless 'taking for granted' attitude of their customers and retired. I would argue that society cannot afford this kind of waste, which implies starting up the whole process again unless the product really can be forgone.

Dr. Miller should have probably added that when the quality deteriorates to an extent that such deterioration becomes readily apparent to the consumer, the low quality commodity will be rejected. The provider of a higher quality would have to emerge then, for the product or the service to survive.

3 Abuse by professionals and professional organisations

Generally speaking, the professions can point with some pride to their performance in subjugating their personal interests to those of their clients and the public. There have been, however, exceptions which were so glaring that they have forced disciplinary action, some of it exposed to public view, but by any statistical analysis these exceptions are of minor importance.[32]

It is much easier, however, to destroy than to build, and such abuse even when it is of minor statistical importance alters the thinking of those who are on the receiving end, and urges them towards losing their trust. Quoting from Alexander Pope:

But when to mischief mortals bend their will
How soon they find fit instruments of ill.

Interestingly, what was predicted in the first edition of this book regarding the adverse effects of price competition in professional fees is now happening, at least in the United Kingdom, as evidenced by the news item in a recent publication.[33] It was reported that consulting engineers were planning to press the government to reintroduce fee scales, fifteen years after these scales were outlawed. The Association of Consulting Engineers said that reinstating fee scales was vital, as the industry was on the brink of collapse in the face of mounting skill shortages and that unless fees can be set at higher levels than those determined by market forces, salaries will stay depressed and skill shortages will worsen.

It was also reported that the chairman of the Association had stated that firms across the consulting sector must increase earnings by between 50% and 60% and raise salaries by up to 100% if they are to recruit and retain the staff they need and that 'The situation is unsustainable and market failure is a very real danger. . . . The crisis point for industry is now. Crisis point for the UK is in 15 years' time, when

---

32 This passage is taken from Dr. Miller's paper referred to earlier in n. 25. However, in reference to the above by professionals and professional organisations, one may be able to quote from the Sunday Times of 10 February 1985, under the heading of *Law Society sued over way it handled complaint*; and from 'Lawyers can Seriously Damage Your Health', by Michael Joseph, Solicitor, 1984.

33 *New Civil Engineer*, Magazine of the Institution of Civil Engineers, London, 17 January 2002.

there won't be managers in the industry capable of running jobs. We are squandering our future.'

The same news item reported that the President of the Institution of Civil Engineers confirmed the direct link between low fees and the inability of firms to attract skilled staff. It was also confirmed that undercutting is still aggressively practised and in rare cases reaches levels as high as 70%.

It is easy therefore to see that the problems that existed then for the construction industry remain. They extend to involve the other two disciplines with which construction is closely associated and these are Law and Insurance, as summarised below:[34]

– The risks inherent in the construction process are not clearly understood by all concerned. Furthermore, they are neither fully identified nor always allocated in accordance with a satisfactory criterion.[35]

– There is a lack of understanding by some of those in the legal profession, the lawmakers, and others involved in the preparation of standard forms of contract, of the theory of 'risk' and the essential difference in the proper treatment of risks of loss and damage as compared with the other risks of economic and time loss.[36] There is even lack of understanding of the inherent characteristics of construction and the features of civil engineering projects which distinguish them from manufactured articles.[37] In this connection, the construction industry has not done enough to clarify its role nor to explain the practical boundaries beyond which the industry cannot tread.

– There is a lack of understanding by society at large of what the various sectors of the construction industry can be expected to do and be able to survive at the same time.

– The apparent wish of society is to lay blame on someone whenever something goes wrong irrespective of whether or not negligence had occurred. Thus the concept of liability is now being enlarged from simple liability to a strict one. The embryonic appearance of strict liability is an example of this trend. Such a trend must emanate from lack of understanding by one strong sector of society of the responsibilities and liabilities of the other. Thus, the relationship of trust (honest communication to the best of the knowledge of the professionals) which should exist between a professional and his client is being adversely affected by the financial considerations and the litigious tendencies of present-day society and the construction process is being distorted in reaction to unjust legal pressures.

– The existing lack of appreciation by some owners of the importance of total project lifetime costs as opposed to construction costs has resulted in dissatis-

34 'The International Situation', by Nael G. Bunni, Symposium on Liability, The Institution of Structural Engineers in association with the International Association for Bridge and Structural Engineering, London, May, 1985. Also from a discussion paper under preparation by FIDIC's Task Committee on Construction, Insurance and Law.

35 See page 48 below under the title 'Risks inherent in Major Projects', for a list of the inherent characteristics.

36 See for example 'FIDIC's New Suite of Contracts – Clauses 17 to 19: Risks, Responsibility, Liability, Insurance and Force Majeure', by Nael G. Bunni [2001] ICLR 523.

37 See page 189 later in this connection.

faction, which is further eroding the important relationship of trust between owners and professionals.

- The role of insurance in providing indemnity against liability is not understood by society in general and not even by other professionals. Principles of insurability must be maintained at all times. Furthermore, some sectors of the insurance market are plagued by irresponsible competition due to the lack of professional know-how of many participants and due to the manner in which benefits are sought without risk-sharing and risk acceptance. Irresponsible competition has led these participants to disregard statistical evidence and to grant ill-defined, over-extended coverage at such inadequate premiums that an erosion of security in the complex insurance/reinsurance system will ensue. The situation is further complicated by the manner in which certain types of construction-related insurances are transacted in practice and which leads to the unavailability of the statistics necessary to quantify the elements of risk. An insured may therefore find it difficult to be indemnified on demand in a major calamity.

- There seems to be a lack of understanding by society at large that there are many facets of competition, some of which are ugly and do lead to the detriment of the very aims which are sought in the name of competition. It is saddening that, whilst experience in one part of the world stands a witness against these ugly facets, other countries seem to head for the very same disastrous route. An example is the price competition in consultancy fees in the United States leading to abandonment of some of the duties attached to the consulting engineer simply because they cannot afford to retain these duties and remain in business.

- Society's expectations of science are confused with those of engineering and its expectations of insurance are slowly but surely departing from the principles of insurance. The principle of loss-sharing is changing into a perception that often an event causing a loss should yield a windfall.

2

HAZARDS AND RISKS

Instincts v. reasoned decisions

The great decisions of human life have as a rule far more to do with the instincts and other mysterious unconscious factors than with conscious will and well-meaning reasonableness. . . . Each of us carries his own life-form – an indeterminable form which cannot be superseded by any other.

Carl Gustav Jung, *Modern Man in Search of a Soul*, 1933.

In its basic form, life on earth is dependent on reflexes and instincts. At its most developed form, it revolves around the ability and quality of decision-making. Life in its basic form can be exemplified by the life cycle of a certain species of beetle, which seeks a mimosa tree and, ignoring all others, climbs to a branch near whose end it starts to cut a longitudinal groove with its mandible. It lays its eggs under the groove and since it knows that the larvae cannot survive in live wood it retreats a certain distance and starts cutting all around the branch. The branch dies and falls onto the ground thus providing the food for the next-generation beetle. The mimosa tree itself, with this process of annual pruning, lives for a century, but without pruning it can only live for a fraction of that period. In this reflex, approaching decision-making, one recognises the choice of a tree, the active carpentry work, the manner in which the work is carried out, the selection of the correct position of the groove and finally the cut.

Moving from animal instinct to human instinct, one recognises the process of decision-making coming into play. At the other end of the scale, the human brain can exemplify decision-making at its best with its capacity and ability for feeling, thought, deliberation, reasoning, application, imagination, invention and speech. In fact, for mankind, living is a decision-making process and the more complicated the pattern of life, the more complicated is the decision process. To achieve the desired result in each step taken is the goal to which one aspires, but the intention is not always pursued through a well-conceived plan and a designed strategy. On an ascending scale, but with unacceptable results, one may act without thinking of the consequences, or may give little thought and reasoning to actions which deserve more care and analysis, or may think but ineffectively. The result is sometimes unacceptable or intolerable, and always inferior.

If, for one reason or another, the decision made does not result in the anticipated outcome, either knowingly or unknowingly, a set of conditions is generated. It remains dormant with the potential for initiating an adverse event, commonly referred to as *'accident'*. The set of conditions, called a *'hazard'*, materialises into an event when an activating agent triggers the change, affecting not only the decision-maker, but also others around him. A hazard is not always a man-made event resulting from a decision taken and events may occur due to sets of conditions beyond the control of mankind. These occur in the form of natural events such as earthquakes, rainfall, floods, volcanic eruption, typhoons, etc., referred to as natural hazards.

Modern society and awareness of hazards

The day-to-day activities of some six billion human beings, each different in character, produce an incredible number of hazards and exposure to hazards which are as complex to identify or predict as is the large number of combinations of all the variable characteristics and the decisions taken. The inability of human beings to foresee the extent and nature of the effects they create with each decision leads to the two words 'hazard' and 'accident' being associated with chance. The dictionary definition of hazardous is 'dependent on chance'.

Awareness in modern society of the hazards around it has reached a significant level in recent times. This awareness occurred as a result of some major accidents in various parts of the world, the consequences of which had a powerful impact on the public mind. These accidents involved high technology industries in the oil, nuclear and chemical fields which fired the imagination of people as to what could happen if the desires of certain sectors of society went ahead unchecked. In more recent times, radio and television, by ease of communication, help to make people dramatically aware of hazards and their effects on people. Society, in general, does not place much significance on single-death incidents whereas it does on potential events in which hundreds of people might die.

The definition of hazard is given in British Standard BS No. 4778 as

> A situation that could occur during the lifetime of a product, system or plant that has the potential for human injury, damage to property, damage to the environment, or economic loss.[1]

Hazards are classified in the same British Standard into four categories:

(a) negligible;
(b) controlled or marginal;
(c) critical;
(d) catastrophic.

---

1 An extract from British Standard BS 4778: Part 3; Availability, reliability and maintain-ability terms. Section 3.1 Guide to concepts and related definitions: 1991, Quality Vocabulary. The British Standards Institution, Linford Wood, Milton Keynes, MKI4 6LE, UK, where complete copies of the standards can be obtained.

These categories are so designated by the effect produced once the hazard materialises. Therefore, it is usually assumed that a *catastrophic hazard* would result in loss of life, personal injury, financial loss, physical or tangible damage and loss of time; a *critical hazard* would result in personal injury, material damage, financial loss and loss of time requiring immediate attention to prevent further damage; a *marginal hazard* would result in financial loss, loss of time and malfunction that could be corrected; a *negligible hazard* would result in only slight damage which could be disregarded.

A scale based on the above grading may be used to quantify the resultant effect and to establish a basis for any calculations.

A hazard or a risk?

The two words 'hazard' and 'risk' are generally used interchangeably and are sometimes confused with each other. However, there is, in fact, an important and subtle difference between the two words and if the definition of 'hazard' is as given earlier, then 'risk' is defined in British Standard No. 4778 as

> A combination of the probability, or frequency, of occurrence of a defined hazard and the magnitude of the consequences of the occurrence.[2]

However, a wider definition and probably more correct is given by Australian/ New Zealand Standard on Risk Management, AS/NZS 3951: 1995, which defines risk as inclusive of not only loss or damage, but also gain.[3] Therefore, the word 'hazard' in the definition given by the British Standard, quoted above, is replaced by a more neutral word such as 'event' that may result in a positive or negative consequence.

Etymologically, the origin of the English word 'risk'; or '*risque*' in French; and '*rischio*' in Italian is uncertain. In Arabic, there is even confusion as to its real meaning – since some say it simply means 'danger' – ('*Khattar*' in Arabic), others refer to '*Rizkk*', which signifies what destiny bestows in the future for someone, positive or negative, good or bad; whereas others use it wrongly to mean '*Gharar*', a forbidden transaction under Islamic law.[4] Of course, the latter version cannot be right because if it were so, then a construction contract, which is known to be exposed to a large number of risks, would not be permitted under Islamic law. This simply does not make any sense nor can it be right. The Latin word '*resecum*' meaning 'danger' or

---

2 An extract from British Standard BS 4778: Part 3; Availability, reliability and maintainability terms. Section 3.1 Guide to concepts and related definitions: 1991, Quality Vocabulary. The British Standards Institution, Linford Wood, Milton Keynes, MKI4 6LE, UK, where complete copies of the standards can be obtained.

3 'Australia/ New Zealand Risk Management Standard', by Roger Keey, part of a book entitled *Owning the Future*, edited by Mr Elms and published by the Centre for Advanced Engineering, University of Canterbury, Christchurch, New Zealand, 1998.

4 '*Gharar*' is an Arabic word which is difficult to define and many differ regarding its true meaning. For further reading on the topic, see *Unlawful Gain and Legitimate Profit in Islamic Law*, by Nabil Saleh, 2nd edition, Graham & Trotman, London 1992, page 63; and *The Law of Business Contracts in the Arab Middle East* by Nayla Comair-Obeid, Kluwer Law International, 1996, page 56.

'rock' may throw some light on the origin of the word 'risk', but the Chinese '*wej-ji*' with the characters representing 'opportunity' and 'danger', is more illustrative of the concept of risk as it applies to the construction industry. This concept has evolved with these two notions embodied in it. It encompasses not only the danger of a loss but also the opportunity of a possible consequent gain when a decision is made.

Thus, various cultures and languages viewed risk differently and whereas common usage of the word 'risk' in English is reserved for adverse events, from a scientific point of view, it is more sensible to follow the Chinese understanding of the word 'risk', as does the Australian/New Zealand Standard on Risk Management, AS/NZS 3951: 1995.

Where decision-making is concerned, it often, if not always, involves risk-taking. However, the well-informed decision-maker will be aware of the risks associated with any decision and will endeavour to eliminate or reduce all foreseeable risks and their consequences to an acceptable minimum. However, it should be remembered that eliminating risk may mean a huge cost penalty and therefore it should be exercised carefully. In making a decision, an individual may deal with any one of the following possibilities:

(a) 'Pure risk' where only negative deviations from the desired outcome are possible and therefore danger of loss is predominant; or
(b) 'Speculative risk' where both negative and positive deviations are possible and therefore there is a danger of loss as well as a chance of gain; or
(c) Only positive deviations are anticipated and therefore only a chance of gain exists. Such events do not form part of the notion of risk.

Based on the definition of the words 'hazard' and 'risk' given above, Risk may be expressed in the form of a mathematical equation, as follows:

Risk=Probability, or frequency, of the occurrence of a defined event×
Consequences of the occurrence of that event; or

$$R = P \times C$$

There are a number of points which flow from the above mathematical expression which are as follows:

(a) An undesirable event may have a number of different causative factors, of which one or any combination could lead to its occurrence. For example, if the undesirable event is the collapse of a cofferdam at a construction site, such collapse may have been caused as a result of bad ground conditions, material failure, defective design, or a combination of some or all of these factors. All these factors could be referred to as hazards.
(b) An event can, therefore, be construed as a dormant potential for gain or for inconvenience, loss, damage to property, damage to the environment, moral damage, injury or loss of life. To eventuate, it is triggered by a particular incident, which may be referred to as a 'triggering incident'. A triggering incident is usually necessary for an event to take place and materialise into a positive gain or a negative undesirable consequence. For example, if the cofferdam collapse mentioned in (a) above was due to a defective section in its wall, the triggering

incident could be the imposition of an additional loading beyond the limit sustainable by that defective section;

(c) The event may result in different levels of magnitude of consequence depending on the particular circumstances and timing of the event itself. For example, the consequence of the collapse of the cofferdam in (a) above may be a financial loss in the form of cost of repair to the cofferdam, or it may extend to a critical delay in the completion of the project, or it may go beyond the financial loss and delay into personal injury and death

(d) Expressing risk in a mathematical formula permits a comparison of the magnitude of the various risks to which a project is exposed. Such a comparative analysis may then be used to decide whether to accept a particular risk or take measures to eliminate it or, at least, to mitigate its effect.

In summary, when an event materialises into an undesirable event, the schematic sequence can best be expressed as follows:

Event → Triggering Incident → Desirable or Undesirable Consequences → Assessment & Evaluation.

Risk is, therefore, a measure of the combined effect of the likelihood and the consequence of an event. For an example of the difference between an undesirable event, a hazard, and risk, let us consider the case of casualties from the hazard of road accidents in the United Kingdom, where the figures were as follows:[5]

The hazard of road accidents occurred	265,861 times
The number of those killed as a result	6,614 persons
The number of those seriously injured as a result	81,681 persons
The number of those slightly injured as a result	259,766 persons
The population for the year in question was	55,852,000 persons

The probability of any one person living in the United Kingdom being killed as a result of the hazard of road accidents in the particular year was:

6,614 death per year/55,852,000 persons$=1.18 \times 10^{-4}$ death per year.person

The probability of serious and slight injury was 14.6×10^{-4} and 46.5×10^{-4} respectively.

On the other hand, the risk to a person venturing on to the road in the United Kingdom in that particular year is a combination of the severity of the effect multiplied by the probability of occurrence. To highlight the meaning of risk in the above example, the severity of the effect of the hazard can be given values along a scale of say 0 to 1000, as suggested earlier. Therefore, if slight injury is given a grading of 200, serious injury 600 and death 1000, the risk of slight injury can be calculated as $46.5 \times 10^{-4} \times 200 = 0.93$. Similarly, the risk of serious injury and death can be calculated as 0.876 and 0.118, respectively.

---

5 *Facts in Focus*, statistics compiled by the Central Statistical Office, U.K. and published by Penguin Books in association with HMSO, 5th edition, pages 17 and 78, the figures quoted are for 1977.

The risk therefore can be represented by the following mathematical expression:

Risk=Event×Probability of Occurrence, where the event is measured in terms of its intensity.

Thus if the probability of occurrence of a catastrophic event is extremely low, the risk may still be acceptable, whereas if the probability of a marginal event is extremely high, the risk may not be acceptable.

The difference in meaning between hazard and risk can be further explained by the following simple example. A statistical analysis compiled in the United States for a particular area showed that, due to various hazards, the risk of meeting with accidental death annually was given by the following figures:

Annual deaths as a result of car accidents	1:4,000
Annual deaths as a result of accidental falls	1:10,000
Annual deaths as a result of fires	1:27,000

The magnitude of the hazard in all these different events is catastrophic and the same grade of severity applies, i.e. death. The risk is therefore dependent on the value of the probability and a comparative analysis can be made based on the value of the probability of occurrence.

Therefore, if one lives in that particular community the risk of dying in a car accident is two-and-a-half times greater than as a result of a fall and six-and-three-quarter times greater than dying in a fire. The triggering incidents leading to the hazard in all these situations are many but are not mentioned in the analysis. They could be any of a number of circumstances causing car accidents such as another car, a pedestrian, ice on the road, etc. In the case of the hazard of a fall, the triggering incident could be an unsecured ladder, a slippery surface, etc. In the case of a fire, it could be an electrical fault, a welding operation, a cigarette etc.

In the construction industry, statistical evidence in the United Kingdom shows that the four most serious hazards on sites were as given in Tables 2.1 to 2.3[6] below:

Table 2.1 Percentage of fatal injuries to workers by kind of accident[7] 1996/97 to 2000/0l p[a]

Injury	96/97	97/98	98/99	99/00	00/01 p
Falls from a height[b]	56%	58%	60%	52%	44%
Struck by moving vehicle	11%	6%	12%	6%	17%
Struck by moving/falling object	12%	15%	12%	21%	8%
Trapped by something collapsing/overturning	7%	5%	5%	2%	17%
Other	14%	16%	11%	19%	14%
Total injuries	90	80	65	81	106

Notes: [a]Reported to all enforcing authorities
[b]Falls from a height include falls from up to and including 2 m, over 2 m and height not known.

6 Tables taken from Section 1.166 of the 'Health & Safety Statistics', HMSO, 2002.
7 Presumably, the word 'accident' here has the same meaning as 'hazard'.

Table 2.2 Percentage of major injuries to employees by kind of accident 1996/97 to 2000/01 p[a]

Injury	96/97	97/98	98/99	99/00	00/01 p
Slips, trips or falls on the same level	35%	37%	37%	36%	37%
Struck by moving vehicle	19%	19%	20%	21%	21%
Struck by moving/falling object	3%	2%	3%	2%	2%
Injured while handling, lifting or carrying	21%	20%	18%	18%	18%
Other	14%	13%	13%	13%	14%
Total injuries	3,227	3,860	4,289	4,386	4,268

Notes: [a]Reported to all enforcing authorities

Table 2.3 Percentage of over-3-day injuries to employees by kind of accident 1996/97 to 2000/01 p[a]

Injury	96/97	97/98	98/99	99/00	00/01 p
Falls from a height[b]	12%	12%	14%	14%	14%
Slips, trips or falls on the same level	17%	17%	17%	18%	19%
Struck by moving vehicle	1%	1%	1%	2%	1%
Struck by moving/falling object	19%	18%	18%	19%	18%
Injured while handling, lifting or carrying	36%	36%	35%	34%	34%
Other	15%	16%	15%	13%	14%
Total injuries	8,637	9,756	9,195	10,159	9,427

Notes: [a]Reported to all enforcing authorities
[b]Falls from a height include falls from up to and including 2 m, over 2 m and height not known.

Risks in construction

Based on the statistics gathered in the past three decades on topics, such as disputes in the construction industry and international arbitration,[8] accidents at work[9] and

8 Annual Reports of the International Chamber of Commerce, Paris; the section on the ICC International Court of Arbitration indicates a constant flow of international construction disputes. Litigation and arbitration cases around the world involving issues of professional negligence add to the list of disputes. Insurance and reinsurance loss statistics and reports complement this picture. See also 'Collection of ICC Arbitral Awards 1974–1985', Sigvard Jarvin and Yves Derains, ICC Publishing SA, ICC Publication No. 433, Paris, 1990; and 'Collection of ICC Arbitral Awards 1986–1990', Sigvard Jarvin, Yves Derains and Jean-Jacques Arnaldez, ICC Publishing SA, ICC Publication No. 514, Paris, 1994.
9 Figures published annually by Central Statistics Offices around the world indicate a high, if not the highest exposure at work to bodily or fatal injuries in construction. See also Health and Safety Statistics, HMSO UK and *Facts in Focus*, Statistics compiled by the Central Statistics Office, UK and published by Penguin Books in association with HMSO, UK.

exposure to natural hazards around the world,[10] it can be concluded that construction projects are sensitive to an extremely large matrix of hazards and thus to risks. This sensitivity is due to some of the inherent characteristics of construction projects, which are summarised as follows:

(a) The time required to plan, investigate, design, construct and complete a construction project spans such a lengthy period that it is often greater than the period of cyclical recurrence, known as the 'return period', of many of the hazards to which such projects are exposed. For example, the hazard of rainfall has usually a return period of one year depending on the time for the rainy season. Therefore, the risks associated with rainfall on a particular project would have to be assessed and managed for the number of years taken to complete it. Any reduction in the period of construction introduces its own risks.

(b) The number of people required to initiate, visualise, plan, finance, design, supply materials and plant, construct, administer, supervise, commission and repair any defects in a construction project is enormous. Such people usually come from different social classes and in international contracts, from different countries and cultures.

(c) Many civil engineering projects are constructed in isolated regions of difficult terrain, sometimes stretching over extensive areas and exposed to natural hazards of unpredictable intensity, frequency and return period.

(d) The materials selected for use generally include a number of new products of unproved performance or strength. Advanced and complex technology is also necessary in some construction projects.

(e) Extensive interaction is required between many of the firms involved in construction, including those engaged as suppliers, manufacturers, subcontractors and contractors, each with its own different commitments and goals.

(f) Construction projects are susceptible to risk cultivation by the parties themselves or by others associated with them or advising them.

It is therefore extremely relevant for the construction industry and those involved in it to understand the concept of risk and to know how to properly manage the risk matrix generated when a construction project is initiated.

The subject of Risk, its assessment, allocation and management in construction projects has been developed and applied on an increasing scale over the last twenty years. The Health and Safety at Work Regulations introduced in a number of

---

10 Publications of the Munich Reinsurance Company and the Swiss Reinsurance Company are a valuable source of reference in this regard. These publications cover topics such as earthquakes, windstorms, flood and inundation, volcanic eruption and hailstorm. In 1978, the Munich Reinsurance Company published a world map of natural hazards, which was updated in 1988. It indicates the intensity, frequency and reference period of various natural hazards (over 670 in number), catalogued in a chronological order and location, with the consequences in terms of loss of life and cost. These worldwide records go back in time to the tenth century. The map and the accompanying publications are extremely useful in risk management calculations and in any attempt at predicting future exposures through extrapolation from retrospective exposure.

countries and in particular those recently imposed in the European Union gave the subject of Risk in construction an even greater significance.[11] Amongst the requirements introduced in the European Union through the Construction (Design and Management) Regulations 1994, 'CDM', there is a requirement to carry out risk assessment of planned work and to take reasonable measures to deal effectively with any significant risk.

However, there is little uniformity of approach to the topic of risk by those involved in the construction industry and, surprisingly, only a few useful general applications of the topic of risk have been developed in the area of planning and management of construction projects. The lack of uniformity relating to risk extends even to the definition of 'Risk' and what is meant by it.

Analysis of hazards

If one moves from intricate day-to-day activities into the zone of a specific decision relating to a particular action, fortuity must give way to reason and lack of thought into a planned scientific approach. Thus it is wise to try, if possible, to identify any hazards that might exist and to either eliminate or mitigate them into a more acceptable class of effect. This upgrading may be done through the provision of a corrective measure or the addition of some act. If, in the implementation of a particular idea, a certain hazard exists and if, neither eliminated nor mitigated, it is found to be totally unacceptable, the whole idea may have to be abandoned.

Once the hazards are identified, the risk of each of them materialising should be evaluated and assessed. The risk itself may again be either accepted or mitigated, weighing the various factors affecting the level of probability of occurrence and its acceptability by those exposed. A flow chart expressing the logical sequence of analysis is given in Figure 2.1.

Risk assessment

As defined earlier, a risk is the combination of the probability, or frequency, of occurrence of a defined hazard and the magnitude of the consequences of the occurrence. Risk assessment is, therefore, the integrated analysis of the likelihood of an event, its effect in terms of extent and also in terms of significance. The likelihood, the extent and the significance of an event can be assessed either from previous experience or from calculations using the theories of probability.

However, not all are initiated in the science of statistics and the theories of probability and an individual's perception of social hazard and risk is usually based on trial and error, and the methods of risk assessment are based on vague and unscientific hunches. Moreover, the incredible number of ways in which human beings differ from each other clouds further this perception in a shroud of bias based on their likes and dislikes. Thus, evidence suggests that higher risks are accepted by people when incurred voluntarily and especially so if they are combined with elements of

11 EU Health and Safety Directive: (Design and Management) Regulations 1994, 'CDM', Health and Safety at Work Regulations.

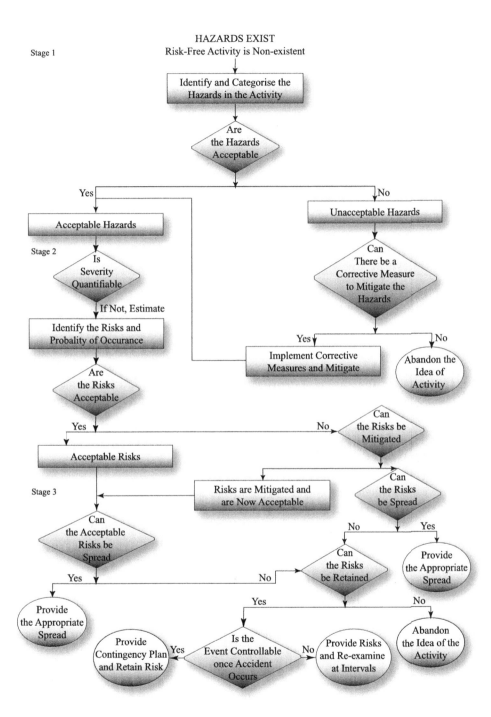

Figure 2.1 Hazards exist.

control and enjoyment. Statistical figures indicate that the level of risk accepted voluntarily is a thousand times that of an imposed risk. Therefore, an individual will start to get concerned at a much lower level of risk about an activity in which he participates but over which he has no control or personal interest. An individual will be prepared to ride a motorcycle if he enjoys it despite statistical evidence that the risk of accidental death thereby is twelve times higher than when driving a car.

It is accepted, however, that some people would be deterred if they were to perceive the high level of risk involved in a particular activity. This suggests that if hazards and risks were identified and quantified, the process of decision-making would be clearer and more accurate in achieving a certain target. On a social level, what is probably needed is an awareness of the various levels of risk in all types of activities, processes, actions, etc.

Risk management

Risk management is defined in British Standard 4778, 1991, as

> the process whereby decisions are made to accept a known or assessed risk and/or the implementation of actions to reduce the consequences or probability of occurrence.

It is defined in the Australian/New Zealand Standard, referred to earlier, as

> the systematic application of management policies, procedures and practices to the tasks of identifying, analysing, assessing, treating and monitoring risk.

By definition, risk management is therefore concerned with the mitigation of those risks deriving from unavoidable hazards through the optimum specification of warning and safety devices; and risk control procedures, such as contingency plans and emergency actions.

If a decision is made to accept a risk, a further decision must be made on whether the risk should be retained or whether it should be shared and if so with whom. Before such decisions can be made, it is necessary to go through a systematic process, which involves the analysis of the possible hazards to which the project may be exposed and the evaluation of their intensity, frequency and return period. In this regard, the following terms and definitions from BS 4778 are relevant:[12]

Hazard analysis	The identification of hazards and the consequences of the credible accident sequences of each hazard.
Risk quantification	The estimation of a given risk by a statistical and/or analytical modelling process.
Risk evaluation	The appraisal of the significance of a given quantitative (or, when acceptable, qualitative) measure of risk.
Risk assessment	The integrated analysis of the risks inherent in a product, system or plant and their significance in an appropriate context.

12 See note 1 above.

Risk criteria   A qualitative and quantitative statement of the acceptable standard of risk with which the assessed risk needs to be compared.

It is interesting to note that Risk Analysis is defined in the Australian/New Zealand Standard, referred to earlier, as

> a systematic use of available information to determine how often events may occur and the magnitude of their likely consequences.

When hazards are identified, assessed and analysed, their management must be allocated to one or more of the various parties in order to keep them under control, prevent the occurrence of any harmful consequences and thus reduce the extent of any risk of harm. Such allocation is part of the risk management process and should be carried out in accordance with appropriate rules rather than haphazardly. The rules for allocation of risks in a construction project may simply revolve around the ability of a party to:

(a) control any arrangements which might be required to deal with the hazard or any triggering incident relating to it;
(b) control the risk or to influence any of its resultant effects;
(c) perform a task relating to the project, such as, obtaining and maintaining insurance cover; and
(d) benefit from the project.[13]

Although these rules were contemplated in 1983, they remain valid, as their application was approved in 1998 by a major report commissioned by the Government of Hong Kong SAR on the allocation and management of risk in the procurement of construction projects.[14] These rules and others were later debated at a conference in Hong Kong which focused on the report and related issues.[15]

On the other hand, the rules may revolve around an already established policy in a large organisation or a governmental agency, such as, that contemplated by some authoritative writers in the following terms:[16]

---

13 'Risk Management', Max W. Abrahamson [1983] ICLR 241, also published as Appendix J to a discussion paper on 'Construction, Insurance and Law' published by FIDIC, 1986, page 49; and 'Defects: A Summary and Analysis of American Law', Justin Sweet, a paper published in *Selected Problems of Construction Law, International Approach*, by Peter Gauch (Switzerland) and Justin Sweet (USA), University Press, Fribourg, Switzerland, Sweet & Maxwell, London, 1983, page 97.

14 'The Grove Report: Key Terms of 12 Leading Construction Contracts Are Compared and Evaluated', published in September 1998 and is available on the web site of Thelen Reid & Priest at (www.constructionweblinks.com).

15 A conference in Hong Kong held in November 2000, details of which were given by Humphrey LLoyd in the International Construction Law Review, [2000] 2 ICLR 302.

16 'Price under Common Law System', I.N. Duncan Wallace; and 'Defects: A Summary and Analysis of American Law', Justin Sweet; pages 149 and 79 of the publication *Selected Problems of Construction Law, International Approach*, quoted in note 13 above.

> ... while an event may be foreseen ... employers may see advantages in a contract which requires him (the contractor) to assume that risk, and to include for the cost of dealing with that situation in his tendered contract price. Where the risk is uncertain, this logically requires that a contingent may possibly not be required. If so, the employer will have agreed to an unnecessarily high price but may regard that as preferable ... than a lower price subject to post-contract upward adjustment at [a] late stage should the risk materialise ... whether or not a particular risk should be so included in the price is in essence a question of policy and not of 'fairness', 'morality', or 'justice',

and

> This (American) system ... also seeks to preserve a large pool of competent contractors and obtain low contract bids by absorbing particular risks and seeking to assure the contractor he will be treated fairly

If risks are not allocated in a contract between certain parties and a dispute arises between them, as to whom a particular risk is allocated, then an arbitrator or a judge would most likely examine the following criteria for risk allocation and determine the dispute accordingly:

1   which party could best foresee that risk?
2   which party could best control that risk and its associated hazard or hazards?
3   which party could best bear that risk?
4   which party most benefits or suffers when that risk eventuates?

In the risk management process, an extensive matrix of general risks can be identified for most, if not all, construction work to which one can add further risk matrices for specific projects. They can be identified from past experience since actual examples from real life are plentiful.[17] These risks are traditionally shared between the parties involved, in accordance with the provisions of two contracts usually agreed: first, between the employer/owner and the design professionals involved; and second, between the employer/owner and the main contractor. From the latter agreement, flows another line of risk sharing between the main contractor, on the one hand, and subcontractors, suppliers, manufacturers, insurers and others, on the other hand.

If these risks are analysed on the basis of the effect they generate once they eventuate, two basic types of risk can be identified. The first type incorporates the risks which could lead to damage, physical loss, or injury and the second type incorporates risks which could lead to lack or non-performance of the contract, delay in completion of the works and/or cost over-run of the constructed project.

Examples of the first type of risk which involves damage, physical loss or injury include defective design, defective material, defective workmanship, acts of God, fire, human error and failure to take adequate precautions. Examples of the second

17   See Chapter 3 below.

type include late possession of the site, delay in receipt of information necessary for timely construction, changes in design, and variations to the original contract.[18]

Once hazards are identified, an analysis must be carried out to work out their probability of occurrence and the effects they would have if they materialise into events. Each hazard and risk must be considered from the point of view of whether or not it can be mitigated. If it can be mitigated, is it feasible to do so? Such a question must be answered, with the following considerations in mind, by those who, entrusted with decision-making, have to judge the benefits, the alternative ways of using the available resources and any other aspect peculiar to the idea being considered:

(a) technical considerations;
(b) time available;
(c) financial resources; and
(d) special features to be maintained.

This type of analysis is expressed schematically in Figure 2.1. The risks are divided into three categories: unacceptable and the idea has to be abandoned; unacceptable but can be mitigated to an acceptable level; acceptable and of such value that the idea considered remains viable and a decision can then be made to implement it. The next stage of the analysis is an attempt to spread the accepted risks. This spreading can be done in various ways and some sectors of society have found ways and means of effectively diffusing the risks to which they are exposed. Thus, in the insurance sector, insurers divide the risks they accept into small portions and make arrangements either with other insurers or with larger reinsurance companies all over the world to accept one or more of these portions against part of the premium they receive (see Chapter 7).

Hazards in construction

Moving away from day-to-day activities into those connected with construction, we find that all categories of hazards exist producing the possibility of loss of life, personal injury, moral damage, material damage, financial loss and loss of time. Terrible combinations of these six consequences may result in the more hazardous areas of construction. At one end of the scale, hazards to water-retaining structures are the most catastrophic. In identifying the hazards associated with the construction of a large dam project, one may visualise that in the case of a breach in the dam an entire river valley can be flooded affecting probably hundreds of thousands of people, destroying whatever towns lie downstream and reducing agricultural land to barren areas. An example of such a failure occurred on 19th July 1985, when the embankment of a dam in the Fiemme valley, North Italy, collapsed, releasing 250,000 cu m of water. More than 260 people died. Although the probability of occurrence of such a hazard is rather low it must be seriously considered due to its devastating effect. Slightly lower down the scale, structural failures can be just as catastrophic, such as the event of the Kansas City Hyatt-Regency

18 'Construction, Insurance and Law', Nael G. Bunni, a paper delivered at a Conference on Structural Failure, Product Liability and Technical Insurance, Technische Universität, Vienna, 1989, and published subsequently in *Forensic Engineering*, vol. 2, 1990, Nos 1/2, page 163.

Hotel walkways failure in July 1981. One hundred and eleven people were killed, 188 were injured and lawsuits, with damages exceeding $1.5 billion, were the result of that disaster.

On the other end of the scale, negligible hazards can produce minor cracking of a cosmetic nature which can be disregarded.

Hazards in construction exist not only after the structure or the project is completed, but they also exist and, in fact, in larger numbers and probability of occurrence, during the construction period. Once again, water-related incidents, including those involving dams and water-retaining structures, feature at the top of the hazard scale.

An example of a natural hazard can be related to the events of 31 July 1976, at Big Thompson Canyon, Colorado, USA when an intense thunderstorm caused a precipitation of more than 250 mm in less than three hours. The topography of the area yielded a runoff of such magnitude that a wall of floodwater swept everything in its path, causing the death of 143 people and property damage in excess of $14 million. Similar effects can arise from hurricanes, earthquakes and volcanic eruptions.

However, from a work-related point of view, construction compares badly with other major sources of employment in terms of the risks of fatal and personal injury. Tables 2.4 and 2.5 show that in the early 1980s it ranked highest in respect of these risks.[19] The extensive work done by the Health and Safety Executive in the United Kingdom seems to have produced a positive effect on the record of construction activities, since recent figures show an improvement in the position of construction. Tables 2.6 and 2.7 show the number and rate of fatal injuries and the number and rate of major injuries in the United Kingdom during 1998/99 and 2000/01, ranking construction as fourth in the list of industries surveyed.[20]

Although some hazards can be attributed to adverse human behaviour, it is by no means true to assume that this is the only or the major cause. It is probably the area where we have most control, provided an attempt is made to minimise or mitigate the hazards.

Table 2.4 Incidence rates of all injuries sustained by employees per 100,000 employees for each of the years 1981 and 1982, in the United Kingdom

Place of employment	Number of injuries of all typesper 100,000 employees		Risk of injury	
	1981	1982	1981	1982
Construction	4190	4070	1:24	1:25
Gas, water and electricity	2560	2440	1:39	1:41
All manufacturing industries	2550	2350	1:39	1:42
Transport and communications	1450	1330	1:69	1:75
Professional and scientific services	860	840	1:116	1:119
Insurance, banking, finance and business services	210	180	1:476	1:556

19 *Health and Safety Statistics*, Her Majesty's Stationery Office, Norwich NR3 1BQ, England.
20 *Health and Safety Statistics, 2000/01*, Her Majesty's Stationery Office, Norwich NR3 1BQ, England.

Table 2.5 Incidence rates of fatal and major injury sustained by employees per 100,000 employees, for 1981 and 1982, in the United Kingdom

Place of employment	Number of fatal and major injuries per 100,000 employees		Risk of fatal and major injuries	
	1981	1982	1981	1982
Construction	164	203	1:610	1:493
Gas, water and electricity	56	56	1:1,785	1:1,785
All manufacturing industries	71	74	1:1,408	1:1,351
Transport and communications	42	42	1:2,381	1:2,381
Professional and scientific services	36	30	1:2,778	1:3,333
Insurance, banking, finance and business services	1.08	0.85	1:92,600	1:117,650

Table 2.6 Industries with the highest rates of fatal injuries to workers, 1998/99–2000/01 provisional combined, in the United Kingdom

Standard Industrial Classification	Number of fatal injuries	Number of workers[a]	Fatal rate per 100,000 workers
Quarrying of stone, ore and clay	9	87	10.4
Agriculture, hunting, forestry and fishing	128	1,423	9.0
Extraction of coal, oil and gas	12	135	8.9
Construction	252	5,225	4.8
Manufacturing of basic metals and fabricated metal products	58	1,682	3.4
Manufacture of wood and wood products	9	284	3.2
Manufacture of other non-metallic mineral products	12	447	2.7
Manufacturing not elsewhere classified	17	737	2.3
Transport, storage and communication[b]	80	4,017	2.0
Electricity, gas and water supply	6	386	1.6
Manufacture of rubber and plastic products	12	728	1.6
Total	595	15,151	–

Notes: [a]Expressed in thousands of workers.
 [b]Injuries arising from shore-based services only. Excludes incidents reported under merchant shipping legislation.

Identification and classification of risks in construction

Risks may be identified and classified under a number of headings, all of which must be considered and assessed for a complete picture. The following are those considered to be important:

A Classification based on geographic distribution

As soon as someone steps outside his own environment, he meets a new set of risks, which may include situations over which he has no control. Risks involving new

41

Table 2.7 Industries with the highest rates of major injuries to employees, 1998/99–2000/01 provisional combined, in the United Kingdom

Standard Industrial Classification	Number of fatal injuries	Number of workers[a]	Fatal rate per 100,000 workers
Quarrying of stone, ore and clay	363	81	449.7
Manufacture of wood and wood products	1,023	243	420.5
Extraction of coal, oil and gas	511	130	392.6
Construction	12,943	3,301	392.1
Manufacture of food products; beverages and tobacco	4,353	1,418	306.9
Manufacture of other non-metallic mineral products	1,247	412	302.8
Manufacture of basic metals and fabricated metal products	4,690	1,576	297.7
Manufacture of rubber and plastic products	1,898	692	274.4
Transport, storage and communication[b]	8,710	3,365	258.9
Agriculture, hunting, forestry and fishing	1,850	872	212.2
Total	37,588	12,090	–

Notes: [a]Expressed in thousands of employees.
[b]Injuries arising from shore-based services only. Excludes incidents reported under merchant shipping legislation.

culture, customs, materials, methods, different politics and varying rates of exchange fall under this type of classification.

B Classification based on the size and complexity of the project

As a construction project increases in size, the risks inherent in its planning, design and execution do not simply multiply in conformity with the increase in size. Instead, new and peculiar risks emerge which need to be identified and taken care of prior to the commencement of design.

The size of projects measured in monetary values has escalated dramatically in recent years. New names had to be conceived such as Jumbo, Giant and more recently Pharaonic. An example of the last description is the US$18 billion Itaipu Hydroelectric plant located on the Parana River between Brazil and Paraguay. See page 390 under the heading of 'Adequacy of finance'.

C Classification based on legal concepts

In general, the legal concepts accepted in the jurisdiction where the project is constructed produce a certain pattern of risks. Therefore, one may classify risks in accordance with law applicable to that jurisdiction, resulting in four areas of concern: contract; tort; equity or custom (depending on the part of the world being considered); and legislation in the form of statutes.

D Classification based on the effect produced by the risk eventuating

As described earlier, the effect produced by a particular risk eventuating can be measured through the severity and the probability of occurrence. A classification may be carried out on the basis of the magnitude of either of these dimensions. There are recognised scales for each of the two dimensions giving a set grading, which starts at zero and ends at a certain value describing the highest anticipated magnitude of that dimension.

The effect, however, can be expressed in terms of monetary loss, property damage, personal injury, or a combination of any or all. The grading for probability of occurrence shown in Table 2.8 is an example:[21]

Severity may be graded as shown in Table 2.9, having been modified from the reference quoted for the previous dimension.

E Classification based on chronology

Risks may be classified in accordance with the chronological staging of a construction project. Thus, risks are divided into those occurring during the feasibility stage, followed by the design stage and so on including the stage when the project has been taken over in part or wholly by the owner and used.

The classification brings together a spectrum of risks in an extensive matrix. In the following Chapter, the expected risks in a construction project are arranged in a succeeding order beginning with the brief and ending with the actual use of the project, i.e. beginning with risk E.1 and ending with E.3.2. Figures 2.3, 3.1 to 3.3, 3.6, 3.9 and 3.11 show flow charts with a listing of some of the risks that can be

Table 2.8 An example of grading for probability of occurrence

Probability of occurrence	The event
0	It is certain that no loss would occur – loss is not possible
0.1	Very remote possibility
0.2	Remotely possible
0.3	Slight probability of occurrence
0.4	A little less than equal chance of occurrence
0.5	Equal chance of occurrence or non-occurrence
0.6	Fairly possible
0.7	Probable
0.8	More than likely to occur
0.9	Strong probability of occurrence
1.0	Certain to occur

21 This grading is modified from the original form introduced in 'Risk Management', by Max W. Abrahamson, Appendix J of the discussion paper, 'Construction, Insurance and Law', published by the International Federation of Consulting Engineers, Switzerland, March 1986.

Table 2.9 An example of grading for severity of consequences of events

Category of loss	The consequences of a hazard eventuating	Percentage of costs of project
0	No loss	Nil–0.09%
1	Nuisance-type small losses	0.1%–0.49%
2	Small losses	0.5%–0.99%
3	Medium Losses which can be borne by the individual concerned	1%–4%
4	Manageable	5%–9%
5	Large losses	10%–19%
6	Probable maximum loss in the range of the largest previous losses of similar projects	20%–40%
7	Serious and exceeding any previous events	41%–50%
8	Very serious Very serious	51%–70%
9	Most serious	71%–80%
10	Catastrophic – total loss	81%–100%

envisaged. Similar charts can be drawn for each of the classifications mentioned above showing the same risks rearranged accordingly.

It must be emphasised, however, that the classification set out here is given in general terms and must, therefore, be looked at with 'scrutiny' when a specific project is concerned. Furthermore, the situation may differ from one part of the world to another as culture, custom and legal concepts change.

F Classification in construction contracts

This is the method used in most standard forms of construction contracts where specific risks are allocated to one party and all the other risks are allocated to the other party in the contract.[22] In general terms, the risks are classified and allocated to the respective parties on the basis of the criteria of control of the risks and their consequences, if and when they eventuate. Some of the consequences are then insured whilst others, although insurable, are not required to be insured. There remains a set of un-insurable risks that cannot be insured because in one way or another they do not conform to the concept and principles of insurance. These un-insurable risks must therefore be allocated to one of the parties in the contract on the basis of the benefit realised from being involved in the project. Of course, whenever risks and their consequences are shifted to an insurer, the principles of insurance must be complied with so that insurance is not degenerated into lottery. This method of classification is discussed further in Chapter 5.

Figure 2.2 shows a flow chart setting out the different methods of classification.

22 In the Engineering and Construction Contract of the ICE, 2nd Edition, 1995, reprinted with corrections in May 1998 (the 1st edition was referred to as the New Engineering Contract), there is a mix of 'Employer's risks' under Clause 80; 'Contractor's risks' under Clause 81; and 'Compensation events' under Clause 60. In all these, the events are specified rather than allocated in accordance with a certain criterion. However, it must be noted that no direct link was established between what is referred to as risks of loss and damage and those others resulting in pure economic and time loss.

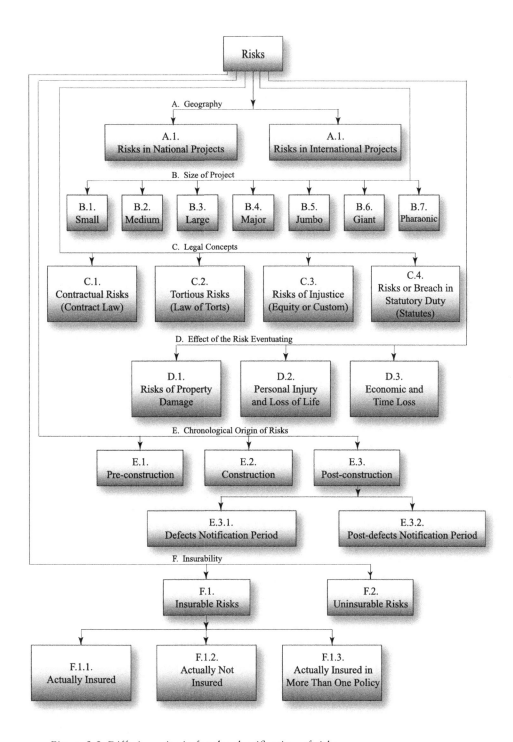

Figure 2.2 Differing criteria for the classification of risks.

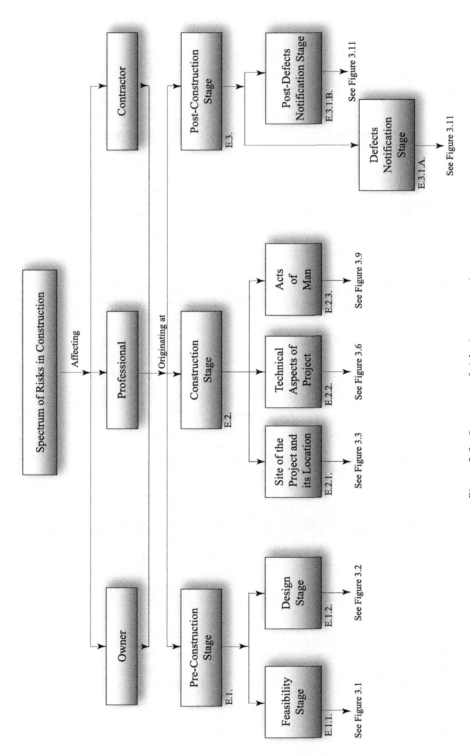

Figure 2.3   Spectrum of risks in construction.

Allocation of risks

When the risks are identified, they should be allocated to the various parties involved in the contract. This allocation should be based on a sound appraisal of the interplay between the parties and the risks. The most appropriate method may be to allocate the risks on the basis of control over their occurrence and the effect they cause when they eventuate. It may be more appropriate in certain circumstances to allocate the risks, or a specific risk on its own, on a different basis. Such a basis may be the optimum ability of one party to execute a specific task related to the project, or it may be the inability of any party, except the owner who initiated the project, to accept the risk.

Examples of the risks, as they are allocated in accordance with the above headings, can be drawn from the spectrum of risks in the previous section. Thus, as the owner is in charge of briefing his design professionals, he should be allocated that risk, No. E.1.1.2, Figure 3.1.

Similarly, as the design professional is in control of his design office, the risks outlined as No. E.1.2.1, E.1.2.2, E.1.2.3, etc., Figure 3.2, should be allocated to him. Similarly again, as the contractor is in charge of the site, such risks as outlined under items E.2.1.1, E.2.1.2, etc., Figure 3.3, should be allocated to him.

Under the basis of optimum ability to carry out a certain task, we have as an example the risk outlined as No. E.2.2.13, Figure 3.6, which should be accordingly allocated to the contractor. Under the basis of inability to accept risks, the example outlined as No. E.1.1.7, Figure 3.1, and E.2.1.8, Figure 3.3, must remain with the owner. Figure 2.4 shows a schematic diagram of the options available under the heading of allocation of risks.

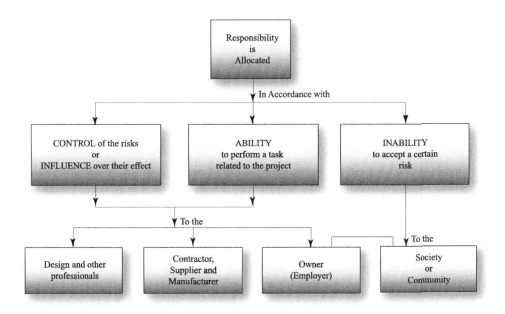

Figure 2.4 Allocation of risks.

It is accepted that such an allocation must be made as the probability of dispute between the parties reduces proportionately with the reduction in the number of unallocated risks. But what happens if the allocation is carried out incorrectly? Generally speaking, if risks are allocated incorrectly and if they eventuate causing loss or damage, as they often do, then disputes arise as to the liability in respect of the cost of such loss or repair.

Risks inherent in major projects

As a construction project increases in size, the risks inherent in its planning, design and execution do not simply multiply in proportion to the increase in size. Instead, new and peculiar risks emerge which need to be identified and taken care of prior to the commencement of design. The following is a list of the reasons for the more acute risks which can be identified in connection with major projects:

1   There is scant experience available to too few people, because major projects are few and scattered around the world. The only information available is through the common denominator of all these schemes, i.e. insurers and reinsurers. However, due to the manner in which insurance is transacted, insurers have not been seriously involved in the dissemination of information and the publication of lessons to be learnt from past experience, which is the fastest way of acquiring knowledge.

2   The financial resources required to complete a major project are more likely than not to be so large that more than one financier is usually involved. As the number increases, so does the level of risk. In some cases, private resources are insufficient or ineffective in producing the necessary conditions for the success of the project envisaged. In such situations, governmental involvement is imperative but such involvement brings with it a new and different set of hazards, increasing the level of risk.

3   The financial aspects of major projects also provide a high element of risk for the contractor. If a major contract proves to be unprofitable, the outcome, from the contractor's point of view, can be disastrous if the number of similar contracts is very low or nil. In other words, a contractor with a turnover of one billion, made up of two contracts of a half billion each, is much more exposed than another contractor with the same turnover but made up of ten contracts each of 0.1 billion.

4   The period of time necessary for the planning, design and execution of a major project is of such length that it increases the probability of occurrence of the risks to which the project is exposed. Compressing that period in any way also produces its own risk-sequence.

5   The long-term economic viability of a project is a course fraught with high risk for all the participants, be they designers, builders or others. At the end of the long period of construction, changes may occur in areas such as technology, product, inflation, the need of society or currency fluctuation, each of which may render the project obsolete.

6   The teams required to visualise, envisage, plan, design, administer, service, finance, assess and execute a major project are many and include widely differing dis-

ciplines. They may include such sectors of discipline as insurers, bankers, finance experts, risk managers, tax advisers, management experts, marketing specialists, commercial advisers and political counsellors.

7   The top personnel must be competent in more than one of the disciplines mentioned. They must be dedicated people willing to remain with the project from inception to completion and to give up a large portion of their time and career to the project. Such personnel are not easy to find.

8   The project is usually in difficult terrain; climate; and location, subject to such natural phenomena as would increase the hazards and level of risk. In certain circumstances, the project is spread over a large area.

9   Major projects involve generally many contractors and subcontractors and, similarly, suppliers and sub-suppliers. This situation is prone to bottlenecks some of which are very critical to the progress of work on or off the site. The level of risk due to this aspect escalates drastically as the number of these critical hurdles increases.

10   High labour content, with multi-million man-hours per month,[23] new and advanced technology in one field or another connected with a major project and unproven performance of materials or their strength, combine to bring with them high and concentrated levels of risk.

11   Risk-cultivation is a phenomenon produced by human greed. The risk level, should such an event materialise, is greater in major projects where pecuniary awards are highest.

Assessment of risks in construction

Risks in construction occur as a result of the existence of dormant sets of conditions, which possess the potential for initiating in most cases adverse events (accidents). They are usually in a dormant state but all that is required is an activating agent to trigger the change from a dormant to an active state. To name and identify the hazards and risks in a project is therefore the first step in the process of managing it to success.

Having identified the spectrum of risks in any particular project, it is important to carry out an integral analysis and assessment of the two elements which identify the effect of the risks, i.e. severity and probability of occurrence. As stated earlier, such analysis can be done either through knowledge from previous experience or from calculations using the theories of probability. There are, however, constraints in both of these methods which may render any analysis erroneous, as can be illustrated from the following light-hearted quotation:[24]

> The young man could open either door he pleased. If he opened the one, there came out of it a hungry tiger, the fiercest and most cruel that could be

23   An example of such a project is a refinery in India, which was hit by a cyclone in June 1998. It was reported in *Schaden Spiegel*, No. 1, 1999, that at the time, erection work was in full swing with about 50,000 workers on site.

24   *Witches, Floods, and Wonder Drugs: Historical Perspectives on Risk Management* by William C. Clark, Laxenburg, Austria, 1979. The quotation referred to is taken from *Engineering Risk*, a publication of the Institution of Professional Engineers of New Zealand, 1983, page 26.

procured, which would immediately tear him to pieces. But if he opened the other door, there came forth from it a lady, the most suitable to his years and station that His Majesty could select among his fair subjects. So I leave it to you, which door to open? The first man refused to take the chance. He lived safe and died chaste. The second man hired risk assessment consultants. He collected all the available data on lady and tiger populations. He brought in sophisticated technology to listen for growling and detect the faintest whiff of perfume. He completed checklists. He developed a utility function and assessed his risk averseness. Finally, sensing that in a few more years he would be in no condition to enjoy the lady anyway, he opened the optimal door. And was eaten by a low probability tiger. The third man took a course in tiger taming. He opened a door at random and was eaten by the lady.

The constraints in using the first method of analysis, i.e. the knowledge from previous experience of similar projects, are based on the fact that each project has its own individual identity and characteristics and, as such, no two projects are exactly the same. Therefore, previous experience has only a limited influence in assessing correctly the risks and their significance and can only be used as a guideline.

The second method, which is based on the theories of probability, has more serious constraints which can be best illustrated by reference to the well-established example of dice-throwing, an example which typifies the calculations of probability. They can be summarised under four categories:

1   The theory of probability is best suited to situations involving large numbers. However, large numbers of similar events do not usually occur in construction; therefore, risk assessment involving the likelihood of occurrence has to be based on only a few occurrences. Calculations of probability have to be modified to suit the situation of low number of events. To illustrate the difference between the probability of occurrence in a large number of events and that in a small number of events, let us consider the probability of any of the numbers 1 to 6 appearing if a dice is thrown. In a large number of throws, the probability is 1 in 6, but if only six throws are contemplated, all of the six numbers may not appear. In fact the probability of this happening is very low and experimentally it is only 1 in 100.

    If the example of a dice thrown is translated into the construction of a civil engineering scheme, the calculations applicable to the high frequency of occurrence would have to be modified if they were to apply accurately to the low frequency events in the construction of a project: a comparison of 1 in 6 to 1 in 100.

2   The theory of probability assumes a known number of possibilities. Thus in a six-sided dice, the number of possibilities (the extent) is six and that in an eight-sided prism is eight. However, in an engineering context, the number of possibilities is not quantifiable, especially in the case of an event with a complicated sequence-pattern.

3   The theory of probability assumes that it is possible to construct a model which can be used to verify the calculations. This is not always the case in construction projects where the situation is similar to that of the dice with faces having concealed numbers.

4   The theory of probability assumes that the magnitude is capable of assessment and such is the case in dice-throwing, in that whatever appears on the top of a thrown dice signifies the magnitude of the throw. In construction, the significance of the event is not always capable of assessment.

Whichever method of assessment is used, the object of course would be to try and eliminate or mitigate the risks and end in a safer project and one with the least exposure to hazards. Such a project would carry with it the minimum of events that could lead to responsibilities and liabilities.

3

THE SPECTRUM OF HAZARD AND RISKS IN CONSTRUCTION

It is only by risking our persons from one hour to another that we live at all.

(Williams James, 1897)

No Risk is the Highest Risk of All.

(A. Wildavsky, *American Scientist*, 1979)

The ideas captured by the above quotations have formed the pivotal factor in many aspects of human life. In insurance, the 1601 Marine Insurance Act of the English Parliament is an example.[1] In law, voluntary assumption of risk results in contributory responsibility, see page 14, and it may be observed in many authoritative works on sociology and philosophy. Nevertheless, few people involved in construction perceive its importance within the work done by the parties involved. It should be perceived that the risks inherent in construction should be specifically allocated in the contract(s) between the parties involved and that they may also be offloaded from one party to another by agreement. They may also be offloaded to a third party such as an insurer or a banker but that choice has to be based on proper principles relevant to the transaction and should also be referred to in their contract(s). The agreements made must be based on clarity and understanding of the respective risks, responsibilities and liabilities which are allocated to each. It is who bears the risk that is important.

As a party enters into contractual obligations freely, it accepts certain risks that are allocated to it and promises to bear these risks if and when they eventuate. In this way, the contracting parties are able to plan ahead with calculable certainty their schemes and arrange their business affairs. There are, however, specific risks that are beyond the capacity of a party to accept. In such circumstances, it would be better to name these risks and specify the method of dealing with and managing them.

In this chapter an attempt is made to classify the risks inherent in construction in such a way that a spectrum may emerge identifying the risks that one might

1 See page 4 above.

expect.[2] If these expectations are clearly understood, attention may then be focused on which risks should be expected on a particular project and how they could or should be dealt with.

Classification based on chronology

In order to produce as wide a spectrum as possible, the classification used in this chapter is that based on chronology as expressed in Figure 2.3. Of course, any other classification could be utilized, but then a different spectrum would emerge. Figures 2.3, 3.1 to 3.3, 3.6, 3.9 and 3.11 are drawn as flow diagrams which arrange, in chronological order, the various stages of a construction project and list in general terms the risks considered important. As it is most difficult, if not impossible, to perceive all possible risks and thus present a complete spectrum, the figures referred to above show a wide but not complete range to which others could be added once a project is accurately defined or new risks are envisaged.

The spectrum of risks in construction

Figure 2.3 on page 46 above displays the spectrum of risks by showing that the risks included are those affecting, in principle, the construction trinity of owner, professional team and contractor and, in certain circumstances, the community at large. It then divides the spectrum, in a chronological manner, into seven stages, allocating to each stage the relevant risks. Figure 3.1 deals with the feasibility stage and Figure 3.2 with the design stage. Figures 3.3, 3.6 and 3.9 refer to the construction stage and Figure 3.11 to the post-construction stage.

E.1.1 Risks associated with the feasibility stage (Figure 3.1)

Figure 3.1 shows the risks in the feasibility stage during which the idea of a particular project on a specific site or in a certain area is born and a decision is then made to either proceed with it or abandon it. In some cases, the choice of a professional team precedes the choice of the site. In others, it is the other way round, relieving the professional team of the responsibility of site selection. For the purposes of this spectrum, we shall assume the first situation whereby the professional team is selected first and the choice of site is decided upon with assistance drawn from that team's recommendation. In the following sections, an example of each of these envisaged risks is given from real life.

E.1.1.1 Owner's choice of professional team and advisers

On 27 March 1981, the reinforced concrete roof of the Harbour Bay Condominium in Cocoa Beach, Florida, collapsed during concreting operations, bringing down with it the whole building and resulting in the death of eleven construction workers and the injury of twenty-three others. The building, a five-storey structure, was

2 'The Spectrum of Risks in Construction', FIDIC's Annual Report of the Standing Committee on Professional Liability, June 1985.

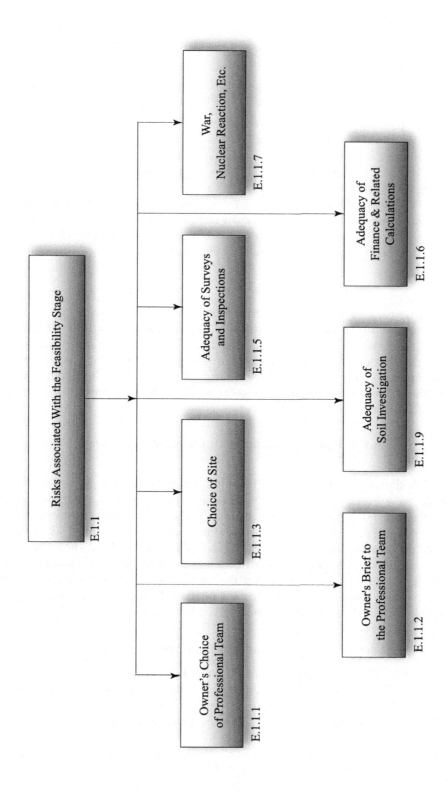

Figure 3.1 Risks associated with the feasibility stage.

242 ft long×58 ft wide and designed as a cast *in situ* reinforced concrete flat slab and columns by two retired structural engineers. One of the two engineers was employed as a consultant by the contractor to carry out design calculations, prepare documents and inspect the construction as it proceeded. The second was brought in by the first engineer as a sub-consultant to carry out the necessary calculations. Out of the seventy-nine sheets of calculations submitted to the Cocoa Beach Buildings Department only one was done by the first engineer.

The cause of the collapse was attributed to punching shear forces around the columns, which is a feature of flat slab design.[3] Despite the fact that it is a basic item of calculation in this type of design, the engineers did not consider it in their calculations. At full loading, the stresses exceeded those allowable by 469%. Even at dead load alone, the shearing stresses significantly exceeded the allowable stresses at all but two of the columns and were sufficient to cause the collapse.

The investigations that followed the collapse showed a variety of design and construction errors, some of which were extremely serious. For example, the weight of the external masonry walls was not taken into account in the calculations for weights on foundations. The columns as constructed were much smaller than designed but, even if they were not, they would have been too small to carry the load safely. Slab deflections were neither calculated nor checked against those permissible in similar type design.

This case serves to show the importance of selecting the right consultant for the particular project or even for the particular discipline.

E.1.1.2 Owner's brief to the professional team

The owner's brief is one of the most important documents and yet that to which the least attention is generally paid. It is a document which sets out the basic definition of the project and the boundary within which it must evolve. It is sometimes referred to as the terms of reference (TOR) and should include such topics as would be necessary for the particular project. They would, however, refer to geographical, financial and technical limits; objectives, capacities and key components; period of service, life span of project and level of input.

Misunderstandings in this area have caused and continue to cause failures in the construction field. Examples include failures due to:

1   Change in use of the project not envisaged by the designer and yet contemplated by the owner;
2   TOR of the mechanical consulting engineer not including the design of a ventilation system capable of handling the corrosive atmosphere in a particular environment; and
3   Greatly differing understanding by the owner and the designer of their duties and responsibilities.

---

3   Professional Liability Loss Control Program prepared by National Program Administrator Inc. in cooperation with Simcoe & Erie General Insurance Co., Canada, Bulletin No. 57, January 1982.

E.1.1.3 Choice of site

A school was recently planned in a mountainous area of the United States of America where many feet of snow are a normal, annual, winter event.[4] The site designated for this project was located at the base of several large mountains and in a bowl-shaped hollow.

Designs were made and contract documents were prepared for a very costly school. The contract for the construction was entered into and work on the foundations proceeded satisfactorily, until the spring when the snow started to thaw. Large quantities of run-off water accumulated in the hollow chosen as the site for the school, forming a shallow but large lake submerging the foundations. The architect who had participated in the selection of the site was blamed and his professional liability insurer had to step in to rectify the situation.

Similar problems exist all over the world wherever natural hazards exist. When a site is chosen for a certain project, its natural characteristics must be investigated. The professional team should not depend on the owner's undertakings or assurances as to the suitability of the site. The defendants in the case of *Bowen v. Paramount Builders* in New Zealand relied on such an assurance from the owner/employer, but they were still held liable to a third party who purchased the property which subsequently suffered from subsidence due to unsuitability of the site.[5]

The site may also be unsuitable due to subsoil conditions, as in the case of *W.C. Kruger and Associates v. Robert D. Krause Engineering Co. and Albuquerque Testing Laboratory, Inc.*, in New Mexico.[6] The facts of the case are that an architect retained a structural engineer to design the structural elements of a post office facility. He also retained a soil-testing consultant to perform a soil investigation based on a proposal which failed to include a partial basement with footings substantially below the depth of the soil investigation made. The soil consultant submitted his report making the recommendation that further settlement computations should be carried out.

The additional computations were not made and the work progressed. During excavations, it was discovered that the site was not suitable for the foundation design recommended by the soil consultant. An additional investigation had to be performed and the foundation had to be redesigned resulting in an additional cost of $54,363. The architect settled the claim made by the owner and brought an action to recover from the structural engineer and the soil consultant on the basis of negligence.

The trial court found that, had the computations recommended by the soil consultant in his report been carried out, the problem that was encountered would not have arisen. On this basis, the court held that the architect, the structural engineer and the soil consultant were all negligent in allowing the work to progress without performing the additional settlement computations. The court, however, rejected the architect's claim against the structural engineer and the soil consultant and did not

---

4 *Guidelines for Improving Practice*, a publication of Victor O. Schinnerer & Co., Inc., in cooperation with the American Institute of Architects, The National Society of Professional Engineers in Private Practice and CNA Insurance, Washington, DC, USA, vol. 11, no. 5.
5 *Bowen v. Paramount Builders* [1977] 1 NZLR 394.
6 *Guidelines for Improving Practice*, op. cit., see note 4, vol VI, no. 1.

permit any recovery on the basis that he had contributed by his own negligence to any damages suffered.

E.1.1.4 Adequacy of soil investigations

Having established the risks to which the site is exposed from nature, perhaps the most controversial, widely litigated and arbitrated subject revolves around what lies below the surface of the ground. One of the reasons is that matters of opinion before construction become matters of fact afterwards and those involved tend to make judgment with the benefit of hindsight. On the other hand, another reason may be that there are many questions of a technical, economic and legal nature which keep changing and developing with time and location. These questions may take the form of:[7]

> . . . to what extent does the employment of 'professional' site investigators entitle the owner to say that he has reasonable grounds to believe that their report was true, even when it proves to be quite misleading?

> Is there a difference between an owner who chooses site investigators of good reputation, supervises their work properly and gives tenderers the full information received from them (or perhaps from a series of site investigations over many years) and an owner who employs the cheapest investigators that can be found, with an inadequate brief, and provides only a part of their conclusions?

> . . . Conclusions about the site drawn from a few boreholes may be worded as fact, but are based not only on the factual results in the columns of ground that have been bored, but on the opinion that those results are typical of the ground between the bores.

Is this a reasonable assumption?

There are also questions about who should bear the responsibility for this risk. Since no two projects or sites are alike, the terminology used in conditions of contract is by necessity quite vague, in order to resolve part of this problem. Clauses 12 of three sets of Conditions of Contract are quoted below which differ slightly in detail but both use the words 'physical conditions', 'artificial obstructions', 'reasonably', 'foreseen', 'experienced contractor' and 'Unforeseeable'.[8]

ICE, 7th edition

> 12 (1) If during the carrying out of the Works the Contractor encounters physical conditions (other than weather conditions or conditions due to

---

7 'Risk Management', by Max W. Abrahamson, [1984] ICLR 241. It was also a paper presented at the International Construction Law Conference organised by the Master Builders Federation of Australia, held in Sydney, October 1982.

8 'Unforeseeable' is a defined term in the New Red Book of 1999. It is defined as 'not reasonably foreseeable by an experienced contractor by the date for submission of the Tender'.

weather conditions) or artificial obstructions which conditions or obstructions could not in his opinion reasonably have been foreseen by an experienced contractor the Contractor shall as early as practicable give written notice thereof to the Engineer.[9]

FIDIC, 4th edition of the Red Book

12.1 If, however, during the execution of the Works the Contractor encounters physical obstructions or physical conditions, other than climatic conditions on the Site, which obstructions or conditions were, in his opinion, not foreseeable by an experienced contractor, the Contractor shall forthwith give notice thereof to the Engineer with a copy to the Employer. On receipt of such notice, the Engineer shall, if in his opinion such obstructions or conditions could not have been reasonably foreseen by an experienced contractor. . . .[10]

FIDIC, The New Red Book

4.12 . . . If the Contractor encounters adverse physical conditions which he considers to have been Unforeseeable, the Contractor shall give notice to the Engineer as soon as practicable.

This notice shall describe the physical conditions, so that they can be inspected by the Engineer, and shall set out the reasons why the Contractor considers them to be Unforeseeable. The Contractor shall continue executing the Works, using such proper and reasonable measures as are appropriate for the physical conditions, and shall comply with any instructions. . . .

If and to the extent that the Contractor encounters physical conditions which are Unforeseeable, gives such a notice, and suffers delay and/or incurs Cost due to these conditions, the Contractor shall. . . .[11]

Following an interesting Australian case in 1972, it could be construed that, where the owner did not accept the task and therefore the responsibility of providing full and accurate information about the site and the contractor had not therefore relied upon such information, the risk was the contractor's.[12] The contract was to deepen a harbour using blasting operations which proved to be slower than the contractor expected. This was attributed to the existence of underground mine workings which were assumed to dissipate the effect of blasting. The contractor alleged that the

9 ICE Conditions of Contract and Forms of Tender, Agreement and Bond For Use in connection with Works of Civil Engineering Construction, Measurement Version, 7th edition, The Institution of Civil Engineers, London, 1999.

10 Conditions of Contract For Works of Civil Engineering Construction, 4th edition, Federation Internationale Des Ingenieurs-Conseils, Lausanne, 1987.

11 Conditions of Contract for Construction for Building and Engineering Works Designed by the Employer, 1st edition, Fédération Internationale des Ingenieurs-Conseils, Lausanne, 1999.

12 *Dillingham Construction Pty. Ltd. & Others v. Downs* (1972), 13 BLR, Supreme Court of New South Wales.

owner knew of the existence of these mine workings but did not make it known in the tender documents. It is no wonder that even eminent authors, arbitrators and judges do not seem to agree on what to do with this risk.[13]

It may, however, be interesting to quote from actual case histories concerning site investigation. It seems that, in 1826, the responsibility of a designer in respect of carrying out his own investigations was established through the case of *Moneypenny v. Hartland*.[14] The designer, in that case, accepted the results of borings taken by someone else who had been previously engaged by the owner. The design based on these borings was found later to be inadequate and the designer failed to recover his remuneration.[15]

In 1981, in *Eames London Estates Ltd. and Others v. North Herts District Council and Others*, the judge stated:[16]

> I consider it normal practice for an architect to draw his client's attention to the need for ground conditions to be investigated. Also, that the client be advised of the possible need to carry out a detailed site investigation, if the architect was uncertain in any way of the type and bearing capacity of the ground.

Commercial decisions on the type of investigation to be carried out or the type of foundation to be designed carry with them responsibility. The owner/employer must, therefore, be involved in those decisions if he were to take part of the risk that might benefit him financially. This was illustrated in the case of *City of Brantford v. Kemp & Wallace-Carruthers & Associates Ltd.*[17] Another view of the relationship between architect, engineer and owner was treated in the case of *District of Surrey v. Church* in Canada where the engineer was appointed by the architect rather than by the owner.[18] The engineer recommended to the architect that a soil investigation be carried out but the latter refused to accept the recommendation due to lack of money in the budget and the presumption that the owner would not approve such an investigation and that he would accept the building to be designed to a certain bearing pressure. Neither the architect nor the engineer approached the owner to verify these statements. It was contended that, had such an approach been made, the owner would have approved the required soil investigation which would have revealed that a layer of marine clay below the surface did not have the necessary bearing capacity to support the building.[19]

In the particular circumstances, serious differential settlement occurred and the owner sued both the architect and the engineer who were held jointly and severally

---

13   See judgments referred to in the article quoted in note 3 above.
14   *Moneypenny v. Hartland*, 1826, 2 C. & P. 378.
15   *Hudson's Building and Engineering Contracts*, by I.N. Duncan Wallace, 11th edition, Sweet & Maxwell, 1995, London.
16   *Eames London Estates Ltd. and Others v. North Herts District Council & Others* (1981) 259 EG.
17   *City of Brantford v. Kemp & Wallace-Carruthers & Associates Ltd.* (1960) 23 DLR.
18   *District of Surrey v. Church*, 1977, 76 DLR.
19   Ibid., See *Guidelines for Improving Practice*, op. cit., see note 4.

liable to the owner. The architect was held liable for breach of contract and the engineer in tort (since he had no contract with the owner) as he was held to have had a duty of care to inform the owner directly of the need for deep soil investigation, notwithstanding the lack of a contractual link.

E.1.1.5 Adequacy of surveys and inspections

There is no substitute for actually visiting the site and physically walking between its extremities and even beyond, keeping one's eyes open for any sign which might require special attention.

In *Balcomb and Another v. Wards Construction (Medway) Ltd and Others,* the engineer was held liable to his client, the builder in this case, for failing to exercise professional skill which would have alerted him to the presence of trees on the site in the immediate past.[20] He was also found liable in tort to the owners of the house which, in this case, was damaged by the heave of the clayey subsoil.

This principle of site inspection goes beyond the boundary of the actual building site, as happened in the case of *Batty and Another v. Metropolitan Property Realisation Ltd. and Others.*[21]

A development company and a builder had inspected land on the side of a valley. They both passed the site as suitable for development, but had they looked across the valley and on adjoining property, they would have seen what should have alerted them to the necessity of carrying out a soil investigation. In the event, three years after the construction of a house on one of the plots, which was located over a steep slope, a landslide occurred below the garden of the house damaging the fence and part of the garden. Although the house itself was undamaged on that occasion, the Court of Appeal in England held that it was doomed to failure and in imminent danger. The developer and builder were held liable to the house owner in tort and, in addition, in contract between him and the developer.

E.1.1.6 Adequacy of finance and related calculations

Getting paid for work done without any strings attached is a real risk. The construction industry cannot function if John Heywood's quotation given below is frequently applied:

> Let the world slide, let the world go; A fig for care, and a fig for woe! If I can't pay, why I can owe, And death makes equal the high and low.

However, in some cases, the calculations made of what is reasonable to pay to complete a project may be erroneous and the finance allocated may prove to be insufficient. In other cases, unforeseen events may result in the owner becoming incapable of honouring his commitments. Thus it is always wise to consider this risk

---

20   *Balcomb and Another v. Wards Construction (Medway) Ltd and Others, and Pettybridge and Another v. Wards Construction (Medway) Ltd and Others* (1981) 259 EG 765.
21   *Batty and Another v. Metropolitan Property Realisation Ltd and Others* 1978 2 All ER 445.

and the consequences which may flow from its occurrence. A few of the most spectacular occurrences of cost overrun are listed below:

The Concorde Project: In 1959, the project was estimated to cost £95 million. The total development cost was finally £1,140 million.

The Sydney Opera House: The Sydney Opera House was originally estimated in 1967 to cost $A6 million and, when it was completed in 1973, the cost had risen to $A100 million.

The North Sea Oil Fields: The cost of 20% of North Sea fields was up to 200% overrun; 30% of North Sea fields were up to 100% overrun and 50% of North Sea fields were up to 50% overrun.

The Channel Tunnel: The Channel tunnel, which is one of Europe's largest infra-structure projects ever, is 31 miles long and, on average, 150 ft under the seabed, started at an estimated cost of £4 billion and ended in the mid 1990s at £15 billion.

E.1.1.7 War, nuclear reaction, etc.

The consequences of hazards such as war, nuclear reaction and such similar events are so devastating, if a project is exposed to them, that the owner must consider them on their own. The risk in such a hazard materialising must be assessed by experts in the relative field or ignored completely. The decision to proceed with a project must be taken by the owner once the balance of probability is considered by him.

E.1.2 Risks associated with the design stage
(Figure 3.2)

Once a project passes from the feasibility stage to the design stage, the decision-maker, amongst other things, must have assessed the implications of the various risks indicated in Figure 2.1 and passed them as acceptable.

Lord Edmund-Davies said in the House of Lords case of *Independent Broadcasting Authority v. EMI Electronics & BICC Construction*, the following:[22]

> . . . The project may be alluring. But the risks of injury to those engaged in it, or to others, or to both, may be so manifest and substantial and their elimination may be so difficult to ensure with reasonable certainty that the only proper course is to abandon the project altogether. Learned Counsel for BICC appeared to regard such a defeatist outcome as unthinkable. Yet circumstances can, and have at times arisen, in which it is plain common-sense and any other decision foolhardy. The law requires even pioneers to be prudent.

22 *Independent Broadcasting Authority v. EMI Electronics and BICC Construction* (1980) 14 BLR 1.

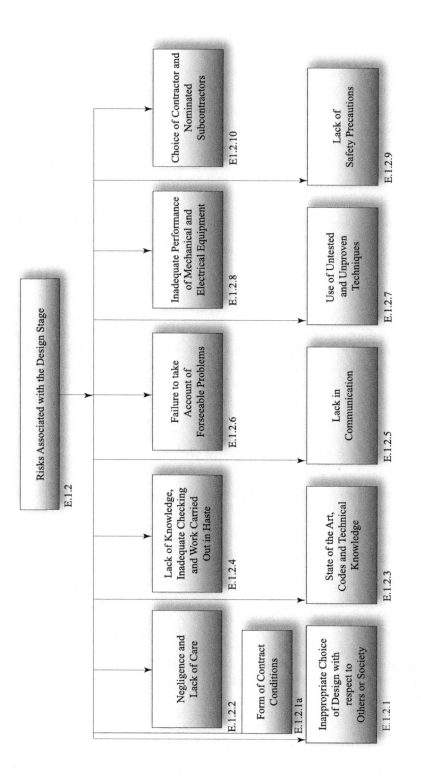

Figure 3.2 Risks associated with the design stage.

Assuming that there are no changes in the identity of the professional team, the first risk that must be considered during this stage is that of inappropriate conceptual design and its suitability, not only in respect of the project itself but also for third parties and for society in general.

Statistics based on an analysis of 10,000 building defects in France during a period of ten years indicate that, where design faults are concerned, conceptual design is responsible for 18% of all causes in terms of cost and 14% in terms of frequency of occurrence.[23] It was also shown in that analysis that the major culprit in design faults was poor detailing. It accounted for 59% of all causes in terms of cost and 78% in terms of frequency of occurrence.

Errors in calculation, to the surprise of some, were only responsible for 13% of design faults in terms of cost and only 3% in terms of frequency of occurrence. The last category of causes was unsuitable material which accounted for 10% of all causes in terms of cost and 5% in terms of frequency of occurrence.

The statistical evidence in respect of the cost and frequency of poor detailing demonstrates that not only is architecture an art but so, too, is engineering. Dr. Peter Miller, in his presidential address to FIDIC stated:[24]

> Somewhere down the track, I think engineering took a wrong turning. It allied itself with Science in public perception when, in fact, it is very much more of an art form. Very few structures have ever failed because of stress analysis faults. They mostly fail because of detailing faults, and detailing is a process in which an engineer applies his experience I suggest to you, his art.

The high number of defects due to poor detailing as compared with defective calculations shows also that, where design is concerned, it is more difficult to master art than science. The following quotation is of interest in this context.[25]

> However, in the rush to master that field (science) we must not lose sight of the difference between the function of the engineer and that of a scientist. The latter's function is to know, but the function of the former is more extensive in that it is not only to know but more importantly, it is to do.
> . . .
> The scientist adds to the reservoir of knowledge; the engineer brings this knowledge to bear on the practical aspects of problem solving. . . .

Let us now look at the risks in their individual capacity.

23 'The Structural Engineer in Context', A.C. Paterson, Presidential Address to the Institution of Structural Engineers, 1984, *The Structural Engineer*, November 1984.
24 'The Future', Peter O. Miller, a presidential address to the International Federation of Consulting Engineers, Annual Conference, Florence, Italy, 1983.
25 'The Civil Engineer – A New Role in an Old Industry', Nael G. Bunni, The Institution of Engineers of Ireland, Special Conference entitled 'A Future for the Civil Engineering Industry', November 1984.

E.1.2.1 Inappropriate choice of design with respect to others and to society

If the design carries a risk of it being injurious to the environment, or to others living in the vicinity, or, in some circumstances, even in faraway places, the risk must be assessed carefully before it can be passed as acceptable. If not acceptable, the design should be altered or changed completely. Examples of such issues, which have become topical in recent years, are acid rain and dumping of nuclear waste. In a similar vein, an example of what can happen occurred in Bhopal, the capital of the State of Madhya Pradesh in Central India, on 2 December 1984.

In 1975, the Indian Government granted Union Carbide, an American-based company, a licence to manufacture pesticides. The plant was located in Bhopal which had, at that time, a population of 900,000 and has since increased by a further 100,000. The site chosen for the plant was on the outskirts of Bhopal, despite its proximity to the densely populated areas of the city and despite an existing local regulation requiring that manufacturing plants producing dangerous substances should be sited at least 15 miles outside populated areas.

The original design of the plant was based on producing the pesticide by the use of an extremely effective imported material called methyl isocyanate (MIC) to produce sevin carbaryl. However, in 1980, in accordance with India's industrial self-sufficiency policy, the plant was modified to produce MIC itself. Thus, carbon monoxide was mixed with chlorine to form the deadly phosgene gas which was in turn combined with methylamine to produce the methyl isocyanate.

At around 2 a.m. on 2 December, 1984, a massive leak occurred in one of the three storage tanks, each holding 40 tonnes of MIC. Due to its very low boiling point, the chemical, which is stored in liquid form, turned into gas as it leaked into the atmosphere.

The needle of the tank's pressure gauge had moved into the danger zone before it was noticed by the night-shift worker. He notified his supervisor who sounded the alarm but it was already too late. The noxious white gas had started to escape from the tank and was spreading with the northwesterly winds in cloud formation towards the densely populated areas of Bhopal. In the thirty minutes which elapsed before the tank was sealed, 5 tonnes of the gas had escaped. With a safe content of only two parts per million, the gas cloud was a deadly one.

The alarm sounded by the supervisor activated the plant's siren and, thinking that a fire had started, hundreds rushed towards the plant straight into the path of the deadly gas.

Methyl isocyanate is so unstable and so dangerous that even professional toxico-logists are reluctant to study it in the laboratory. It belongs to a family of toxins for which there is no antidote and no treatment. But the first effect of exposure is water-ing of the eyes and damage to the cornea rendering its cells opaque. Subsequent effects resemble those of nerve gas. When inhaled, it reacts with water in the lungs, often choking the victim to death instantaneously. It can be just as lethal when absorbed through the skin. The survivors may be left with permanent disabilities such as blindness, sterility, kidney and liver diseases, tuberculosis and brain damage.

The catastrophe in Bhopal is probably the worse industrial accident in history causing the death of over 2,500 and injuring as many as 100,000. Bhopal, in minutes, had turned into a city of corpses. Muslims were piled on top of each other in hurriedly dug graves and Hindu funeral pyres burned around the clock because of fear of a cholera epidemic.

Besides the human toll, the tragedy of Bhopal is without precedent in terms of insurance and financial implications. Some lawyers, in the tradition of champerty, travelled without delay to Bhopal and initiated and filed suits in the US District Court in Charleston, W. Va., for $5 billion in punitive damages on behalf of 'all injured and deceased residents'.

Another aspect of unacceptable design occurs if there is an infringement of patent such as took place in the following case, whose ending was not a typical outcome of similar cases:

An engineer was retained by an owner to design a water treatment plant.[26] When the designs and documents were completed, tenders were invited for the supply and installation of the special equipment from contractors experienced and capable in this field of activity. The contract was awarded to the lowest tenderer A and the others, including tenderer B, were informed accordingly.

When the fabrication of the equipment was half-completed, Company B advised the owner that the equipment as designed infringed its patents. The owner in turn claimed from the engineer and the latter claimed from contractor A who advised him that fabrication would be discontinued until the matter was resolved and he (contractor A) was fully indemnified against patent infringement. The building contractor could proceed no further and he joined the battle and brought the project to a halt.

The engineer's design so closely matched the features depicted in Company B's bulletin that performance by Company A of its contract with the owner involved infringement if Company B's patent was valid. The engineer faced delay claims by the building contractor, plus either Company A's damages for breach of contract should the owner decide to take such proceedings against Company A for non-performance, or payment to Company A for modifications to avoid patent infringement. Alternatively, the engineer had the choice to pay damages to Company B for the infringement.

However, the engineer was satisfied that his design did not constitute an infringement because there was no such allegation at the tendering stage. He obtained a copy of Company B's patent in order to either establish invalidity, or design some modifications to avoid the infringement.

The patent had been applied for on 4 May 1956 which meant that, in order to invalidate it, it was necessary either to establish a use of its content prior to that date anywhere in the world, or to show that the content was obvious to anyone skilled in the art as of 4 May 1956.

A search of the *American Waterworks Journals* showed that a similar device to the subject of the patent had been used as far back as 1939. With this information, the patent for the earlier device was obtained and, when compared with that of Company B, it was found to be sufficiently close to make a strong case for making Company B aware of the consequence of their intended action. Faced with that risk,

---

26 'Investigate, Don't Capitulate', Report of the Standing Committee on Professional Liability, FIDIC, Item 3.2.1., 1984, Switzerland.

Company B withdrew their claim and indicated that they did not intend to pursue the allegation.

E.1.2.2 Negligence and lack of care

Negligence of a professional person has been defined in common law jurisdictions through various legal decisions. In *Bolam v. Friern Hospital Management Committee*, Mr Justice J. McNair stated:[27]

> How do you test whether this act or failure is negligent? In an ordinary case it is generally said that you judge that by the action of the man in the street. He is the ordinary man. In one case it has been said that you judge it by the conduct of the man on the top of the Clapham omnibus. He is the ordinary man. But where you get a situation which involves the use of some special skill or competence, then the test as to whether there has been negligence or not is not the test of the man on top of the Clapham omnibus, because he has not got this special skill. The test is the standard of the ordinary skilled man exercising and professing to have that special skill. A man need not possess the highest expert skill; it is well established law that it is sufficient if he exercises the ordinary skill of an ordinary competent man exercising that particular art.

The courts have since relied on this test. In the recent case of *QV Ltd and QV Foods v. Fredrick F. Smith and Others*, this test of an ordinary competent skilled building designer was applied to show the standard of care owed by the first defendant in carrying out the design of the building in question.[28]

To whom can one be negligent? This is an important consideration. In most legal systems, the law of negligence has developed to such an extent that the risk is of major importance, and one may be liable not only in contract but also in tort. The construction trinity may, therefore, be liable to each other and also to third parties who have no interest in the construction project. The length of the period during which one is exposed to liability has also been extended through the tort net. Specialist reference books in the relevant jurisdiction should be consulted when answers to these questions are sought.

E.1.2.3 State of the art, codes and technical knowledge

Innovation and technological advancement in all facets of construction must continue in order to improve standards and reach beyond present achievements. The results, if successful, can be expressed in terms of either cost benefit or the production of something new for the benefit of human existence or luxury. If the results prove to be unsuccessful and cause loss or damage, then as the risk of such an event

---

27 *Bolam v. Friern Hospital Management Committee* [1957] 2 All ER 118.
28 (1) *QV Ltd (formerly Holbeach Marsh Co-operative Ltd.)*; (2) *QV Foods Ltd (formerly QV Ltd.) v. (1) Fredrick F. Smith (Trading as Fredrick F. Smith Associates)*; (2) *D.A. Green & Sons Ltd (Defendants) and Eternit UK Ltd. (Formerly Eternit TAC Ltd) (Third party)*, (1998) QBD Official Referees' Business.

occurring is high, it is only just that it should be borne by those benefiting, providing they were given the opportunity to decide for themselves whether or not the innovation was to be pursued.

This principle in common law can be traced back to 1853 in the case of *Turner v. Garland and Christopher* where a designer was asked to prepare plans for the erection of model lodging houses, using a new patent concrete roofing which was cheaper than the alternatives available.[29] The patent concrete roofing was not a success and had to be replaced. The owner claimed in negligence from the designer but the judge told the jury that, although failure in an ordinary building was evidence of want of competent skill, yet if, out of the ordinary course, a designer is employed in some novel concept in which he has no experience and which has not the test of experience, failure may be consistent with skill.

In more recent times, however, another design at the frontier of professional knowledge ended in collapse and was the subject of a court case with a different outcome. It was the case of *Independent Broadcasting Authority v. EMI Electronics Ltd. and BICC Construction Ltd.* (quoted in note 22). The case was decided in the United Kingdom by the House of Lords in 1982, but the events occurred in 1969 when, on 19th March, the 1,250 ft high cylindrical television mast at Emley Moor in Yorkshire collapsed. The collapse occurred after a flange at a height of 1027 ft. above ground level fractured due to vortex shedding induced by wind and the asymmetric loading of ice on the mast and the stays. The cause of the fracture was attributed to defective design, which at the time was accepted as being at and beyond the frontier of professional knowledge. The designers had assumed that excessive deposits of ice would crack and fall away in the wind and this did not happen.

The statement quoted below is relevant to the discussion here:

> What is embraced by the duty to exercise reasonable care must always depend on the circumstances of each case. They may call for particular precautions: *Redhead v. Midland Railway Co.* (1869). The graver the foreseeable consequences of failure to take care, the greater the necessity for special circumspection: *Paris v. Stepney Borough Council* (1951). Those who engage in operations inherently dangerous must take precautions which are not required of persons engaged in the ordinary routine of daily life: *Glasgow Corporation v. Muir* (1943). The project may be alluring. But the risks of injury to those engaged in it, or to others, or to both, may be so manifest and substantial and their elimination may be so difficult to ensure with reasonable certainty that the only proper course is to abandon the project altogether. Learned Counsel for BICC agreed to regard such a defeatist outcome as unthinkable. Yet circumstances can, and have at times arisen, in which it is plain commonsense and any other decision foolhardy. The law requires even pioneers to be prudent.

Had the owner been informed of the features of the design, would the above statement have been different?

---

29 *Turner v. Garland and Christopher* (1853), *Hudson's Building Contracts*, 4th edition, vol. 2, page 1.

E.1.2.4 Lack of knowledge, inadequate checking and work carried out in haste

Although a professional may be qualified and experienced to carry out the design of a certain project, he may still lack knowledge of a particular aspect of the design. The problem is that if he does not realise his limitation, he may proceed without executing his duties properly. This occurred in the case of a firm of consulting engineers commissioned to design a steam power station for which the various pieces of equipment were ordered directly from the manufacturers who supplied, independently of each other, in accordance with the specification.[30] Unsuitable relays were, however, ordered and installed for the safety and protection system of the generator of a 32 MVA turboset. During commissioning, it failed to operate and respond properly when a fan blade broke accidentally and was thrown into the stator winding head. The blade fragments caused an earth and short circuit with arcing. The unsuitable relays in the safety devices responded but only after a thirty-second delay causing considerable damage to the whole turboset.

Due to either economic restraints or shortage of time, this type of risk increases, reaching unacceptable levels and situations, which may produce problems later during construction. The level of this type of risk also increases if economic restraints or shortage of time exist. This may occur during construction as in the following example.

An engineer was engaged for the design of retaining walls as part of a site stabilisation plan for a large city building.[31] A system of ground anchors was chosen to stabilise the retaining walls, and due to the short period allocated to the construction, the engineer permitted work to proceed without carrying out preliminary tests to establish the load capacity of the anchors. It was not until the work was well advanced that stressing of the anchors was first attempted; it was discovered then that their capacity was below the design load. Work on the contract came to a halt until the matter was resolved by adding further anchors throughout the whole wall.

The owner, who had to pay for the delay and the additional anchors, started to prepare a case against the engineer who, in allowing the work to proceed, was only giving the owner a commercial advantage and benefit.

It was fortunate for the engineer in this case that, by virtue of a sympathetic and reasonable report by an independent consulting engineer, the claim was averted.

E.1.2.5 Lack of communication

Communication has been identified as the most important requisite of success, and lack of it is perhaps the most significant factor in human failure. Its recognition as a cause of failure goes back to the first construction project, that of Babylon, as recorded in the Revised Standard Version Common Bible, Genesis, Chapter 11, which reads:

> Now the whole earth had one language and few words. And as men migrated from the East they found a plain in the land of Shinar and settled

30 *Schaden Spiegel*, a publication of the Munich Reinsurance Company, No. 1, 1982, Munich.
31 'Lessons to be Learnt', FIDIC Standing Committee on Professional Liability Report, 1983, page 14, Switzerland.

there. And they said to one another, 'Come, let us make bricks and burn them thoroughly.' And they had bricks for stone and bitumen for mortar. Then they said, 'Come let us build ourselves a city and a tower with its top in the heavens and let us make a name for ourselves lest we be scattered abroad upon the face of the whole earth.' And the Lord came down to see the city and the tower which the sons of men had built. And the Lord said, 'Behold they are one people and they have all one language; and this is only the beginning of what they will do; and nothing that they propose to do will now be impossible for them. Come let us go down and then confuse their language, that they may not understand one another's speech.' So the Lord scattered them abroad from there over the face of all the earth and they left off building the city. Therefore its name was called Babel.

On a more recent note on communication or lack of it, a number of disputes would not have arisen had the parties involved explained to each other, in clear language, what risks and responsibilities each has been allocated; see Chapter 2.

E.1.2.6 Failure to take account of foreseeable problems

An engineer advised a city water authority to close certain valves in its water supply system to allow leakage tests to take place.[32] During the tests and whilst the valves were closed, a fire broke out in a factory and the water pressure in the system was not sufficient to fight the fire.

The court ruled that the engineer, as an expert, was aware that closure of the valves in one district would reduce the water pressure in the adjacent district and greatly increase the risk of fire damage. The engineer was held liable for the fire damage sustained by the factory.

E.1.2.7 Use of untested and unproven techniques

Robert C. McHaffie Ltd recommended the use of a material to produce lightweight concrete, which material proved to be unsuitable.[33] The court in the case of *Sealand of the Pacific v. Robert C. McHaffie Ltd* held that the respondents should not have relied on manufacturer's literature in recommending the material for use. Furthermore, if they wanted to use the material, they should have carried out their own tests and examinations. Accordingly, they were held liable.

E.1.2.8 Inadequate performance of mechanical and electronic equipment

More and more designers are using the electronic equipment readily available in today's design offices for analysis, design and drafting. In doing so, they are using hardware equipment designed and manufactured by others and software written and checked by yet another party. In order to guard against unauthorised use of copying, the software is secured in such a way that the user cannot check or disassemble the

32 'Lessons to be Learnt', op. cit., see note 31.
33 *Sealand of the Pacific v. Robert C. McHaffie Ltd* (1974) 51 DLR, Canada.

steps used in the design of the software. He is, therefore, unaware of the assumptions made and the methods utilised in the solution of the problems. The risk of incorporating incorrect computer results in construction is a very real one and can only be mitigated by meticulous and critical checking using common sense and experience.

E.1.2.9 Lack of safety precautions

An employee of a seed grain drying plant went to sweep up grain around the hatches on top of one of the grain bins.[34] He fell through a hatch and was fatally injured. His widow was awarded US$280,000 by the courts which held that the designers of the plant and the builders were jointly liable. The fact that there were no code recommendations concerning such hatches did not absolve the designer and the builder of the responsibility to protect users.

E.1.2.10 Choice of contractor and nominated subcontractor

There is an implied warranty in a construction contract that good material and workmanship will be used. The risk that defective material and workmanship are used in a construction contract can only be mitigated through careful selection of the contractor and any subcontractors to be named in the contract.

E.2.1 Risks during construction associated with the site of the project and its location (Figure 3.3)

Statistics based on an analysis of 10,000 building defects, recorded through the decennial liability insurance in France between 1968 and 1978, showed that construction faults ranked highest in frequency of occurrence.[35] The analysis showed that 51% of all faults were due to construction, 37% to design, 7.5% to faulty maintenance and 4.5% to defective material.

In terms of cost of repair, the analysis showed that design faults and construction faults each accounted for 43% of the total cost of repair. Faulty maintenance accounted for 8% and defective material for 6% of the cost.

These faults did not all occur during the construction period. In fact Figure 3.4 shows the distribution of when faults occurred or were detected against time, indicating that only 11% occurred during the construction period. The figures may be different if civil engineering or if building defects in another country were to be considered.

But let us look first at examples of the risks commonly referred to as 'Acts of God'.

E.2.1.1 Excessive Rainfall

Water pipeline in Africa: Trenching had already been completed for the entire 60 km pipeline that was to be joined with couplings.[36] Some of the pipes had already been

---

34 'Lessons to be Learnt', op. cit., see note 31, page 25.
35 'The Structural Engineer in Context', op. cit, see note 23.
36 *Schaden Spiegel*, op. cit., see note 30.

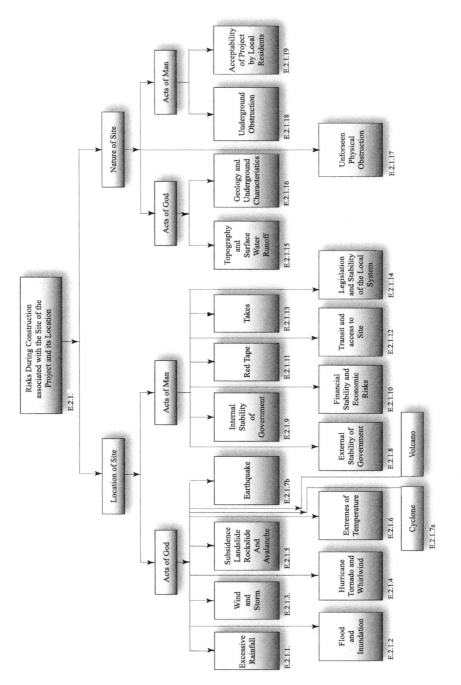

Figure 3.3 Risks during construction associated with the site of the project and its location.

laid in their final position. However, the couplings, for which adequate room between the pipes had been left, were still not available. When sudden, intensive rainfall started, the ditch was flooded and the pipes became filled with mud. During the repair operation, it was found that only some sections of the pipes could be salvaged by cleaning. The rest had already been rendered useless by the solidified mud, causing a total loss of DM 1 million. This example demonstrates how important it is to limit the length of open trench. (Figure 3.5)

E.2.1.2 Flood and inundation

It is well recognised that projects exposed to the effect of water may encounter some of the most hazardous conditions during construction. Damage can result in any of the following modes:

A Inundation and/or damage caused by rainfall;
B Flooding caused by excessive surface run-off either immediately after rainfall or consequent upon melting snow;
C Flooding due to failure of water conduits or storage structures;
D Flooding and inundation due to a combination of natural forces such as windstorms, tides, etc;
E Inundation due to underground water.

The damage which may occur due to this risk can be summarised as one or more of the following:

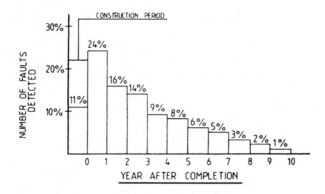

Figure 3.4 Building defects occurring or detected in the first ten years.

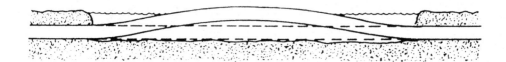

Figure 3.5 Flotation of a pipeline due to rainfall.

(a) Damage due to erosion and wasting away.
(b) Damage due to under-design of temporary works because of their short period of exposure.
(c) Damage due to lack of, or insufficient, flood warning in areas with certain topographical properties.
(d) Damage due to non-compliance with the construction programme.
(e) Damage due to insufficient waterway.
(f) Damage due to improper organisation which includes method of execution, arrangement of storage space, etc.
(g) Damage caused directly by rainfall.

There are many examples of damage relating to these risks during construction. Perhaps the more spectacular are those involving bridges and dams, but the most unexpected are floods occurring in the deserts. The following accident took place on a road project under construction, 105 km long, in a desert area where a 1,000 m high mountain range had to be crossed.[37] The topography of the mountains made it necessary for the road to follow the course of a wadi over a distance of 35 km. In this section of the road, thirty-one prefabricated viaducts had to be built as well as a number of retaining walls and culvert structures for side wadis.

The tender specifications did not contain much information on the amount of rainfall to be expected, as this desert area was only just being developed and, so, no statistics were available. During the two-year period of road construction, unexpected and heavy rainfalls occurred which flooded the wadi nine times. The floods lasted only one to two hours each but reached flow velocities of nearly 40 km/h and water discharge rates of up to 1000 m³/h.

The highest flood lasted about 45 minutes. Within the first two minutes the water rose by 1 m. The maximum of 4 m was reached after 15 minutes. The quick rise of the flood and the vast amount of water caused severe damage on the construction site. All nine floods caused extensive damages, including:

• Construction equipment, such as a screening and crushing plant, was destroyed or washed away.
• Excavators, bulldozers, loading gear, etc. which could not be removed from the wadi in time, filled with mud or silt.
• Culverts were washed away or clogged.
• Falsework and formwork material were lost or damaged.
• Foundations for abutments, piers or retaining walls were covered with mud. Slopes and benches were scoured.

E.2.1.3 Wind and storm

The strength and lateral stability of an uncompleted structure are in most cases much lower than those for a completed one. Severe damage can result on building sites exposed to wind forces, in particular to roofs; uncompleted walls; formwork and temporary buildings.

37 Ibid., July 1977.

Perhaps the most vulnerable to wind forces during construction is the large steel storage tank, as the next example demonstrates.

A project for a gas liquefying plant in the Middle East included the erection of nine tanks for the storage of final liquefied products: butane and propane.[38]

The tanks were erected by the usual method: pre-curved plates were welded together one after the other around the entire tank circumference. Thus, the tank grew in height ring by ring. Seven tanks had been completed in this way. When the workers were welding the top section of the eighth tank 30 m above ground, a hailstorm occurred with hailstones measuring 1 cm in diameter and wind velocities up to 100 km/h. The storm lasted for about 10 minutes. There was no time to secure the tank with guys, as had always been done before extended breaks and during the night.

As the tank had not yet been covered by its mushroom-type roof, suction forces were created inside the tank in addition to the great pressure forces applied by the storm on the outside. As a result, an area of 400 m² imploded. The working plat-form around the top tier of the tank was torn from its mountings and crashed to the ground. Under the influence of these forces, the buckled side of the tank was lifted off the foundation, leaving a gap of 10 cm at the bottom.

The first repair operation was to pull out the imploded section using winches and tractors. Then material samples from the buckled areas were examined in the Ismaning Institute's laboratory to see which plates could be reused and which had to be replaced. The damage was unforeseeable, since the storm had arisen within a few minutes. Fortunately, through a careful and well-conceived repair programme, the storage tank was slowly put back into horizontal alignment on its foundation which made possible a successful repair, thus avoiding a total loss. The total amount of the damage was accordingly limited to US$750,000.

E.2.1.4 Hurricane, tornado and whirlwind

As in the previous hazard of wind, the low strength and lateral instability of uncom-pleted structures make this type of hazard devastating in its effect. Hurricanes, by definition, have a speed greater than 120 km/h.[39] Whirlwinds, which are basically a column of air rotating rapidly with low atmospheric pressure at its centre, can be of three types: the tropical storms which develop in tropical sea areas having large diameters; the dust devils which develop in desert regions having much smaller diameters; and tornadoes which are the most destructive, reaching a velocity of 400 km/h and perhaps more. At that velocity, lightweight structures and industrial build-ings would be completely demolished, whilst reinforced concrete and steel framed structures could suffer serious damage.

The Midwest part of the United States is the area most exposed to tornado activity and the Wichita Falls/Mexico tornado of 10 April 1979 produced one of the most expensive single tornadoes on record with about US$200 million of insured damage.

38 Ibid., no. 2, 1981.
39 Ibid., no. 2, 1981.

E.2.1.5 Subsidence, landslide, rockslide and avalanche

Subsidence is one of the risks the causes of which are extremely complex to assess. This is basically due to the inexact nature of the science of soil mechanics, which covers the problem of subsidence. Subsidence may occur due to any of the following reasons or a combination of them:

- Lack of or insufficient site investigation;
- Incorrect assumption of distribution of stresses in or under foundations and the supporting soil layers;
- Improper support to sides of excavations;
- Changing the properties of the surrounding soil;
- Deterioration of the foundation material due to presence of aggressive substances in the soil such as sulphate salts, etc.;
- Inferior properties of the foundation soil;
- Defective design of the foundations.

Catastrophic accidents have occurred because of this type of peril. In many ways, it is connected with the previous hazard and therefore it is sufficient to add that instability of soil strata can only be averted by very careful handling of the design and construction of projects which are dependent, in one way or another, on the soil. Even then, accidents will happen.

E.2.1.6 Extremes of temperature

The effect of extremes of temperature on some materials and processes can be demonstrated by the following incident, which occurred during the erection of a plant for the production of coffee extract. A loss occurred in the most important section which houses the deep-freezer room where coffee is quick frozen at a temperature of −45°C and then it is ground and screened.[40] This room was insulated with 30 cm thick, expanded polystyrene slabs. Four suspended air coolers were installed to produce the low temperature required for the freezing process. The automatically controlled defrosting units attached to the coolers were designed to be operated at intervals of 24 hours for approximately 8 minutes.

   The loss occurred during installation of the defrosting units. When the wiring had been completed, the defrosting units (each having a power input of 15 kW) were switched on manually for testing purposes. At the beginning of their lunch break, the workers left the deep-freezing room and shut the door, without switching off the units. A technician happened to re-enter that deep-freezing room only 20 minutes later. The room temperature had already risen to more than 60°C and around the coolers it was even higher. The entire polystyrene insulation had shrunk considerably; at certain exposed areas the material had even become liquid and was dripping off. The damage was so extensive that the insulation had to be replaced throughout the upper third of the deep-freezing room and new lamps, switches and cables were required.

---

40 Ibid., October 1976.

If the technician had not discovered the error when he actually did, the damage would have been even more extensive as the temperature would have increased further and the air coolers suspended from nylon elements would have fallen down into the expensive coffee preparation plant.

This loss shows that, in the case of erection projects, it is imperative to obtain permission from the site management for any commissioning or testing operations, even if only minor plant items are involved. In fact, it is only in this way that it is possible to cope with the hazards inherent in the provisional operation of individual plant sections prior to completion of the entire project.

E.2.1.7a Cyclone

Cyclones pose a threat to nearly all coastal areas. An example of the devastating effect of such an event occurred on 9 June 1998, at the relatively sparsely popul-ated peninsula of Kathiawar in the north-west Indian state of Gujarat when it was hit by a tropical cyclone (Cyclone 03A) with peak wind speeds of 170 to 180 km/h. On the coast there was a storm surge with a height of 2 to 4 m. The toll was as follows:[41]

- probably more than 10,000 fatalities and 30,000 homeless;
- economic losses of about US$ 1.7 billion; of which
- the insured losses exceeded US$ 400 million.

Cyclone 03A developed on 6 June 1998 from a tropical low pressure system in the south-east of the Arabian Sea and steadily gathered in strength on its way north. On 8 June, the storm attained its maximum intensity with peak wind speeds of 240 km/h, but was already much weaker when it finally hit land on 9 June. Judging by the wind speeds and the duration of the storm, the waves must have been 5 to 6 m high. The astronomic tide reached its maximum that day about three hours after the eye of the cyclone had passed. If these two events had occurred simultaneously, the water levels would have been even higher.

In the cyclone track there were two refineries under construction some 20 to 30 km from each other and in one of which the erection work was in full swing. The total investment value of these two plants exceeded US$ 1.7 billion. Both of these refineries had erection all risks insurance cover. One of them was also insured for advance loss of profits. There were about 50,000 workers at one of these two con-struction sites in June 1998, making it one of the world's largest construction sites in the industrial sector at that time.

The severest damage was to the temporary installations like the site offices, workers' barracks, stores and the power supply.

The cyclone tore off roofs and hurled the corrugated metal sheets through the air like sheets of paper. Brick walls were smashed down. As a result of the rain that followed, there was considerable damage to office equipment, including computers and stored materials. The damage to the tank farm was likewise catastrophic. More

41 Taken from a paper by Andreas Gerathewohl, Martin Jenne, Ernst Rauch, Werner Teichert, published in *Schaden Spiegel*, op. cit., see note 30, No. 1 1999.

than a quarter of the 200 tanks, up to 92 m in diameter, were damaged or destroyed. The damage mainly affected tanks that were in the course of being erected and were therefore not sufficiently secured.

Installations directly on the coast were also destroyed. A pumping station, which had been built to provide seawater for desalination and supply the refinery itself with cooling water, was destroyed by barges that had broken loose and had to be rebuilt. This proved to be particularly critical for the advance loss of profit insurance cover. However, it was not only on land that Cyclone 03A left its mark of devastation but also at sea. Total losses and partial damage to over fifty ships off the coast of Gujarat generated marine insurance losses totalling hundreds of millions of dollars.

E.2.1.7b Earthquake

Earthquakes occur much more frequently than most people realise and on average ten potentially catastrophic earthquakes occur per year around the world. In addition, many hundreds occur causing serious local damage, many thousands can only be felt and modern seismographs record even a larger number.

Between 1974 and 1984, thirty-three major earthquakes occurred in twenty-two countries with a magnitude greater than 6 on the Richter scale. The number of resultant fatal injuries depends largely on the density of population in the affected area and varied between five dead in Guerrero (Mexico) and 242,000 in Tangshan (China). The property damage in the latter quake was estimated at US$5600 million. The heaviest property damage during that period occurred in Irpina in Italy on 23 November 1980, reaching an estimated amount of US$7200 million and resulting in the death of 3,114 persons.[42]

The total injury and damage that has resulted in the ten-year period is 330,000 dead and US$25.5 billion loss.

Despite the obvious catastrophic nature of this hazard, the risk of exposure to earthquake damage can be reduced through knowledge of its facets.[43] These are:

1 Zone

Site location determines the probability of occurrence of earthquake and, whilst it is not easy to generalise, one may observe that 80% of all earthquakes occur in the Circum-Pacific belt which follows the west coast of South America, Central America, North America, the arc of islands in the northern part of the Pacific, Japan, Taiwan, the Philippines and a section of Indonesia.

About 17% of earthquakes are observed in a belt which extends from the Azores in the Atlantic to the southern part of Europe and part of North Africa, Turkey, the Near East, including Iran, part of Arabia, Afghanistan, Pakistan, India and Burma. The probability of occurrence of earthquakes obviously varies greatly for each of the regions within these belts.

---

42 *Schaden Spiegel*, op. cit., see note 30, No. 1, 1984.
43 'S.R. Focus – A Short Guideline to Earthquake Risk Assessment', a publication of the Swiss Reinsurance Company, 1982, Zurich (H. Tiedemann).

2 Subsoil conditions

The type of subsoil and its stratification, the depth of layers and the position of the water table determine to a large extent the resultant effect. As a general rule, the harder the subsoil material on which a structure is founded, the smaller is the damage. Groundwater and sloping ground increase the damage.

3 Building materials

It is usual that various materials are used in any one building, each of which responds differently to earthquake forces. Therefore, the relative and cumulative response of these materials must be considered carefully. Elastic materials such as steel respond better than brittle materials. Stiff design incorporating shear walls capable of resisting bending moments is more effective in resisting earthquake than soft design. Prefabricated elements are in general more prone to damage due to earthquake forces than cast *in situ* parts.

4 Shape of buildings

The shape of buildings plays an important part in the type of response displayed. Deviations from absolute symmetry introduce different oscillations and vibrations which increase the probability of damage.

5 Sensitivity of machinery and plant

Most machinery and plant are usually susceptible to damage by falling debris, tilting bases and cracking foundations. The behaviour of the building elements and the type of damage they are expected to sustain due to earthquake forces are important factors for consideration when the building is designed to house machinery and plant.

6 Tolerances:

Permissible deviations from acceptable standards of design, material and work-manship are much lower in circumstances where earthquake forces are to be accommodated.

E.2.1.8 to E.2.1.14 Acts of man as related to location of site

These are a group of risks connected with the political, financial, sociological and status of the country in which the site is located. They can be enumerated as follows:

- External stability of government;
- Political risks;
- Internal stability of government;
- Financial stability and economic risks;
- Red tape;
- Transit to site and condition of infrastructure;
- Taxes; and
- Legislation and stability of the legal system.

E.2.1.15 Acts of God in relation to nature of site – topography and surface water run-off

'See E.2.1 Risks'.

E.2.1.16 Adverse geological and underground characteristics

The hazard of adverse geological and underground characteristics and the risks attached to it form a major topic in engineering, particularly when these conditions are not foreseen and not discovered during soil investigations that are carried out prior to construction.

Unforeseen adverse ground conditions have been described in an interesting article on this topic as one of the most notorious causes of disputes under engineering contracts.[44] Their incidence may have far-reaching effects on the course of the works and their resolution can often have a most serious effect on the economic balance under the contract. Given their importance, one would expect to find a rational scheme of transfer and placement of this risk, but unfortunately, experience suggests that the effect of the relevant provisions in standard forms of contract is anything but rational. This is perhaps due to the fact that these provisions pose a test which is related in part to what has immediately occurred, but is also dependent on conditions which existed at the date of the tender, and which are likely to be obscure and highly susceptible to dispute at the date of the occurrence. The important point is that the contract terms make no attempt to define the occurrence of risk in terms which can be applied directly or readily.

In the article quoted above, John Uff explains that unforeseen conditions typically produce large contractual claims, which could remain in dispute even after completion of the project. The principal reasons for this are, first, the lack of any clear criteria for determining whether the relevant events are established, and second, the qualified right to be reimbursed in respect of all additional cost if the event is established. The effect of these clauses in practice is, therefore, not to transfer risk but to provide a vehicle for making a claim for additional payment in the event that the relevant facts can be established subsequently. Indeed, it may be said that the contractual provisions embody two risks namely:

1   that circumstances will arise which allow the contractor to bring a claim; and
2   that an arbitrator might subsequently find the claim proved.

The article concluded that a risk event should not be regarded as equivalent to a claim situation created at the will of one of the parties. While the contractor may legitimately expect proper compensation for variations and imposed delays, risks should be dealt with so as to preserve a proper incentive to minimise their incidence.

---

44   'Contract Documents and the Division of Risk', by John Uff, part of a book entitled *Risk, Management and Procurement in Construction*, published by the Centre of Construction Law and Management, Kings College London, edited by John Uff and A. Martin Odams, 1995.

The tunnel collapse in September 1994 of the second phase of Munich's U-Bahn U2 underground extension is a dramatic example of an incident resulting from an unforeseen adverse ground condition. Two people died and thirty were injured when a crowded bus was sucked into a large hole that suddenly opened up in the middle of a main suburban road.[45]

The Munich area has a high water table, just 4 m below ground level. The tunnel was being driven with a cover between 3 m and 1.5 m of marl above the soffit, but it appears that the drive hit an unforeseen local depression in the marl stratum after less than 50 m of progress, prompting the collapse. It was reported that the collapse was preceded by an influx of water at the tunnel face and caused by a breach of the tunnel soffit.

Workers in the tunnel had enough warning to escape before the tunnel was flooded with tonnes of water and gravel for 20 m along its length, but nothing could prevent the crowded bus from plunging into the void that was created. The collapse also undermined the foundations of an adjacent apartment block and a shop, which were temporarily stabilised after the hole was plugged with concrete and supported with crushed stone fill. The bus, which became partially embedded into the concrete that was poured in, was later cut in two and lifted by a crane.

E.2.1.17 Acts of man in relation to nature of site – underground obstructions

Underground obstructions in the form of man-made cables, pipe ducts, and other conduits are susceptible to damage causing not only physical loss but also consequential damage which in many cases exceeds the cost of repair to the items directly affected. An investigation into the causes of damage of underground cables and pipes in the course of construction work carried out in a European country revealed the following statistics which give an insight into the risk:[46]

- In 30% of all cases, the contractor had not procured any plans showing the positions of telephone and electric cables or gas pipes.
- In 11% of the cases concerned, the plans received by the contractor were not available at the construction site.
- In 10% of these cases, the plans submitted to the contractor were incorrect.
- The telephone office was not notified in 40% of all cases in which its cable network had been damaged; the gas works in 18% and the electricity supply company in 13% of these cases.
- 84% of the cases were accounted for by construction machines.
- 60% of the cases occurred outside built-up areas.

The above findings can probably be regarded as generally valid in other countries also.

Due to the severity of this risk, some insurers recommend the inclusion of the following clause in covering major earthwork contracts:

---

45 *New Civil Engineer*, Magazine of the Institution of Civil Engineers, London, 29 September 1994.
46 *Schaden Spiegel*, op. cit., see note 30, No. 2., 1981.

The Insurers shall indemnify the Insured only in respect of loss of, or damage to, existing underground cables and/or pipes or other underground facilities if, prior to the commencement of works, the Insured has inquired with the relevant authorities as to the exact position of such cables, pipes or other underground facilities. The indemnity shall in any case be restricted to the repair costs of such cables, pipes or other underground facilities, any consequential damage being excluded from the cover.

There are many examples of this type of hazard of man-made obstructions. In February 1995, tunnelling work on the Jubilee Underground Line extension in London caused Blackwall Tunnel to sink 3 mm. The incident occurred when preparation for the chamber for the tunnel-boring machine was being carried out by hand digging. The workmen hit a pocket of peaty ground, which caused water to gush out into the westbound tunnel on the £71 million contract. Three miners were lucky to escape injury.[47]

The construction of the underground tube line at the Blackwall Tunnel in 1966 had included a steel sheet piled cofferdam to enclose the peaty soil that was encountered rather than remove it. The Jubilee tunnel required to be cut through the cofferdam and breaking through it released the retained soft material and pore water causing it to fall into the pit leaving a huge void. The existence of the cofferdam was known, but its purpose was unclear.

Excavation work was stopped immediately. Emergency procedures were then employed, and frantic remedial measures, including an immediate injection of grout into the ground to stabilise the weak material, averted a major ground collapse.

E.2.1.18 Acceptability of project by local residents

A concrete sewage pipeline had to be laid, extending 700 m into the sea.[48] The complete pipeline was assembled and laid on land ready to be dragged into the sea. It was then that overzealous environmental protectionists paid a visit and ignited three explosive charges destroying over 150 m of pipeline. The cost of repair amounted to DM800,000.

E.2.2 Risks during construction associated with the technical aspects of the project (Figure 3.6)

E.2.2.1 Extended duration of construction

It is evident that the longer the period of construction, the greater is the probability of occurrence of the hazards to which a project is exposed. However, in certain circumstances, there are seasonal hazards which occur at specific times of the year and thus require special consideration if the period of construction is to be extended. These hazards include rainfall, temperature changes, flood, storm and wind. To illustrate this point, the example of Diyala Bridge in Iraq may be cited. Designed as a

---

47 *New Civil Engineer*, op. cit., see note 45, 16 February 1995.
48 *Schaden Spiegel*, op. cit., see note 30, September 1975.

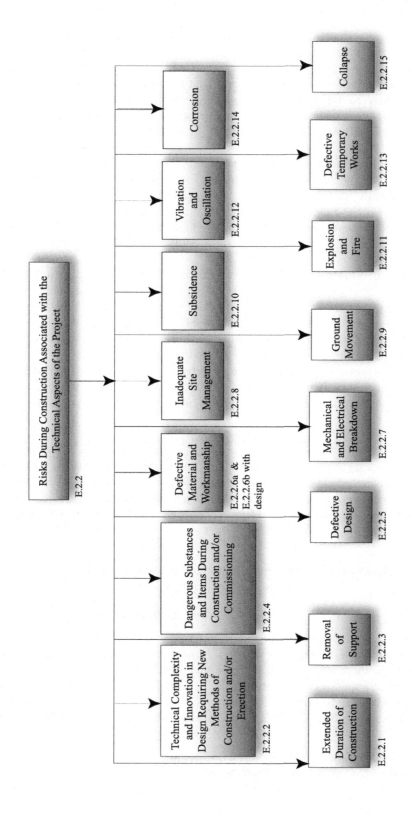

Figure 3.6 Risks during construction associated with the technical aspects of the project.

prestressed concrete multi-span structure, it crossed a river known to flood during the month of April. The bridge was constructed using closely spaced formwork supported on the riverbed. The prestressing operation of the deck was scheduled to be completed prior to the flood season, but due to the permitted tolerances in the deck level being exceeded by the contractor in construction, the prestressing operation was delayed. The contractor attempted to rectify the levels, but in doing so he spent more time than he originally allocated for the construction of the deck. When the floodwaters finally arrived at the site, the restricted area of the river caused increased water velocities under the bridge, which resulted in severe erosion below some of the formwork supporting members under the end span of the bridge. In a very short period of time, the floodwaters swept the formwork downstream of the river. The end span collapsed into the riverbed. As usual, there were other factors that contributed to the occurrence of this episode, nevertheless one could focus on the issue related here as being the main cause.

E.2.2.2 Technical complexity and innovation in design requiring new methods of construction and/or erection

When traditional materials or methods are used in construction, the familiarity of those involved with the design or the work itself may permit an occasional ambiguity in the drawings or specifications without them being misinterpreted. It may even provide correction of a mistake. However, in a novel or relatively new design, material or construction method, what is needed is precise and thorough communication between the designer, manufacturer or contractor, as the case may be, and others involved in the construction process.

Examples cram the literature on failures. The brittle fracture of high tensile structural steel to British Standard 968:1962 was little known in the construction industry until load-bearing members designed for the Kleinwort-Benson building cracked one day in September 1965 as they lay on the ground in the fabricator's welding yard.[49] It was fortunate that this phenomenon was discovered at an early stage of using this material before a major disaster could take place. The most disturbing aspect of brittle fracture is that the metal breaks without warning at very low stresses with cracks spreading at a speed of 1,000 metres per second and often one may find it associated with welding.

In the precast concrete industry in the United Kingdom, the roof collapse of Camden School for Girls and of the reading room in Leicester University, within hours of each other, in June 1973, showed that the bearing distance between precast members and their supports must be of a minimum dimension greater than that allowed for in these two buildings.[50] This minimum distance must be chosen not only for practical reasons of placement of steel reinforcement but also for accommodating movements in the various elements of the structure and for their lateral stability. These two buildings were in fact built in 1956 and 1965, respectively, prior to compilation of codes of practice in respect of the use of such precast concrete

49 'How the Umbrella Technique Failed', *Sunday Times*, 4 July 1965, London.
50 'Hundred Schools Checked after Roof Fall', *Sunday Times*, 17 June 1973, London.

sections. However, the reading room in Leicester University was built after the collapse of four precast concrete buildings under construction in the United Kingdom and one of the reasons given in the technical statement made by the Building Research Station was 'Bearing area of beam to column inadequate'.[51]

Another serious consequence of the use of new materials and methods is that such use may create a problem in another area of design as is the case in high-rise buildings. The fire risk increased dramatically in the 1950s and 1960s due to the use of new materials in their construction. Prior to that period the infill elements of old skyscrapers were much more fire-resistant and their masonry was non-combustible. There were no large glass surfaces which, when burning, would have permitted a fire to spread to other storeys from the outside.

Catastrophic fires were the result of the use of the aluminium curtain walls, plastic façade elements, large areas of glass, suspended fitted ceilings concealing large undivided areas and finally the use of synthetic materials. An example in this context is the large fire which occurred in São Paulo in a thirty-one storey building in 1972.[52]

E.2.2.3 Removal of support

The risk of removal of support has usually very serious consequences, even in minor parts of the work, as can be seen from the following example.

A four-man gang was engaged in backfilling a trench excavated to lay a 229 mm diameter saltglazed pipe as part of a contract for kerb laying and surface water drainage on a road project. The sidewalls of the trench where the pipes were to be laid were supported by 19 mm-thick plywood sheeting held in position by cross-struts. As the men prepared to take a mechanical roller along the bottom of the trench, they removed the cross-struts holding the plywood in place. Once the struts were removed, part of one of the walls collapsed, burying one of the labourers. The 21 year old worker died.[53]

E.2.2.4 Dangerous substances and items during construction and/or commissioning

The following example from Japan highlights the effect of such substances on construction work. Shortly after commencement, the construction of a water reservoir had to be stopped when the concrete cubes cast for testing purposes did not meet the required compressive strength and neither the concrete nor the cement manufacturers were able to give any explanation for the failure of the test cubes.

The mystery was solved when it was discovered accidentally that the blossom and leaves of nearby Jacaranda trees had fallen into the concrete aggregate.[54]

---

51 'Technical Statement by The Building Research Station of the Department of Scientific and Industrial Research on the Collapse of a Precast Concrete Building under Construction', 19 December 1963, United Kingdom.
52 *Schaden Spiegel*, op. cit., see note 30, October,1974.
53 *New Civil Engineer*, op. cit., see note 45, 30 November 1978.
54 *Schaden Spiegel*, op. cit., see note 30, October 1976.

Usually, concrete reaches 90% of its compressive strength within 28 days after manufacture. In this case, however, certain substances contained in the Jacaranda trees, although of minor concentration, apparently retarded the curing of the concrete. In another series of tests carried out later the compressive strength of the test cubes proved to be satisfactory.

E.2.2.5 Defective design

In September 1994, a ship-to-shore walkway in the Port of Ramsgate in the United Kingdom collapsed, killing six people and seriously injuring a further eight. The contractor was responsible for the design and construction of the walkway, but the cause of the accident was seen to be relating to the design of the steel structure.

The walkway structure was hinged at the shore end and spanned two intermediate vertical supports fixed to a floating pontoon. The third span connected the walkway to the vessel. The construction of the walkway was in three sections, which were shipped to the site along with hinges and other fixing materials, where the whole structure was then assembled. The support bearings had to accommodate vertical and lateral movement caused by tidal changes, vessel roll and motion of the pontoon. It was reported that the two shore-end bearings and a third corner bearing at the first internal support allowed longitudinal movement along the direction of the span and rotation around a pin.[55] The bearings on the fourth corner had two pins, one vertical and one horizontal, allowing rotation about two axes. Thus, the vertical pin at the fourth bearing provided the only restraint against longitudinal movement.

The accident occurred when the horizontal pin joint failed, thus disconnecting the bearing, and leaving no restraint to prevent horizontal movement. The structure's integrity relied on a single 55 mm diameter steel pin, a 5 mm butt weld and a 7 mm fillet weld, one of which failed causing the bearing to fail.

E.2.2.6a Defective workmanship and material

The warranty of incorporating or using only good workmanship and material is implied in construction contracts. Despite that warranty, one finds that as long as quality means perpetual care and high cost, this risk of defective workmanship and material will always exist. Even the smallest defect can sometimes cause a disastrous effect, as happened in the case described below.

The main distillation column of a new oil refinery became a total loss in an accident which occurred during erection.[56]

The column, which was approximately 50 m in height and weighed over 120 tonnes, had been shipped by a cargo vessel from the factory to the refinery pier. It was moved to the erection site on low loaders. Two cranes with a capacity of 250 tonnes and 200 tonnes, respectively, were used to hoist the column into a vertical position and place it on its foundation.

---

55 *New Civil Engineer*, op. cit., see note 45, 22 September 1994.
56 *Schaden Spiegel* op. cit., see note 30, No. 1, 1980.

During the initial phase of the joisting operation a third crane guided the base of the column. The foundation had been covered up with wooden planks to protect the anchor bolts while lifting work was in progress. When the column was suspended vertically a few centimetres clear of the foundation, the plank cover was removed. At this point the column made a slight turn, shifting out of position and sagging a little.

A weld in the cross-strut in the top section of the jib of the 250 tonne crane had failed, causing the failure of the welded joint and ultimately of the strut. The jib became distorted causing the column to turn and sag as described.

The operator of the 200 tonne crane, warned by the sudden movement, released the brake for the hoisting cable which deposited the column on the platform but leaning to one side. As a result, the 250 tonne crane was unable to carry the additional load and both cranes and the column crashed on to the ground sustaining irreparable damage. Other equipment on the site suffered damage which, when added to the cost of the column and cranes, amounted to $1.2 million. This collapse resulted from failure of the weld in the cross-strut of the crane.

Sometimes, however, such defects arise out of lack of knowledge rather than lack of care or intentional acts, as happened in the following incident. Aggressive material was shipped in 5,000 plastic sacks each containing a weight of 50 kg.[57] The sacks were heat-sealed at one end but, when they arrived at their port of destination, they were found to have burst open.

The cost of salvage and removal operations was very expensive due to the aggressive nature of the material being shipped. The sacks were examined and found to be made of a plastic material of a thickness of 0.25 mm. They were loaded on pallets with up to twenty sacks on top of each other. A chemical analysis carried out with the help of an infra-red spectroscope revealed that the sacks were made of polyethylene film with a density of 0.94 g/cm^3. A tensile strength examination showed that near the heat-sealed ends the strength of the material was between 20% and 40% lower than elsewhere. This phenomenon, which is well known in the packaging industry, is caused by the fact that the films grow thinner in the area around the seams due to the heat-sealing.

The next step was to examine the way the sacks had been loaded on top of each other in the light of the above-mentioned inherent weakness. Calculations showed that, with twenty sacks placed on top of each other, the reduced tensile strength around the seams would already be exceeded under static load conditions. Considering the shocks and bumps hardly avoidable during loading and unloading, not more than a maximum of ten sacks should have been placed on top of each other.

E.2.2.6b Defective design, workmanship and quality control

During the construction of the rail link from London's Paddington Station to Heathrow Airport, the Heathrow Express line, three partly built station tunnels caved in during the early hours of Friday 21 October 1994 and continued to collapse over a number of days. Fortunately, no one was injured or killed in the

57 Ibid., October 1976, Munich.

accident, but the failure, which was estimated to have cost Heathrow Airport operator BAA around £50 million, brought chaos to the heart of the airport.[58]

The station complex was to comprise three large caverns 9 m in diameter which form the central concourse area and two up-and-down-running platforms together with a complex network of tunnels and escalator shafts to link the station to the surface and to the main airport terminals.

At about 1 a.m. on the morning of the collapse, ground-monitoring equipment measured movements 'of the scale' which alerted workmen in the down-platform tunnel to the impending disaster. Twenty-five people were evacuated to the surface moments before the roof of the new station complex collapsed. Chaos ensued as the contractor's and consultant's engineers tried to contain and arrest the collapse and prevent further damage.

It was discovered that the collapse started at the base of the main shaft at the connection to the down-platform. With overburden material pouring into the fractured tunnel the semi-complete cavern was severely breached and the ground above swiftly sank. As the ground around the shaft slid into the down-platform, abnormal stresses were induced in the linings of both the concourse and the up-platform tunnels. These quickly began to fold up around the junction with the main shaft, causing further movement in the ground above.

To prevent further damage, the first task was to secure the stability of the main shaft. Structural concrete, at a rate of 27 truckloads per hour, was pumped into the shaft. This formed a 9 m plug at the bottom covering the tunnel accesses completely. Despite this, and despite pouring thousands of cubic metres of structural and light-weight foamed concrete, the ground was still sinking into the hole and eventually the site headquarters building tilted on its foundations and crumpled towards the hole.

Some 24 hours later, a rotary piling rig was employed to drill from the surface into the concourse and the down-platform caverns and more concrete was pumped through these shafts to plug the failed area. Access was gained to construct concrete bulkheads to seal the damaged tunnels. The flow of concrete into the hole continued the whole time until ultimately these measures were successful and ground movement around the failed area was stabilised. It was stated that by the end 18,500 cu m of concrete was pumped in.

The designers of the tunnel and the contractors were prosecuted under the Health and Safety Act in England for failing to ensure that their conduct during the construction of the tunnel did not expose the construction workers and members of the public to risk. The contractors pleaded guilty after an expert report, which was carried out for them in 1998, showed that a weak tunnel invert resulting from poor construction was the cause of the collapse. During the 27-day trial, it transpired that various warnings of an impending collapse were given, but these warnings were not heeded.

The final report of the Health and Safety Executive, published in 2000, referred to the accident as the worst civil engineering disaster in the UK in the last quarter

---

58 'HEX Collapse Report Slates Poor Risk Management', report by Anthony Oliver, *New Civil Engineer International*, 1 August 2000, UK. See also various earlier reports in the *New Civil Engineer*, 27 October; 3 November; 10 November; 1 and 8 December 1994; and 26 January 1995.

of the twentieth century resulting from a catalogue of design and management errors, poor workmanship and quality control.[59] The Executive claimed that the designers were responsible for monitoring the behaviour of the lining during construction and failed in their duty to issue warnings when data from their monitoring instruments showed that a collapse was imminent in the weeks preceding the collapse. The designers claimed that an 'unpredictable and unpreventable' landslide in the clay above the tunnels triggered the collapse and that even with a defective lining caused by poor workmanship, the tunnels could not have collapsed without an outside influence.

The report also stated

> Such accidents must be prevented through effective risk management. The industry cannot simply rely on good fortune. Risk assessment should be a fundamental step in the procedures adopted by all parties: it is inappropriate wholly to leave the control of risk to contractors.

The jury found both the designers and the contractors guilty of the charges against them. The judge in the case, Mr Justice Cresswell stated that the contractors should bear the greater responsibility for the collapse, as they fell seriously short of the 'reasonably practicable' test. He stated that it was a matter of chance whether death or serious injury resulted from the breaches committed. The contractors were fined £1.2 million, whereas the designers were fined the lesser sum of £500,000 for their 'less culpable role'.

A material factor in the collapse was the nature of the contractual arrangements, the contract management, and all engineering questions relating to the New Austrian Tunnelling Method (NATM) process in soft ground being devolved to the contractor with self-certification as part of a competitive contract.

E.2.2.7 Mechanical and electrical breakdown

Site operations are becoming more dependent on plant and equipment, the breakdown of which forms a major risk element. An interesting study was made of 409 failures of diesel and natural gas engines reported in the period from 1975 to 1979 with damage amounting to or exceeding US$2500. The study covers only such cases where the cause and development of the failure were clearly determined.[60] Failures due to 'unknown causes' were not included in the study.

Table 3.1 shows the distribution of failure in terms of the application to which the engine is used. Table 3.2 shows the distribution in terms of the component mainly affected and Table 3.3 in terms of primary cause.

In some cases, the damage to the piece of equipment or machinery is minor when compared with the damage or risk of damage to the project itself, as occurred in the following case where the loss amounted to DM1.7 million.

59 *The Collapse of NATM Tunnels at Heathrow Airport*, [2000] published by HSE Books, UK, which could be viewed from www.hse.gov.uk web site.
60 *Schaden Spiegel*, op. cit., see note 30, No. 2, 1981.

Table 3.1 Distribution of failures of diesel and natural gas engines by fields of application

Application	Number	% of total
Earthmoving machines	209	51.1
Power generation	58	14.2
Watercraft	42	10.3
Construction site vehicles	32	7.8
Fork-lift trucks	29	7.1
Railbound vehicles	19	4.6
Compressors, pumps	11	2.7
Other	9	2.2
	409	100.0

Table 3.2 The principal failure areas in diesel and natural gas engines

Failure areas	Number	% of total
Pistons and connecting rods	191	23.3
Crankshafts	138	16.8
Bushings	132	16.1
Bearings	123	15.0
Casings	81	9.8
Cylinder heads	63	7.7
Controls	48	5.9
Other	45	5.4
	821	100.0

A 2.3 km underwater pipeline with a diameter of 0.6 m was to connect a refinery on land with a planned tanker jetty.[61] The pipeline was winched out from the shore by means of a steel cable. The winch stood on a moored pontoon. Then the cable became entangled and the winch was ripped apart. The pipeline, already partly under water, had to be salvaged. A cyclone then caused a tidal wave which pushed the pipeline some 200 m from its correct position to where it could be brought back only after a great deal of effort. Apart from this, construction equipment was also damaged. While a second attempt was being made to tow out the pipeline, the winch broke. At that point a length of pipeline measuring 1.3 km was in the water. The resistance caused by friction on the seabed had obviously been underestimated. It was not possible to repair the winch in the country itself and a reserve machine was not available.

Following this, tugs belonging to the harbour authority were used to tow the pipeline. Only approximately 400 m had been positioned in this way before the tugs were forced to give up. The next attempt was made with the help of a 16,000 tonne tanker used to tow the pipeline on its own. This also failed as, when the heavy ship

61 Ibid., September, 1975.

Table 3.3 Primary causes of damage in diesel and natural gas engines

Product faults	Number	% of total
Faulty assembly	23	5.6
Faulty design	20	4.9
Faulty material	17	4.3
Faulty repair	13	3.3
Poor workmanship	12	2.9
	85	21.0
Operational faults	*Number*	*% of total*
Maintenance faults	177	43.2
Faulty handling	102	24.8
	279	68.0
Outside influences	*Number*	*% of total*
Foreign bodies	25	6.1
Sabotage and other extraneous causes	20	4.9
	45	11.0
Total	409	100

started to move, the cable was torn by the force of the sudden jerk and the pipe sprang back and bent over.

These constant misfortunes led to a considerable delay in the laying operations. In addition, assembly became more difficult when the monsoon period began. The pipeline was finally towed with the original winch which had meanwhile been repaired abroad.

E.2.2.8 Inadequate site management

A company contracted to build a section of motorway procured the necessary stones from a nearby quarry.[62] The rock was blasted into fragments and loaded onto dump trucks. The hydraulic excavator had a loading shovel with a capacity of 4.5 cu m and was driven by a 500 hp (DIN) diesel engine.

During operation, fire (probably caused by a short circuit in the 24 volt electrical system) broke out in the excavator. The flames consumed 1,000 litres of diesel fuel and an equal quantity of hydraulic oil in two tanks at the rear of the excavator.

Although fire brigades from the neighbourhood quickly reached the scene, it took an hour to extinguish the fire. The losses amounted to US$300,000. The fire could well have been brought under control at the outset if adequate fire-fighting equip-

62 Ibid., October 1976.

ment (manual fire extinguishers) had been available. The loss, in that case, would have been minimal.

E.2.2.9 Ground movement

Ground movement could take place from a number of causes, including landslides, frost heave, earth slips and ground pressure leading to collapse. Two examples are given here. The first occurred in a sewerage plant which was damaged during construction by an earth slip.[63] Due to heavy rainfall the earth on the slope above the building site slipped down 10 m. The soil pressed against a shaft structure made of precast concrete elements until it collapsed. Consequently, surface water and silt were able to get into a sewer at the point where it had been connected to the shaft. The sewer had already been completed and was ready for use but then became filled with mud along a length of 2500 m.

The second incident occurred during construction work for new loading and landing piers which included the driving of steel sheet pile walls into about 8 m deep water and anchoring the walls on the landside by steel anchors. These anchors were held by a smaller sheet pile wall driven about 20 m further inland. When the driving operation was completed, the space between the two sheet pile walls was gradually filled with liquid soil. At the same time steel piles were driven along the waterside of the outer sheet pile wall, which were to be connected later by a solid concrete slab to form the final quay.

The ultimate fill height had nearly been reached when the inland sheet pile wall started to move. Deprived of its backward anchoring, the sheet pile wall on the waterside also gave way and collapsed over a length of 100 m pulling the inland sheet pile wall with it. Large amounts of fill material poured into the bay, tearing down several steel piles standing in the water.

The damage amounted to about US$2 million.[64]

E.2.2.10 Subsidence

In 1975, an international consortium of contractors were awarded the contract for the construction of the terminus station in Hong Kong Island for the Hong Kong Mass Transit Railway Corporation.[65] The station, basically a large underground concrete box, almost half a kilometre long and approximately 27.5 m deep, was built in the central business district only a few metres away from surrounding properties. One of these buildings was the premises of the Supreme Court of Hong Kong which was built around 1910 on wooden piles in very poor ground.

During the diaphragm wall construction and after the sides of the excavation has been stabilised with bentonite, unexpected ground behaviour and dewatering influences caused the building to subside and tilt. Serious cracks appeared and in July

---

63 Ibid., October 1974.
64 *Schaden Spiegel*, op. cit., see note 30, special issue, 1998.
65 'Settlement at Court or Why the Judges Sought an Adjournment!', *Risk Review*, a publication by Stewart Wrightson, Insurance Brokers, No. 9, October 1984.

1978, when the learned judges had become concerned at lumps of plaster falling on their heads, the building was evacuated.

In 1984, the loss, which was calculated to be well into seven figures, was settled out of court, by the insurers. A single insurance policy had been arranged to cover the employer, contractor, subcontractors, and 'all other parties engaged to provide goods or services'. No subrogation recovery procedures were initiated and the insurers accepted responsibility.

E.2.2.11 Explosion and fire

Even the best-organised construction sites are, by their very nature, prone to fire hazards. Inflammable construction materials such as timber, shuttering, packing material, plastic foils, fuel, paints and other hazardous material are generally found on site. The temporary nature of many items on site such as camps, stores and temporary heating and cooking facilities adds to the fire hazard. Moreover, only a few sites maintain complete and efficient fire-fighting equipment and many civil engineering projects are remote from public firefighting facilities. A project concentrated in one location can be threatened in its entirety by fire and the risk involved increases with the progress of construction.

Welding operations in an enclosed environment constitute a major fire risk both during and after the welding operation. The following incident of a fire that occurred during welding illustrates what could happen. An amusement park under construction within a hotel and shopping complex was almost completely destroyed by fire.[66] The roof of the multi-storey 'theme park' was to be spanned by a 200 m×60 m glass dome. Among the attractions was a presentation of the Arabian tale of Sindbad the Sailor. The artificial rock walls used for the show consisted of glass-fibre reinforced polyester resin and were covered with a refractory coating on the front side only. The fire broke out during flame-cutting operations on pipework situated under the ceiling. It was thought that welding beads must have dropped on to the back of the artificial rock walls, which were at the time unprotected, and they caught fire immediately.

The workmen tried to combat the fire with portable extinguishers, but dense smoke and the toxic gases it contained soon forced them to give up. Large amounts of combustible material and the presence of a great many shafts for transporting installations, lifts, and escalators between the individual storeys accelerated the spread of the fire up to the glass dome.

Several hours elapsed before the fire brigade managed to extinguish the fire. Four of the approximately 300 workmen present in the building when the fire started suffered minor injuries. The fire caused considerable damage to the interior of the building, including its structural components, surfacing slabs, wall panelling and floors. Protective coatings covering the steel structure were affected by the heat and smoke and serious damage was inflicted to the mechanical and electrical installations, to lifts and to loudspeaker systems. The panels of the glass dome had to be cleaned or replaced. The material damage covered by CAR insurance amounted to the equivalent of about US$3 million.

---

66 Taken from an article by H. Maier, published in the Special issue of *Schaden Spiegel*, op. cit., see note 30, 1998.

In a similar incident, fire caused severe damage to a thermal power station designed to house three 400 MW units. The fire occurred many hours after the end of a day's work.[67] At the time of the accident, the structural steelwork of the 29 m high machine hall was nearing completion and the equipment for the first unit was being installed. Concreting work for the third unit was under way. The foundations and steel columns had been completed and work was concentrated on the completion of the reinforcements and scaffolding for the turbo-generator platform. Concrete was to be poured the following morning and completion work continued late into the night.

In the early hours of the morning, a watchman on an inspection round discovered flames coming from the formwork. He triggered the fire alarm and fire engines were called from a nearby industrial area and the nearest town. The works fire brigade and another seven fire engines fought the fire, which was finally brought under control after one hour.

The turbo-generator platform was completely destroyed. The flames lashing up high and the enormous heat had caused serious damage to the turbine house and a neighbouring building. The entire roof structure and two cranes parked above the fire area had to be replaced.

The damage amounted to about US$3.5 million, some 10% of the sum included for the removal of debris. As no cause for the fire could be established, it was assumed that the accident was caused by flying sparks from the welding of steel reinforcement, which ignited timber and combustible wastes.

E.2.2.12 Vibration and oscillation

A serious loss amounting to DM3.5 million occurred during the erection of one of the world's largest blast furnaces with a daily output of 8,800 tonnes of pig iron.[68]

A self-supporting steel-plate, brick-lined chimney with an overall height of 140 m was to be erected for discharging the waste gases. The lower section of the chimney was 35 m high and cone-shaped, tapering from 9 m to 6 m in diameter. The upper chimney section consisted of a cylindrical tube having a length of 105 m and a diameter of 6 m. The material used for the steel plate was mild steel and the thickness of the plate varied from 12 to 30 mm. After the two chimney sections had been erected, the brick lining work was started. When the lining had reached a height of only a few metres, technicians discovered a crack, measuring 1 m in length, around the periphery of the chimney. The crack was at a height of 35 m, just below the joint between the lower conical section and the upper cylindrical section (see Figure 3.7). Within a period of seven hours, the crack extended to a length of 8 m. The prevailing wind at the time was force 6 on the Beaufort scale and the risk of the chimney toppling over and crashing down on to the furnace air preheater unit could not be ignored. It was decided, therefore, after consultation with the insurers, to blast off the chimney approximately 15 m above the crack. This was done successfully, but how had the crack originated?

When checking the fracture and the design of the chimney, it was found that, due to severe oscillation of the structure, excessive stress had been exerted at the point

---

67 *Schaden Spiegel*, op. cit., see note 30, No. 1., 1982.
68 Ibid., October 1976.

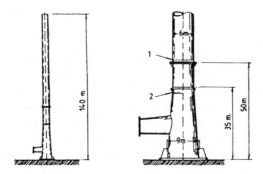

Figure 3.7 Brittle fracture in a steel chimney due to oscillation.

where the conical and cylindrical sections met.[69] Eventually, this had resulted in a brittle fracture of the steel plate. Wind tunnel tests, which could have uncovered the weakness, had not been carried out during the design stage.

E.2.2.13 Defective temporary works and their design

A steel girder, used as falsework to support formwork, and incomplete bridge span sections, collapsed during construction of the second span of a six-span 300 m long continuous, prestressed concrete box-girder railway viaduct (Figure 3.8).[70]

At the time of the collapse, all piers and one span had been completed. The steel girder was supporting the completed span and the moving formwork for the concrete box-girder into which approximately 40 cu m of fresh concrete had been

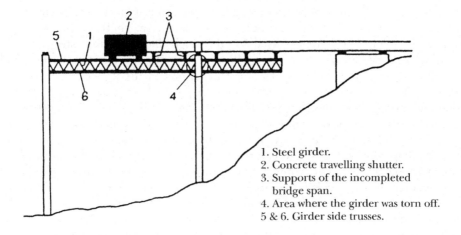

1. Steel girder.
2. Concrete travelling shutter.
3. Supports of the incompleted bridge span.
4. Area where the girder was torn off.
5 & 6. Girder side trusses.

Figure 3.8 Defective design of a temporary girder.

69 Ibid., October 1976.
70 Ibid., October 1976.

placed for the first 10 m length of the second span. The formwork was felt to drop suddenly and buckling was noticed in the web members of the lattice trusses which made up the steel girder. Buckling progressed slowly over a period of 30 minutes until the concrete box-girder section, which was two-thirds of its final length, was torn off near the pier and collapsed, together with the formwork. The completed piers and span were undamaged by the collapse.

The cause was traced to buckling of tubular web members in the steel girder side trusses. The girder, imported from overseas, had been originally designed to position and support precast concrete bridge sections slung beneath it. It had been substantially redesigned for use on the viaduct project, which necessitated a different construction method with the girder mounted under the viaduct span, and involved heavy loads and long spans. The affected side truss web members had not been strengthened and were loaded beyond their safe limit. There were no injuries in the collapse, but the financial loss sustained was approximately DM250,000.

E.2.2.14 Corrosion

A 63 mm stop valve was connected to a fire water supply line by means of an aluminium flange with a screw thread about 80 mm in length. The stop valve had a brass body of the material $CuZn_{39}Pb_2$ whilst the flange was made of aluminium alloy $G\text{-}AlSi_{10}Mg$. The quality of each of the two materials by itself was not in question, but when used together they result in galvanic corrosion when in contact with moisture and therefore leakage. This is precisely what happened on this project.

The water leaked through the localised corrosion in the joint and saturated the wall in the basement. To repair the damage, the soil outside the wall had to be removed and then filled in again afterwards. The wall also had to be dried and painted.

The loss, for which the water damage insurer paid, amounted to approximately US$15,000 and could only have been prevented by an electrochemical separation of the materials inside the valve. However, since as the valve could not be constructed in such a way, a different material should have been chosen for the flange. In this case, recourse action was taken against the plumbing firm responsible for this configuration.[71]

E.2.2.15 Collapse

Total collapse is the most catastrophic of all hazards. It rarely gives any warning and it therefore carries with it the risk of injury. Such an event occurred in Kuwait in 1976 when twenty-one workers were killed.[72] The chain reaction, which resulted in the total collapse of a garage building under construction, lasted for just five seconds. Six parking levels collapsed like a house of cards. While the slabs fell on to each other to form a 'sandwich', the columns broke like sticks at each level and all that remained was a pile of wood, steel and concrete. What had happened? The formwork and reinforcement for the sixth floor of the building had already been

71 Ibid, No. 2, 1998.
72 Ibid., October 1976.

completed. As a total of more than thirty floors had already been made in the same way, concreting seemed to be just a routine affair. The pouring of concrete for the sixth floor was thus started half an hour before midnight. The concrete was being pumped up through a riser. Some 70% of the slab had already been concreted around 6 a.m. when the timber structure supporting the formwork suddenly collapsed. As a result, the concrete, some of which had cured but other areas had not yet hardened, fell on to the slab below from a height of 3.5 m. The mass of falling concrete weighed no less than 450 tonnes.

The floor slab below had only been completed fifteen days earlier, but the formwork had already been removed. The floor was not able to support the weight of the collapsed floor and the dynamic load of the collapsing concrete masses. The columns buckled, and both floor slabs fell to the next level together. This induced a chain reaction, causing all of the six floor slabs to collapse right down to the basement level. The accident occurred while some workers for the next shift were still asleep on the lower floors and were thus crushed to death underneath the rubble. The workers doing the concreting fell to the ground together with the collapsing structure.

The investigations that followed indicated that the lateral support of the formwork of the floor slab being concreted was insufficient. In fact, the timber structure used for this purpose had been previously used a number of times and was worn considerably and did not have sufficient stability. Moreover, additional horizontal forces were exerted on the weakened structure by the riser through which the concrete was being pumped up. The actual reason for the collapse was, without doubt, the fact that there was not enough sound timber material in the formwork used. So, used and reused parts were patched together to serve as supports, but these were no longer able to bear the imposed loads. In addition, the specification for concrete curing was disregarded. There can be no doubt that tough competition and piecework contracts often force contractors to make full use of the materials they have at their disposal and to really exploit their schedules to the utmost. However, once economic limits and sound practical engineering are reached and sometimes even exceeded, the failure of just one minute detail may be sufficient to cause a disaster. The loss described above amounted to a total of about DM3 million, not to mention the loss of life. However, it is unusual for one single cause to be identified with such a collapse. More usually, one particular factor can be found to have contributed most in bringing about the final collapse.

E.2.2.16 Collapse of temporary works

On 5 August 1999, the £300 million Grand Bridge in South Korea partially collapsed and thirty-seven precast concrete bridge segments from one partly complete and two completed spans crashed to the ground.[73]

The collapse occurred during construction of the 5.82 km section of the precast concrete segmental twin box girder southern viaduct. The viaduct, with 60 m long spans, was founded on twin 2.5 m diameter *in situ* concrete piers. The accident occurred as work was approaching the fifty-fifth pier and the 80 tonne deck segments were being assembled.

73 *New Civil Engineer*, op. cit., see note 45, 19 August 1999.

The segments were precast on site and transported along a previously completed bridge-deck before being positioned on a steel launching truss, which spanned adjacent piers and supported the segments making a span until all the segments were in position. Epoxy adhesive was then applied to the segment faces and external post-tensioning was carried out stressing the segments together. The launching truss then slid forward to span the next pair of piers.

It was thought that the launching truss failed causing the incomplete span to collapse and to over-load the two preceding spans, causing their collapse. It was fortunate that no one was killed or injured in the collapse.

E.2.3 Risks during construction associated with Acts of Man (Figure 3.9)

E.2.3.1 Human error

It is now generally accepted that human error is in some way or another the cause of a large percentage of the accidents in the industry in the sense that actions by people either initiated or contributed to the accident or that people might have acted better to avert them. Recent data indicate that approximately 80% of industrial accidents, 50% of pilot accidents and 50–70% of nuclear power accidents are attributable to human error.[74]

Construction is part of these statistics. In particular, such accidents result in accidental death, personal injury, property damage or combinations of them. Workers may feel that safety measures, such as wearing protective equipment, are cumbersome or it is not manly to follow them. A person may rationalise the idea of risk, believing that it would not happen to him/her, or deviate from safety procedures to gain some personal benefit. In the context of the intense time pressures typical of construction work, workmen may even cut corners in the belief that they are acting in the interests of their employer in finishing a particular task earlier or on time. However, simply put, the truth is that to err is human, and humans are fallible and liable to make mistakes or behave unpredictably for many reasons.

It has been suggested that modern technology has advanced to the point at which improved safety can only be achieved through attention to human error mechanisms.[75] Therefore, human control must remain and must be exercised to intervene when unplanned events occur. In fact, there is evidence to suggest that introducing safer technology can lead to more risky behaviour because people feel uncomfortable with the 'low' level of risk they experience and try to 'compensate' for this by behaving in an unsafe manner, often referred to as risk compensation.

The importance of the human element in reducing the risk in construction projects means that there ought to be a successful management of construction workers' occupational health and safety (OHS) behaviour. It is therefore important to

---

74 *Human Reliability Analysis*, E.M. Dougherty, and J.R. Fragola, John Wiley, 1988, New York; '*Human Factors*', R.S. Jensen 1982, 'Pilot judgment: Training and evaluation', 34, 61–73; and *Cognitive Systems Engineering*, J. Rasmussen, A.M. Pejtersen and L.P. Goodstein, John Wiley, 1994, New York.
75 *Human Error*, J. Reason, Cambridge University Press, 1990, Cambridge.

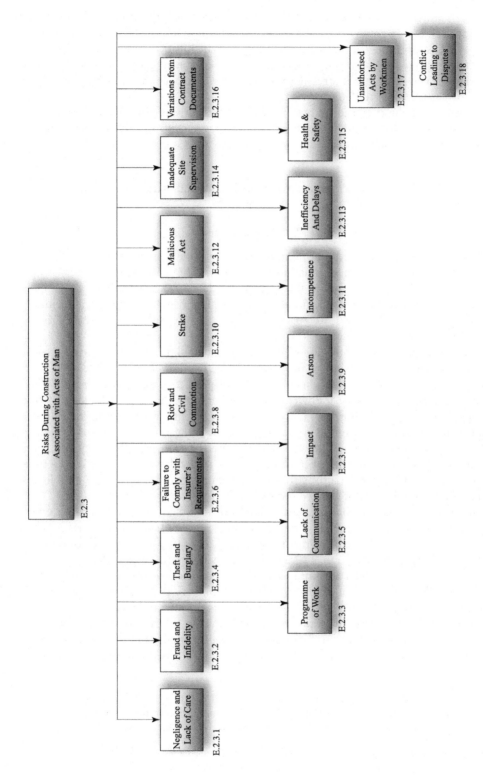

Figure 3.9 Risks during construction associated with Acts of Man.

understand the psychological processes which result in behaviour that leads to mistakes and accidents. Although detailed study of this topic is outside the scope of this book, it is interesting if not important to understand the types of error that may occur. The Health and Safety Executive in the United Kingdom classified the types of error in the following manner.[76]

(a) Lapses of attention

Intentions and objectives are correct and the proper course of action is selected but a slip occurs in performing it. This may be due to competing demands for attention. Paradoxically, highly skilled operatives may be more likely to make slips because they are not used to carefully thinking about every minor detail.

(b) Mistaken actions

Doing the wrong thing under the impression that it is right. For example, the individual knows what needs to be done but chooses an inappropriate method to achieve it.

(c) Misperceptions

Misperceptions tend to occur when an individual's limited capacity to give attention to competing information under stress produces 'tunnel vision' or when a preconceived diagnosis blocks out sources of inconsistent information.

(d) Mistaken priorities

An organisation's objective, particularly the relative priorities of different goals, may not be clearly conveyed to, or understood by, individuals. A crucial area of potential conflict is between safety and saving cost or time. Misperception, described above, may be partly intentional as warning signs are ignored in the pursuit of competing objectives. Where top management's goals are not clear, individuals at any level in the organisation may superimpose their own.

(e) Wilfulness

Willfully disregarding safety rules is rarely a primary cause of accidents. Sometimes there is only a fine line between 'mistaken priorities' and 'wilfulness'. Managers need to be alert to the influences that, in combination, persuade employees to take and condone 'cutting corners' where safety rules and procedures are concerned, in the belief that the benefits outweigh the risks.

An example of a human error of the penultimate type, which could have easily been fatal, is the incident that happened during excavation works at Heathrow

---

76 *Human factors in industrial safety*, UK Health and Safety Executive, HSE HS (G) 48, 1989, London: HMSO.

Express trial tunnel project. A truck driver escaped death when his 30 tonne vehicle plummeted down the access shaft to the tunnel. Four men who were working at the bottom of the shaft escaped injury by pressing themselves against the side of the 10.67 m wide and 25 m deep shaft as the truck landed, roof first, on a protective steel mesh 3 m from the bottom.[77]

The driver, who had to be cut from the wreckage, explained that he lost traction in deep mud and his laden truck plummeted down the shaft. The shaft perimeter was only protected by a pedestrian barrier consisting of scaffold tubes. The Health and Safety Executive stated there were no regulations stipulating the installation of vehicle barriers, but the standard practice is to protect the top of such shafts with lining segments, which were to be added on the next day.

An example of a human error of the last type, which did end in a fatal accident, occurred on a road in the Far East during asphalting operations.[78] The asphalt, which had to be mixed with a diluting substance to increase its viscosity, was heated in batches to a temperature of between 110 and 135°C in a special tank lorry. The tank had a built-in churning device, which stirred the contents until they had formed a homogeneous mixture which was then carried to the laying point by the lorry.

The diluting substance when used on site should normally be a non-volatile hydrocarbon with a high boiling point. In this case, however, the mistake was made by using, on a site operation, kerosene – a volatile hydrocarbon with a low boiling point with similar properties to petrol.

On the day of the loss event some of the asphalt/kerosene mixture suddenly spilled over the edge of the opening, ran down the outside of the tank, and ignited immediately upon contact with the heating elements. The flames surged back into the tank and within seconds the fire had engulfed both the tanker and a forklift truck next to it that was used to carry the kerosene from the storage area. An unsuccessful attempt was made to extinguish the fire using CO_2 extinguishers. The local fire brigade was alerted, but in spite of arriving on the scene almost immediately, it could not assist in the fire fighting because the only extinguishing agent it had was water, which is not suitable for fighting hydrocarbon fires.

Ultimately, the fire was extinguished by throwing earth and sand onto the flames using shovel dozers. Two workers died in the fire and the material damage totalled more than US$110,000.

E.2.3.2 Negligence and lack of care

For the extension of a power plant, a 500 m pipeline with a diameter of 1.7 m was needed to supply additional cooling water from a nearby lake.[79] Owing to the very gradual slope of the lake shore and the great fluctuation in the water level, the pipeline was to be laid in a trench both on land and under water. The method chosen for laying was to weld together the 6 m pipe lengths on land into several lengths of

---

77 *New Civil Engineer*, op. cit., see note 45, 10 February 1994.
78 An article by Maqbul Ahmad, Dacca taken from *Schaden Spiegel*, op. cit., see note 30, No. 2, 1997.
79 *Schaden Spiegel*, op. cit., see note 30, September 1975.

pipeline and temporarily seal them. The sections were then floated into position by means of pontoons and were mechanically coupled. In order to ensure even lowering of the pipeline, the inside was fitted with sealing discs dividing the pipes into separate chambers, each chamber having its own valve. Thus, it was possible to flood the chambers individually, giving maximum control over the rate of sinking.

After the complicated sinking procedure had been completed and the pipeline lowered into its channel, the temporary partition walls between the pipe section were to be removed by specially installed tackle lines. One of the walls then jammed as it was being extracted. A diver was engaged to rectify the problem. To make work easier for him, it was decided to let some air into the pipeline. Unfortunately, too much air was pumped in and the pipe began to lift. A whole section of the pipeline was pushed to the surface, but its ends had already been secured by anchorages. Consequently, the pipeline fractured and sank back into the water. The damaged pipeline was useless and not worth salvage. A new ditch was prepared parallel to the original one and a second pipeline had to be laid. The loss amounted to DM1.9 million.

Another instance of lack of care during construction can be taken from the mechanical engineering field when, after completion of the construction of a modern iron plant containing a 74 m long, 4.6 m diameter, rotary kiln, it found that the pneumatic clutch of one of the two 35 kW auxiliary drive motors had not engaged properly.[80]

The coupling discs between the two main reduction gearboxes and the spur-gear drive shafts were found to be fractured near the key slots. Investigation of the damage revealed that the auxiliary drives, which were both connected to the two main reduction gearboxes, had acted in contra-rotating directions, thus causing the 1,400 tonne kiln to lift by a couple of millimetres fracturing the coupling discs and then failing down on its bearings.

The impact also caused the dislocation of bricks of the refractory lining of the kiln which had to be removed and rebuilt. The entire kiln bearings had to be checked, gears and drive shafts had to be dismantled and checked, and the drive shafts were found damaged in the key slots. The total claim was estimated to be DM100,000.

According to the electrician, all power cables were connected in the same sequence of red/white/blue, on the motor side as well as in the switch room. It was found, however, that whilst the connection for the right auxiliary drive motor had been changed in the switch room to red/blue/white to give the motor the correct direction of rotation, the connection for the left auxiliary drive motor had remained un-changed. This was noted neither during individual testing of the motors nor during testing of the auxiliary drives by rotating the kiln.

E.2.3.3 Fraud and infidelity

Although these risks are not common, they are nevertheless real, albeit in some cases rare, possibilities. This risk is best illustrated by the legal case of *Applegate v. Moss* which went to the English Court of Appeal in December 1970.[81]

---

80 Ibid., October 1974.
81 *Applegate v. Moss and Archer v. Moss* [1971] 1 All ER 747.

The facts of the case were that in February 1957, Mr Moss, an estate developer, agreed to build for Mr Archer and Mr Applegate two houses already in the course of construction. He had employed a builder, Mr Piper of Piper Ltd, to build the houses in accordance with plans and specifications approved by the local authority, on condition that the foundation should be a reinforced concrete raft with a specific concrete mix. The reason for this condition was that the site formed a sloping ground on wet clay soil.

Messrs Applegate and Archer occupied the houses towards the end of 1957. In 1965, when one of them tried to sell his house, experts for the would-be purchasers found serious cracking. On investigation, it was found that there was no foundation raft but instead only a simple concrete footing of inadequate dimensions and concrete strength. The cracks were so serious that the houses were deemed unsafe, uninhabitable, beyond repair at a reasonable cost and only fit for demolition. They were, therefore, evacuated in 1966. Both Mr Applegate and Mr Archer sued Mr Moss for breach of contract despite the fact that they were outside the limitation period of six years. Mr Moss denied liability, basing his denial on an exclusion of liability clause in the original contract and also on the assumption that the action was time-barred. In reply, Messrs Applegate and Archer claimed that their rights of action were concealed by the fraud committed. The judge in the case, Paull J., held that both the builder and the developer had concealed the right of action by fraud and thus the action was not time-barred. He awarded as damages the cost of the houses in 1957, plus interest from that date.

The case went to the Court of Appeal. The developer appealed against the finding of concealment by fraud, and a cross-appeal was made by the owners as to the date from which damages should be assessed. The appeal was dismissed and the cross-appeal was allowed, awarding damages of the value of the house when the problem was discovered in 1965 and interest from the date the houses were evacuated.

It is interesting to note that the exemption of liability clause was considered to be inapplicable to a situation created by a fundamental breach of contract and fraud, of which this case was a good example.

E.2.3.4 Programming the work

As an example of this risk, see page 81, Risk E.2.2.1.

E.2.3.5 Theft and burglary

On large sites, this risk can be quite substantial particularly if events lead to losses of a repetitious nature. Housing schemes present a good example of projects exposed to this risk.

E.2.3.6 Lack of communication

Lack of communication during the construction period can be a major risk no matter how small the project. Inaccurate communication, or lack of it, can also be a cause of misunderstanding between the various members of the professional team and also between members of the team and either the contractor or the owner. An example of this risk can be shown in the English case of *Sutcliffe v. Chippendale* where an architect was held liable to the owner for certifying payment in respect of defective work.[82] The contractor subsequently went into liquidation and could not make good the defective work. The cause of the problem in this case was a failure of communication between the architect and the quantity surveyor who drew up the certificates on behalf of the architect. The architect knew that the work was defective but he simply did not pass that information to the quantity surveyor.

However, one of the most catastrophic examples of poor communication and organisational failure during construction and erection of an engineering project occurred on 28 January 1986 when the space shuttle *Challenger* was given the clearance for ignition.[83] The space-shuttle-rocket-booster exploded after lift-off and all seven crew-members perished. The flight began in the late morning at 11:28 and ended 73 seconds later in an explosive burn of hydrogen and oxygen propellants that destroyed the external fuel tank and exposed the space shuttle to severe aerodynamic forces that caused complete structural break-up. Although the technical cause of the *Challenger*'s explosion was the result of a faulty design of the O-ring seal, which failed at the launch, the Presidential Commission, which was established to investigate and enquire into the cause of the disaster, found that the underlying cause 'was rooted in organisational failures and poor communication'.

The explosion was found to have been caused by hot combustion gases that escaped from a booster via a failed field-joint seal. The design of the joint included two O-rings that did not function correctly at launch due to the low ambient temperature that prevented them from responding correctly to the rising pressure after ignition and rotational movement within the joint.

For a number of years prior to the tragedy, engineers had been concerned about the behaviour of the seals at low temperatures and such temperatures were forecast for the morning of the launch. Analysis of the records, showed that of the previous twenty-three launches in which the field-joints had been examined following booster recovery and where data was held, seven showed damage to the O-ring seals.[84] This damage had only occurred at ambient temperatures below 24°C and it occurred in all cases where the temperature was below 18°C. The lowest recorded temperature was 12°C. However, various factors, including the management structure of the project, and ultimately time pressures to maintain the space shuttle programme, created a situation where launch proceeded despite technical advice to the contrary and at an ambient temperature near to freezing, where seal damage was likely to occur.

The Presidential Commission Report (Bermingham 1999, pers. Comm.) traced the technical cause of the accident to hot gas escaping, known as *blow-by*, following the failure of the O-ring pressure seal in a joint of the casing of the booster. The failure was due to a faulty design, which was unacceptably sensitive to a number of factors, including the effects of temperature, physical dimensions, the character of the seal materials, as well as the reaction of the joint to dynamic loading. The shuttle's solid

---

82 *Suttcliffe v. Chippendale and Edmondson* (1971), 18 BLR, 149.
83 Quoted from *Management of Engineering Risk*, by Roger B. Keey, Centre for Advanced Engineering, University of Canterbury, New Zealand, April 2000.
84 *Engineering Ethics: Balancing Cost, Schedule and Risk*, by R.L.B. Pinkus, L.J. Shumann, N.P. Hummon and H. Wolfe, Cambridge University Press, Cambridge, 1997.

rocket boosters were made up of several sub-assemblies; the nose cone, solid rocket motor, and the nozzle assembly. Marshall Space Flight Centre was responsible for the solid rocket boosters, while Morton Thiokol was the contractor for the solid rocket motors. The boosters are one of a set of 'elements' that make up the complete craft.

Prior to the launch of this flight, the procedures of the Flight Readiness Reviews (FRR) were carried out in accordance with normal procedures. However, concerns of Level III NASA personnel, and element contractors, regarding the joint seals of the Solid Rocket Motors were not adequately communicated to the NASA Level I and II management responsible for the launch. The management structure of the Shuttle Programme had four levels: Level I was responsible for policy, budgetary and top level technical matters; Level II was responsible for overall supervision of the Shuttle programme; Level III was responsible for development, testing and delivery of hardware to launch site; and Level IV was responsible for design and production of hardware. The managers at Levels I and II were unaware that the O-rings had been designated a 'Criticality 1' feature – a term denoting a failure point, without back-up, that could cause a loss of life or vehicle if the component fails. This component had previously been designated 'Criticality 1R' – the R implying redundancy. The R was removed when it became understood that the secondary O-ring was unlikely to seal if the primary O-ring failed.

The managers at Levels I and II were also unaware that since July 1985 a launch constraint had been imposed and then for six consecutive flights waived. The crucial factor seems to have been that *neither the management of Thiokol nor the Marshall Level III manager believed that the O-ring blow-by and erosion risk was critical.* The testimony and contemporary correspondence show that Level III believed that there was ample margin to fly with the extent of O-ring erosion that was being experienced, provided the leak check was performed at an increased pressure. The fact that the increased test pressure was a contributor to the increased failure rate in service seems not to have been recognised. *What is clear is that the NASA Level III managers, and Thiokol management, had no such understanding or a least had a different perspective of the failure mechanism to that held by Thiokol's engineers.*

The Mission Management Team (MMT) postponed the launch scheduled for 27 January due to high crosswinds. The MMT met again at 14:00 on that day and concerns were raised about the effect of the forecast low temperatures on such facilities as drains, eye wash and shower water, and fire suppression systems, but not about the O-rings. When the situation was relayed to the engineers at Morton Thiokol they were adamant about their concerns over the low temperature: '. . . way below our database and we were way below what we qualified for . . .' They contacted Morton Thiokol's liaison officer at the Kennedy Space Center, expressed their concern, and requested more forecast temperature data. He recognised the significance of the concerns and ensured that a teleconference was set up. This was in turn followed by a second.

At the second teleconference Morton Thiokol engineers presented the history of O-ring erosion and blow-by. Their recommendation was not to launch until the O-ring temperature reached 53°F (12°C). A long-detailed, and reportedly, not acrimonious discussion followed. Thiokol's Vice-President of Engineering was asked for a recommendation and he replied that he could not recommend launch. The Deputy Director, Science and Engineering at Marshall, was reported to have said he was 'appalled' at the recommendation not to launch. The Manager SRB (Solid Rocket Booster) Project at Marshall was said to have asked, 'My God, Thiokol, when do you want me to launch, next April?' Under this pressure, Thiokol management asked for a recess to consider their recommendation further and a Thiokol management-level discussion took place. One of the managers is said to have remarked that he 'took off his engineering hat and put on his management hat'. The Thiokol managers seem to have concluded that, although blow-by and erosion was to be expected, there was not sufficient evidence to predict joint failure. In the absence of such evidence Thiokol engineers described it:

> This was a meeting where the determination was to launch and it was up to us to prove beyond a shadow of a doubt that it was not safe to do so. This is in total reverse to what the position usually is in a pre-flight conversation or a flight readiness review

and the launch subsequently took place, with fatal results.

E.2.3.7 Failure to comply with insurer's conditions and requirements

All insurance policies are based on full disclosure by the insured of any relevant information to the insurer and almost all insurance policies make it a condition that the insured is to abide by any other conditions of the insurance contract. *Williamson and Vellmer Engineering v. Sequoia Insurance Company* is a case that illustrates the importance of this risk.[85]

A mechanical and electrical services consulting engineer received a quotation valid for thirty days for professional indemnity insurance through his broker on 15 May 1973. Because of financial problems he did not act on or respond to the quotation until 2 August 1973 when he sent to the broker a cheque for the quoted premium. During the intervening period, problems arose in connection with the mechanical design of an air conditioning, heating and ventilating system in the library project which was designed by him in 1968.

The insurer, upon receipt of the premium, requested a new application form to be completed and submitted by the consulting engineer. The broker copied the original application and sent it to the engineer with instructions to return it noting any changes. The engineer returned the application without change and the insurer issued a one-year insurance policy effective from 10 August 1973.

In 1974, the design problems in the library project resulted in a legal case against the engineer who turned to his insurer for indemnification. The insurer refused to defend or indemnify the engineer. The legal action initiated by the engineer against his insurers failed and the case was appealed to the California appellate court which affirmed the original judgment. Reference was made to the pertinent questions in the application form which asked the applicant to describe any claim made against the applicant and to set out whether he is aware of any circumstances that could result

---

85 *Guidelines for Improving Practice*, op. cit., see note 4, vol. VIII, No. 1.

in a claim against him. The engineer's response to these questions gave no indication of problems with the library project.

In contrast with the aforementioned case, the insurer may unjustly fail his insured, using this condition as a basis for his refusal to defend or pay a claim. An example of this occurred in Canada where one firm of consulting engineers was refused coverage by their insurer under their professional indemnity policy. The insurer refused to defend a claim made against the insured firm alleging late notification of the claim and failure to disclose relevant information. The firm faced with this dilemma had to, in the end, defend the claim and pay the legal costs. The defence was successfully made and the insured firm pursued the insurer for the legal costs it paid in defending the claim. It was awarded judgment against its insurer by the trial court. The insurer still refused to honour the insurance contract and appealed against the judgment. The appeal court dismissed the appeal and approved the judgment, ordering the insurer to pay the legal costs incurred in the defence of the original claim as well as the non-legal costs incurred in defending the claim, and the legal costs of the actions, trial and appeal.[86]

E.2.3.8 Impact

As part of construction work for a quay, 30 m long reinforced concrete piles were driven into the seabed.[87] They were to be connected by a solid reinforced concrete platform. During a windstorm, a 350 GRT pontoon, moored some hundred metres away from the construction site, broke away from its moorings and drifted against the piles. As these were still standing free and unconnected, they were unable to absorb any appreciable horizontal force, so two rows of piles were bent over in the impact. A number of piles had to be removed and replaced. In order to provide the necessary access for the pile driver, even undamaged piles had to be extracted and re-driven, which considerably increased the cost of repair. The damage was estimated at DM300,000.

E.2.3.9 Riot and civil commotion

Riot is defined as a tumultuous disturbance of the peace by three or more persons assembled together without lawful authority, with intent to assist each other, if necessary by force.

Civil commotion may be defined as public disorder. Riot or civil commotion may occur within the boundary of the construction site or outside it, and it may involve the employees of the contractor or members of the public or both. The responsibility for the risk in these different circumstances is assessed differently and thus may be allocated to different parties.

---

86 'Insurance, An Ultimate Solution or a Failing Expectation', by Gerald Beaumont, Workshop on Risks and Liability, FIDIC Annual Conference, Vienna, 1985. The case referred to by Mr. Beaumont was *Stevenson et al. v. Simcoe & Evie General Insurance Company* (1982) *Insurance Law Reporter* 5462, Alberta Court of Queen's Bench, upheld by the Alberta Court of Appeal by judgement dated October 1982.
87 *Schaden Spiegel*, op. cit., see note 30, July 1977.

E.2.3.10 Arson

Arson may be defined as the wilful and malicious damage to or destruction of property by the setting of a fire.[88] Experience has shown that the following are the usual causes:

- Vandalism;
- Covering up a crime or diverting suspicion;
- Greed for profit, insurance fraud;
- Terror, intimidation, sabotage;
- Mental defects.

It has been established that the target of arson is usually unattended and isolated premises with little or no security. Construction projects during construction or after completion fit that description and are often the target.

E.2.3.11 Strike

Strike is the usual term given for a simultaneous and concerted cessation of work by an employer's employees, or a substantial group of them, normally in pursuance of an industrial dispute. Strikes, however, may be political aimed at coercing the government.

Damage to property or injury to people may occur as a result of a strike action within the site involving the employees of the contractor or outside the site involving others. Allocation of responsibility for these two different risks is usually established in the General Conditions of Contract.

E.2.3.12 Incompetence

Mistakes are sometimes made by the most qualified and experienced of people. However, more frequent by far are the mistakes made by those who have not previously had the experience of similar work. This risk becomes more acute in rapidly developing countries where many projects of a wide variety are under construction without the necessary level of expertise either on the drawing board or on site.

E.2.3.13 Malicious acts

A 14 km pipeline with a diameter of 900 mm was laid to convey water between a reservoir and waterworks.[89] The bitumen-lined steel pipes were to be laid in a shallow trench and backfilled after testing. However, prior to the testing and backfilling of a section of the pipeline, it was exposed with one end remaining open.

Unknown persons emptied into the trench a 200 litres barrel of diesel oil which flowed into the open end of the pipeline. The manner in which this had been carried out indicated an act of sabotage. The diesel oil caused a chemical reaction with the

---

88 *Arson*, a publication of the Munich Reinsurance Company, Munich, 1982.
89 *Schaden Spiegel*, op. cit., see note 30, September 1975.

bitumen and the lining was ruined. Forty-two pipes, each 9 m long, had to be replaced and the repair, necessitating electric welding, was extremely expensive. It was necessary to separate the pipes from one another, give them a new lining, re-weld them and then lay the section of pipeline once again. The loss amounted to DM300,000.

E.2.3.14 Inefficiency and delays

Time is money, with the exception that one cannot help spending it. Inefficiency and delays cost time and in the end additional expenditure in one form or another, directly and indirectly. This risk is recognised as one of the major factors in most project overruns and in essence it can also apply to other cost overruns experienced by the contractor and the design professional.

E.2.3.15 Inadequate site supervision

This is a multi-faceted risk affecting all the parties to a construction contract. The owner should realise the importance of full-time site supervision and allocate sufficient funds to provide for a suitable and qualified individual, or a team of inspectors and supervisors. If the project is the owner's first and he is unaware of the importance of this risk, then it is up to the professional adviser to acquaint him with the problems and benefits. He should understand that if inexperienced, poorly trained and underpaid personnel are employed on site, they would be no match for the contractor's team and in most cases they would not earn the essential respect and cooperation.

The contractor also has to supervise his own workers and ensure that they carry out the work properly and to the requirements of the contract documents.

The professional designer is also expected to carry out his part of the supervision, which can be best explained by quoting from the document prepared by the International Federation of Consulting Engineers (FIDIC) for the purpose of assisting individuals discussing the subject of the various aspects of supervision. It is entitled 'FIDIC Policy Statement on the Role of the Consulting Engineer During Construction', and states in part the following:[90]

> A full professional service by a Consulting Engineer to a Client for a project comprises five main stages, as follows:
>
> (1) investigation and report,
> (2) detailed design and preparation of contract documents,
> (3) arranging a contract,
> (4) services during construction,
> (5) acceptance of Works, commissioning of systems, and resolution of final account.
> . . .
>
> A Consulting Engineer who undertakes only some of the services comprised in a full professional service, is not in a position to take responsibility for the performance or consequence of those which are not entrusted to him.
> . . .

90 'FIDIC Policy Statement on the Role of the Consulting Engineer During Construction', a FIDIC Policy Statement, 1984, Lausanne Switzerland.

Therefore, FIDIC recommends as follows:

(1) The Consulting Engineer should recommend to his Client the advantages of a full professional service providing continuity from inception to completion of a project.

   If the Client does not accept this recommendation, the Consulting Engineer should analyse and agree with his Client, before accepting an appointment for partial services, on the allocation of responsibilities for the different services respectively, and the procedures to be adopted for any independent checking or repetition of previous services that may be required.

(2) The Consulting Engineer undertaking services-during-construction should recommend to his Client the advantages of the Consulting Engineer undertaking entire services-during-construction with delegated authority to exercise comprehensive powers under the construction contract, as the agent of the Client, and authority to act as independent arbiter on matters properly referred to the Consulting Engineer for decision under the construction contract.

   If the Client does not accept this recommendation, the Consulting Engineer should analyse and agree with his Client, before proceeding with the services, on the allocation of responsibilities for the various duties respectively, between the Client and the Consulting Engineer, between the Client and the Contractor, and between the Consulting Engineer and the Contractor, all of which should be recorded in writing.

(3) Remuneration for services-during-construction should comprise two main parts:

   (i) payment for all services other than resident site staff, on the basis of a retainer per month, or on the basis of a percentage of the cost of the Works.

   (ii) payment for resident staff at man-month rates plus mobilisation payments.

In addition, payment to the Consulting Engineer should include reimbursement of . . .

This risk is so intense that the technical publications are full of horror stories relating to the inspection of work which is either carried out improperly or not at all. The extension of the law of negligence in the past decade or two has made the parties involved in the construction contract, and others too (see Chapter 5), responsible to third parties for any lack of care or negligence in the process of supervision. Some design professionals have already decided to lessen their exposure to this risk either by not undertaking the task of supervision at all, or by withdrawing from the site. Some lawyers, cultivating this risk, are advocating the idea that the professional involved in supervision should not be the same person responsible for design. Such an idea can only multiply the number of disputes and increase the magnitude of this risk because, if a problem arises during construction, the person best qualified to deal with it is the designer.

The American Society of Civil Engineers, concerned with the problem of construction inspection, established in 1967 a task committee on inspection. The Committee gathered a wealth of information through replies to two questionnaires; the first was sent to owners and their representatives and the second to contractors. The replies showed the following problem areas as being conducive to claims and extra payments:[91]

- Inspectors who are too young and inexperienced;
- Personality conflicts with contractors' personnel;
- Unfamiliarity of the owner's representatives with the plans and specifications;
- Poor documentation;
- Owner's representatives directing the contractor's operations;
- Demanding a higher quality of work than is necessary;
- Owner's representatives exceeding their authority;
- Unnecessary delay by contractors;
- Unfamiliarity with construction practices.

One example of this risk eventuating can be taken from the case of the *Governors of the Peabody Donation Fund v. Sir Lindsay Parkinson and Co. Ltd and Others.*[92] Architects and engineers were retained to design a housing complex of 245 houses on a hillside site in inner London. The owners, Peabody Donation Fund, were required, under the London Government Act 1963, to install a suitable drainage system for the development. The system was to be to the satisfaction of the local authority and had to conform to the requirements of the drainage by-laws.

The site, being hilly, presented problems and had to be terraced. The subsoil was London clay which tends to expand and contract seasonally, thus giving rise to movement. For this reason, the professional team of architects and engineers designed the drainage system using flexible joints.

Early in 1973, the contractor, Sir Lindsay Parkinson & Co. Ltd, was ready to start work on the drainage system. The architect's representative on site was a young trainee architect who was responsible for supervision of the works. The local authorities instructed a drainage inspector to carry out inspections of the drainage works and on 2 February 1973 he met on site the trainee architect supervising the work and agreed with him to abandon the planned flexible jointing system in favour of a rigid pipe jointing system. The latter accordingly instructed the contractor, whose agent had attended the meeting on site. However, neither the inspector nor the trainee architect informed their respective principals of the change in design to which they had agreed.

Tests carried out in late 1975 and early 1976 revealed that many of the drains laid with rigid joints had failed. Reconstruction was necessary at a cost of £18,000 and the completion of the development was delayed for about three years with consequent loss of rents for the owner, who was also faced with substantial claims by the contractor for additional payments.

91 Report on Construction Inspection, ASCE Paper No. 9192, Summary Report, published in the Proceedings of the America Society of Civil Engineers, September 1972.
92 *Governors of the Peabody Donation Fund v. Sir Lindsay Parkinson & Co. Ltd. & Others* [1984] 3 All ER 529.

The owner started legal proceedings against the contractors for faulty workmanship, against the architects for failing to supervise properly and for the change in the design of the joints, and against the local authorities for knowing that rigid joints were being installed yet failing to require that they be flexible.

In the event, the case against the architects was compromised but it continued against the contractor and the local authorities. It was held that,

> although there had been some faulty workmanship on the part of the contractors, this was not the cause of the failure of the drains, and that the cause of the failure was the design change, instructed by the supervising trainee architect, from flexible joints to rigid joints.

The court did not have to deal with the claim against the architects since it was settled. The local authorities were judged liable in damages to the owner on the grounds of failure to take steps to ensure that the drainage system, as installed, complied with the design originally approved by them.

The local authorities appealed against the latter part of the judgment. The Court of Appeal allowed the appeal and reversed the decision of the trial judge. Subsequently, the owner appealed, though without success, to the House of Lords.

This decision was followed in the case of *Investors in Industry Commercial Properties v. South Bedfordshire District Council & Others.*[93] The defendants appealed the earlier decision of the court's finding of negligence in the defendant's approval of plans and inspections of two warehouses with inadequately designed foundations. It was held that the local authorities owed no duty of care in its supervisory powers to the original building owner, who had the advice of architects, etc.

E.2.3.16 Variations from contract documents

On a Friday evening in July 1981, over one thousand people were crowded on to the main floor of the lobby of the Hyatt-Regency Hotel, in Kansas City, USA and on three walkway bridges spanning it, to watch and participate in a dancing contest. Shortly after 7 p.m., a loud cracking noise was heard and the fourth-level walkway was seen to buckle and fall on to the second-level walkway two storeys below, causing it to collapse and dump some 60 tonnes of debris, along with the spectators from both walkways, on to the crowded dance area. The death toll was 111, and 188 were injured.

Lawsuits were quickly launched, seeking compensation damages exceeding US$1 billion and punitive damages of more than US$500 million. Legal fees involved were estimated to be in the order of US$100 million. Several technical investigations were undertaken, including one by several consultants retained as engineering counsel on behalf of the steelwork fabricator.

The technical facts of the case are altogether simple. The failure initiated in the fourth-level walkway at the hanger rod connections to the floor beams. Each floor

---

93 *Investors in Industry Commercial Properties v. South Bedfordshire District Council & Others* [1986] 1 All ER 787.

beam consisted of a pair of light channels joined together, toe to toe, with weld beads placed along the outside of the joints only, except for inside passes 30 mm long at each end. Such welds have no code status. Figure 3.10 shows the general arrangement of the walkway steelwork.

Investigations showed that the as-built steelwork was different to that shown on the original design drawings. The original connection detail showed each of the 32 mm steel hanger rods passing continuously from the ceiling through the fourth-level floor beam and terminating at the second-level floor beam, with an ordinary round washer and a nut at each floor beam bearing point. In order to simplify fabrication and erection, the fabricator submitted an alternative detail, Figure 3.10, incorporating two half-length hanger rods, each terminating at the fourth-level floor beam. The

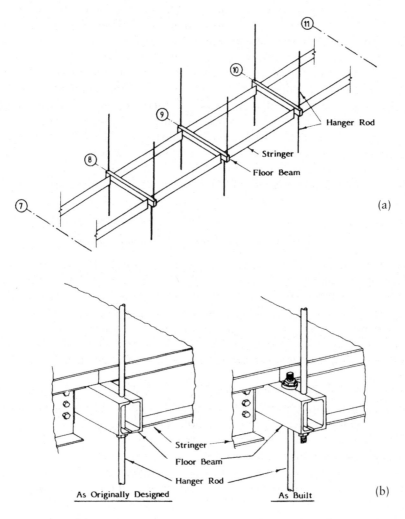

Figure 3.10 (a) General arrangement of framing of the walkway; and (b) Floor beam detail of the Hyatt-Regency Hotel, Kansas City, USA.

fabricator's shop detail drawing showing this hanger rod connection detail was seemingly reviewed and authorised by the design engineer as attested by his stamp. The hanger rod connection failed.

The change in detail of the connection resulted in doubling of the load applied against the lower flange of the upper floor beam. The walkway failure cycle began when one of the upper hanger rods pulled through the bottom flange of a fourth-level floor beam.

It is interesting and significant to note that whilst the original hanger rod connection detail at the upper floor beam was adequate to support the loading subjected at the time of the collapse, its capacity was far below that required by the Kansas City Building Code.[94] Furthermore and in comparison, the as-built connection detail that failed reached its capacity under the weight of only the dead load.[95]

E.2.3.17 Illegal activities

A shopping mall in Donguan in the Chinese province of Guangdong collapsed as workers were engaged in illegally adding two floors to the single-storey reinforced concrete-framed building. The disaster occurred on Friday 1 December 2000, as a result of which eleven people were killed, 40 were injured and 120 were trapped for a number of hours. It was thought that the weight of the additional floors caused the building to collapse.[96]

E.2.3.18 Risks associated with dispute resolution

Construction contracts are prone to disputes, the resolution of which is highly uncertain as to duration, costs and outcome. In particular, these elements cannot be known in the early stages of the dispute resolution proceedings and risk management is of the utmost importance if optimum results are to be achieved.

Experienced construction lawyers and specialists, however, can usually provide very preliminary, but realistic estimates by drawing up a step-by-step schedule of activities of the dispute resolution mechanism. Important dates should be laid down, for example, dates for the exchange of witness statements; dates for the exchange of expert reports; and hearing dates. The duration and costs of these activities should then be assessed together with the amount of the outcome on the basis of probabilities of optimistic and pessimistic boundaries. Risk strategy can then be developed for each of these boundaries.

This process should be continued throughout the dispute resolution mechanism to provide a continued assessment and flexibility of reaction.

Of course, it should be noted that in arbitration, the arbitrator in consultation with the parties sets the above-mentioned dates, which are unknown at the early evaluation stages. In order to evaluate fully the extent of the risks involved in

---

94 'Hyatt-Regency Walkway Collapse: Design Alternatives', by George F.W. Hauck, ASCE *Structural Engineering*, vol. 109, No. 5, May 1983.

95 'Some Liability Aspects of Steelwork Design and Construction', by Jackson Durkee, The IABSE Henderson Colloquium on Liability, Cambridge, 1984.

96 *New Civil Engineer*, op. cit., see note 45 above, 7 December 2000.

arbitration, as well as the objectives sought, sufficient information should be continually examined, including the following:

(a) An assessment of the estimated or likely sum recoverable if the case proceeded to a hearing and an award;
(b) An estimate of the percentage of costs that is likely to be recoverable if a party is met with success or failure at arbitration; and
(c) Whether or not an offer of settlement is made.

The answers to the above are combined to form compound figures for decision-making in relation to the risks involved in dispute resolution.

E.3.1 Risks associated with the post-construction stage (Figure 3.11)

A project is born with the completion of the construction period and it is then expected that it should carry out whatever functions for which it has been conceived. Accordingly the risks to which a project is exposed during the post-construction stage differ from those which exist prior to its completion. They can be categorised however, as shown in Figure 3.11, under the following headings:

- Safety
- Serviceability
- Resistance to fire and arson
- Resistance to natural and other hazards
- Resistance to man-made hazards
- Fitness for its purpose
- Operation
- Resistance to wear and tear during its designed life span.

There is little difference between the risks during the Defects Notification period, and those in the period that follows. The difference mainly lies in the fact that the contractor may be present on the site of the project during the earlier period in fulfilment of his obligations under the contract. Each of these risk categories is discussed separately. The term 'maintenance period' has been renamed as the 'Defects Notification Period' in the new suite of FIDIC Forms of Contract published in September 1999 and the 'Defects Correction Period' in the old Red Book of FIDIC and the ICE forms of contract. In this book, the term Defects Notification Period will be used.

E.3.1.1 Risks associated with safety

The combined quality and performance of the design of the project, the material used and the workmanship employed in its construction makeup the level of its safety. If one considers each in a scale where white represents perfection and black represents fault, then there are as many combinations as there are shades of grey. Lack of safety in construction projects continue to cause concern all over the world, but naturally more in some parts than in others. The concern in the United States

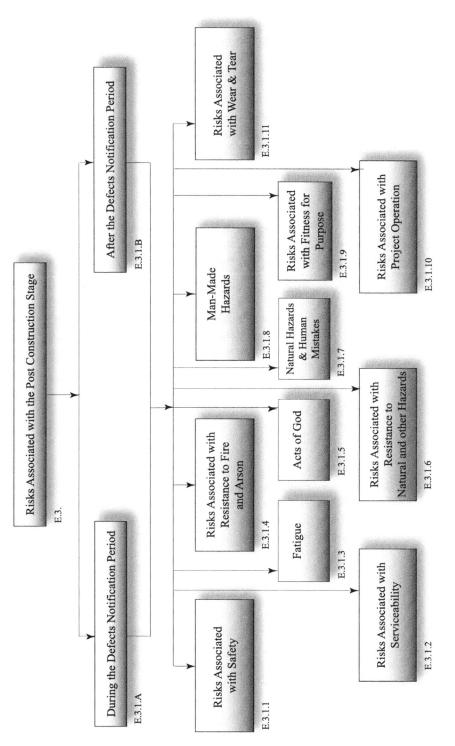

Figure 3.11 Risks associated with the post-construction stage.

about the apparent increase in the number of structural failures in the 1980s prompted the US House of Representatives Committee on Science and Technology to investigate these failures. The Committee's report, submitted in February 1984, discusses the findings and provides some recommendations as a result. Part of the report is quoted here for its relevance to the problem of safety:[97]

> The Committee found six significant factors that are critical in preventing structural failures. The Subcommittee also found five factors to be of heavy-to-moderate impact and eleven factors to be of lesser significance.
>
> The six critical factors are:
>
> (1) communications and organisation in the construction industry;
> (2) inspection of construction by the structural engineer;
> (3) general quality of design;
> (4) structural connection design details and shop drawings;
> (5) selection of architects and engineers; and
> (6) timely dissemination of technical data.
>
> The five moderately significant factors are:
>
> (1) overall accountability for structural integrity;
> (2) impact of 'cost cutting' on design;
> (3) impact of 'cost cutting' on construction;
> (4) potential for improving quality during construction; and
> (5) selection of construction contractors.
>
> The eleven least significant factors are:
>
> (1) adequacy of building codes;
> (2) impact of higher strength materials;
> (3) adequacy of state and municipal design reviews;
> (4) increased use of speciality contractors;
> (5) support by testing laboratories;
> (6) certification of buildings;
> (7) role of lending institutions;
> (8) adequacy of seismic codes;
> (9) impact of construction manager type Organisation;
> (10) impact of fast track scheduling; and
> (11) need for legislative changes.

The Subcommittee also made formal findings and recommendations for the six factors which it concluded as being of critical importance to structural failures. The subcommittee believed that 'these factors warrant the greatest and most immediate attention by both government and the building industry'.

97 *Structural Failures in Public Facilities*, Report by the Committee on Science and Technology of the United States House of Representatives, United States Government Printing Office, Washington, 1984.

The Subcommittee then elaborated on the six critical factors as follows:

(a) Communications and Organisation in the Construction Industry

There is no set pattern for organising design/construction projects. Virtually all projects are different, and the companies involved have different capabilities. The data received by the Subcommittee, however, through its hearings and investigation strongly indicate the existence of loosely structured organisations, unclear definitions of responsibility, and poor lines of communication between the participants in construction projects, all of which can contribute to the occurrence of a structural failure.

(b) Inspection of Construction by the Structural Engineer

For a variety of reasons the structural engineer of record or his designee is often not present on the job site during the construction of principal structural components. The absence of the structural engineer has permitted flaws and changes on site to go unnoticed and uncorrected. One reason for the structural engineer's absence is the possibility that he or his firm could be subject to lawsuits which have no relationship to their project responsibilities.

(c) General Quality of Design

The quality of design is being compromised by client desire to speed construction and reduce overall project costs. This can lead to the subsequent elimination of essential engineering services, such as peer reviews, that provide an important check on the commission of errors in the design process and help ensure that high quality designs are produced.

(d) Structural Connection Design Details and Shop Drawings

Structural connections, which are critical components of structural design, are often not designed by a structural engineer. Instead, they are frequently left to persons who do not have a sufficient understanding of the interaction of all stresses between structural members. Moreover, shop drawings are not sufficiently reviewed by the structural engineer. As a result, errors in design can be made and go unnoticed.

(e) Selection of Architects and Engineers

The selection of architects and engineers is generally made on a 'low bid' basis. Even when peer reviews, design scope, and on-site inspection are included in the bid request, there is a tendency to unrealistically reduce the price when price is known to be the primary basis for the contract award. Nearly exclusive use of this 'low bid' procedure has frequently resulted in insufficient funds allocated to a project to adequately verify the accuracy of design and to thoroughly check plans before construction. In other words, selection of an architect or engineer solely on a price-competition basis provides the potential for reductions in quality due to initial underestimation of the costs and resources required to adequately perform the work.

(f) Need for a National Board to Investigate Structural Failures

The records in many cases of litigation involving structural failures have been closed to public review as a result of settlement agreements between the parties. Consequently, little information has been made public about the technical causes of several recent major structural failures. Structural engineers and others have thus often been prevented from learning from the experiences and mistakes of others. This failure to disclose information provides the opportunity for others to commit the same mistakes in the future. A national investigative body is needed to obtain information about structural failures and provide individuals involved in the building industry with much needed data on failures around the country.

Examples of the risk of failure in safety are plentiful and some have been quoted earlier in this chapter. Others can be seen in the technical press and those which are dramatic have been the subject of investigation by various committees formed in various parts of the world for that purpose.

E.3.1.2 Risks associated with serviceability

Serviceability is the second essential and basic performance requirement of any construction project throughout its intended life span.

The serviceability requirement can be very stringent in that it restricts the acceptable and yet inevitable movement and deformation of the various elements of a project to a maximum limit. Such limit is usually fixed so that any movement or deformation is within a boundary beyond which the intended use of the project is rendered less effective.

The movement and deformation may take any of the following forms:[98]

- Settlement of foundations
- Deflections due to loads including wind
- Strain and creep deformation
- Temperature and shrinkage movement
- Movement due to varying moisture content
- Movement due to natural forces such as those resulting from earthquakes
- Cracking
- Vibration

The most dramatic examples of serviceability failures are perhaps due to the first type of movement, i.e. settlement of foundations. The building for the Société Minoteries Tunisiennes at Tunis is but one example of such failure with others easily located in the technical press.[99]

---

98 'Guide to the Performance of Building Structures', a publication of the Institution of Structural Engineers, December, 1984, London, UK.
99 *Building Failures*, Thomas McKay, McGraw-Hill Book Co., 1962, page 114.

E.3.1.3 Risks associated with fatigue

In June 1989, a 90 m diameter radio telescope at Green Bank, Virginia, USA, collapsed after twenty-six years of service. It collapsed due to fatigue following progressive cracking and then failure of a key joint. The fatigue was caused by very high cyclic loading derived from secondary forces as the radio telescope swept back and forth across the sky.[100]

The shallow dish consisted of a light aluminium surface supported on a bolted steel frame of matching shape, all supported on a diamond shaped truss. The collapse was precipitated by failure of a gusset plate on one side of the diamond truss at a connection between three main members. A crack spread from two punched bolt-hole positions to extend over the whole 1.2 m width of the plate. Secondary cracking was found around the rogue boltholes, indicating that 'severe working produced by the punch could have left an initiating small crack'. Cracking was also found on corresponding gusset plates.

E.3.1.4 Risks associated with fire and arson

A construction project should be capable of containing a fully developed fire within a restricted and prescribed area and in such a way that it does not spread to adjacent compartments. Furthermore, it is a requirement in many parts of the world that collapse should not occur as a result of fire before a prescribed period of time has elapsed which depends on the type of the project.

Examples of fire disasters can best be studied from the files of insurance and reinsurance companies and the following is one of these examples. A 114-year-old department store in Singapore, one of the landmarks of the city, was destroyed by a large fire on 21 November 1972.[101] Nine people were killed, the property damage and consequential loss amounting to about DM18 million.

The three-storey building was of brick, masonry and timber construction with a tile roof. The whole department store covered an area of about 8,000 sq m which was not separated into individual fire compartments and not protected by sprinklers. The building was air-conditioned and had three passenger lifts.

On the day when the fire occurred, the store opened at 9 a.m., as usual. Shortly afterwards, the air-conditioning system was put into operation by an electrician. At 9.45 a.m., some employees on the ground floor noticed a smell of smoke which they thought had been caused by a short circuit. The electrician was called and, as he arrived on the ground floor, all the lights suddenly went out. Just as repairs at the main switchboard were to be started, the fire alarm rang and, shortly afterwards, fire was noticed at the rear of the ground floor. At first, some employees tried to bring the fire under control using portable extinguishers. However, they soon gave up as fire and dense smoke spread very rapidly from the ground floor and also took hold of the upper storeys of the building.

When the fire brigade arrived at the scene, the fire had already reached such pro-portions that they had to limit their action to protecting the adjacent buildings.

---

100 *New Civil Engineer*, op. cit., see note 45, 8th June 1989.
101 *Schaden Spiegel*, op. cit., see note 30, October 1974.

The building, where about 300 employees worked, had several staircases but these were difficult to find, especially for the customers. In addition, the lifts were inoperative due to the failure of the power supply and the combination of these unfavourable factors led to the loss of nine lives.

The risk of fire in certain structures, such as grain mills, forms part of their function. Grain mills and silos are essential structures for human food programmes. To guarantee sufficient food supplies for the world's population, the main needs are grain, powdered milk, and oil seed. In order to store and provide these basic substances, harbours and areas with intensive, large-scale grain production are all equipped with silos, trans-shipment terminals, and facilities to store grain for export and import.

Dust explosions in mills and grain elevators occur frequently despite the accumulated knowledge of their cause and devastating effect. In the 1960s and 1970s, for example, the Corn Belt in the United States was frequently the scene of devastating dust explosions. Major incidents then occurred in the harbour at Bremen in 1979 and in the harbour at Metz in 1982, with many fatalities in each case.

One of these explosions happened recently in France.[102] In the harbour area of a small town a mighty dust explosion destroyed the silos used to store wheat, barley and maize. It was not until the dust cloud settled that the rescue teams realized the total extent of the catastrophe:

- 29 of the 44 silos had vanished completely, together with the conveying equipment installed on top;
- The head house, 52 m in height, had buried offices, a grain-loading station, and the packing unit beneath it;
- The maize drier in the rear section was badly damaged;
- The oil and soda containers at a neighbouring tank farm were riddled with holes; and
- Eleven persons died.

The fire brigade, which took only three minutes to arrive on the scene, could do nothing but fight a relatively small fire and make the ruins safe. A close watch was kept on the three remaining parts of the structure so that a timely warning could be given if there was any likelihood of them collapsing.

It was assumed that the explosion originated in the central dust extractor because of its large-scale destruction. Mechanically induced sparks, friction heat or self-ignition could have triggered the event.

The mechanism of such an explosion is well known. Grain seeds rub against each other during transportation and release extremely fine dust. Unlike the grain itself, this organic dust is highly inflammable and can burn so fast that there is an explosion. Such a sudden combustion can only be caused by finely dispersed combustible dust, such as a dust cloud, coming into contact with oxygen in the air and an ignition source, which may be, for instance, an electrical discharge or a hot piece of metal.

102 An article by Robert Schmid, Munich, taken from *Schaden Spiegel*, op. cit., see note 30, No. 2, 1999.

The initial explosion usually swirls up further dust clouds, which ignite on the first flame front and release a much larger explosion. The chain reaction continues until such time as there is no dust left.

The energy released in this sudden combustion manifests itself just as suddenly in an expansion of gases. The expanding gases finally produce a shock wave. In this incident, the shock wave was so strong that it was able to pulverise solid concrete, hurl large elements into the harbour basin, and catapult a car some 50 m through the air. Subsequent explosions led to the collapse of the head house.

Obviously, risk prevention measures are taken very seriously in such structures. In principle, atmospheric oxygen and combustible materials are always present in any silo. Risk prevention therefore primarily involves eliminating ignition sources and limiting combustible grain dust as far as possible, which means, in practical terms, sucking it off.

E.3.1.5 Acts of God in relation to nature of site – topography and surface water run-off

On 9 October 1963, one of the worst reservoir disasters in history, killing nearly 3,000 people, took place when over 240 million cubic metres of hillside slid into the Vaiont Reservoir in Italy.[103] The dam 265.5 m, then the world's highest thin arch, survived.

The dam, completed in 1960, blocks the Vaiont gorge, a mile above its confluence with the Piave, in the Italian Alps, 90 km north of Venice. The disaster was entirely caused by dangerous geological conditions, accentuated by groundwater changes due to the filling of the reservoir.

The area consists of a thick succession of sedimentary rocks, dominantly lime-stone with frequent shaley partings and sequences of thin limestone and marl, ranging in age from Lias (Lower Jurassic) to Senonian (Upper Cretaceous). These strata are folded into a syncline, the valley coinciding with the axis, and the north limb of the syncline (right bank) is cut by a fault bringing Upper Cretaceous against Dogger (Middle Jurassic). Minor faulting and close jointing have created blocky rock masses, and solution cavitation of the limestones has occurred.

Topographically, the reservoir area consists of an outer U-shaped glaciated valley and an inner steep-sided post-glacial gorge. Morainic deposits occur in the outer valley, and on the rock surface of both valleys there are accumulations of talus, slope-wash and old landslide material. Large-scale landslips are common in the Vaiont Valley.

The rocks of the outer valley are affected by an older set of stress-relief joints parallel to the surface, and a younger set occur parallel to the walls of the inner gorge.

The dam itself is built on the Middle Jurassic, which contains some thickly bedded massive limestone.

The events before the disaster were that in 1960 a slide of some one million cubic metres occurred on the left bank of the reservoir near the dam; a pattern of cracks developed upslope from the slide and continued eastwards.

---

103 *Case Histories in Engineering Geology*, by J.G.C. Anderson and C.F. Trigg, Elek Science, London, 1976. This case is reported here almost as it appeared in the Dams and Reservoirs section of this book.

In 1960–1, a 2 km by-pass tunnel was driven under the right wall to enable water from upstream to reach the outlet works of the dam in case of future slides. As a further precaution the top water level (TWL) was limited to 680 m, about 40 m below the dam-crest, and a grid of geodetic stations installed throughout the potential slide area. Drill holes and an adit did not detect a major slide plane, but an analysis after the disaster showed that they were too shallow. Gravitational creep continued to be observed during the 1960–3 period.

During the spring and summer of 1963, scattered observations of the eventual slide area showed an average creep movement of 1 cm per week. About mid-September, numerous geodetic stations were moving at 1 cm a day, but it was believed that only individual blocks were involved, not the whole area.

Heavy rains, beginning about 28 September and continuing until after 9 October, increased groundwater recharge and run-off, and the reservoir TWL rose to 690 m. Early in October, the Mayor of Casso, a town above the right bank, posted a warning to the townspeople. About 8 October, engineers realised that all the geodetic stations were moving on an unstable mass, and on that date they began to lower the water level through two outlet tunnels, although heavy run-off reduced effectiveness.

On 9 October, accelerated movement was reported and in spite of the open gates, the reservoir level rose; movement must by that time have been reducing reservoir capacity.

The disaster occurred late in the evening of 9 October when over 240 million cubic metres of the hillside slid from the left bank into the reservoir in less than 30 seconds. The speed of the mass movement was 15–30.5 metres per second. Over a length of 2 km, the entire reservoir piled up as a vast curving wave for ten seconds. A terrific updraft of air created by the slide and accompanying the wave sucked water and rocks up to about 270 m above reservoir level. Both the blast and the subsequent decompression added to the destruction. The water swept over the dam to a height of some 100 m above its crest. A wave 70 m high overwhelmed Longarone, 1.6 km down the Vaiont Valley from the dam, at the confluence of the Vaiont and Piave. Two kilometres up the latter valley, the wave was still 5 m high, and the main volume swept for many miles downstream. Nearly 3,000 people perished.

Seismic (L waves only) tremors were recorded as far away as Vienna and Brussels. The records showed that these were entirely due to the kinetic energy of the sliding mass and that no tectonic earthquake triggered the movement.

It was established that the Vaiont disaster was caused by a combination of adverse geological conditions (dip slope of tectonically jointed Mesozoic limestones affected by later relief joints and interbedded with weaker shale), change in environment due to the reservoir itself and excessive groundwater recharge from heavy rain. Geological assessment, not only of a dam site, but of a whole reservoir area, should be directed not only to present conditions but to past events (e.g. former sliding in the Vaiont Valley) and to likely future changes. Rock masses in a changed environment can weaken rapidly, particularly once creep (for which there is often surface evidence) gets under way; acceleration to collapse can occur very quickly.

E.3.1.6 *Risks associated with natural hazards*

Natural hazards are many. By the very function they perform, construction projects are exposed to the effects of natural hazards such as wind, hurricane, typhoon, landslide, earthquake, rainfall and flood. The degree of protection afforded by a particular design to a specific project depends on the probability of occurrence beyond which the project is expected to be unaffected. Where the hazard is of a probability of occurrence lower than that for which the project is designed, damage would be expected to occur. Table 3.4 provides a list of loss to property, which resulted from natural hazards within the period of 1970 to 1980, and shows the extent and the disastrous nature of these hazards.[104]

Since 1980, many natural hazards have occurred, some of which would be more appropriately designated as natural catastrophes. In particular, the more significant of these natural catastrophes with devastating effects include: storms, typhoons, hurricanes, floods, tornadoes and last, but not least, earthquakes. Examples of the latter include the earthquake in Armenia on 7 December 1988, where 25,000 people died and 65,000 were injured; in Kobe, Japan, on 17 January 1994, where 1,000 people died; in Los Angeles in January 1994; in Istanbul, Turkey on 17 August 1999, where 30,000 people lost their lives; and in Gujarat, India, on 26 January 2001, where 100,000 people lost their lives.

It is noteworthy that there seems to be an upward trend in the number and the devastation caused by these natural hazards. The number has increased by a factor of 2.3 when comparing the 1980s with the 1960s; and by a factor of 3.3 when comparing the 1990s with the 1960s. The economic losses suffered have increased by a factor of 2.8 when comparing the 1980s with the 1960s; and by a factor of 8.6 when comparing the 1990s with the 1960s. The insured losses, which form only a small portion of the economic losses, have increased by a factor of 3.6 when comparing the 1980s with the 1960s; and by the huge factor of 17.0 when comparing the 1990s with the 1960s.[105]

Insured losses form a superior basis for analysis of these natural hazards since they can be established precisely. When the insured losses resulting from a single event pass the significant threshold of US$1 billion, there would usually be a large number of people killed, and even larger number made homeless. A list of insured losses exceeding that threshold is quoted below in Table 3.5 for the years between 1980 and 2001.[106] The figures given represent the original losses recorded without taking into account inflation.

In construction, as explained below, the worst effects occur when natural hazards are combined with standards of construction that are lower than acceptable.

104 *Reinsurance Principles and Practice*, by Dr Klaus Gerathewohl, Verlag Versicherungswirtschaft e. V., Karlsruhe, vol. II, 1982, page 114.

105 'Annual Review: Natural Catastrophes 2001', Munich Re Topics 2001, published by the Munich Re Group, Munich, Germany, 9th year 2002, page 15.

106 The list is reproduced from page 17 of the previous reference marked Copyright to 2002 Munich Re NatCat*SERVICE*'.

Table 3.4 List of loss to property from natural hazards within the period 1970 to 1980

Date	Location	Approximate overall value of the loss
May 1970	Earthquake/Chimbote, Peru	US$510m
Aug. 1970	Hurricane Celia	US$450m, of which US$330m is an insurance loss
Oct. 1970	Flood/Northern Italy	US$200m
Feb. 1971	Earthquake/California, USA	US$535m, of which US$50 m is an insurance loss
June 1971	Flood/India	US$530m
June 1972	Hurricane Agnes/east coast of USA	US$3,100m, of which US$100 m is an insurance loss
July 1972	Flood/Philippines	US$35m
Nov. 1972	Winter gale/Lower Saxony/ Central Europe	US$420m, of which US$200 m is an insurance loss
Dec. 1972	Earthquake/Managua, Nicaragua	US$800m, of which US$100m is an insurance loss
Jan. 1973	Lava from the Helgafjell volcano/ Iceland	US$200m
April 1973	Flood/St Louis, USA	US$500m
Jan. 1974	Tornado Wanda, floods/Australia	US$300 m
April 1974	Tornadoes/USA	US$1,000m, of which US$20m is an insurance loss
Sept. 1974	Hurricane Fifi, floods/Honduras	US$500m, of which US$20m is an insurance loss
Dec. 1974	Tornado Tracy	US$500m
Jan.–Aug.1975	Severe rainfalls/Japan	US$400m
Sept./Oct.1975	Typhoons Phyllis, Rita and Cora/Japan	US$360 m
Sept. 1975	Hurricane Eloise/Florida, USA	US$420m
Jan. 1976	Winter gale Capella/Europe	US$1,300m, of which US$500m is an insurance loss
Feb. 1976	Earthquake/ Guatemala	US$1,100m, of which US$55m is an insurance loss
May 1976	Earthquake/Friuli, Italy	US$2,000m
May 1976	Typhoon Pamela/ Guam, Pacific	US$120m, of which US$66m is an insurance loss
March 1977	Earthquake/Rumania	US$800m
June 1978	Earthquake/Japan	US$1,800m, of which US$2m is an insurance loss
Aug. 1979	Hurricane David/ Caribbean and USA	US$2,000 m, of which US$250 m is an insurance loss
Sept. 1979	Hurricane Frederi/USA	US$2,300m, of which US$750m is an insurance loss
May 1980	Volcanic eruption, Mt St Helens/ USA	US$2,700 m, of which US$27 m is an insurance loss
Aug. 1980	Hurricane Allen/ Caribbean and USA	US$1,400 m, of which US$50 m is an insurance loss
Oct. 1980	Earthquake/Algeria	US$3,000m
Nov. 1980	Earthquake/Italy	US$10,000m, of which US$40m is an insurance loss

Table 3.5 List of insured losses from natural hazards after 1980 of US$ 1 billion and above

Rank	Year	Event	Region	Insured losses US$m	Economic losses US$m
27	1983	Hurricane Alicia	USA	1,275	3,000
10	1987	Winter storm	Western Europe	3,100	3,700
6	1989	Hurricane Hugo	Caribbean, USA	4,500	9,000
5	1990	Winter storm Daria	Europe	5,100	6,800
26	1990	Winter storm Herta	Europe	1,300	1,950
15	1990	Winter storm Vivian	Europe	2,100	3,250
25	1990	Winter storm Wiebke	Europe	1,300	2,250
4	1991	Typhoon Mireille	Japan	5,400	10,000
18	1991	Oakland forest fire	USA	1,750	2,000
1	1992	Hurricane Andrew	USA	17,000	30,000
20	1992	Hurricane Iniki	Hawaii	1,650	3,000
19	1993	Snow storm	USA	1,750	5,000
33	1993	Flood	USA	1,000	16,000
2	1994	Earthquake	USA	15,300	44,000
11	1995	Earthquake	Japan	3,000	100,000
29	1995	Hail	USA	1,135	2,000
22	1995	Hurricane Luis	Caribbean	1,500	2,500
16	1995	Hurricane Opal	USA	2,100	3,000
21	1996	Hurricane Fran	USA	1,600	5,200
28	1998	Ice storm	Canada, USA	1,200	2,500
34	1998	Floods	China	1,000	30,000
24	1998	Hail, severe storm	USA	1,350	1,800
7	1998	Hurricane Georges	Caribbean, USA	4,000	10,000
30	1999	Hail storm	Australia	1,100	1,500
23	1999	Tornadoes	USA	1,485	2,000
14	1999	Hurricane Floyd	USA	2,200	4,500
8	1999	Typhoon Bart	Japan	3,500	5,000
13	1999	Winter storm Anatol	Europe	2,350	2,900
3	1999	Winter storm Lothar	Europe	5,900	11,500
12	1999	Winter storm Martin	Europe	2,500	4,000
32	2000	Typhoon Saomai	Japan	1,050	1,500
31	2000	Floods	Great Britain	1,090	1,500
17	2001	Hail, severe storm	USA	1,900	2,500
9	2001	Tropical Storm Allison	USA	3,500	6,000

E.3.1.7 *When risks of natural hazards are added to human mistakes*

The earthquakes in India and Turkey provided examples of such disasters where the number of casualties is usually highest. In Turkey, many buildings collapsed because of 'soft storey' failure after walls in the ground floor were removed to accommodate shop fronts. This reduces the stiffness of the lower floors and weakens the structural resistance to torsion leading to shear failure at the beam to column connection or at the base of the column. Much of the destruction in the nearby Golcuk was through liquefaction, where whole floors disappeared into the ground, since the water table is so close to the ground level that the buildings could be considered as floating. Where buildings were engineered properly, only a small amount of damage occurred.

In the Indian earthquake, poor construction, poor supervision, and lack of adherence to design codes, were at the heart of the disastrous consequences. Eighty percent of Gujarat city was reduced to rubble within two minutes. The estimated costs of rebuilding the city are over £3.5 billion.

E.3.1.8 Risks associated with man-made hazards, including political risks

The twentieth century has witnessed great wars and disastrous destruction, as well as very significant progress in science and technology, which has not only increased the quality of human life and the respect for human rights and the freedom of the individual, but also unfortunately permitted the possibility of more sinister atrocities. Acts of terrorism and acts of force against the rights of individuals or of a whole nation are but only two of the hazards that have emerged and face humanity as a whole. Construction artefacts are in particular vulnerable to these two types of risk.

No example of the first of these two risks, terrorism, can be more striking in atrocity than the events that took place on 11 September 2001 in New York, USA. The events are lucidly counted in an insurance publication on these events.[107]

On 11 September 2001, an American Airlines Boeing 767 flew right across the Manhattan peninsula at low altitude, heading southwards. The aircraft had a wingspan of almost 48 m, weighed approximately 180 tonnes, and had 92 passengers and crew on board. The aircraft had taken off in Boston shortly before and was hijacked en route to Los Angeles. At 8:45 it slammed into the North Tower of the World Trade Centre, between the 96th and 103rd floors. A major explosion immediately followed the impact, and the entire building was shrouded in black smoke. The steel columns of the façade were severed over a width of roughly 50 m. The heavy aircraft probably also severed a number of steel columns in the inner core. The aircraft had an almost full complement of fuel, so that over 90,000 litres of kerosene poured into the interior of the building, ran down through the vertical elevator shafts to the storeys below and ignited.

A second Boeing 767, operated by United Airlines, with 65 people on board was also hijacked en route from Boston to Los Angeles. This aircraft approached the World Trade Centre in a long drawn-out curve from the seaward side and struck the South Tower at an angle roughly between the 73rd and 77th floors at 9:03, little more than a quarter of an hour after the first impact. Whether by coincidence or through perfidious planning, the kerosene in the wing tanks was distributed over several storeys by the oblique impact of the 48 m wide aircraft, thus accelerating the fire with fatal consequences. A huge fireball on the outer façade and dense black smoke from the building's interior heralded its imminent demise.

Both towers were now ablaze. Before long, the fire reached temperatures of over 800 °C and as much as 1,400 °C according to some experts. The fireproof coating of the steel trusses in the core area was designed to withstand at best a local fire, such as burning archives. At temperatures of only 600 °C, steel loses around 75% of its strength. Despite their coating, the columns consequently gave way or melted completely.

---

107 '11th September 2001', Central division: Corporate Communications, Münchener Rückversicherungs-Gesellschaft, 2001.

In the case of the South Tower, the aircraft had struck the building lower down and also severed the columns of the outer façade near one of the edges. Due to the higher load of the thirty-five or so floors above, reputedly around 100,000 Mp, the upper half of the tower initially buckled. Then, at 10:02, almost exactly an hour after the collision, the tower completely collapsed in a huge cloud of dust.

Although the North Tower had been struck first, the aircraft hit the building higher up and the fire raged longer there before the weakened steel columns in the floors finally caved in abruptly. Due to the dynamic force of this sudden failure of the load-bearing structure, the upper storeys hit the undamaged floors below with their full weight. The lower floors were not designed to withstand such loads and likewise collapsed. As a result, the North Tower caved in like a telescope at 10:28, almost an hour and three-quarters after the collision.

The third building to succumb was the 47-storey 7 World Trade Centre on Vesey Street. Severely damaged by flying debris from the twin towers, it collapsed floor by floor, almost in slow motion, at 17:40. Subsequently the other four buildings of the World Trade Centre collapsed one after the other too.

It was estimated that up to 50,000 people worked in the two towers every day and that the number of visitors could exceed 100,000 on peak days. The number of parties affected by the attack is therefore high. Those directly affected include, in addition to the owners and lessees of the towers, above all the firms domiciled there: telecommunications companies, banks, insurance companies, brokers, hotels and public authorities. The interruption or even discontinuation of their business activities has led to considerable losses of rental value as well as loss of business income and extra expense.

However, as an indirect consequence, the collapse of the two towers following the outbreak of fire resulted in another fifty buildings being severely damaged or even collapsing in Manhattan, with its dense concentration of high-rise buildings. This is not surprising, considering the dynamic force and energy released during the collapse of the two towers, the resultant pressure waves, and the masses of falling structural components and flying debris that were spread over the district.

The entire area of Lower Manhattan was closed off as a result of the catastrophe. Over 3,000 people lost their lives. Over 150,000 people lost their jobs temporarily or permanently because thousands of smaller businesses and offices were forced to close due to limited access. This in turn led to a breakdown of the entire infra-structure. Bridge and tunnel operators are suffering from the loss of toll fees, whilst subways, ferries and other public transport companies have had to suspend opera-tions, and there are no passengers for the taxis.

The second act of flagrant disregard for human rights is the economic embargo against a whole nation by a certain group in response to the acts of one person or a group of people from that nation. Examples of this type of political risk that affect the maintenance and care taking of all types of construction and engineering projects are beyond the scope of this book, but must be mentioned.

E.3.1.9 Risks associated with fitness for purpose

Although it is unusual, there is always the risk that when a construction project is completed, it is found to be unfit for a specific purpose, either because the purpose was not made clear to the designer prior to the design stage or by virtue of some changes in circumstances.

The latter situation can be illustrated by the example of the sewage sludge incineration plant, which was completed in 1975, but never used and abandoned by the owner.[108] It was found to be too expensive to run and the decision to abandon the project was taken despite the fact that it cost £2.5 million to construct. This risk has assumed particular importance recently since the introduction of standard forms of contract that placed the responsibility for design on the contractor. The new suite of contracts introduced by FIDIC in September 1999 include such contract forms where the level of liability for design is that for fitness for purpose.

E.3.1.10 Risks associated with project operation

Investigations have shown that in mechanical and electrical plant, operational faults form the largest number of incidents causing failure and damage. In combustion engines, 68% of all incidents are attributed to operational faults.[109] In construction projects, due to their nature, operational faults are not as significant as other considerations. However, the following incident is an example of what can happen.

The rain continued to fall in the Brazilian province of São Paulo and, in the forty hours preceding the morning of 21 January 1977, 230 mm fell down in a storm of a probability of occurrence of one in ten thousand.[110] The runoff swelled the Rio Pardo, a tributary of the Rio Grande and exposed four dams to risk. The uppermost dam, Graminha, was saved by opening the gates. The nearby 60 m high earth fill Caconde dam also survived. The release from the Graminha resulted in a rise of 7 m in the flow of the Rio Pardo above its normal level. This flood wave struck the Euclides da Cunha, another 60 m high earth fill dam, 40 km downstream, overtopping its embankment. The flood destroyed about one-third of the dam and filled the machine hall with water, wrecking the generating equipment and so locking the gates in their part-opened position.

The flood wave thus continued its course and struck the lower 41 m high Armando Salles de Oliveira dam, 10 km downstream and destroyed half of its length and the power house. The two dams held 13.6 and 25.4 million cubic metres of water respectively. Further downstream, the flood waters destroyed a small village and inundated several towns. It was reported that 4,000 homes were washed away and the loss was estimated at $40.5 million.

It was understood that the dam operators hesitated over instructions to open the gates of the lower two dams because they were afraid to flood downstream farmland. When failure of the two dams occurred by overtopping, the gates were only part open.

E.3.1.11 Risks associated with wear and tear during the project's designed life span

The life span of materials and components incorporated in construction projects is limited and inversely proportional to the deterioration, wear and tear which take

---

108 *New Civil Engineer*, op. cit., see note 45, 1 January 1976.
109 *Schaden Spiegel*, op. cit., see note 30, No. 2., 1981.
110 *New Civil Engineer*, op. cit., see note 45, 27 January and 3 February 1977.

place usually for many reasons; some are natural and others are artificial. To increase the life span, one must reduce the deterioration, wear and tear and increase the durability of the various elements. Such a result could only be achieved by a strict programme of inspection and maintenance.[111]

The risk of lack of or faulty maintenance programme is a grave one and could affect not only the owner of the project but also others involved in its execution. The latter event usually occurs when things go wrong and the liability for any injury or damage is then held to be shared between the owner and others.

---

111 'Guide to the Performance of Building Structures', Institution of Structural Engineers, London, December, 1984.

4

THE RISKS AS CLASSIFIED IN STANDARD FORMS OF CONSTRUCTION CONTRACTS

As discussed in the previous two chapters, the criterion used for allocation of risks in standard construction contracts is that of control, either of the probability of occurrence or the consequences of the risk, if and when it eventuates. Once allocated a risk, it would be natural and logical for that party to bear its resultant consequences. However, it is also logical to be in control of a risk, whilst having its consequences borne by another party, provided that the liability for these consequences is shifted to that other party through an indemnity provision. In doing so, there must be trust between the two parties, or explicit conditions imposed, to the extent that the party liable for the consequences must be able to trust that the other party which is in control of the risk would do all that is possible to manage it properly. Otherwise, if the party in control does not properly manage the risks, then economic and/or time loss would most likely be the result, and would fall on the party bearing the consequences. Whilst this might be obvious, it would be worthwhile to illustrate the point by a practical example usually found in any construction insurance policy.

The contractor, who is an insured party under a CAR policy in a construction project, is allocated certain risks of which he is in control under the contract with the employer. Through the CAR policy, the contractor shifts the liability for bearing the consequences of these risks, if and when they eventuate, to the insurer. The insurer, in turn, places certain conditions and requirements in the insurance policy establishing an obligation on the insured to take care of the insured works and to take all precautions to prevent accidents and to mitigate losses, if and when accidents occur.

Not only would badly managed risks result in economic and/or time loss, but so too does incorrect allocation of risk. In a review of major international construction contracts, carried out by Mr. Jesse B. Grove on behalf of the Government of Hong Kong, he stated that one aspect of the philosophy of risk allocation must be that

> [T]he ultimate goal of optimal risk allocation is to promote project implementation on time and on budget without sacrifice in quality, that is, to obtain the greatest value for money. The goal for a repeat employer should be to minimise the total cost of risk on a project, not necessarily the cost of

130

either party. A study in the U.S.A. has shown that 5 percent of project cost may be saved by choice of the most appropriate terms of contract alone. The question is therefore what is 'most appropriate' and how can it be recognized? There is a variety of answers.[1]

This review was followed in November 2000 by a timely conference held in Hong Kong.[2] A number of the papers presented at the conference were subsequently published in the *International Construction Law Review*.[3]

In a section entitled 'Application of Philosophy', Mr Grove referred to four criteria for allocation of risks:

- *The fault standard:* cost and time impacts of risks caused (or not avoided) through the fault of a party should be borne by that party;
- *The foreseeability standard:* He who is best able to foresee the risk is allocated that risk;
- *The management standard:* He who is best able to control and manage the risk is allocated that risk;
- *The incentive standard:* risks should be placed on the party most in need of incentive (presumably already with the ability) to prevent and control them.[4]

He drew on conclusions made in an earlier report on the topic by Thompson and Perry.[5] He stated the following:

9 *Application of Philosophy*

9.1 It is not enough to say that there should be a 'balance of risk' or 'efficiency in risk allocation' because all of us will never agree on what is a fair and reasonable balance between the contractor and the employer or which terms are most efficient for either of them.

9.2 When studying the views of the proponents of, and commentators on, the various philosophies of risk allocation, one is tempted to conclude that the same principles underlie them all. Certainly there seem to be the following common considerations:

- Which party can best control the events that may lead to the risk occurring?

---

1 'The Grove Report: Key Terms of 12 Leading Construction Contracts Are Compared and Evaluated', published in September 1998 and is available on the web site of Thelen Reid & Priest at (www.constructionweblinks.com).
2 A conference focusing on the report commissioned by the Government of Hong Kong SAR and prepared by Mr Jesse B. Grove III of Thelen Reid & Priest LLP, New York, to carry out a 'fundamental review of the General Conditions of Contract, in particular the allocation and management of risk in the procurement and work projects . . .', For the report itself, see the web site of Thelen Reid & Priest LLP: >www.constructionweblinks.com<.
3 'The Grove Report', by Humphrey LLoyd [2000] 2 ICLR 302.
4 These criteria are mentioned on page 38 above.
5 'Engineering Construction Risks', by Thompson and Perry, Science and Research Council, UK, 1992.

- Which party can best manage the risk if it occurs?
- Whether or not it is preferable for the employer to retain an involvement in the management of the risk.
- Which party should carry the risk if it cannot be controlled?
- Whether the premium charged by the transferee is likely to be reasonable and acceptable.
- Whether the transferee is likely to be able to sustain the consequences if the risk occurs.
- Whether, if the risk is transferred, it leads to the possibility of risks of different nature being transferred back to the employer.

If these considerations are applied, it should be possible to achieve clear and realistic terms that are acceptable to the employer and on which contractors are prepared to tender at prices which do not contain contingencies for unclear terms or for significant risks which are not possible to estimate with some certainty or which are unlikely to materialize.

9.3 In my opinion, Max Abrahamson has come the closest to laying down an acceptable 'formula' for risk allocation, as follows:

[A] party should bear a construction risk where:

1   It is in his control, i.e., if it comes about it will be due to wilful misconduct or lack of reasonable efficiency or care; or
2   He can transfer the risk by insurance and allow for the premium in settling his charges to the other party . . . and it is most economically beneficial and practicable for the risk to be dealt with in that way; or
3   The preponderant economic benefit of running the risk accrues to him; or
4   To place the risk on him is in the interests of efficiency (which includes planning, incentive, innovation) and the long term health of the construction industry on which that depends; or
5   If the risk eventuates, the loss falls on him in the first instance, and it is not practicable or there is no reason under the above four principles to cause expense and uncertainty, and possibly make mistakes in trying to transfer the loss to another.

The job of trying to balance the five principles in practice is the hard one . . . But at least it is best to work from declared principles rather than undeclared and perhaps unconscious prejudices.[6]

Irrespective of the above criteria, however, the allocation of the risks in construction contracts is traditionally based on a sharing between the parties involved, in accordance with the provisions of two contracts usually executed between the parties: the first is a contract for the provision of services between the employer/owner and a professional, be it design or advice, etc. The second is between the employer/owner and a main contractor. From the latter agreement flows another line of risk sharing

---

6   In paragraph 9.3 of this section of his Report, Mr. Grove quoted from an article by Mr. Max Abrahamson which was referred to earlier in note 13 of Chapter 2 above.

between the main contractor, on the one hand, and subcontractors, suppliers, manufacturers, insurers and others, on the other hand.

If these risks are analysed on the basis of the effect they generate once they eventuate, two basic types of risk can be identified. The first type incorporates the risks which could lead to personal injury, death and/or physical damage and the second type incorporates risks which could lead to economic and/or time loss, see Figures 4.1 to 4.3. In both types of risk, those that are allocated to the employer/owner are explicitly specified and all others are allocated to the contractor.

Examples of the first type of risk, which involves personal injury, death and/or physical damage include defective design, material, or workmanship; Acts of God; fire; human error and failure to take adequate precautions. Examples of the second type include late possession of the site; delay in receipt of information necessary for timely construction; changes in design; and variations to the original contract.[7]

The treatment of these two types of risk in construction contracts differs in that the first type encompasses risks that might be insurable, whereas the second type involves, in principle, uninsurable risks.

It is most important that one appreciates that the treatment of the identified risks in most of the forms of construction contracts is dealt with in two different ways.

First, there are the risks that lead to death, bodily injury and/or physical loss or damage, which are specified separately in certain parts of the contract and which might be insurable. So, in the 4th edition of FIDIC's Red Book, 1987 to 1992; the 7th edition of the ICE Form, Measurement version; and the 2nd edition of the ICE Form 'Design and Construct', these risks and the respective insurance provisions are dealt with in Clauses 20 to 25. In the ICE Form 'Engineering and Construction Contract', these risks and the respective insurance provisions are dealt with in Clauses 80.1 to 80.7. In the 1981 RIBA Building Contract, these risks and the respective insurance provisions are dealt with in Clauses 20 to 22.

Second, there are the risks that lead to economic and/or time loss, which are dealt with throughout the remaining part of the contract conditions, but whilst the employer's risks are explicitly specified only some of the others that are allocated to the contractor are specified. If the above divisions and allocations of risks are not understood, many problems could arise.[8]

Irrespective of the method chosen to allocate risks in a construction contract, a most important question arises in relation to any unidentified risk, if and when it eventuates: to whom should the consequences be allocated? At common law, the contractor is liable for the consequences of all the risks that are not specifically allocated to the employer. This is the approach taken in FIDIC's standard forms of contract and in the Engineering and Construction Contract. Against that view, the contractors would argue that if and when such risks eventuate they should best be borne by the party who gains in the long run the benefit of the project, namely the employer. The American Institute of Architects adopts that view and accepts that all

---

7 'Construction, Insurance and Law', Nael G. Bunni, a paper delivered at a Conference on Structural Failure, Product Liability and Technical Insurance, TecvolV. 2, Nos. 1/2, page 163, 1990.

8 'FIDIC's New Suite of Contracts – Clauses 17 to 19: Risk, Responsibility, Liability, Indemnity and Force Majeure', by Nael G. Bunni, ICLR, Vol. 18, Part 3, July 2001.

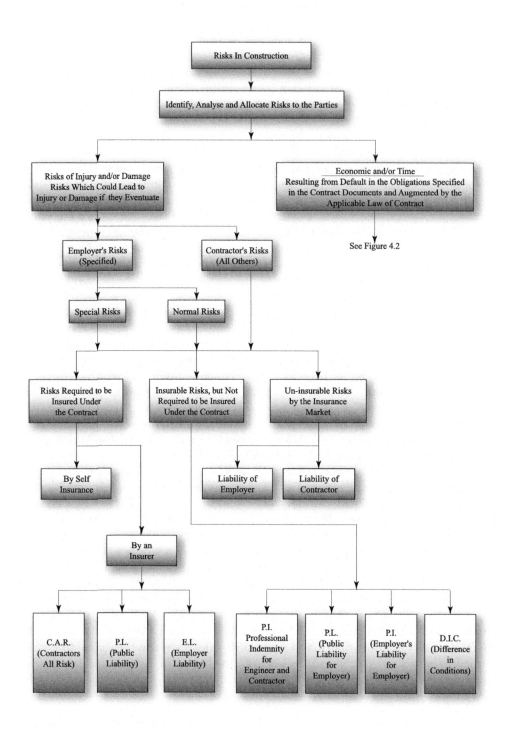

Figure 4.1 Risks of injury and/or damage.

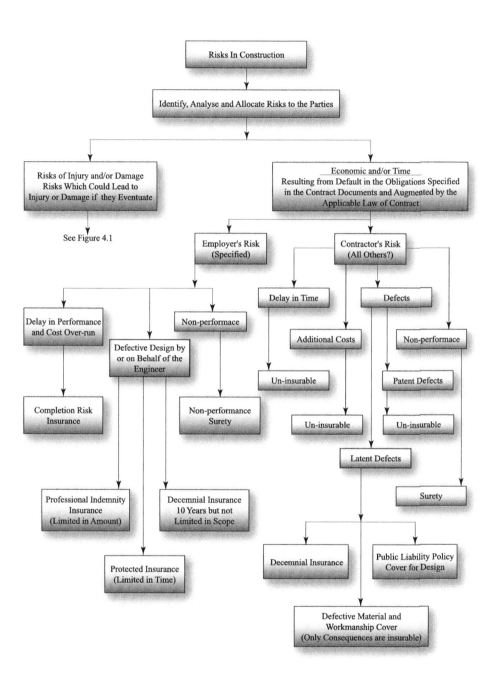

Figure 4.2 Risks resulting in economic and/or time loss.

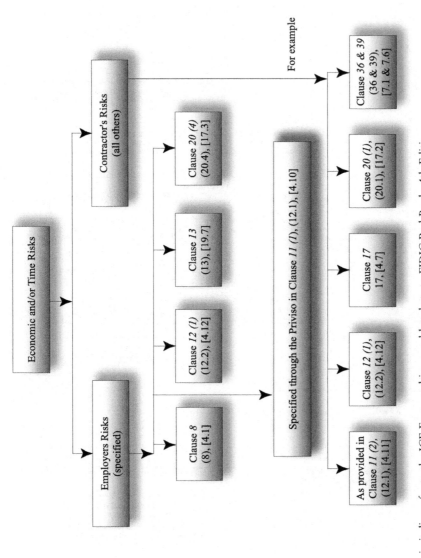

Numbers in italics refer to the ICE Form and in round brackets to FIDIC Red Book, 4th Edition.
Numbers in square brackets refer to FIDIC New Red Book, 1999.

Figure 4.3 Risks as specified in FIDIC's Red Book and the ICE Form.

risks belong to the employer when no other party can either control the risks or prevent the loss.[9] In other words, they adopt the principle that unidentified risks can neither be controlled nor can the resulting loss be prevented.

Perhaps, the least confusing from the above point of view is the Engineering and Construction Contract, where the phrase used in Clause 81.1 is '. . . the risks which *are not carried by the Employer* are carried by the Contractor'.[10]

Examples of problems resulting from lack of or incorrect allocation of risks can be very instructive. An owner may increase risk exposure by applying unreasonable monetary and time restraints, or by not implementing appropriate maintenance or operating procedures once the project is completed.

Unfortunately, the division of risk as referred to above is not clearly and explicitly explained in a number of the well-known standard forms of contract, a problem that has resulted in major misunderstandings. As an example, the wording of Sub-clause 17.1 of the new suite of FIDIC's forms of Contract, published in September 1999, should have started by explaining that the risks included under Clause 17 of the Conditions are only those risks of loss and damage and not the whole matrix of the risks to which the project and the contracting parties are exposed. The term 'Employer's Risks' in the context of this clause should have been replaced by 'Employer's Risks of Loss and Damage', since these risks are confined to those which lead to some form of accidental loss or damage to physical property or personal injury, which in turn may lead to economic and/or time loss risks, directly or through the other clauses of the contract.

As this explanation is not stated in the Conditions, the mistake of referring to the risks under Clause 17 as 'Employer's Risks' could lead to serious error in that the reader, and of course the user, would conclude that having identified in Clause 17 the Employer's Risks, all the other risks belong to the contractor, including the economic and/or time risks in the remaining provisions of the contract. This problem can be highlighted by reference to Clause 17 of the Orange Book,[11] where the draftsman fell into that trap and stated expressly in Sub-clause 17.5 that 'The Contractor's risks are all risks other than the Employer's Risks listed in Sub-Clause 17.3'. This mistake has led to many instances of misunderstanding, conflict and at least one serious arbitral proceedings, where the employer pointed out that by Sub-clause 17.5 he bears no risks under the contract other than those specified in Sub-clause 17.3.[12]

When the risks are allocated, the consequential flow referred to earlier of responsibility, liability, indemnity and insurance would apply.

---

9 See the Grove Report, referred to above in note 1.
10 The Institution of Civil Engineers (United Kingdom), *The Engineering and Construction Contract*, 2nd Edition (1995, reprinted with corrections May 1998) ('ECC,' formerly the 'NEC').
11 The draftsman of the new suite of FIDIC's Forms of Contract, published in 1999, applied the same format of the 1995 Orange Book to these new forms and ended up with the same problem.
12 'FIDIC's New Suite of Contracts – Clauses 17 to 19: Risk, Responsibility, Liability, Indemnity and *Force Majeure*', op. cit., see note 8.

Summary

It may be helpful to briefly summarise the above.

Sophisticated forms of construction contract identify the risks to which the project is exposed, analyse them and allocate them to the parties in the contract. These risks are divided into two sets, the first of which incorporates the risks that lead to personal injury, death and/or physical damage; and the second set incorporates the risks that lead to economic and/or time loss. This division is necessary because the consequences of the first set of risks could be, as a general principle, insurable; whereas the consequences of the second set could not, as a general rule, be insured.

In most of the international forms of contract, the general principle of control of the risk or management of the consequences is applied in the allocation of risks to the contracting parties. Moreover, the employer is allocated in specific and explicit terms certain of each of these two sets of risk leaving the remaining risks to the contractor.

Figures 4.1 to 4.3 illustrate the above. Figure 4.1 shows the flow from risks of personal injury, death and/or physical damage to insurance; and Figure 4.2 shows the risks that lead to economic and/or time loss. Figure 4.3 identifies the clauses of various forms of contract where the employer's economic and/or time losses are explicitly specified, and provides examples of clauses where some of the remaining risks are allocated to the contractor.

In the next chapter, the liabilities that arise if and when risks eventuate are discussed.

5

RESPONSIBILITY AND LIABILITY IN CONSTRUCTION

Logic

The spectrum of hazards and risks in construction, as displayed in the previous two chapters, covers such a variety that it can only serve to show the cautious manner with which one should approach this field. Risks are born as soon as the decision is taken to proceed with a project. They increase and grow as the project advances from the feasibility stage to the design and, later, construction stages. In the latter stage, risks reach their maximum intensity at the completion of the project when they begin to diminish but they never cease.

When and if hazards mature into events and the unexpected happens resulting in one or more of the four categories of effect, damage to property and/or injury to persons are inflicted. Figure 5.1 diagrammatically illustrates the result. Thus, on a building or civil engineering site, a loss of life or a personal injury might occur to employees, to others associated with the project and to third parties. The result might also be loss or damage to material on or off the site, affecting the permanent and temporary works, the plant and machinery used by the contractor and property, whether it belonged to the employer, the contractor, others associated with the project, or third parties.

Therefore, when risks eventuate, it is wise and, more often than not, essential for those involved in construction to be prepared for such events and to forestall the undesirable results which usually accompany them. Such action can only be taken through forethought, risk analysis and risk management. For, in order to be effective, the latter term must include the allocation of each of the risks, or group of risks, to one of the parties involved. Once risks are allocated, each party becomes responsible and/or liable for certain risks and in control of certain events and/or their consequences. The liability does not necessarily follow the responsibility and one may be responsible for a certain risk but not liable for its consequences, see page 147.

Risks should be allocated at the outset of a project as part of the contractual agreement between the parties involved when amicable relationships, and perhaps friendship, exist. It is too late to try and make that allocation after the event when the responsibilities and liabilities become the focal point and disputes arise. Of course, disputes may still arise even if risks are already allocated, mainly for the reason that no one can envisage all the possible situations that might occur. Even then, the analysis might be incorrect. leading to risks being offloaded from those

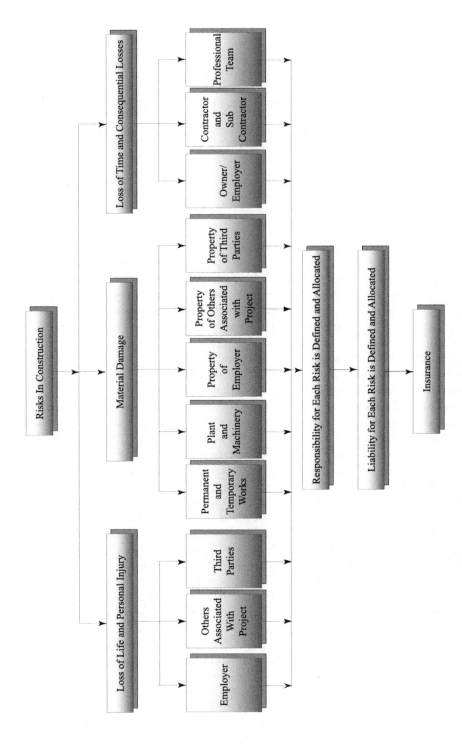

Figure 5.1 Flow of risk into responsibility, liability and insurance indemnity.

who should bear the responsibility for them to others who should not. However, such disputes will be fewer if risks are properly allocated.

Allocation of risk and responsibility

When a construction project is being considered, one of the principal decisions to be made is the type of contract to be executed between the parties concerned.[1] The contract conditions can only be chosen if a decision is made on the allocation of risks and the responsibility which stems from that allocation. That responsibility is not only for the events which may follow but also for the avoidance of risks and their mitigation, should they mature into real events.

At one end of the scale, it might be agreed that there should not be any sharing of risk between the parties and that one party alone, usually the contractor, should take the responsibility and the liability for all the risks to which the project is exposed. At the other end of the scale, it might be decided that, as a matter of principle, risk sharing is to be used extensively and that such sharing should be based on appropriate criteria leading to contract conditions with a balanced risk allocation. Alternatively, the decision might be that, although risk-sharing should form the basis of the contract, the allocation of risks should follow some predetermined pattern rather than being fairly balanced.

The standard forms of conditions of contract have moved steadily from the first alternative of no risk-sharing towards full implementation of the idea of risk-sharing on a balanced footing and more recently into a choice between the three alternatives. Thus, until 1992, with every revision of the major standard forms of contract, more and more of risk-sharing ideas were injected into their conditions, whereas subsequently and in particular after the launch of the 1999 FIDIC suite of contract forms, various options were introduced. These are exemplified by the new Red Book with its balanced allocation of risks, compared with the new Yellow Book with more risks allocated to the contractor and the new Silver Book where the risks are mainly allocated to the contractor. As an example, the risk of unforeseen adverse ground conditions can be cited to illustrate this shift in policy. The early version of the provision in the contract conditions put all the risk on the shoulders of the contractor. FIDIC's old Red Book of 1992 allocated that risk to the owner of the project and the new 1999 Red Book kept it with the owner, whereas FIDIC's 1999 Silver Book shifted it back to the contractor.

Of course, if the risk is shared, the contract price should decrease in accordance with the extent of risk offloaded and that retained. A price tag commensurate with the burden carried by the risk-taker will be attached as a compensation for each allocation of risk. Some will argue, however, that such a method of risk-sharing does not necessarily result in an ultimate reduction in price since this price differential would be minimal in the face of competition between contractors to obtain work,

---

1 The allocation of risks between the contracting parties in a construction contract is one of four criteria usually applied for the choice of the type of contract conditions to be used. The other three are: the allocation of functions that exist in the construction process; the choice of the preferred method of remuneration; and the allocation of the management functions.

whereas large increases in price will result if risks eventuate. The difficulty in reaching agreement on a fair and reasonable compensation for the additional work after contracts are awarded is another argument often made against risk-sharing. This may be true if the criterion for selecting a contractor is based on price alone. However, in that case, the owner is taking the risk that materials and workmanship may not be the best of their respective kind, that corners may be cut and ultimately a different kind of risk may be taken. It is generally agreed, however, that work profitably executed is usually of the best standard.

The type of contractor interested in a contract where risk-sharing is not accepted is different from the one who accepts full sharing. In the former type of contract it must be ascertained whether the additional burden imposed on the contractor can be carried without any risk of financial collapse. The same assurance should be sought even in a contract with loss-sharing, but to a lesser degree.

Where risk-sharing is incorporated in the contract, the next step is to establish the basis on which this sharing is to be allocated and the position at the end of the contract, when the risks pass from one party to another.

Risks can be allocated, as discussed earlier on page 47, illustrated by Figure 2.4 and summarised below, in accordance with the following criteria:

A   Control over the risk to be allocated: This is probably the most effective criterion for risk allocation which can be done on the basis of any one, or more, of the headings set out below.

   1   Control over the risk eventuating;
   2   Control over mitigating the risk; and
   3   Influence over the effects resulting from the risk eventuating, thus being in control of mitigating the losses which would generally flow from the event.

With Acts of God risks, such as rainfall or storm, where the first element of control mentioned above cannot be exercised, the second and third elements can be used as the applicable criteria. With risks such as political risks, where control over the first two elements cannot be exercised, the third element becomes the criterion.

B   Ability to perform a task related to the project: this is another important criteria used in standard forms of contract, for example, the case of the risks attached to transport to site, which are allocated to the party responsible for that transport.

C   Inability of all the parties to accept a certain risk: Such risks are usually allocated to the party that most benefits from the project, the owner or society in general. An example of this type of risk is war or external instability of government.

Definitions

Before proceeding further to establish the logical transition from decision-making, to risk, to responsibility, to liability, to indemnity (see Figure 5.2), it is important to define these terms: responsibility, liability and indemnity, with some precision, due to their relevance to the arguments presented later.

The word 'responsibility' is defined in *The Oxford Companion to Law* as follows:[2]

2   *The Oxford Companion to law*, by David M. Walker, Clarendon Press, Oxford, 1980.

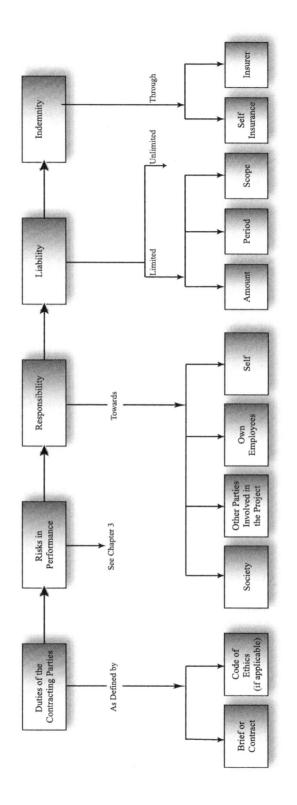

Figure 5.2 Logic in the interaction between law, insurance and construction.

Responsibility. A word used in several senses. A person may be said (1) to be responsible if he generally displays care and forethought and considers the possible results of his actions. He may also be said (2) to be responsible for certain events if his conduct has been a material factor in bringing them about; thus a reckless driver may be said to be responsible for an accident. In this sense the word means little more than that he has caused the events and does not necessarily imply accountability. An animal or a snowfall may be said to be responsible for causing a happening. A person may also be said (3) to be legally responsible when of such an age and in such a state of mind and body that he is deemed to be capable of controlling his conduct rationally and such that he can fairly be held accountable and legally liable for the consequences of what he does. Conversely, a person mentally ill or under the influence of drink or drugs may be held legally irresponsible. Responsibility in this sense is fundamental to liability to punishment. Responsibility in the third sense has a substantial moral flavour, but moral responsibility or blameworthiness and legal responsibility are not wholly equivalent. A person may by law be held responsible in cases where he has not been personally blameworthy at all. Thus under the principle of vicarious liability a person is held responsible for, and legally liable for the wrongs of, his employees, though not personally in fault at all. . . .'

Similarly, the word 'Liability' is defined in the same reference as follows:

Liability. The legal concept of being subject to the power of another, to a rule of law requiring something to be done or not done. Thus a person who contracts to sell goods is liable to deliver them and the buyer is liable to pay the price. Each is required by law to do something, and can be compelled by legal process at other's instance to do it; the other is empowered to exact the performance or payment. It is sometimes called subjection. The correlative concept is power.

A person is said to be under a liability when he is, or at least may be, legally obliged to do so or suffer something. Thus, one may be said to be liable to perform, to pay, to be sued, to be imprisoned, or otherwise to be subject to some legal duty or legal consequence. In general, liability attaches only to persons who are legally responsible; an insane person does not generally incur any liability.

Liability may arise either from voluntary act or by force of some rule of law. Thus, a person who enters into a contract thereby becomes liable to perform what he has undertaken, or to pay for the counterpart performance, or otherwise to implement his part of the contract. If he acts in breach of contract, he becomes liable by law to pay damages in compensation for the breach. Similarly, if a man acts in breach of any of the general duties made incumbent on him by statute or common law, such as to refrain from injuring his neighbour, or to maintain his tenant's house in reasonable repair, or to exercise diligence in administering property of which he is trustee, he incurs legal liability to make good his omission or default.

Liability is commonly distinguished according to its legal grounds into civil liability, whereby one is subject to the requirement to pay or perform something by virtue of rules of civil law, and criminal liability, whereby one is subject to being fined, imprisoned, or otherwise treated by virtue of rules of criminal law. Civil liability may arise from many grounds, from the natural relations of the family, from undertaking or contract, the commission of a harm, from trust, statute, or decree of court. In respect of liability arising from harm done, it arises from intentional harm, harm brought about in breach of duty, and in some cases there is strict liability if harm befalls despite care taken. Criminal liability arises from the admitted or proved commission of some kind of conduct declared by the rules of criminal law to be a crime inferring punishment. At common law criminal liability normally also requires that the conduct has been done intentionally or recklessly but not merely negligently or accidentally, and sometimes proof of a particular intent is a necessary ingredient of a crime, but many cases under statute have been held to impose strict or absolute liability, i.e. liability irrespective of the actor's state of mind . . .

The word 'Indemnity' is defined from the same text:

Indemnity. An undertaking to compensate for loss, damage, or expense; . . .
Indemnification. The making good of a loss which a person has suffered in consequence of the act or default of another.

Responsibility in construction

Once the risks are allocated, the responsibilities follow in accordance with that allocation. Briefly, responsibility as defined is generated from actions and duties allocated to the various parties bound by the project and, through these, responsibility is owed. But, to whom are the parties responsible for their action?

- *Towards society*
  Construction work is closely related to the environment and society. In most cases the results of designs prepared by the professional team, financed by the employer and implemented by the contractor, can be seen and felt for a long period of time, thereby shaping the environment and, to some extent, the society. In some cases, it involves the 'act of directing the great source of power in nature for the use and convenience of man . . .'.[3]

  The responsibility in this area tends towards the owner and his professional design team. Some therefore argue that the professional team's first duty and responsibility is towards the environment and society in which the project is located. To fail society is to fail one's self. It is not very difficult to succeed in this objective if there is in existence legislation to control development, in which case it is the duty of the designer to abide by the rules set out in such legislation. Should he fail to do so, he will be in breach of a statutory duty and/or will have committed

---

3 Description of Civil Engineering in the Charter of 1828 of the Institution of Civil Engineers (UK).

a crime. A problem arises, however, where legislation is either non-existent or deficient, in which case a conflict of interest will almost certainly arise between the duty and responsibility towards the owner and that towards society. Should such a conflict arise, the responsibility would tend acutely towards the employer.

It must be recognised here that the first responsibility must lie with the legislative authority in not specifying what might and what might not happen with society and its environment. Professionals in the forefront of technology in their appropriate fields must be given the task of drafting such legislation.

- *Towards other parties involved in the project*
  The parties to a construction contract owe a responsibility towards each other to do certain things at certain times or, in some aspects, all the time. These responsibilities are set out in the agreement, conditions of contract or in codes of ethics which control the behaviour of professional people. This responsibility can be divided into four sections:

1 Responsibility of the professional team towards the owner, which can be stated, in part or as a whole, as follows:

    (a) to design the project in a comprehensive, skilful and cost-effective manner;
    (b) to provide the necessary documentation to obtain a reasonably priced tender from a competent contractor and to administer the contract to a successful conclusion;
    (c) to supervise the construction and to administer the contract to ensure that the work is carried out in accordance with the design and specifications provided; and
    (d) to perform his statutory obligations.

2. Responsibility of the professional team towards the contractor centres generally around the principle that the warranty of authority conferred upon the team under the contract should not be breached. This authority, whilst intended to serve the owner, may influence the contractor or his behaviour if breached. Such authority might include any or all of the following:

    (a) dealing with ordinary variations from the contract;
    (b) dealing with measurement and valuation of variations;
    (c) dealing with any necessary or appropriate instructions;
    (d) intervening in the process of construction. However, interference with the rights and function of the contractor to carry out his own construction operations as he thinks fit does constitute a breach of that warranty; and
    (e) complying with all the imposed legal requirements, including health and safety aspects.

3. Responsibility of the contractor towards the owner and the professional team revolves generally around the quality and rate of performance of the works and may be stated under the following headings:

    (a) to ensure that the works are properly planned and managed so as to achieve a high standard of construction;

(b) to co-ordinate and plan the execution so that all labour, materials and plant are available when needed to perform the work at the rate demanded by the programme;

(c) to comply with all the imposed legal requirements such as employment, safety and health Acts; and

(d) to provide a quality of materials supplied, workmanship performed and design (if required) carried out to a standard at least equal to that required under the contract; and

(e) to complete the project.

4. Responsibility of the owner towards the contractor and the professional team is generally concentrated on providing the financial and other resources required to perform the work. These can be summarised as follows:

(a) to choose only competent parties to act within the construction scene, so that a relationship of trust may exist;

(b) to give possession of site when investigation work is to be performed and later when construction work is to proceed;

(c) to supply instructions promptly so as not to interfere with the progress of the works;

(d) to permit the contractor to carry out the whole of the work;

(e) to pay the professional team adequately and on time to perform the tasks required of them; and

(f) to pay the contractor as agreed and on time.

- *Towards one's employees*
  Although the responsibility in this area was basically a moral and contractual one, it is now to a large extent part of legislation. Thus, to fail in the latter area would be to breach statutory requirements.

- *Towards oneself*
  This area of responsibility is perhaps the most ignored by all parties. It has basically two demands: the first is for the party to be able to remain in existence and the second is for that party to be able to uphold its standards and reputation.

  To be able to remain in existence means that the organisation must be profitable and the responsibility of owner, contractor and professional team towards themselves and/or their shareholders (as the case may be) is a very real one.

  Profitability must also be accompanied by a high standard, otherwise it would be short-lived. The responsibility of maintaining such a reputation of high standard is fundamental to construction.

Liability emanating through construction

The two words 'Responsibility' and 'Liability' are, more often than not, confused with each other and are used loosely to indicate a state which the former could express more precisely than the latter, or vice versa. As can be seen from the definitions provided earlier, one could be responsible and also liable as a result of a certain action, one could also be responsible for an action but not liable for the damages that result and, finally, one could be liable but not responsible. An example of the first situation would be the professional person who acts negligently causing damage. He is responsible for his actions and is also liable for the damages towards the party who has suffered the loss. Both the second and the third situation can be illustrated by the example of the employer who nowadays is held liable for the negligent acts of his employee, even though the employer himself may have been completely without blame and thus not responsible for the negligent act. Another example, which may help to clarify this point, is that of an insurance contract that imposes on the insurer a liability to indemnify the insured in respect of an occurrence for which the former has no responsibility.

The two words also differ in respect of the authority needed to establish their existence. Therefore, whilst someone could be empowered to define the responsibility of a certain event, it is left to arbitral tribunals and/or the courts, but sometimes only to the courts, to establish liability and its apportionment. An arbitral tribunal or a court of law can, of course, establish both.

In any case, the more usual situation is that where both responsibility and liability are attached to the same individual, or organisation, or to groups of either, or both. This attachment always spells out affliction, harm, mishap, etc. It never has a happy content, and thus one is said to be liable to err or to go astray. One is liable to be blamed, or censured, or criticised, etc.

Developments in the laws of liability

As stated earlier, one of the fundamental precepts of law is that ignorance of the law is no defence; but how much of the law is a non-lawyer expected to know? A pertinent example can be cited from the case *B.L. Holdings Ltd. v. Robert J. Wood & Partners* (1979), see page 150 below. The legal systems of various countries differ and are so complicated that only those who study the subject professionally understand its impact.

The situation in common law jurisdictions is even more difficult because one has to know the most up-to-date position at all times. As construction is a subject bridging most national borders, this is true not only within the jurisdiction in which one practices but also in all other common law jurisdictions. The reason is that, although a legal argument setting a precedent in one jurisdiction is not binding on courts in another, it is nevertheless understood to be persuasive. In practical terms, this means that, unless there is a good reason to ignore it, the same argument will be used in reaching a decision in other jurisdictions.

The matter is further complicated by the fact that a change in the law may occur any day anywhere there is a new event presenting a suitable cause for change. It is perpetual change which the late Lord Denning aptly described as 'from precedent to precedent'. He also stated:[4]

> Of all the developments in the 20th Century – by the Judges – the greatest has been in the law of negligence.

4 *The Discipline of Law*, by the Rt. Hon. Lord Denning, Butterworths, London, 1979, page 227.

The law of negligence for those involved in construction has, therefore, been chang-ing fast, and if one were to keep abreast of that change, a certain minimum amount of knowledge of the law is necessary and more is preferable.[5] Furthermore, if the construction industry is to be in control of its destiny, then those involved in it must take an interest in the law affecting construction and must influence its path. This path has become riddled with liability issues that are adversely affecting construc-tion.

Levels of liability

Under contract, liability is basically generated either pursuant to the contractual provisions or in breach of them. However, in tort, the standards of liability differ between the different parts of the world but, in general, there are three standards in common law jurisdictions and only two in jurisdictions which follow the Civil Code.

In the common law systems, the three levels or standards of liability are:

- Liability based on lack of care and negligence;
- Strict liability; and
- Absolute liability.

Liability based on lack of care and negligence is the most usual and, for one to be liable in negligence, it is essential for the claimants to establish a proof of the respon-dents' negligence. A higher and more stringent standard of liability exists in the case of strict liability which is the standard sometimes set by statute where liability arises if the harm to be prevented takes place irrespective of whether or not care and precautions have been taken. In such a case, the onus of proof shifts from the claimants having to prove negligence to the respondents having to prove non-negligence and even then liability may attach.

Strict liability occurs also under the principle of *Rylands v. Fletcher* where harm or damage is caused by the escape of a danger from one's own land.[6] A more stringent level of liability is the absolute liability which is imposed under certain statutory provisions and is incurred by reason of the intentional occurrence of an event of a kind deemed prohibited, without regard to care or precautions taken and without need for proof of negligence or fault. This principle was discussed in some detail in

---

5 'FIDIC's view of Design Liability', Nael G. Bunni, a paper delivered at a Colloquium on Liability, 1984, IABSE in association with the Institution of Structural Engineers, Cambridge, UK.

6 *Rylands v. Fletcher* (1868) LR, 3 HL, 330, where the defendant employed an independent contractor to construct a reservoir on his land. When it was filled, water flowed into disused mine workings underneath, which communicated with and, consequently, flooded the plaintiff's mines. Despite the fact that there was no proof of negligence, it was held that a person who used his land in a non-natural way, as where he, for his own purposes, brought on his land, collected and kept there anything likely to do mischief if it escaped, must keep it in at his peril. Such a person was *prima facie* answerable for all damage which was the natural consequence of its escape, unless he excused himself by showing that the escape was due to the plaintiff's default, or was the consequence of *vis major*, or act of God: *The Oxford Companion to Law* by Professor David M. Walker.

the more recent case of *Cambridge Water Co. Ltd v. Eastern Counties Leather plc*, *[1994] 1 All ER*.

The standard of liability under the Civil Code is generally based on whether negligence is a criterion, but there one has to also consider gross negligence.

Liable or not?
Developments in contract law

The two important areas of the law concerning construction are contract and tort. Recent developments in the former have been few and the essential principles have remained the same for some time.

In contract, the terms used can be either expressed or implied. The expressed terms are those agreed by the parties and, if they are clearly stated, there can be little dispute about their meaning. Therefore, most developments have occurred in relation to implied terms in a contract, and these can be either implied by reason of statute, by reason of an established procedure through previous contracts, or by the courts in resolving a dispute. A case from the last century which still holds true in respect of risk-sharing as implied in a contract is *The Moorcock* in which Bowen L.J., in 1889, stated:[7]

> . . . I believe if one were to take all the cases, and there are many, of implied warranties or covenants in law, it will be found that in all of them the law is raising an implication from the presumed intention of the parties with the object of giving to the transaction such efficacy as both parties must have intended that at all events it should have. In business transactions such as this, what the law desires to effect by the implication is to give such business efficacy to the transaction as must have been intended at all events by both parties who are businessmen; not to impose on one side all the perils of the transaction, or to emancipate one side from all the chances of failure, but to make each party promise in law as much, at all events, as it must have been in the contemplation of both parties that he should be responsible for in respect of those perils or changes.

In general terms, the cost varies inversely with the level of risk inherent in a number of alternative solutions to a particular situation or problem. In such a situation it is essential for the designer to discuss the alternatives and their advantages and disadvantages with the owner and to leave the decision-making to him. The decision will most probably be made on a commercial basis relative to the risk to be taken and, therefore, ought to be taken by the owner. The same applies if a cheaper but a higher risk design is followed; see the Canadian case of the *City of Brantford v. Kemp and Wallace-Carruthers & Associates Ltd.* (1960).[8]

In the case *of B.L. Holdings Ltd v. Robert J. Wood & Partners* in 1979, an owner employed design professionals to design an office development on a site in Brighton,

---

7   *The Moorcock* (1889) LR 14 PD 64.
8   *City of Brantford v. Kemp and Wallace-Carruthers & Associates Ltd.* (1960) 23 DLR (2d), 640 (Canada).

UK.[9] At that time, an Office Development Permit (ODP) was required to be issued for a development larger than 10,000 sq ft. When the design professionals approached the local planning authorities, they were told that car parking should be provided within the development, but that car parking was not included in the calculation for the purposes of the ODP.

They were also informed that if a self-contained residential unit were added at the top of the development, its area also would not be counted for the purposes of ODP.

Despite their surprise, the design professionals neither investigated the matter further nor informed the owner of their discussion with the planning authorities. Their finished design measured 16,100 sq ft, out of which only 10,000 sq ft were allocated to office development and thus no ODP was obtained. The building was completed in 1972 and when a prospective tenant enquired into the matter of the ODP, it was found that the permit was, in fact, required. The building remained unoccupied until 1976 when the limit for such a permit was raised to 50,000 sq ft.

The owner claimed that the design professionals were negligent in not ascertaining that an ODP was required. The judgment was against them and the judge stated:

> . . . it may be thought by some to be hard to require of an architect that he knows more law than the planning authority . . . I am left with the clear conviction that I have not on the facts set too high a standard of care of judgment for an ordinarily competent architect who in 1970 was undertaking to advise in planning matters relating to office development. Indeed, I am convinced that the standard which the law sets, namely that the ordinarily competent and skilled architect certainly requires of (the designer) that he should at least have given that advice and warning to his clients.

The judgment was appealed and the Court of Appeal allowed the appeal reversing the original judgment. Commenting on this case, D.L. Cornes, in his book *Design Liability in the Construction Industry*, states:

> The one lesson that can be clearly drawn from the case is that where circumstances arise where a decision needs to be made by a designer and where that decision involves risk attaching to the employer, the designer would be well advised to draw the risk to the attention of his employer and that where appropriate advise that independent legal opinion should be sought.

Duty of care in design

The duty of care and skill owed by the professional designer to the owner in common law has evolved in the following manner and sequence. Starting from 1957 with the judgment by McNair J. in *Bolam v. Friern Hospital Management Committee,* who stated:[10]

> How do you test whether this act or failure is negligent? In an ordinary case it is generally said that you judge that by the action of the man in the street.

9 *B.L. Holdings Limited v. Robert J. Wood & Partners* (1979) 12 BLR 1.
10 *Bolam v. Friern Hospital Management Committee* [1957] 2 All ER 118.

He is the ordinary man. In one case it has been said that you judge it by the conduct of the man on the top of the Clapham omnibus. He is the ordinary man. But where you get a situation which involves the use of some special skill or competence, then the test as to whether there has been negligence or not is not the test of the man on top of the Clapham omnibus, because he has not got this special skill. The test is the standard of the ordinary skilled man exercising and professing to have that special skill. A man need not possess the highest expert skill; it is well established law that it is sufficient if he exercises the ordinary skill of an ordinary competent man exercising that particular art.

The *Bolam* test is of general application and the law requires that a person, in doing such an act, shall exercise the skill of an ordinary competent human being of the same calling.

There is no implied term in the contract between owner and professional designer permitting the delegation of the duty of design. In *Moresk v. Hicks,* an architect subcontracted the design of the structural frame of the building to a contractor who supplied and erected the frame.[11] The frame, however, proved to be defective in two major design issues. The judge in that case stated:

In my view if a building owner entrusts the task of designing a building to an architect he is entitled to look to that architect to see that the building is properly designed. The architect has no power whatever to delegate his duty to anybody else. Certainly not to a contractor who would in fact have an interest which was entirely opposed to that of the building owner.

In order to be relieved of the responsibility of design of a special part, a contract must be established between the owner and the specialist designer. Care must be taken, however, in making any recommendation to the owner proposing the new design. If that recommendation is made negligently, the principle of *Hedley Byrne* will apply, see page 170, as happened in the Canadian case of *Nelson Lumber Co. Ltd. v. Koch* (1980).[12]

The designer is also under a continuing duty to check that his design will work in practice and to correct any errors which may emerge.[13] This duty can only be exercised if the design professional is involved in the supervision as well as the design of the project. The risk of something going wrong increases; therefore, when the design professional is either not engaged in supervision or another professional is brought in, instead of him, to carry out the supervisory duties.

If the relationship of trust between the owner and the design professional breaks down, the owner may even interfere with the design professional's work. He would then attract the liability for any errors made due to his interference. *In Kitchens of Sara Lee Inc. v. A.L. Jackson Co. et al.,* Illinois, USA, the owner employed, in 1962, a firm of architects and engineers to design an immense holding freezer as part of his

11 *Moresk v. Hicks* [1966] 2 Lloyd's Rep. 338.
12 *Nelson Lumber Co. Ltd. v. Koch* (1980) III DLR (Canada).
13 *Brickfield Properties v. Newborough* [1971] 3 All ER 328.

headquarters and bakery.[14] In 1964, insulation panels in the freezer started to fail and the owner filed a suit against the designers and the suppliers of material for $680,000 in alleged damages for repairs and later the figure was increased by $1.5 million. The designers were accused of negligence in recommending and permitting the use of inappropriate materials in the holding freezer, in failing to design the holding freezer to prevent the ingress of moisture and free water and in failing to supervise construction properly.

In defence, the designers were able to show from their records that, during the design period, the owner's personnel overruled them on many occasions. Amongst a number of items on which they were overruled were the ceiling design, material selection, freezer doors, installation of the insulation, vapour-barrier quality and extent of use and the employment of the insulation subcontractor. They were also able to show that they had informed the owner of their opinion that the design and material desired by him would not work. The decision of the court was in favour of the designers on the basis that the owner was contributorily negligent and the owner recovered nothing. ('In Illinois, the law does not compare degrees of negligence, and it therefore precludes recovery to one who contributed in any measure to his own damages.')

Reasonable skill, care and diligence v. fitness for purpose

In construction contracts, there are two levels of legal liabilities attached to the work and services supplied by contractors and professionals. These are:

(a) Reasonable skill, care and diligence; and
(b) Fitness for purpose.

Fitness for purpose is a greater obligation than that of reasonable skill, care and diligence. It is an absolute obligation independent of negligence. Therefore, negligence does not have to be proved where there is an obligation to provide fitness for purpose. On the other hand, where the duty is simply to use reasonable skill and care, the employer or any claimant against a designer must show that the designer has been negligent in order to establish liability. This aspect of onus of proof forms an important element in the cost of dispute settlement, if and when disputes arise.

In general terms, the first level of liability applies to professional persons providing services to an employer, whereas the higher level of fitness for purpose applies to contractors. Therefore, designers are under a duty to use reasonable skill and care, but the word 'reasonable' has to be interpreted as appropriate in circumstances where the level of skill varies.

The statement, quoted below, made in the American case of *Cagne v. Bertran,* has remained as the criterion in the United States since 1954:[15]

> . . . those who sell their services for the guidance of others in their economic, financial, and personal affairs are not liable in the absence of negligence or intentional misconduct.

---

14 *Kitchens of Sara Lee Inc. v. A.L. Jackson Co. et al.* (1972) Lake County, Illinois, USA, GFIP Vol. III No. 2.
15 *Cagne v. Bertran* (1954) 43 Cal. 2d 481, 275 p. 2d15.

That judgment was used as a basis and was followed in the case of *Allied Properties v. Blume* and also in others.[16] *In Xerox Corporation v. Turner Construction Co. et al.*, the court noted:[17]

> In the absence of an express agreement to the contrary, the duty of an architect, in performing his duties to his employer, is to exercise reasonable care and diligence, to use ordinary and reasonable skill usually exercised by one in that profession.

And more recently, in *Gravely v. The Providence Partnership*, in 1977, the plaintiff was a guest in a hotel suite which had a spiral staircase connecting the bathroom at an upper level to the lower level.[18] During the night, he went upstairs and opened the door of the bathroom but discovered that it opened towards him. As he stepped back, he fell and was seriously injured. The architect who designed the facility was sued in negligence and in breach of warranty. The court decision was in favour of the architect. The plaintiff then appealed. The Federal Appellate Court assessed the duty as being:

> An architect, in the preparation of plans and drawings, owes to his employer the duty to exercise his skill and ability, his judgment and task reasonably and without neglect. . . . Even if there were a warranty of expertness, it would not be actionable by the plaintiff, for it would not run to the public generally but only to the architect's employer. This absence of mutuality we know as want of privity of contract.

It should be noted that an error of judgement may not necessarily be regarded as negligence. This was highlighted in the medical negligence case in England, *Whitehouse v. Jordan.*[19]

A fundamental distinction between the use of reasonable skill and care and an obligation of fitness for purpose in connection with design activities is that in the former case negligence has to be shown whereas in the latter case there is an absolute obligation, which is independent of negligence. Therefore, it is not necessary to prove negligence where an obligation of fitness for purpose exists. However, as can be seen from the case of *Greaves v. Baynham*, referred to on page 156 below, where a particular purpose is made known to a designer, that designer has an obligation to produce a design that is fit for that particular purpose, and this is so whether or not there is negligence. Moreover, a designer may be in breach of his obligation to use *only* reasonable skill and care if he knew of a particular purpose to which a specific part of his design was to be put; and did not sufficiently take into account the relevant particulars, as then this could amount to negligence.

---

16 *Allied Properties v. Blume* (1972) 25 CA 3d, 848.
17 *Xerox Corporation v. Turner Construction Company, et al.* (1973), GFIP vol. IV No. 4.
18 *Gravely v. The Providence Partnership* (1977), Federal Appellate Court, USA, G.F.I.P. Vol. VIII No. 1.
19 *Whitehouse v. Jordan* [1981] 1 WLR 246; 125 SJ 167; [1981] 1 All ER 267.

The skill of an ordinary competent person would normally be judged by reference to the state of the art of the design when it was carried out and not at a later date.

It is accepted that the precise scope of the duties undertaken by the engineer in the contract with his client, the employer, would determine his responsibilities and, consequently, his liability. Obviously, should the terms of the agreement between the employer and the engineer include detailed expectations of the design, the standard of care would be elevated to that of fitness for purpose as was concluded in the Canadian case *Medjuck & Budovitch Ltd. v. Adi Ltd.* in 1980.[20] In that case, it was concluded that there was a 'common intention that the building should be fit for its purpose'. It was held that 'this gave rise to a term implied in fact that if the structure was completed in accordance with the design it would be reasonably fit for use as a . . . store'.

Therefore, the liability of the engineer can become higher than simply having to use skill and care when he is made aware of the purpose for which the design is intended.[21]

It is also worth noting that the designer is under a continuing duty to check during the construction period that his design will work in practice and to correct any errors that may emerge.[22]

Fitness for purpose

The obligation for fitness for purpose in construction contracts has to be viewed from three separate angles: fitness for purpose for the supply of materials; for workmanship; and for design.

In common law jurisdictions, there are generally two main sources for the legal obligation of fitness for purpose in a construction contract: statutory and common law sources. From a statutory point of view, the applicable Sale of Goods and Supply of Services Act is relevant to the application of the principles concerned.[23]

The Sale of Goods and Supply of Services Act generally provides the legal principles that apply in the sale of goods and in the supply of services. Where the supply of materials is concerned, there is usually an implied warranty on the part of contractors that the materials they supply for use in the work will be reasonably fit for the intended purpose and are of good quality. The leading case in that connection in England is *Young and Marten Limited v. McManus Childs* in 1969.[24]

On this topic, if an owner, relying on a manufacturer's information, specifies an item in his contract with a contractor, then a separate and distinct contract between the owner and the manufacturer is created. This latter contract also implies fitness for purpose; see *Shanklin Pier Ltd. v. Detel Products Ltd.* (1951).[25]

---

20 *Medjuck & Budovitch Ltd v. Adi Ltd* 33 NBR 2nd 271 (80 Apr. 271, paragraph 110).
21 See also the statement of Lord Denning in *Greaves (Contractors) Limited v. Baynham Meikle & Partners* below.
22 *Brickfield Properties v. Newton* [1971] 1 WLR 862 and also *Eckersley, T.E. and Others v. Binnie & Partners & Others*, Court of Appeal, (1988) 18 Con LR 1.
23 See also the standard form of contract of sale under the United Nations Convention on Contracts for the International Sale of Goods. This convention has been ratified by many jurisdictions worldwide and applies to contracts of sale in the international field.
24 *Young & Marten v. McManus Childs* [1969] 1 AC 454; [1968] 3 WLR 630.
25 *Shanklin Pier Ltd. v. Detel Products Ltd.* [1951] 2 All ER 471.

However, where the building owner relies on his own judgment rather than on that of the contractor, the warranty as to fitness for purpose will not be implied and the contractor will have no responsibility to supply materials which are fit for their purpose. In such circumstances, the warranty of quality will still usually be implied. A designer who selects materials and specifies them, so that the contractor has no choice as to what materials to buy, will remove the employer's cause of action against the contractor in respect of an implied warranty for fitness for purpose. The employer can still usually rely on the implied warranty as to good quality, at least in cases where the defect complained of is one of quality rather than fitness for purpose.

A similar implication would apply to workmanship where the contractor is obliged to ensure that finished work is reasonably fit for the purpose intended.

Where a professional is responsible for design, the Sale of Goods and Supply of Services Act generally provides that there will be an implied term for the supplier to use *due* skill, care and diligence. (Note the use of the word 'due' instead of 'reasonable'.) However, it is important to recognise that this provision does not prejudice any rule of law which imposes a duty stricter than that of skill and care, and therefore the Act does not of itself prevent the implication of a fitness for purpose term, where appropriate. Four cases are relevant and helpful to the understanding of the difference between that obligation and the higher obligation of fitness for purpose. These are:

(a) Greaves (Contractors) Limited v. Baynham Meikle & Partners;[26]
(b) IBA v. EMI and BICC;[27]
(c) Norta Wallpapers (Ireland) v. Sisk and Sons (Dublin);[28] and
(d) George Hawkins v. Chrysler (UK) Limited and Burne Associates.[29]

However, it must be noted that all these cases were on particular facts and were not cases that arose out of the commonly found relationship between designers, contractors and employers. So, a brief summary of the facts in these cases could be instructive.[30]

Where the contractor is responsible for the design of a construction project, the employer relies on the contractor, not only in respect of the selection of the materials and the proper workmanship, but also in respect of design. In the first case referred to above, which related to a package deal contract, Lord Denning MR said:

Now, as between the building owners and the contractors, it is plain that the owners made known to the contractors the purpose for which the building was required, so as to show that they relied on the contractors' skill and judgment. It was, therefore, the duty of the contractors to see that the finished work was reasonably fit for the purpose for which they knew it was required. It was not merely an obligation to use reasonable care, the

26 *Greaves (Contractors) Limited v. Baynham Meikle & Partners* [1975] 1 WLR 1095.
27 *Independent Broadcasting Authority v. EMI Electronics and BICC Construction* (1980) 14 BLR 1.
28 *Norta Wallpapers (Ireland) v. Sisk and Sons (Dublin) Limited* [1978] IR 114.
29 *George Hawkins v. Chrysler (UK) Limited and Burne Associates* (1986) 38 BLR 36.
30 *Design Liability in the Construction Industry*, by David L. Cornes, 4th edition, Blackwell Scientific Publications, 1994, Oxford.

contractors were obliged to ensure that the finished work was reasonably fit for the purpose. That appears from the recent cases in which a man employs a contractor to build a house *Miller v. Cannon Hill Estates Limited* (1931); *Hancock v. B.W. Brazier (Anerley) Limited* (1966). It is a term implied by law that the builder will do his work in a good and workmanlike manner; that he will supply good and proper materials; and that it will be reasonably fit for human habitation.

That statement of Lord Denning was made where the liability of the package deal contractor to the building owner had been admitted and it is not, therefore, binding, although it is of great persuasive authority. The same applied to statements made in the House of Lords in England in *IBA v. EMI and BICC* (1981), referred to in (b) above, where Lord Scarman said:

> In the absence of any term (express or to be implied) negativing the obligation, one who contracts to design an article for a purpose made known to him undertakes that the design is reasonably fit for the purpose.

Lord Scarman had equated the position to that of a dentist making a set of false teeth where it has been held that there is a an implied term that the false teeth will be reasonably fit for their intended purpose: *Samuels v. Davis* (1943). In *Samuels*, Lord Justice Du Parcq said:[31]

> If someone goes to a professional man . . . and says: 'Will you make me something which will fit a particular part of my body?' and the professional gentleman says 'Yes', without qualification, he is then warranting that when he has made this article, it will fit the part of the body in question. . . . If a dentist takes out a tooth or a surgeon removes an appendix, he is bound to take reasonable care and to show skill as may be expected from a qualified practitioner. The case is entirely different where a chattel is ultimately to be delivered.

The situation could also be surmised from Lord Denning's statement in the *Greaves v. Baynham* case, as follows:

> The law does not usually imply a warranty that he (the designer) will achieve the desired result but only a term that he will use reasonable skill and care. The surgeon does not warrant that he will cure the patient. Nor does the solicitor warrant that he will win the case. But, when a dentist agrees to make a set of false teeth for a patient, there is an implied warranty that they will fit his gums.

These words were approved by Lord Scarman in 1980 in the *IBA* case when he distinguished between the dentist, using reasonable care in taking out a tooth, and

---

31 *Samuels v. Davis* [1943] 1 KB 526.

the more onerous task of providing false teeth, but he went one step further in allocating a duty of fitness for purpose to a designer who contracts to design 'an article' for a purpose made known to him: 'Such a design obligation is consistent with the statutory law regulating the sale of goods.'

The above cases were followed in 1985 by the case of *Viking Grain Storage Limited v. T.H. White Installations Limited and Another*,[32] where there is now little doubt on the applicability of fitness for purpose on design and build projects. This case concerned a preliminary issue before the court as to whether the following terms were to be implied:

(a) that the design and build contractor would use materials of good quality and reasonably fit for their purpose; and
(b) that the completed works be reasonably fit for their purpose.

The works in this case were related to a grain drying and storage installation. The Official Referee had little difficulty in concluding that there was reliance by the plaintiff owners on the skill and judgment of the defendant contractors and it followed that those two implied terms contended for were implied.

The question remains as to whether an employer might consider restricting the contractor's liability for design from the standard of 'fitness for purpose' to the lower standard of 'due skill and care'. However, whatever the acceptability of such an action from a legal point of view, the practical aspect of its advisability should be carefully considered. This is in the context that by restricting the liability to a narrower application, one of the main advantages of such a method of procurement of works, the single source of liability in design and build contracts, disappears. The employer is told that in design and build contracts, he does not have to enquire into distinctions between defects arising from material, workmanship or design. Neither would he have to show that negligence is the source of a complaint when problems arise.

The role of insurance

It is accepted that where supply of material is concerned, the party responsible for selection is to some extent covered through that selection by a manufacturer's warranty or by the insurance cover of that manufacturer. Where design is concerned, however, the designer does not provide such warranty and the professional indemnity insurer does not extend to cover fitness for purpose.

Contractors are therefore exposed to liability for which they have no indemnity, whether through suppliers of such service or through insurance. The suppliers of design services do not usually have assets to provide any real comfort to a contractor undertaking design and build contracts. The insurers of professional indemnity resist the provision of a cover against fitness for purpose for reasons of insurability. They argue that such a cover would be against the principles of insurance. Similarly,

---

32 *Viking Grain Storage Limited v. T.H. White Installations Limited and Another* (1985) 33 BLR 103.

contractors argue that their liability should be restricted to the lower standard of due skill and care.

The above problem exists all over the world. FIDIC, the International Federation of Consulting Engineers, which consults widely throughout its over sixty national associations world-wide concluded when publishing the Orange Book in 1995 that it is necessary to impose a fitness for purpose liability on contractors in respect of design in addition to that usually imposed in respect of material and workmanship.[33] Fitness for purpose is also imposed in the recently published Silver Book: *Conditions of Contract for EPC Turnkey Projects* and the new Yellow Book: *Conditions of Contract for Plant and Design-Build for E & M Plant, and for Building and Engineering Works designed by the Contractor*. For both of these forms of contract, the relevant provisions are in Sub-Clause 4.1.[34] This leads to the conclusion that unless expressly restricted, there is liability of fitness for purpose imported into contracts of design/build nature.

Against that background, it is necessary to mention that contractors in the United Kingdom have secured a restriction in the design liability imposed under the ICE 1992 Design and Construct Conditions of Contract, where Sub-Clause 8(2)(a) states as follows:

> (2)(a) In carrying out all his design obligations under the Contract . . . (and including the selection of materials and plant to the extent that these are not specified in the Employer's Requirements) the Contractor shall exercise all reasonable skill care and diligence.

In fact, in considering the precise meaning of the words quoted above, it could be argued that the contractor has in fact a lower standard of design liability than that which would attach to a professional designer, if the latter were made aware of the intended purpose of the design he has been asked to undertake.

It may therefore be concluded that unless expressly restricted in the contract, the applied liability of a contractor for design, including any duty of selection of materials and methods of workmanship, is one of fitness for purpose.

If a professional designer is made aware of the intended purpose of a particular aspect of a project, the liability could be elevated from one of due care and skill to one of fitness for purpose.

A restriction in the liability for design from fitness for purpose to skill and care would lead to an increase in the matrix of risks allocated to the employer in a particular contract. Such an increase in the employer's risks, if contemplated, must be made clear to the employer. On the other hand, the risks allocated to the contractor would be much more intense if he is allocated the liability for fitness for purpose.

If the increase in risk is to be allocated to the employer, the decision to accept that increase must be taken by the employer and not by the professional adviser on his behalf.

---

33 The Orange Book (FIDIC's standard design and build form of contract), see Clause 4.1.
34 See Chapter 10 below.

Selection of site staff

The custom that a consulting engineer is empowered to have a say in the selection of site staff to be employed by the owner has been settled in law since 1890. The case of *Saunders v. Broadstairs Local Board* established that the design professional could not rely on the incompetence of site staff as a defence to a claim against him for breach of a duty to supervise knowing of such incompetence.[35]

Similarly, the matter of delegation of supervisory duties assumed by a design professional has been settled, since 1911, in the case of *Leicester Guardians v. Trollope.*[36] In that case, extensive dry rot was discovered in a building four years after completion of the works. The cause of the dry rot was traced to the pegs supporting the joists of the ground floor which were driven into the ground and through which moisture was passing. The concrete layer, which was supposed to have been laid prior to the joists, was missing. The architects denied any liability for discovery of that omission which, in their opinion, should have been discovered by the clerk of the works. The trial judge Channell J. stated:

> The defence is that this dry rot was the fault of the clerk of the works. That is so in one sense. It clearly was the duty of the clerk of the works to attend to the laying of concrete in accordance with the design, but does that relieve the defendant? To my mind there is little difficulty in deciding the point. The position of the architect and of the clerk of the works was made quite clear. The details were to be supervised by the clerk of the works. The architect could not be at the works all the time, and it was for that reason that the clerk of the works was employed to protect the building owner . . . If the architect had taken steps to see that the first block was all right, and had then told the clerk of the works that the work in the others was to be carried out in the same way, I would have been inclined to hold that the architect had done his duty; but in fact he did nothing to see that the design was complied with. In my view, this was not a matter of detail which could be left to the clerk of the works. It may have been natural to leave it to him, but in my judgment it was an omission to do that which it was his duty to do.

A resident engineer who is an employee of the owner and receives a salary from him is a servant of the owner and not an independent contractor. If members of the resident site staff are employees of the design professional, then he is responsible for their actions not only in cases of negligence but also in fraud.[37] *See Lloyd v. Grace, Smith & Co.*[38] and in 1953 *London County Council v. Cattermoles (Garages) Ltd.*[39]

---

35 *Saunders v. Broadstairs Local Board* (1980), Reported in *Hudson's Building Contracts*, 4th edition, Vol. 2.
36 *Leicester Guardians v. Trollope* (1911) 75 JP 197.
37 *Morren v. Swinton and Pendlebury Borough Council* [1965] 1 WLR 576.
38 *Lloyd v. Grace, Smith & Co.* [1912] AC 716.
39 *London County Council v. Cattermoles (Garages) Ltd.* [1953] 2 All ER 582.

Supervision

The design professional's duty of care and skill in supervision has recently been the centre of controversy and confusion. This is mainly due to the different interpretations drawn by the courts of law from the meaning of the word 'supervise' and the definition of the duty that emerges from its implementation. Some argued in accordance with the spirit of what was intended by the profession, whilst others argued along the strict meaning of the word 'supervision', ignoring the realities of what can take place in practice. The confusion was encouraged by the fact that the profession itself had no precise definition to offer. A recent statement by the International Federation of Consulting Engineers goes some part of the way in explaining the role of the engineer during construction.[40] What is not stated in that document, however, is that the owner cannot afford and the design professional cannot provide a person supervising each person working on the site, which seems to be the interpretation of some.

The various views can be seen from the following quotations. In the case of *East Ham Corporation v. Bernard Sunley & Sons Ltd.* (1966):[41]

> As is well known, the architect is not permanently on the site but appears at intervals, it may be of a week or a fortnight, and he has, of course, to inspect the progress of the works. When he arrives on the site there may be many very important matters with which he has to deal: the work may be getting behind-hand through labour troubles; some of the suppliers of materials or the subcontractors may be lagging; there may be physical trouble on the site itself, such as, for example, finding an unexpected amount of underground water. All these are matters which may call for important decisions by the architect. He may, in such circumstances, think that he knows the builders sufficiently well and can rely upon them to carry out a good job; that it is more important that he should deal with urgent matters on site than he should make a minute inspection on site to see that the builder is complying with the specifications laid down by him ... it by no means follows that, in failing to discover a defect which a reasonable examination would have disclosed, in fact the architect was necessarily thereby in breach of duty to the building owner so as to be liable in an action for negligence. It may well be that the omission of the architect to find the defect was due to no more than an error of judgment, or was a deliberately calculated risk which, in all the circumstances of the case, was reasonable and proper.

In 1976, the interesting judgment in *Oldschool v. Gleeson* gave another example.[42]

> Plainly it is the consulting engineer's duty to produce a suitable design for the works which will achieve what the building owner requires, and it is further his duty to ensure that that design is carried out. The difference of opinion

---

40 'FIDIC's Policy Statement on the Role of the Consulting Engineer During Construction' discussed earlier on page 108.
41 *East Ham Corporation v. Bernard Sunley & Sons Ltd.* [1966] 3 All ER 619.
42 *Oldschool v. Gleeson* (1976) 4 BLR 103.

between the experts – Mr H on behalf of the first defendants and Mr M on behalf of the second – defendant is as to the extent of the consulting engineer's duty in regard to the manner in which the contractors execute the work in order to achieve the required result. Here may I pay tribute to the two experts, both of whom are consulting engineers of high qualification and considerable experience and both of whom in my estimation gave this evidence in support of their respective opinions in a manner deserving of the highest praise. Mr H obviously is, if he will forgive me saying so, of the older school. Although he was disposed to agree at one stage that a consulting engineer's duty was to design and see that the design was properly carried out, but otherwise to leave the contractors to get on with the job and not give instructions as to how the work was to be done, he nevertheless maintained that it was still the consulting engineer's duty to see that the contractors executed the work in a competent manner, particularly where the safety of the works was involved.

He regards the consulting engineer as what he described as being 'the father and mother of the job', whose duty it is to direct the contractors as to the manner in which the work is to be done, if he sees that the method which they are employing might endanger the safety of the works, and to stop the work if necessary. He considers it to be the consulting engineer's duty to ensure that the contractors carry out the work in a manner which will not endanger the safety of the works and thereby to assume responsibility for insisting that the contractors should undertake the work, if necessary, in a manner and sequence different from that which they may have planned or may be proposing to follow. Mr M, on the other hand, was equally insistent that the manner of execution of the works is a matter for the contractors. He considers that the consulting engineer is in no position, for instance, to require the contractors to comply with any particular sequence of works; he has no right let alone duty, to involve himself in the work of the contractors. Of course he would interest himself in their work, would offer advice to assist the job to go better and would certainly not turn his back on a situation that he could see was likely to give rise to danger to life. Equally he would intervene if he could see imminent damage to property. Those are matters of common sense; but that is a very different matter from assuming responsibility for the method of work to be adopted by the contractors. In my judgment, Mr M's view is the right one.

These views are contrasted by a third made as part of the judgment in *Eames London Estates v. North Hertfordshire District Council and Others* in 1980, where it is reported that Judge Edgar Fay stated 'that the blameworthiness of the policeman who fails to detect the crime is less than that of the criminal himself'.[43] One would normally fail to see how such blame could be attached to the policeman.

---

43 *Eames London Estates Ltd and Others v. North Hertfordshire District Council and Others* (1981), reported in *Design Liability in the Construction Industry*, by D.L. Cornes, Granada, 1983, page 69.

The duty of care was extended, in 1974, to certification. It was thought until then that when a design professional certifies, he is in the position of a quasi-arbitrator and therefore cannot be held liable for errors committed in the certificate. This changed with the decision in *Sutcliffe v. Thackrah* [1974] 1 All ER 859, which imposed a new dimension on the duty of care, as postulated by D. L. Cornes in his book on Design Liability, referred to earlier on page 151. He stated:

> The employer engaged architects to act in respect of the design and erection of a house. Eventually, a 1963 Edition JCT Contract was entered into between the employer and a contractor. The contractors were slow and in due course the building contract was determined. The architect's interim certificates had been issued without deduction in respect of defective work. The employer could not recover his losses from the contractor, who was now insolvent, and sought damages from the architect. It was held that the architect, in issuing interim certificates, was not acting as a quasi-arbitrator and was not, therefore, immune from liability. An architect issuing interim certificates has a duty to act fairly between the employer and the contractor. The employer can therefore recover his loss from an architect who has negligently issued interim certificates.
>
> This case is of fundamental importance to the position of architects acting under the JCT Standard Form and engineers acting under the ICE form of contract. The finding in the *Sutcliffe* case that a certifier is not immune from action means that contractors may be able to bring actions against certifiers where they certify negligently. This could arise, for example, where an architect, following his inspection of the work, reduces the sum to be paid to the contractor under an interim certificate in respect of alleged defective workmanship, when that workmanship is not, in fact, defective. In the *Sutcliffe* case Lord Reid, having described many of the decisions that an architect makes which affect the amount of money a contractor will receive, said: 'and, perhaps most important, he has to decide whether work is defective. These decisions will be reflected in the amounts contained in certificates issued by the architect. The building owner and the contractor make their contract on the understanding that in all such matters the architect will act in a fair and unbiased manner and it must, therefore, be implicit in the owner's contract with the architect that he shall not only exercise due care and skill but also reach such decisions fairly holding the balance between his client and the contractor'.

The conclusion reached in this quotation is obviously deduced with the spirit of the quotation 'If one step, why not fifty?', see page 167 below. The difference between over-certified amounts, which subsequently cannot be recovered, and under-certified interim valuations, which can later be adjusted, is immense. If an architect or engineer suspects, in carrying out the function of a supervisor, that defective work had been carried out, it is not only his entitlement, but also his duty to investigate and establish whether or not it is in fact defective. The suggestion that the contractor might then sue him if the work is found to be not defective could only lead to a breakdown in the very function a supervisor performs. Instead of the supervisor getting on with his work he 'would be for ever looking over his shoulder to see if someone was coming up with a dagger'.[44] It is essential that the relationship of trust between the parties to a construction contract prevails unhampered.

In the United States of America, a number of legal decisions in the early sixties decided that the contractual right of the designer to supervise construction created a duty to supervise. They defined supervision as the right to control the construction work and thus created the ridiculous situation that, if a design professional had a contractual right to supervise and negligently failed to control the actual construction work, even though this was the contractor's responsibility, he could be found liable for any injuries that might follow.

The American Institute of Architects and the National Society of Professional Engineers changed their standard contract documents in such a way that the word 'supervise' would not be used to describe the architect's, or engineer's, responsibilities during construction. The word 'supervise' was replaced by words such as 'inspect' and 'oversee' resulting in recognition by the American courts that, amongst other things, it is not the duty of the design professional to warn the contractor of a hazard of which he should be aware, provided the revised standard forms are used; see *Vonesck v. Hirsch and Stevens Inc.* (1974).[45] However, when the new standard conditions were not used, designers got caught in the web of this word 'supervise'; see *Geer v. Bennett*.[46]

Developments in contract law: contractor's duties

In a contract between owner and contractor, there is an implied warranty that the contractor undertakes:

1 to do work with care and skill in a workmanlike manner;
2 to use materials of good quality; and
3 that both the work and materials will be reasonably fit for the purpose for which they are required.

The implied warranty in this format was established and confirmed through a number of legal cases in the late sixties.[47]

The contractor is also held to have an implied duty to inform the employer/owner of potential defects in the project, as held by the Alaska Supreme Court in *Lewis v. Anchorage Asphalt Paving Co.* (1975).[48] It was stated:

. . . in building or construction contracts whenever someone holds himself out to be specially qualified to do a particular type of work, there is an

---

44 The words quoted were used by the late Lord Denning in his book *The Discipline of Law*, page 243, as part of what he told the jury in the case of *Hatcher v. Black and Others* (1954), which revolved around medical negligence and concerned a doctor.
45 *Vonesck v. Hirsch and Stevens Inc.* (1974), Wisconsin Supreme Court, USA, GFIP, Vol. V. No. 1.
46 *Geer v. Bennett* (1970) Florida, USA, GFIP Vol. V, No. 1.
47 See *Hudson's Building and Engineering Contracts*, 11th edition, by I.N. Duncan Wallace, Sweet & Maxwell, London, 1995, pages??? (pages 274–306 in the 10th ed.).
48 *Lewis v. Anchorage Asphalt Paving Co.* (1975), Alaska Supreme Court, USA, GFIP, vol. VII No. 6.

implied warranty that the work will be done in a workmanlike manner and that the resulting building, product, etc., will be reasonably fit for its intended use. . . . Thus, the contractor is required to bring his expertise into play and to notify even an architect (expert) of reasonably discovered defects.

Where subcontractors are concerned, the main contractor has an implied duty to supervise their work and inform the owner, or the architect or the engineer if potential defects are suspected including design defects carried out by them. This implied duty extends even to nominated subcontractors.

Examples can be drawn from a number of legal cases headed by *Eames London Estates and Others v. North Hertfordshire District Council and Others* in 1980 (referred to earlier on page 162). The excavation subcontractor in that case questioned the adequacy of the soil at the level he was required to excavate to. The architect was informed, but made no change in the design of the foundations or the level at which they were to be founded. The foundations proved later to be defective and caused damage in respect of which the contractor was held to be partly liable.

Two other legal cases may be quoted in this connection, both in 1984: *Equitable Debenture Assets Corporation Ltd. v. William Moss and Others*,[49] and *Victoria University of Manchester v. Hugh Wilson and Lewis Womersely and Others*.[50] In the first case, a nominated subcontractor designed and installed curtain walling, which proved to be defective. Having admitted liability, he went into liquidation, leaving the main contractor to face the courts, which held him liable for the default under the following implied terms:

- The main contractor must inspect and supervise the work of a nominated sub-contractor.
- The architect must be informed if a main contractor discovers that a design carried out by a nominated subcontractor is either defective or impractical. Similarly, if the main contractor discovers that the subcontractor is in breach of a statutory requirement, he should inform the architect.

Two limitations were, however, placed on that duty. The first is in the case of a subcontractor who does not have to warn of potential problems in someone else's work, unless they are likely to cause failure in his own part of the work. Second, once the owner is warned and acknowledges the warning, the contractor is released from liability.

Problems will, however, arise if the architect and the owner ignore the warning of the contractor or deny its validity. The contractor may find himself facing two options, neither of which is expedient. The first is not to proceed with the work at the risk of being found in breach of contract and the second is to proceed at the risk of being held liable for defective work if his warning turns out to be valid.

Recently, changes under contract affecting the contractor's liability have occurred under revisions in the standard forms of the General Conditions of Contract. The

49 *Equitable Debenture Assets Corporation Ltd. v. William Moss and Others* (1984) 2 Con LR 43.
50 *Victoria University of Manchester v. Hugh Wilson and Lewis Womersely and Others* (1984) CILL 126.

transfer of responsibility for some risks from one party to another has affected the respective liabilities, see page 141.

Liable or not? Developments in the law of torts

'I view tort as one would a funnel web spider – fascination, respect, fear.'

Comment at International Construction Conference
in Sydney, Australia (October, 1982).

The legal principles of the law of torts were laid down a long time ago by great jurists of the law of nature. Examples include the Islamic law of tortious liability, best reflected in the maxim *'la darar wa la dirar'*, meaning 'no hardship or the causing of hardship'; and the principles of Roman law which divided torts into the three constituents of intentional interference with the person, intentional interference with property and the rule of the Lex Aquilia covering losses caused by negligence.[51] Jurisdictions using the Civil Code embodied their principles in legal codes such as the French Code Napoléon or Code Civil of 1804. The French law of torts is contained in five articles, of which the first reads: 'Every act whatever of man which causes damage to another binds the person whose fault it was to repair it' (1804, Article 1382).[52]

The common law on the other hand developed as time passed by, but there was little need in the field of negligence before the Industrial Revolution. The increase in exposure to risk of injury and damage to property which accompanied the extensive use of machinery, which followed in every aspect of daily life, necessitated the use of some form of distribution of the effect produced by those risks that eventuated. New policies were gradually injected into the substance of the law and resulted in the rise of the Welfare State in some parts of the world. Professor Dennis Lloyd explains that process as:[53]

> The rise of the welfare state contains the implicit assumption that many of the social and economic risks from the ordinary wear and tear of existence should be as widely distributed as possible and not allowed to fall only on the unfortunate. The idea of an earlier and sterner age of individualism (where misfortune was almost equated with the culpability of the sufferer who was therefore no more than an object of charity to be relieved from destitution but without any legal claim whatever to relief) has been replaced by a partial attempt to provide a legal claim to a reasonable subsistence in relation to many of the main contingencies of human life. Such provision has been made in the case of sickness, industrial injuries, old age, and the death of a breadwinner leaving dependants behind him. Yet it is obvious that however favourably courts of law may desire to react towards this

---

51 'Liability in Islamic Countries', John Beechey, a paper presented at the Henderson Colloquium on Liability, 1984, IABSE in association with the Institution of Structural Engineers, Cambridge, UK.
52 *A Casebook on Tort*, Tony Weir, Sweet & Maxwell, 3rd Edition, London, 1974, page 2.
53 *The Idea of Law*, Dennis Lloyd, Penguin Books, 1983, page 264.

general change in human attitudes it is not for them but for the legislature to introduce far-reaching schemes of social insurance in order to secure the citizens against undeserved misfortune.

Moreover, even in the context of those fields of law where courts have scope to adjust the law to new situations, their ambit of operation remains rather limited. If we take one of the main activities of our present-day courts, which is to try actions for damage arising out of negligent driving of motor-vehicles on the highway, it has to be borne in mind that this whole branch of law has only been rendered tolerable in modern times by legislative intervention imposing compulsory insurance against third-party risks on all drivers. For in the absence of such insurance much of the compensation recoverable in actions of this sort could not be paid. In addition there remains the grave lacuna in the existing law that such liability requires proof of negligence, and so depends upon an assessment of the facts in each case so far as they can be ascertained, the decision frequently turning on an opinion as to the interpretation of these facts which might well differ substantially from one judge to another. It may be argued that what is needed here is some kind of social insurance against the misfortunes of road accidents comparable to that which operates already in the field of industrial injuries. Indeed, a case could be made out against the whole idea of actions for damages amounting in some instances to vast sums which can only be met by large corporations or an insurance company. Social insurance can and should provide a reasonable subsistence in all cases; compensation for serious injuries or loss of earning-capacity might arguably be left as a field for private insurance, as it is in some considerable extent already.

The development of the law of torts and in particular the law of negligence stems therefore from the concept that the purpose of these laws is the adjustment of certain losses that are the inevitable result of living in a common society through compensation by the wrongdoer to the person affected. Three major problems may hinder or prevent such an adjustment of loss through financial compensation. They arise from any one, or a combination, of the following:

- the identity of the wrongdoer cannot be ascertained;
- the fault cannot be proven; and
- the financial means of the wrongdoer are insufficient to compensate the victim.

In the process of trying to overcome these problems and implement the adjustment and sharing of risk and ultimately loss, insurance was brought into play. The law of negligence was extended and strict liability grew to different levels in various parts of the world covered by indemnity through insurance. In certain countries legislation was introduced and in others judgments moved in the direction of wider liability and closer interaction with insurance.[54]

---

54 The German Legislature introduced in 1884 a system of special insurance for industrial accidents, which formed the basis for other systems all over the world.

The beginning of the twentieth century saw that gradual movement with cases led by *MacPherson v. Buick Motor Co.* in 1916 in the United States and *Donoghue v. Stevenson* in 1932 in the United Kingdom. Until then the judgment in the earlier case of *Winterbottom v. Wright* in 1842 applied, when Judge Alderson stated:[55]

> If we were to hold that the Plaintiff could sue in such a case, there is no point at which such actions would stop. The only safe role is to confine the right to recover to those who enter into the contract: if we go one step beyond that there is no reason why we should not go fifty.

It was the last few words which formed the stumbling block against any shift in policy. It was recognised that any change, no matter how small, would create a precedent which would open the door for further change. The sociological developments in the United States and in the United Kingdom were of such magnitude that the inevitable shift in policy was made by the judiciary.[56] The masses of ordinary men and women at the start of this century placed more importance than heretofore on suitable and fitting food, drink, clothing and housing. These items were being manufactured in bulk as a result of the continuing industrial revolution.

Living conditions improved and society expected more equitable distribution of the four essential requisites for agreeable living. But the distribution was not free of charge and these requisites had to be purchased with hard-earned income. The masses aspired therefore to get their money's worth and the maximum in respect of their spending. They also felt that it was not sufficient for them to know what is right or wrong but also to practise it, and thus the principle of love-your-neighbour had to be put into practice to protect the weak from the strong, the consumer from the manufacturer and the unaware from the enterprising. A decision, however, had to be made as to who is your neighbour.

All this happened when, in 1928, Miss Donoghue asked a friend of hers to purchase a bottle of ginger beer from a shop in Glasgow.[57] In pouring her second glass she saw a decomposed snail drop from the bottle. She became ill and suffered

---

55 *Winterbottom v. Wright* (1842).
56 In his judgment in the case of *Dutton v. Bognor Regis U.D.C.* [1972], 1 QB 373, page 397, the late Lord Denning throws a revealing light on policy-making by the judiciary: 'It seems to me that it is a question of policy which we, as judges have to decide. . . . In previous times, when faced with a new problem, the judges have not openly asked themselves the question: What is the best policy for the law to adopt? But the question has always been there in the background. . . . *In Rondel v. Worsley* [1969] 1 AC 191, we thought that if advocates were liable to be sued for negligence they would be hampered in carrying out their duties. In *Dorset Yacht Co. Ltd. v. Home Office* [1970] AC, 1104, we thought that the Home Office ought to pay for damages done by escaping Borstal boys, if the staff was negligent, but we confined it to damage done in the immediate vicinity. In *S.C.M. (United Kingdom) Ltd. v. W.J. Whittall & Son Ltd.* [1971] 1 QB, 337, some of us thought that economic loss ought not to be put on one pair of shoulders, but spread among all the sufferers. In *Launchbury v. Morgans* [1971] 2 QB, 245, we thought that as the owner of the family car was insured she should bear the loss. In short, we look at the relationship of the parties and then say, as a matter of policy, on whom the loss should fall.'
57 *Donoghue v. Stevenson* [1932], AC 562.

nervous shock, and thus she sued Mr Stevenson, the bottler of the ginger beer, for negligence. Miss Donoghue knew that she had very little chance of success because previously, on two similar occasions, the law was not on the side of the consumer. In 1913, a bottle of ginger beer exploded in the hands of a plaintiff and, in 1928, the remains of a decomposed mouse had been found in another but, in both cases, the bottler was not held to be liable.

In this instance, however, the case was taken to the House of Lords and three out of five held that the bottler could be liable, thus marking a change in the principle that a manufacturer can only be sued successfully in contract. After 1932, he became liable in tort and in contract.

Lord Atkin delivered the following statement, which became the guideline in defining the duty of care:

> The rule that you are to love your neighbour becomes in law, you must not injure your neighbour; and the lawyer's question, Who is my neighbour? receives a restricted reply. You must take reasonable care to avoid acts or omissions which you can reasonably foresee would be likely to injure your neighbour. Who, then, in law is my neighbour? The answer seems to be – persons who are so closely and directly affected by my act that I ought reasonably to have them in contemplation as being so affected when I am directing my mind to the acts or omissions which are called in question.

The words used in the above judgment are so fluid that the definition of duty of care is still being developed. In fact, Lord MacMillan, in the same case, foresaw that possibility of development of the law of negligence when he stated:

> . . . the grounds of action may be as various and manifold as human errancy; and the conception of legal responsibility may develop in adaption to altering social conditions and standards. The criterion of judgment must adjust and adapt itself to the changing circumstances of life. The categories of negligence are never closed.

An interesting part of the judgment of one of the two dissenting judges in this case, Lord Buckmaster, is given below, specifically referring to construction and showing an awareness of the repercussions of changing from the existing law of that time:

> There can be no special duty attaching to the manufacture of food apart from that implied by Contract or imposed by Statute. If such a duty exists, it seems to me it must cover the construction of every article, and I cannot see any reason why it should not apply to the construction of a house. If one step, why not fifty? Yet if a house be, as it sometimes is, negligently built, and in consequence of that negligence the ceiling falls and injures the Occupier or anyone else, no action against the builder exists according to English Law, although I believe such a right did exist according to the Laws of Babylon.[58]

---

58 See page 2 of this book in reference to Hammurabi's Code.

It took exactly fifty years, however, to apply the principle established in *Donoghue v. Stevenson* to a construction case.[59] It was the case of *Dutton v. Bognor Regis U.D.C. and Another,* which changed the *caveat emptor* or 'let the buyer beware' position with regard to the builder of a house as established earlier in the case of *Bottomley v. Bannister.*[60] In that case, the duty of care owed by a builder, who was also a vendor, was determined by and limited to the contract of conveyance or lease executed and no further duty existed beyond the privity of contracts. Lord Atkin, in his judgment in *Donoghue v. Stevenson,* mentioned the *Bottomley v. Bannister* case but did not commit himself to the view that a duty of care is owed by builders. In 1972, in judgment on *Dutton v. Bognor Regis,*[61] see page 171, the late Lord Denning referred to the *Bottomley v. Bannister* case when he stated: '. . . But I do not think it is good law today.'

During the intervening period 1932 to 1972, the ground was being prepared for change by such judgments as that of Lord Denning, dissenting, in the case of *Candler v. Crane, Christmas* in 1951;[62] (later overruled by *Hedley Byrne & Co. Ltd. v. Heller & Partners*) the judgments in *Gallagher v. McDowell Ltd.* in 1961;[63] in *Clay v. A. J. Crump & Sons Ltd.* in 1964;[64] in *Hedley Byrne & Co. Ltd. v. Heller & Partners Ltd.* in 1964;[65] in *Dorset Yacht Co. Ltd. v. Home Office* in 1970;[66] and in the *Ministry of Housing and Local Government v. Sharp* in 1970,[67] all in the United Kingdom. In the United States of America, amongst a group of legal cases, it is worthwhile mentioning the case of *Nelson v. Union Wire Rope Corporation* in 1964.[68]

In the first case, the late Lord Denning, dissenting, felt that the accountant who negligently prepared accounts for a company should have been held liable when a third party, induced to invest in that company on the strength of the accounts, lost the sum of money invested. In the event, he was not held liable, as the rest of the court disagreed with Lord Denning.

In the second, *Gallagher v. N. McDowell Ltd.,* the Court of Appeal in Northern Ireland held that a contractor who built a house negligently was liable to a person injured by his negligence.

In the third, the case of *Clay v. Crump* was decided against an architect who approved of leaving an existing wall standing without support on a site where demolition was taking place. The architect apparently accepted the demolition contractor's opinion that the wall was safe and did not examine the wall, despite the fact that he visited the site subsequently. The building contractor assumed that the wall was safe. The wall collapsed and injured a workman. The Court of Appeal in the United

---

59　See quotation on page 168 of Judge Alderson, where he predicted that if one step is taken from the situation at the time, fifty would follow.
60　*Bottomley v. Bannister and Otto v. Bolton & Norris* [1936] 2 KB 46.
61　*Dutton v. Bognor Regis Urban District Council.* [1972] 1 QB 373.
62　*Candler v. Crane, Christmas,* [1951] 1 All ER 426.
63　*Gallagher v. N. McDowell Ltd.* [1961] NI 26.
64　*Clay v. A.J. Crump & Sons Ltd.* [1964] 3 All ER 687.
65　*Hedley Byrne & Co. Ltd. v. Heller & Partners Ltd.* [1964] 2 All ER 575.
66　*Home Office v. Dorset Yacht Co. Ltd.* [1970] 2 All ER 294.
67　*Ministry of Housing and Local Government v. Sharp* [1970] 1 All ER 1009.
68　*Nelson v. Union Wire Rope Corporation* (1964) 199 N.E. Rep. (2d) 769.

Kingdom held that the architect, the demolition contractor and the building contractor were all liable.

The most important case of the sixties is probably that of *Hedley Byrne & Co. Ltd. v. Heller & Partners Ltd.*, where the liability for negligent statement was finally established. The facts of the case were that, before entering into a contract with a customer, advertising agents Hedley Byrne required a banker's reference from their own bankers who passed the enquiry to merchant bankers. The latter reported that the customer was creditworthy but headed their report 'without responsibility'. However, it transpired that the customer was not creditworthy and Hedley Byrne lost a considerable sum of money and sued the merchant bankers. The merchant bankers escaped liability by reason of having expressed their report to be without responsibility. The British House of Lords, however, held that a professional man is liable for statements made negligently in circumstances where it is known that those statements are likely to be acted upon and, in fact, were. In simple terms the following was established:

> If someone possessing a special skill undertakes, quite irrespectively of contract, to apply that skill for the assistance of another person who relies upon such skill, a duty of care will arise.

This decision was only the beginning because, in the case of the *Ministry of Housing and Local Government v. Sharp* in 1970, liability for economic loss following negligent statement was established.

The decision in the *Dorset Yacht Co. v. Home Office* in 1970 applied the principle of *Donoghue v. Stevenson* to a situation where a public authority is concerned.

Finally, in the case of *Nelson v. Union Wire Rope Corporation* in 1964 in the United States of America, during the building of a courthouse, a lift plunged down six floors with nineteen workmen aboard. It had been regularly inspected by an insurance company and passed as safe. The insurer's inspector was negligent because he passed the lift as safe when it was unsafe. The Supreme Court of Illinois, by a majority, held that the insurance company was liable for the negligence of the inspector. It was stated that the defendant's liability

> is not limited to such persons as might have relied upon it to act but extends instead to such persons as defendant could reasonably have foreseen would be endangered as the result of negligent performance.

But to return to the *Dutton v. Bognor Regis case,* where the events presented a new situation not previously brought before the courts: the facts were that, in Bognor Regis, there was a rubbish tip which was subsequently filled in and the surface was restored to match the adjoining ground. In 1958, a builder bought the land in that area and developed it into plots for a housing estate. The old rubbish tip became plot No. 28. The builder, having obtained the necessary approvals from Bognor Regis UDC, started work on a house on plot No. 28. He notified the council when the foundations were ready for inspection and an inspector visited the site and approved them. The house was finished towards the end of 1959 and, in early 1960 the builder sold it to a Mr Clark who sold it later, in December, to Mrs Dutton.

The inevitable happened in 1961 and settlement cracks led later to the discovery of the rubbish tip under the foundations. Mrs. Dutton sued the builder and council claiming damages of £2,740 in respect of the cost of repair and diminution in value. The builder's insurers claimed that he was exempt from liability in accordance with the earlier legal decisions of *Bottomley v. Bannister* and *Otto v. Bolton & Norris* [1936], 2 KB 46. Mrs Dutton was advised to settle the claim against the builder for £625 and she complied but continued her action against the council, claiming negligent inspection.

It was held, in 1971, that the council's inspector was negligent and the judge gave judgment in favour of Mrs Dutton for £2,115, the remainder of her claim. The council appealed the decision and it was then that the law in this area was expanded. The Court of Appeal dismissed the appeal, holding that the council was entrusted with the control of the construction of the house under the Public Health Act and that it had, therefore, a duty in common law to such persons as could reasonably be foreseen to be in danger, whether or not they could be shown to have relied on the council's inspection. The court held that the liability extended to a subsequent purchaser of a house with a defect which could not have been detected on any intermediate survey. It was also held that the principle in *Donoghue v. Stevenson* should, as a matter of common law policy, be applied to the duty situation existing between the council and the plaintiff, and that the liability of the council was not limited to damages for physical injury but extended to the damage to the house and might also include economic loss. Lord Denning and Sachs LJ were quoted in the law report giving their judgments that the principle of *Donoghue v. Stevenson*

> . . . as developed in later cases applied to real property as well as to chattels with a hidden defect so that an action in negligence could now be brought against a builder or builder owner who put a defective house on the market. The earlier authorities to the contrary no longer bound the court, page 394, C–D and 402 A–C.

The late Lord Denning, in his judgment, expounded the law as it applied in various areas relevant to the case under the heading of 'Power or Duty'. He stated:[69]

> The reason for this discussion was the case of *East Suffolk Rivers Catchment Board v. Kent* [1941] A.C. 74. The agreement was that if the local authority had a mere power to examine the foundations, they were not liable for not exercising that power. But if they were under a duty to do so, they would be liable for not doing it. This argument assumes that the functions of a local authority can be divided into two categories, power and duties. Every function must be put into one or other category. It is either a power or a duty. This is, however, a mistake. There is a middle term. It is control . . .

Under the heading of the 'Position of the Builder', he stated:

69 *The Discipline of Law*, by the Rt. Hon. Lord Denning, Butterworths, 1979, pages 255 to 261; and case report 1 QB 373.

The distinction between chattels and real property is quite unsustainable. If the manufacturer of an article is liable to a person injured by his negligence, so should the builder of a house be liable . . .

Under the heading of the 'Position of the Professional Adviser', he stated:

. . . it is clear that a professional man who gives guidance to others owes a duty of care, not only to the client who employs him, but also to another who he knows is relying on his skill to save him from harm . . .

He then went on to 'Reliance to Proximity to Economic Loss to Limitation of Action' where he stated:

The Council would be protected by a six year limitation, but the builder might not be. If he covered up his own bad work, he would be guilty of concealed fraud, and the period of limitation would not begin to run until the fraud was discovered: see *Applegate v. Moss* [1971] Q.B. 406.

The *Dutton* case was thus important in establishing, under common law, that tortious liability applied to construction, to the contractor, to the design professional, to the local authority inspector and to valuers.[70] It further established the liability for economic loss.

An earlier case in Queensland, Australia, is worthy of mention in this connection because of the dicta which supported liability either from a breach of contract or in tort. It is in *Voli v. Inglewood Shire Council* (1963) where an architect was held liable for his failure to provide for joists of sufficient strength under the stage of a hall.[71] The joists which were specified in his drawings and specifications were not strong enough to support the minimum live load recommended by the Standard Association of Australia and required by the local Council's by-laws. The stage collapsed under the load of a gathering of a local association and the architect was held liable in negligence for the injuries of the plaintiff. It was stated by J. Windeyer:

. . . for the reasonably foreseeable consequences of careless or unskilled conduct an architect is liable to anyone whom it could reasonably have been expected might be injured as a result of his negligence.

Concurrent tortious and contractual liability towards an employer/owner was not established, however, in the English courts until 1976 in the case of *Esso Petroleum v. Mardon*.[72] Esso wanted to get a tenant for their filling station at Southport and they informed Mr Mardon that the forecast of the estimated annual consumption was 200,000 gallons. Mr Mardon took the tenancy.

Unfortunately, nobody ever told Esso and they didn't bother to find out that a new one-way system had been inaugurated in this particular town after

70 *Dutton v. Bognor Regis U.D.C.* [1972] 1 QB 373, pages 391 to 396.
71 *Voli v. Inglewood Shire Council* (1963) 10 CLR 74.
72 *Esso Petroleum Co. Ltd. v. Mardon* [1976] 1 QB 801.

which it was necessary to drive for nearly half a mile through back streets to get to the filling station.

In the event, very little petrol was sold and Mr Mardon sued for damages under the principle *of Hedley Byrne and Co. Ltd. v. Heller & Partners Ltd.* The Counsel for Esso argued that when negotiations between two parties resulted in a contract between them, their rights and duties were governed by the law of contract and not by the law of tort. There was, therefore, no place in their relationship for *Hedley Byrne* [1964] AC 465, which was solely on liability in tort . . .

Lord Denning, however, disapproved and went on to state:

> . . . in the case of a professional man, the duty to use reasonable care arises not only in contract, but is also imposed by the law apart from contract, and is therefore actionable in tort. It is comparable to the duty of reasonable care which is owed by a master to his servant, or vice versa. It can be put either in contract or in tort . . .
>
> . . . A professional man may give advice under a contract for reward; or without a contract, in pursuance of a voluntary assumption of responsibility, gratuitously without reward. In either case he is under one and the same duty to use reasonable care. . . . In the one case it is by reason of a term implied by law. In the other it is by reason of a duty imposed by law. For a breach of that duty he is liable in damages: and those damages should be, and are, the same, whether he is sued in contract or in tort.

This principle was adopted by other courts in different parts of the world where the common law system applied.[73] So in Ireland, in *Finlay v. Murtagh*, Mr Justice Kenny held:[74]

> When a client retains a professional person, i.e., a Solicitor, an Architect, an Accountant or a Doctor, to do work for reward, there is implied from the retainer a contract between them, one of the terms of which is that the professional person has the competence to do the work and that he will act with that degree of care and skill which is reasonably expected from a member of that profession. If he is negligent in the performance of the work, an action for damages for breach of contract may be successfully brought against him. The professional person, however, owes the client a general duty and not one arising from the contract but from the proximity principle to exercise reasonable care and skill in the performance of the work entrusted to him. This duty arises from the obligation which springs from the situation that he knew or ought to have known that his failure to exercise care and

---

73 'The Architect – Liability in Perpetuity', by David Keane, a lecture delivered at a Conference on Liability, The Royal Institute of the Architects of Ireland, October 1982.
74 *Finlay v. Murtagh* [1979] IR 249.

skill would probably cause loss and damage. This failure to have or to exercise reasonable skill and care is Tortious in origin so a plaintiff in such an action may successfully sue in Contract or in Tort or in both.

To complete the picture, similar judgments from other jurisdictions may be quoted.

From Canada, the case *of District of Surrey v. Church* in [1977] can be quoted where the architect and the engineer were held jointly and severally liable to the owner.[75] The architect was held liable for breach of contract and the engineer (having been appointed directly by the architect and not by the owner) was held liable in tort for negligence.

In *Aluminium Products (Qld) Pty. Ltd. v. David Hill and Others* (1980), the Supreme Court of Queensland held that an action could be brought in tort independent of the right to bring an action in contract.[76] In New Zealand, the case of *Vlado Vulic v. Bohdam Bilinsky and Others* (1982) is one of the authorities in this context.[77]

Leaving aside the common law and turning back to civil code jurisdictions, one finds that they differ in their interpretation and application of the law of torts. For example, the Belgian courts reject the idea of concurrent liability under both contract and tort, whereas certain others accept it.

Date of accrual

One of the important features of tortious liability distinguishing it from contractual liability is the date on which it ceases to exist. This date is linked directly, under both systems, to an inception date which, under contract law is the date of the breach of contract, except effectively in cases of fraud. In construction contracts, it is usually the date of either the Taking-Over Certificate or the Final Certificate, depending on the legal system in the jurisdiction, since until that time, the Engineer may change the entitlement.

The inception date under the law of torts, however, is the date of 'accrual of the cause of action' (see page 14). The definition of these few words within the context of construction has caused much controversy recently after a number of contradicting and confusing judgments. The reason behind the confusion is mainly related to the practical difficulty in establishing when 'accrual' takes place. Is it the time when the faulty act was conceived, or drawn, or specified, or constructed, or inspected, or eventuated into a loss, or its effect felt and noticed?

Until the decision of the Court of Appeal in the United Kingdom in 1976, in the case of *Sparham-Souter and Others v. Town and Country Developments (Essex) Ltd.*, it had always been held that the damage was caused at the date of the negligent construction.[78] In the case of *Bagot v. Stevens Scanlan and Co. Ltd.* [1966] 3 All ER 577, in the United Kingdom, Lord Diplock stated that the damage occurred 'when

---

75 *District of Surrey v. Church et al.* (1977) 76 DLR (3d) 721 and 4 BCLRC 31.
76 *Aluminium Products (Qld.) Pty. Ltd. v. David Hill and Others* (1980) 3 BCLRC 103.
77 *Vlado Vulic v. Bohdam Bilinsky and Others* (1982) NSW Supreme Court No. 17700/78.
78 *Sparham-Souter and Others v. Town and Country Developments (Essex) Ltd.* [1976] 2 All ER 65.

the drains were improperly built'. As late as 1972, this question had been considered in the case of *Dutton v. Bognor Regis,* referred to on page 171 and the afore-mentioned conclusion was reached, giving a date similar, if not identical, to the date when the cause of action also accrued in contract. Four years later, however, Lord Denning reversed his earlier judgment stating:

> But now, having thought it over time and again – and been converted by my brethren – I have come to the conclusion that, when building work is badly done – and covered up – the cause of action does not accrue, and time does not begin to run, until such time that the plaintiff discovers or ought with reasonable diligence, to have discovered it.

It was obvious that his decision could extend the period for claims in tort to perpetuity. Yet, in the subsequent case of *Anns and Others v. London Borough of Merton* (1977),[79] the House of Lords approved *Sparham–Souter* in substance and it was not until December 1982 that the decision was found to be wrong when a further opportunity occurred in the case of *Pirelli General Cable Works Limited v. Oscar Faber and Partners.*[80] It was held that the cause of action accrues on the date when the damage first came into existence. The Lords viewed the decision they had to come to (in view of the law as it existed) with dissatisfaction as they thought it to be 'unreasonable' and recommended that the law be altered by Parliament. The whole matter of the period of limitation was examined by the Law Reform Committee and their report was published in 1984.[81] The facts of the *Pirelli* case were that, in March 1969, Pirelli engaged Oscar Faber to advise in relation to the extension of their factory including the design and construction of a 160 ft high industrial chimney. Although Oscar Faber neither designed nor constructed the chimney, they did specify its order from a specialist nominated subcontractor which subsequently went into liquidation leaving them with the liability. The chimney was built in June/July 1969 utilising a relatively new material called Lytag in refractory inner lining. This material was found later to have been unsuitable for the purpose and the design negligent. Damage, in the form of cracks near the chimney's top, must have occurred not later than April 1970, more than six years prior to the commencement of the proceedings in October 1978. The damage, however, was not discovered by Pirelli until November 1977 and it was found by the court that, with reasonable diligence, this could not have been discovered before October 1972 so that, on the Sparham-Souter 'discoverability' test, the proceedings were brought in time. The chimney had to be partly demolished and replaced.

The Lords held that the cause of action had in fact accrued in April 1970 when the cracks first occurred at the top of the chimney and that, accordingly, Pirelli's claim was outside the period of limitation and therefore must fail. This decision was

---

79   *Anns and Others v. London Borough of Merton* [1977] 2 All ER 492.
80   *Pirelli General Cable Works Ltd. v. Oscar Faber and Partners* [1983] 1 All ER 65.
81   *Latent Damage,* Law Reform Committee 24th Report, HMSO, London, November, 1984.

reached on the basis that the Limitation Acts, as had been passed by Parliament after an earlier judgment in 1963, permitted no other interpretation. It was the case of *Cartledge and Others v. Jobling & Sons Ltd.* where a man whose lungs had become diseased due to exposure to dust was unsuccessful in claiming damages because it was more than six years after his exposure and despite the fact that it took that length of time for the disease to manifest itself. [82] This judgment prompted the British Parliament to change the law but it was only changed in respect of personal injury. It was thus contended by Oscar Faber's Counsel that Parliament had considered the question and had intended the law to change in respect of personal injury only and to remain unchanged for other actions and that, accordingly, the decision in the *Sparham-Souter* case was incorrect.

The House of Lords decision in Pirelli was unpopular and the whole question of limitation in England was referred to the Law Reform Committee. As a result, significant changes were recommended in the case of 'latent' damage, which were incorporated in the Latent Damage Act 1986.

The result of *Anns v. Merton* was revisited by the House of Lords in 1990 in the case of *Murphy v. Brentwood District Council*.[83] The claimant purchased a newly built house. Eleven years later, signs began to appear that there were problems with the foundations of this home. The claimant decided to sell the house in its damaged state for £30,000. His insurers paid him £35,000 for the difference between the value of the undamaged house on the open market and the price received. The insurers then proceeded to recover from the local authority the £35,000 together with the costs they incurred in (a) mending the fractured pipes; (b) refitting carpets; and (c) the sale itself.

The trial judge held that in passing the building plans, the local authority, which relied on advice given negligently by a consulting engineer, had itself been negligent. The defects in the property, having become an imminent danger to the health and safety of the claimant whilst he was still in occupation of his house, had given the claimant a good cause of action against the defendants in tort.

The Court of Appeal dismissed the defendant's appeal, but the House of Lords allowed a further appeal. It was held that the principle in *Donoghue v. Stevenson* applied and a duty was imposed on the builder of a house to take reasonable care to avoid injury or damage, caused by defects in its construction, to the person or property of those whom he ought to have in contemplation as likely to suffer such injury or damage. Nevertheless, the principle only extended to latent defects. Once a defect was discovered and became patent, before any injury to person or health had materialised or any damage to property, other than the defective building itself, had been occasioned, any defect in question amounted to pure economic loss.

The *Pirelli* decision is not expected to be followed by other jurisdictions and, in fact, it has already been rejected in Ireland in the case of *Brian Morgan v. Park Developments Ltd.* in 1983, where Miss Justice M. Carroll stated:[84]

---

82 *Cartledge and Others v. Jobling & Sons Ltd.* [1963] AC 758.
83 *Murphy v. Brentwood District Council*, [1991] 1 AC 398. See also the *Department of the Environment v. Thomas Bates & Sons Ltd.* [1991] 1 AC 499.
84 *Brian Morgan v. Park Developments Ltd.*, judgment delivered on 2 February 1983.

Accordingly, I hold that the date of accrual in an action for negligence in the building of a house is the date of discoverability, meaning the date the defect either was discovered or should reasonably have been discovered. In relation to contract, the date of accrual of the right of action for breach of the building contract in my opinion must depend on the wording of the contract.

Economic loss

The decision in *Hedley Byrne and Co. Ltd. v. Heller and Partners Ltd.* in 1964 was the cornerstone in the extension of the law of torts to include claims in respect of economic loss. It was only a short period of time before it was applied to a construction case when in *Batty v. Metropolitan Property Realisation Ltd.* the owner of a house was awarded damages in respect of the future resale value. Both of these cases have been referred to earlier. The landmark in respect of economic loss in construction is, however, the case of *Junior Books Ltd. v. Veitchi Co. Ltd.*[85]

Veitchi Co. Ltd. were specialist flooring subcontractors nominated to lay a floor in a factory owned by Junior Books Ltd. under a building contract. There was no privity of contract between the two parties. The factory was built in 1969 and 1970, but in 1972 it was averred that the flooring showed defects allegedly due to bad workmanship, or bad materials, or both.

Junior Books Ltd. claimed the cost of replacing the floor (estimated at £50,000), plus the cost of storing books during the period (estimated at £1,000), plus the cost of removing machinery (estimated at £2,000), plus £45,000 for loss of profit during the period in which the business would have to be closed, plus £90,000 for wages to employees during that period, plus £16,000 for fixed overheads and £3,000 for investigations into the treatment required. The House of Lords held that Veitchi were specialists in flooring and must have known that Junior Books Ltd. relied upon their skill and experience. The relationship between the parties was as close as it could be, short of actual privity of contract, and Veitchi must be taken to have known that if they did the work negligently the resulting defects would require remedial measures as a consequence of which there would be financial or economic loss.

There was, on the facts, nothing whatever to restrict the duty of care arising from the proximate relationship. The only criterion made for limiting the damages recoverable for breach of the duty of care enunciated by the House of Lords was that the law had not before permitted recovery for economic loss and, therefore, ought not to do so in the future. It was stated by Lord Roskill that he saw recovery for economic loss as 'the next logical step forward in the development of this branch of the law'. On this basis Veitchi were held liable for both the physical damage to the floor and the consequential economic loss.

The question is: has this case opened the 'floodgates' to a 'liability in an indeterminate amount for an indeterminate time to an indeterminate class?[86] Only time will tell.

---

85 *Junior Books Ltd. v. Veitchi Co. Ltd.* [1982] 3 WLR 477.
86 *Ultramares Corporation v. Touche* (1931) 255 N.E. 170, *per* Cardozo C.J.

Future developments

All changed, changed utterly: A terrible beauty is born.

W.B. Yeats

Of all the different areas of the law of torts, negligence is the most relevant from the construction point of view. It is also the area where developments in the last century have been the most extensive. Negligence may be described as a failure to comply with that standard of care that would correspond to the conduct expected of a reasonable man of ordinary prudence under similar circumstances. Any conduct that falls below that standard of behaviour is considered to be negligent. The law of negligence, which combines that description with the liability emanating from negligent acts, has been developing out of the changes in the meaning of the various elements of the aforementioned description and the definition of to whom that standard of care is owed.

Interaction between insurance and law in areas of negligence other than those related to construction has moved in a direction towards strict liability compulsory insurance, indemnity irrespective of fault and a shift in the old principle of the law of torts of 'the loss lies where it falls' to the new position of 'the loss lies with the community'.[87]

The evolution of industrial accident insurance and motor insurance discussed earlier on pages 166 and 167 took place hand in hand with developments in the law. The indications are, however, that developments in the law of negligence are taking place irrespective of the evolution of compatible insurance schemes. This is happening at a time when even the existing insurance arrangements are not coping with the present legal principles. Two recent articles paint the picture of what is happening in the insurance world, appearing within days of each other. The first article states:[88]

> Statistics have been published and widely distributed showing that the professional liability claim frequency has increased in Canada from one claim for each 20 firms in 1966 to one claim for every three firms in 1983 . . . The sudden withdrawal from a class of insurance [*professional indemnity*] by a number of insurers and the simultaneous shrinkage of capacity will become a cause of major concern for design consultants. . . . We cannot afford a multi-million dollar judgment to a single injured claimant no matter how serious the injury. Members of the public who ultimately foot the bill will soon start asking questions: Are the courts too plaintiff-oriented? Is the legal system such that ways are always found to assign liability to any

---

87 Common law countries have generally adhered to the principle that liability for traffic injuries presupposes a wrongdoer's fault. This means in practice that victims of traffic accidents must prove that their injuries were the result of another's negligence. A large number of civil code countries on the other hand have adopted a rule of strict liability with regard to the operation of motor vehicles, i.e. liability regardless of fault.

88 'Liability Insurance . . . A Crisis in the Making', by Claude Y. Mercier, *Canadian Consulting Engineer*, June 1985.

available deep pocket whenever the truly negligent party (e.g. the contractor) has no insurance and is impecunious?. . .

The second, with a stark title, gives a chilling feeling:[89]

> Premiums have risen broadly and for some types of coverage, are up enormously – by 300%, 500%, even 1000%. The availability of insurance, nevertheless, has sharply contracted. Many Corporations do not have as much coverage as they would like, and for some kinds of potential losses have none at all . . .
> The dread is particularly great because most corporations have grown in recent years to want more insurance, not less. Asbestosis, toxic waste, the general litigiousness of society, the tendency of judges and juries to reinterpret the legal doctrines of negligence and fault – all these have driven home the need for bountiful insurance protection.

These quotations show some of the drift that is taking place in the interaction between law and insurance. But this may grow wider if the embryonic appearance in Canada of strict liability in construction-related professional liability is an indication of what is to come in the future. Evidence exists that others may follow suit.[90] Fitness for purpose is another aspect where the law may interfere with and interpret a contract in a context different from that accepted at present.

None of these developments have yet taken place, but only time will tell.

89 'Naked Came the Insurance Buyer', by Carol J. Loomis, *Fortune*, 10 June 1985. Time Inc. All Rights Reserved.
90 'Professional Liability Insurance – A New Requirement', by Chris Hart, *Engineering Dimensions*, Nov./Dec. 1984.

6

INDEMNITY AND INSURANCE

Indemnity

As defined earlier, on page 145, an agreement to indemnify means an undertaking to compensate for loss or damage suffered. Thus, a contract of indemnity is one where a party **A** undertakes to assume the legal liability which another party B may be held to be under. This liability can be either towards another party C or towards A.

However, if B's liability is towards A and he is also entitled to call on A to indemnify him, A's claim is, in effect, nullified (a situation similar to having an exemption clause in the contract between A and B).

If a risk had been allocated to B, it would be more likely than not that a court of law would find it unreasonable to shift that risk back to A, unless a very clear and precise indemnity clause were provided.[1] This attitude of limiting one's attempts to indemnify oneself against one's own negligence prevails, due to an overriding concern for the best interests of the public and society.[2] Any shift in liability or, in other words, any application of indemnity can take place through either an established legal routing or a contract between an indemnifier and an indemnified.

Indemnity through law

Indemnity through an established legal routing can be cited using one of the insurance law principles known as subrogation. Subrogation is defined as

> the right which the insurer has of standing in the place of the insured and availing himself of all the rights and remedies of the insured, whether already enforced or not, but only up to the amount of the insurer's payment to the insured. This right of subrogation is exercisable at common law after the insurer has paid the claim. However, a condition of the policy may entitle the insurer to exercise the right before the payment is made.[3]

Thus, following a loss suffered by a policyholder, the insurer may seek indemnification from a third party who has either caused or contributed to the loss. However,

1 *Smith v. South Wales Switchgear Co. Ltd.* [1978] 1 All ER 18, and *Photo Products Ltd. v. Securicor Transport Ltd.* [1980] 1 All ER 556.
2 *Guidelines for Improving Practice*, op. cit, see Chapter 3, note 4, vol. 111, No. 5.
3 *Insurance for the Construction Industry*, F.N. Eaglestone, George Godwin Ltd., London, 1979, page 8.

it is important to note that this right is not extended to apply against a joint-insured or co-insured, as can be seen from the judgment made in the case of *Petrofina v. Magnaload*.[4] There, it was held that, where a contractors' all risks insurance policy is issued in respect of a complete contract inclusive of all the property and in the names of all contractors and subcontractors, the whole contract is fully insured. Therefore, an insurer, having settled a claim in respect of one of those named as insured, cannot exercise the right of subrogation against another.

In his judgment, Lloyd J. referred to the Canadian case of *Commonwealth Construction Company v. Imperial Oil*, which was decided by the Supreme Court of Canada in 1976.[5] He gave the facts and drew his conclusions, as follows:

> . . . the facts were that Imperial Oil Ltd. entered into a building contract for the construction of a fertilisers plant with Wellman Lord Ltd. as main contractor. Wellman Lord Ltd. entered into a sub-contract with Commonwealth Construction Co. Ltd. for the construction of the pipework. Imperial took out a policy known as a 'course of construction policy'. The policy was in the name of Imperial, together with their contractors and subcontractors. It covered:
>
>> 'all materials . . . and other property of any nature whatsoever owned by the insured or in which the insured may have an interest or responsibility or for which the insured may be liable or assume liability prior to loss or damage . . .'
>
> There was a fire at the site which was said to have been due to the negligence of Commonwealth. The Insurers paid the loss to Imperial, and then sought to recover in the name of Imperial from Commonwealth under their right of subrogation. It will be seen that the facts are thus almost identical to those in the present case. The matter came before the court on a preliminary issue. The court at first instance rejected the claim on the ground that Commonwealth was fully insured under the policy. The Court of Appeal reversed the trial judge holding that Commonwealth could only claim to be indemnified under the policy to the extent of that part of the work performed under the subcontract, i.e. property belonging to Commonwealth and property for which it was responsible before the loss occurred. The Supreme Court, consisting of the Chief Justice and eight judges, restored the judgment of the trial judge. The Supreme Court stated that the main issue was as follows (at 560):
>
>> 'Did Commonwealth, in addition to its obvious interest in its own work, have an insurable interest in the entire project so that in principle the insurers were not entitled to subrogation against that firm for the reason that it was an assured with a pervasive interest in the whole of the works?'

4 *Petrofina (U.K.) and Others v. Magnaload Ltd. and Others* [1983] 3 All ER 35.
5 *Commonwealth Construction Co. Ltd. v. Imperial Oil Ltd.* [1976] 69 (3rd) DLR (3d) 558.

There was a preliminary question whether the policy was to be regarded as an insurance on property at all or whether it was an insurance against liability. The Supreme Court held that it was property insurance. Indeed there seems to have been no real argument to the contrary. At the beginning of their judgment the court described the policy as a 'multi-peril subscription policy stated to be property insurance which it clearly is'. Later the court said (at 560) '. . . given the fact that the policy is property insurance and not liability coverage the reasoning of the Court of Appeal may be summarised thus . . .

Having decided that preliminary question in favour of Commonwealth, the court went on to consider whether Commonwealth had a 'pervasive interest' in the entire property. The judgment referred to *Waters v. Monarch Fire and Life Assurance Co., Tomlinson v. Hepburn*, and other bailment cases, including a number of American cases, and then said (at 562–563):

> 'In all these cases, there existed an underlying contract whereby the owner of the goods had given possession thereof to the party claiming full insurable interest in them based on the special relationship therewith. Although in the case at bar Commonwealth was not given the possession of the works as a whole, does the concept apply here? I believe so. On any construction site, and especially when the building being erected is a complex chemical plant, there is ever present the possibility of damage by one tradesman to the property of another and to the construction as a whole. Should this possibility become reality, the question of negligence in the absence of complete property coverage would have to be debated in Court. By recognising in all tradesmen an insurable interest based on that very real possibility, which itself has its source in the contractual arrangements opening the doors of the job site to the tradesman, the Courts would apply to the construction field the principle expressed so long ago in the area of bailment. Thus all the parties whose joint efforts have one common goal, e.g. the completion of the construction, would be spared the necessity of fighting between themselves should an accident occur involving the possible responsibility of one of them.'

The *Commonwealth Construction* case is, in my view, indistinguishable from the present case, and is highly persuasive authority.'

Another example of indemnity through a legal routing is where a 'distribution of a loss is made among several wrongdoers, each compelled to pay a proportionate share'.[6] This quotation is taken from an article whose author went on to warn against confusing 'distribution' with 'contribution', with particular reference to the situation in the United States of America:

---

6 'Indemnity – A Bird's Eye View', Philip F. Purcell, in *Guidelines for Improving Practice*, op. cit., see Chapter 3, note 4, vol. IV, No. 9.

Contribution is prohibited in many jurisdictions, but exists in whole or in part in some jurisdictions by judicial decision or by legislation. For there to be indemnity by operation of law, there must be a qualitative difference between the conduct of the party seeking indemnity and the conduct of the party from whom indemnity is being sought. This means, for example, that a defendant may obtain indemnity for his liability if there is a difference in the quality of his negligence (or conduct) and that of the party from whom he seeks indemnity, such other party also having contributed to the cause of the injury for which the defendant was found liable. However, the negligence of the party seeking indemnity must be of a lesser degree than that of the party from whom indemnity is sought. This situation is most frequently referred to by the use of the words 'active' and 'passive' – the passive or lesser wrongdoer being entitled to indemnity from the active or primary wrongdoer.

Indemnity through contract

Indemnity through contract arises out of specific agreements requiring one party to indemnify and hold harmless another from all liability for damages arising out of a specific endeavour. It can be illustrated by referring to indemnity clauses in standard conditions of contract such as Clause 22 of the 4th edition of the FIDIC Conditions of Contract for Works of Civil Engineering Construction. Sub-Clause 22.1 establishes agreement between the contractor and the employer, referred to sometimes as the owner, for the former to indemnify the latter should certain events occur; see page 255 for the text of the Sub-clause.

Sub-clause 22.2 provides a list of events which are excepted from that agreement and where it is the employer who indemnifies the contractor should any of these events take place. A further example of an indemnity contract is illustrated by an insurance policy issued by an insurer promising to take over the task of compensating a sufferer on behalf of an insured wrongdoer or, in some cases, the reparation of any damages sustained.

Insurance as a contract of indemnity

There are circumstances where the extent of the obligation undertaken through a contract of indemnity is larger than the indemnifier can either afford or is capable of fulfilling, through his own resources. In such a case, party A (as referred to at the beginning of this chapter) may choose to shift the liability it has undertaken to yet another party D (usually an insurance company), and the contract of indemnity is then made in the form of an insurance policy.

To provide indemnity through insurance is to ensure that compensation is paid when a certain endeavour fails.

Due to the specific nature of insurance agreements – and construction insurance is no exception – certain legal rules apply, irrespective of the jurisdiction where the agreement is made. These rules can be distinguished throughout the wording of the three documents associated with such an agreement, namely: the proposal form, the policy and any endorsement issued, either with the policy or subsequently.

They can be summarised as follows:

1   *Utmost good faith contract:* When insurance is transacted, the purchaser of insurance makes a proposal to the insurer, sometime by the completion of a 'proposal form', designed by the insurer for the specific perils to be insured and in which the proposer is expected to answer specific questions about these perils. Unlike the purchaser of ordinary goods, who is under no obligation to give any information to the seller, the purchaser of insurance must furnish to the insurance company any information it requires and volunteer any further facts relating to the risks for which cover is being sought.

   Thus, when one purchases equipment, or any consumable item, payment of the price means entitlement to the goods. In insurance, the proposer is expected to divulge to the seller, the insurer, all available knowledge regarding the perils and the risks involved. The insurer relies on the information given to him to assess the level of risks involved and to determine the premium and the conditions which may have to be attached to the insurance policy.

   In fact, the duty of the proposer goes further than just to divulge or not to conceal. Indeed, everything that might affect the insurer's decision regarding whether or not to insure or the amount of premium necessary to cover the insurance agreement must be revealed. For this reason, insurance contracts are called contracts of utmost good faith.

   Problems, however, arise in two main areas: first, in the area of defining what is relevant information and what is not and, second, in the area of defining the difference between known facts and facts which ought to have been known. The first area can be explained by the following case taken from the legal page of the magazine *New Civil Engineer*:[7]

   A woman who was taking the contraceptive pill applied for and obtained a life insurance policy. She died and her husband claimed the £1,500 benefit. The insurers refused to pay on the ground that the woman was on the pill and should have disclosed that information. The Life Offices Association stated that women must tell insurers if they are taking the pill when they apply for a life policy. The dispute went to court and it was decided in favour of the insurers, the court holding that the information withheld was a material fact which should have been brought to the attention of the insurer.

   In the Irish courts, the information that should be given to an insurer when a proposal is made was considered in the case of *Chariot Inns Limited v. Assicurazioni Generali S.P.A. and Coyle Hamilton Hamilton Phillips Ltd.* and the conclusions are given on page 211 below.[8]

2   *No financial profit:* The insurance contract invariably includes a statement establishing and clarifying the basis of claim settlement; see Memo 2, Appendix B, Contractors' All Risks Policy. As a contract of indemnity, the insurance policy is intended to place the insured, after a loss event covered by it, in the same

---

7   'Why Your Insurers Must Know All', *New Civil Engineer*, 17 January 1974, page 32.
8   *Chariot Inns Limited v. Assicurazioni Generali S.P.A. and Coyle Hamilton Hamilton Phillips Ltd.* (Supreme Court 1981).

financial position as that which existed immediately prior to the event. Therefore, except in limited circumstances, profit is not allowed as a result of an
insured event and it is generally accepted that to allow profit would be against
the interest of society.

3   *Insurance interest:* The insured matter must bear some recognised relationship
to the insured party. This relationship must be such that a benefit continues to
accrue as long as the insured subject matter is free from injury.

4   *Subrogation:* The principle of subrogation discussed earlier in this chapter forms
a corner stone of the legal powers of the insurer unless it is curtailed in the
agreement.

5   *Contribution:* If an insured subject matter is covered against a peril for the
benefit of an insured party by more than one policy, and if that peril eventuates
into a loss, the insured cannot recover from more than one insurer. In that event,
an insurer, having paid a claim, can seek a contribution from other insurers
liable for the same loss to contribute towards the payment made.

In certain policies, a condition may exist making it contentious as to which
policy, if more than one is liable for the same loss, is made to pay first; see
Condition 3, Appendix D, Employer's Liability Insurance.

Construction insurance

Construction insurance encompasses all contracts of indemnity within the activities
of the construction industry where insurance is chosen as the medium through which
liabilities are shifted. It involves not only many branches of insurance but also many
disciplines and professions.

Thus, an insurer is expected, when dealing with construction insurance, to understand the intricate and complex problems of building, construction, engineering, and
mathematics in the form of quality measurements, probability and statistical calculations, economics, law and all aspects of insurance. He is also required to contrive
means of providing cover against the perils of construction in an effective and
efficient manner while at the same time achieving a profitable transaction. This is
expected despite 'unsound competition characterised by inadequate knowledge' of
the hazards and risks involved.[9]

A design professional is expected to enter into the controversial lush pastures of
hazards and risks, identify the acceptable and advise on what to do with the
unacceptable. Dealing with risks, whilst essential to all, is nevertheless of special
importance to the design professional in view of his function of designing safe and
serviceable projects with the minimum of risk.

A lawyer is expected to know more than law and to properly understand the
complex problems surrounding construction.

Furthermore, it is expected that all parties involved in construction insurance
should act in unison to provide correct allocation of risks and responsibilities, which
must be reflected in the contractual agreements. These agreements must also encom-
pass the allocation of liabilities and how they are to be dealt with, if they arise. The

9   *Annual Report for the 104th Year of Business 1983/1984,* Munich Reinsurance
Company, Munich, Germany.

provision of indemnity must be considered and, if required, the shift towards insurance should be carried out with the minimum of gaps and overlaps.[10]

The concept of construction insurance stems from four inherent characteristics of the construction contract which are peculiar to it and distinguish it from other types of contract. These are as follows:

1   Construction contracts include the traditional requirement imposed on the contractor to complete the works, in all but few specified circumstances, whatever the difficulties and cost. This requirement is usually stipulated in the relevant conditions of contract.
2   Vast sums are normally associated with many construction projects. In recent years, the size and cost of construction contracts have escalated to such an extent that few, if any, employers, owners or financiers can absorb the financial implications of failure.
3   An artefact of civil engineering and, to a lesser extent, of building construction is a unique object, which cannot be displayed to the buyer prior to purchase. Construction projects are, therefore, different from manufactured products and other consumer articles, see page 190.
4   As explained in Chapters 3 and 4 above, there is a complex matrix of hazards and risks that could lead to personal injury and/or physical damage during the construction period and beyond. Difficulties generally arise in construction projects due to their inherent characteristics. However, when hazards eventuate and risks materialise through events that could result in costly losses, which must be absorbed by the contractor while carrying out his legal obligation to complete the works. These difficulties could and sometimes do cripple the contractor financially and lead to disruption of the construction programme and, in extreme cases, to his insolvency. This also applies to the owner and the design team, but probably to a lesser extent, although equally detrimental.

Characteristics of the construction contract

The four characteristics referred to in the previous section are now considered in greater depth:

1 The contractor is to complete the contract

Most, if not all, construction contracts include an express or an implied undertaking by the contractor that, except in certain specified circumstances, he must complete the works and the project as a whole. With the exception of contracts of professional services, all other construction-related contracts are based on the premise that liability for non-performance of contractual obligations is a strict one. Failure to perform the required duties under such a contract would give rise to a claim for

---

10   There is also an element of risk in the assumption that an insurer will honour its commitment and compensate an insured for loss sustained from a covered peril. Insurers can refuse to pay, rightly or wrongly, or they may even become insolvent.

damages.[11] The rationale for this rule in construction contracts might lie in their common law origin, but in any case, except for specified events in the contract, the contractor is obliged to complete the contract.[12] Accordingly, Clause 8 of the ICE, FIDIC's Red Book and all other standard traditional forms of contract rooted in them stipulate that requirement. For building works, Clause 2 of both of the RIBA Standard Form of Building Contract, used in the United Kingdom, and the RIAI form of Agreement and Schedule of Conditions of Building Contract, used in Ireland, includes this obligation of the contractor to complete the contract. These conditions state or infer that, except in certain circumstances irrespective of cost or difficulties encountered, the contractor must complete the works in order to fulfil his legal obligations under the form of contract.

Where strict liability applies, why a party failed to fulfil its obligation is immaterial, and it is no defence for that party to plead that it had done its best.[13] As a party enters into contractual obligations freely, it accepts certain risks that are allocated to it and promises to bear these risks if and when they eventuate. In this way, the contracting parties are able to plan ahead with calculable certainty their schemes and arrange their business affairs. The contractor is, therefore, not excused from this obligation by pleading that he did not appreciate fully the extent or the nature of the works when tendering. Of course, the contractor may seek reimbursement of costs from the owner under the terms of contract, or may take recourse for indemnity against other parties, but his basic responsibility for completion remains unchanged.

In this connection, however, the few specified circumstances that excuse non-completion of the contract must be noted. Such circumstances include specified risks that are beyond the capability of the contracting parties, in which case the method of dealing with and managing them is also specified. Furthermore, where unacceptable risks become harder to identify or define in an explicit manner in the contract, or where the law recognises that, without default of either party, a contractual obligation has become incapable of being performed because the circumstances in which performance is called for would render it a thing radically different from that which was undertaken by the contract ('It was not this that I promised to do'), the contract is said to be frustrated.

The specified circumstances in standard conditions of contract also include, more often than not, an express term for the relief from this obligation, the most important of these being:

(a) Physical or legal impossibility: An example is Clause 13 of FIDIC's Conditions of Contract for Works of Civil Engineering Construction, known popularly as the old Red Book.[14] Another example is Clause 19.7 of FIDIC's Conditions of

---

11 Where a contract for professional services is concerned, unless the contract provides otherwise, the liability is based on the lesser requirement of exercising reasonable skill and care in the performance of the duties under that contract.
12 See page 326 below in Chapter 10.
13 *Raineri v. Miles* [1981] AC 1050, 1086.
14 Clause 13 of FIDIC's Conditions of Contract for Works of Civil Engineering Construction, 'Work to be in Accordance with Contract', 4th edition of the old Red Book, 1987/1992.

Contract for Construction for Building and Engineering Works designed by the Employer, known popularly as the New Red Book.[15]

(b) If the works are suspended for more than a specified period: An example is Sub-clause 40.3 of the FIDIC's aforementioned old Red Book.[16] A similar provision is included under Sub-clause 8.11 of FIDIC's New Red Book.[17]

(c) If any one, or a combination of a number of the employer's risks resulting in loss or damage is encountered during execution of the contract: An example of these risks can be found in the risks enumerated in Sub-clause 20.4 of FIDIC's afore-mentioned old Red Book.[18] A similar provision is included under Sub-clause 17.3 of FIDIC's aforementioned New Red Book.[19]

(d) In the case of default of the employer: An example is Sub-clause 69.1 of FIDIC's aforementioned old Red Book.[20] A similar provision is included under Sub-clause 15.5 of FIDIC's aforementioned New Red Book.[21]

2 Vast contract sums

It is generally accepted that, due to the vast cost of many major building and civil engineering projects, financial institutions, banks and governments are frequently involved in raising funds for their execution. In many cases, the cost of one project is shared by more than one of these establishments and it is generally accepted that the cost of some schemes is so high that a total loss would be financially unacceptable to any of them. It is logical, therefore, that they require guarantees towards the safety of the capital expended in the creation of the projects being financed. These guarantees would be required not only for apportioning the liability for any possible loss but also for ensuring the availability of any funds necessary to compensate for the loss and to complete the project. Such assurance is achieved through a shift in their liability towards insurance.

A point to remember in this connection is that, in some cases, it is not only important to complete the project but also to do so in a short period of time.

---

15 Sub-clause 19.7 of FIDIC's Conditions of Contract for Construction for Building and Engineering Works designed by the Employer, 'Release from Performance under the Law', The New Red Book, 1st edition, 1999.

16 Sub-clause 40.3 of FIDIC's Conditions of Contract for Works of Civil Engineering Construction, 'Suspension lasting more than 90 days', 4th edition of the old Red Book, 1987/1992.

17 Sub-clause 8.11 of the FIDIC's Conditions of Contract for Construction for Building and Engineering Works designed by the Employer, 'Prolonged Suspension', The New Red Book, 1st edition, 1999.

18 Sub-clause 20.4 of FIDIC's Conditions of Contract for Works of Civil Engineering Construction, 'Employer's Risks', 4th edition of the old Red Book, 1987/1992.

19 Sub-clause 17.3 of FIDIC's Conditions of Contract for Construction for Building and Engineering Works designed by the Employer, 'Employer's Risks', The New Red Book, 1st edition, 1999.

20 Sub-clause 69.1 of FIDIC's Conditions of Contract for Works of Civil Engineering Construction, 'Default of Employer', 4th edition of the old Red Book, 1987/1992.

21 Sub-clause 15.5 of the FIDIC's Conditions of Contract for Construction for Building and Engineering Works designed by the Employer, 'Employer's Entitlement to Termination', The New Red Book, 1st edition, 1999.

3 A Construction project is unique and not comparable to a manufactured product

A construction project is a unique artefact different from a manufactured product in many ways and therefore requires special treatment. It is different for the following reasons:

(a) Every building project and, to a greater extent, every civil engineering project is a unique creation with its own individual characteristics which differ from those of any other.

(b) There are many parties involved in the planning, design, construction and finally the use of a construction project. Each of these parties has a different role to play, a situation which can result in a conflict of interests.

(c) Before a manufactured product is completed, it is possible to carry out an extensive programme of research and development in order to eliminate any defects and errors. Such procedure is not possible in the construction of a project which must be right at the first attempt.

(d) Different materials and manufactured items are normally incorporated in a construction project, each having been produced by a different manufacturer with differing standards and performance specification.

(e) When a manufactured product is marketed, it is usual to produce an exact cost-analysis, which could be kept under control and perhaps guaranteed, at least for a certain period of time. On the other hand, the cost of a construction project cannot be guaranteed.

(f) The period of construction is long, spanning over seasonal changes during which many events may occur with effects beyond the control of the parties involved.

(g) In the design and construction of a project, many new materials and methods of construction may be employed. These materials and methods, because of their recent development and novelty, have no established performance standards.

(h) A prospective purchaser can view a manufactured product. If it does not live up to expectations, is not as promised, or is defective, it can be returned and the purchaser reimbursed its cost. A construction project, on the other hand, is usually built into or on to the ground and is functional at that single location. It cannot be returned.

4 Hazards and risks in construction

The quality of every decision taken is dependent on the quality of thought put into it. The level or risk in a decision being erroneous is a positive figure and can in most cases be assessed mathematically using various theories of probability.[22]

However, in general, as one leads a more active and demanding lifestyle, the risks to which one is exposed become greater. Similarly, in major and more complicated

---

22 This branch of mathematics is credited to the Babylonians and one channel of its development, from that beginning to a sophisticated science, was through commercial insurance calculations of risk probabilities in the early Renaissance period in Northern Italy.

schemes, the risks increase in number, intensity and consequence, despite the fact that the principles remain the same.

As noted in Chapters 3 and 4 above, the risks associated with construction are extensive and come into existence with the decision to initiate a project. They continue to grow on the drawing board and increase numerically and in intensity once work starts on site. They begin to diminish after completion of the project but they never cease. As discussed in Chapter 3 above, hazards do sometimes eventuate into occurrences and the unexpected happens. Murphy's Law is apt in this connection:

- Nothing is as easy as it looks.
- Everything takes longer than you expect, but what is more important is that anything that can go wrong will go wrong at the worst possible moment!

And perhaps, one should also remember Paddy's Laws:

1   Nature always sides with the hidden flaw.
2   The unexpected will always happen.
3   Anything you try to construct will take longer than you thought.
4   If everything seems to be going well, you don't know what is going on.

It is important, therefore, to be prepared for such occurrences and to forestall their undesirable results through good management and foresight.

When the 'unexpected' happens and the responsibilities and liabilities are resolved, the most effective way of meeting this eventuality and forestalling its effect is through insurance, provided that insurance is arranged in a manner that would lead to helping, rather than hindering, the quick resolution of the problem, getting the project back to its primordial state. The following example illustrates this point:

Assume that an uninsured building project of a total cost of £1 million suffers a total loss due to fire at a stage when the project is nearly completed and only a few days prior to handing over to the owner. The financial situation then would be that the owner would have paid to the contractor £0.9 million (assuming 10% retention) and the contractor would have received £0.9 million. He would have spent that amount to reach the completion stage (assuming 10% profit margin) but neither he, nor the employer, would have anything to show for the money spent. In fact, the situation would be even worse when one considers the delay and the consequential losses, such as costs of demolition and clearance of the site, which result from such an occurrence. Nevertheless, the owner would be still entitled, under normal conditions of contract, to get a completed building but that could only happen if the contractor were financially capable of absorbing those losses and remaining solvent.

It is prudent, therefore, when drafting a contract between the parties to a construction project, to establish where the responsibility lies for such risks and for their management. Furthermore, it is necessary to establish who would be liable for any compensation in case of accidents. It is also imperative, if indemnity is to be provided through insurance, to stipulate the cover required and to identify the party responsible for obtaining and maintaining the policies of insurance.

Definition of the word 'accident'

Construction insurance is based on the assumption that events causing damage are not deliberate or certain to occur and are therefore accidental or fortuitous. If injury or damage can be foreseen in a project and can be expected to occur as a matter of certainty, then they should be avoided by intentional planning and not by wishful thinking. The word 'accident' is, therefore, a very important term in understanding this principle of insurance and must be defined as accurately as possible. In everyday usage its precise meaning is abused and many interpretations of its definition have been attempted in a number of legal cases where the word was a pivotal factor. Some of these cases appropriate to construction insurance are quoted below to enable the reader understand the precise legal definition. In *Fenton v. Thorley & Company Limited,* Lord Lindley stated:[23]

> The word 'accident' is not a technical legal term with a clearly defined meaning. Speaking generally, but with reference to legal liabilities, an accident means any unintended and unexpected occurrence which produces hurt or loss. But it is often used to denote any unintended and unexpected loss or hurt apart from its cause; and if the cause is not known the loss or hurt itself would certainly be called an accident. The word 'accident' is also often used to denote both the cause and the effect, no attempt being made to discriminate between them. The great majority of what are called accidents are occasioned by carelessness; but for legal purposes it is often important to distinguish carelessness from other unintended and unexpected events.

This definition has been adopted in leading dictionaries.[24] It has also been applied in court cases, the most recent being in 1998.[25]

In *Trim Joint District School Board of Management v. Kelly,* Lord Loreburn stated:[26]

> A good deal was said about the word 'accident'. Etymologically, the word means something which happens – a rendering which is not very helpful. We are to construe it in the popular sense, as plain people would understand it, but we are also to construe it in its setting, in the context, and in the light of the purpose which appears from the Workmen's Compensation Act 1906 [repealed] itself. Now, there is no single rigid meaning in the common use of the word. Mankind has taken the liberty of using it, as they use so many other words, not in any exact sense but in a somewhat confused way, or rather in a variety of ways. We say that someone met a friend in the street quite by accident, as opposed to appointment, or omitted to mention some-thing by accident, as opposed to intention, or that he is disabled by an

23 *Fenton v. Thorley & Company Limited* [1903] A.C. 443, at page 453.
24 *Judicial Dictionary of Words and Phrases,* 6th edition, by Frederick Strand, Daniel Greenberg, Alandra Milbrook, Sweet & Maxwell, 2000.
25 Social Security Commissioners Decision (No. CI/105/1998) [2000] C.L.4 4845 (SSComm).
26 *Trim Joint District School Board of Management v. Kelly* [1914] AC 667, HL, at page 680.

accident, as opposed to disease, or made a discovery by accident, as opposed to research or reasoned experiment. When people use this word they are usually thinking of some definite event which is unexpected, but it is not so always, for you might say of a person that he is foolish as a rule and wise only by accident. Again, the same thing, when occurring to a man in one kind of employment, would not be called accident, but would be so described if it occurred to another not similarly employed. A soldier shot in battle is not killed by accident, in common parlance. An inhabitant trying to escape from the field might be shot by accident. It makes all the difference that the occupation of the two was different. In short, the common meaning of this word is ruled neither by logic nor by etymology, but in custom, and no formula will precisely express its usage for all cases.

From Canada, in *McCollum (R.D.) Ltd. v. Economical Mutual Insurance Company*, it was stated:[27]

In common parlance when one hears someone relate that there has been an accident it does not . . . follow that there has been no negligence involved at all. For the word 'accident' has in commonplace the significance of being opposed to a willful and deliberate act, or, short of this, one which is of so obviously gross negligence [as to be] the obvious and natural result of a most imprudent and unreasonable act.

From *Makin (F. & F.) Ltd. v. London & North Eastern Ry. Co.*[28] (The Peak Forest Canal Act 1794 empowered a company, of whom the defendant company was successor in title, to construct a canal. Section 15 thereof provided that compensation should be paid to the owner or occupier of any mill, edifice, lands or hereditaments, for injury or damage caused by the breach of any reservoir, or any of the locks or works of the canal, or by the water flowing from the canal or reservoir, or from any other 'accident'.):

When Parliament in 1794 refers to an accidental flowing of water, did it intend to exclude from that category a flowing of water caused by an event which to the ordinary person would be described as accidental, but which to the more technical mind of a lawyer might have to be put for certain purposes under the special label of 'Act of God'? In my opinion, that is a refinement which is not justified by anything in this section. It seems to me that, if Parliament meant damage caused by the accidental flowing of water or the accidental breach of the reservoir, those words are wide enough in their natural meaning to cover something caused by an Act of God. It is to be observed that the language here does not deal with the cause of the breach, nor does it deal with the cause of the water flowing or of the other

27 *McCollum (R.D.) Ltd. v. Economical Mutual Insurance Co.* [1962] OR 850 *per* Lancheville, at page 858.
28 *Makin (F. & F.) Ltd. v. London & North Eastern Ry. Co.* [1943] 1 KB 467, CA *per* Lord Greene, MR page474.

matters. It does not appear to go behind those occurrences. In any event in ordinary language the word 'accident' seems to me without any doubt to cover an Act of God. Whether the bank of the canal is broken by an ordinary or by an extraordinary flood, the sufferer has suffered damage as the result of an accident. Any such sufferer, if asked whether he had received damage as the result of an accident, would scarcely, in the latter case, say: 'No, I have not suffered as the result of an accident. I have suffered as the result of an Act of God, because this particular flood cannot properly be described as an accident'. I do not see why a refinement of that kind should be introduced into clear and simple language of this sort.

From *Mills v. Smith*, it was stated:[29]

The word 'accident' is difficult to define, indeed, one judge very many years ago said the word was undefinable. The dictionary definition given in Murray's *Oxford Dictionary* is (and I leave out immaterial meanings) 'Anything that happens without foresight or expectation or is an unusual effect of a known cause'. The application of that definition depends almost entirely on the point of view from which the particular matter is approached. If, quite unexpectedly, someone coming up to me, hits me in the face and gives me a black eye, the event, so far as I am concerned, is quite unexpected, yet I would not say that I got it by an unprovoked assault. Nor would it be an accident so far as the attacker is concerned. Yet, under the Workmen's Compensation Acts [repealed; cf. now National Insurance (Industrial Injuries) Act 1965, s. 5(1)] an injury by an unprovoked attack has been decided to be an injury by accident for the purpose of those Acts.

The insurability of risks

Not all risks are insurable and while the principle of the equitable contribution of many for the benefit of an individual suffering a loss is the corner-stone of insurance philosophy, certain limitations must, of necessity, be put on that principle to make the insurance transaction viable.

The limitations are as follows:

1   The principle of insurance is based on the theory of probability and, therefore, there must be an element of uncertainty relating to the matter to be insured, i.e. accidental or fortuitous in character (see definition of accident above).

2   An insurable risk should preferably be measurable in quantitative terms and in such a way that the theories of probability and the law of large numbers may be used. Without this stipulation, the premium required to insure the risk could not be scientifically calculated. Insurance becomes lottery in the absence of such calculations. It is, however, important to note that, if the extent of the risk is unquantifiable, it is the assessment of the premium and not the insurability that is in question.

29   *Mills v. Smith* [1963] 2 All ER 1078, *per* Paul, J.P. 1079.

3　An insurable risk should preferably be such that it is acceptable to the insurance market through appropriate risk selection methods. The objects insured must be numerous enough and homogeneous enough to allow sufficient selection.

4　An insurable risk should preferably be such that one can determine whether loss has in fact occurred and the cause of the resultant damage. The extent of the damage should also be capable of assessment.

Fortunately, most risks in construction contracts fall within the limits set out and are, therefore, insurable.

Uninsurable risks

When risks fall outside the limits indicated in the preceding section, they are normally uninsurable. The word 'normally' is used here to indicate that if the premium is high enough and/or if the insurer is adventurous enough then the limits set out above diminish.

It is important to note that the responsibility and liability for damage to property and/or personal injury emanating from uninsurable risks must be clearly defined in any contract. Examples of such risks are given below, classified in accordance with the four categories previously outlined:

1　*Foreseeable risks:* An insurer will argue that if a contractor stores cement in an uncovered condition during a rainy season, then any damage caused is foreseen to be inevitable and, thus, is not the liability of the insurer. On the other hand, if the cement was stored in a watertight shed and the roof of the shed blows away under severe wind, then the contractor will argue that this is unforeseen damage.

2　*Unquantifiable risks:* A consequential economic risk is unquantifiable, even in a certain circumstance. It is, therefore, very rarely covered. However, the word 'consequential' must not be confused with 'consequence' as in risks resulting as a consequence of defective design, material and/or workmanship because these risks are quantifiable and their limit is the value of the contract which is insured. Such damage resulting from, or occurring as a consequence of these defects is insurable and the intention of a good insurer must always be clear in this respect. Insurance policies must be written in clear and precise language at all times but more especially so when dealing with this issue because, otherwise, it could result in a dispute if repair to a resultant damage is costly. See page 349.

3　*Political risks and risks on an international scale:* War is a good example of these risks that are normally uninsurable. The reason is that the principle of the contribution of many for the benefit of an individual suffering loss breaks down in such a situation, unless governmental institutions carry out the insurance.

4　*Causation:* To prove the cause of any damage on a project is to establish the responsibility and liability for it and to establish whether or not the damage is covered through the provisions of the insurance contract. If such a cause cannot be proven for any particular risk, the risk becomes uninsurable.

Figure 6.1 illustrates graphically the insurability element of the risks in a construction contract. It shows at a glance the relationship that exists between the Construction Trinity, risks and insurability.

Insurance policies required in construction

The type of insurance generally required in connection with a construction project can be divided into two basic categories. The first is property insurance; the second is liability insurance. In international construction, marine insurance may also be required. Construction insurance has been traditionally transacted by issuing a number of insurance policies for the benefit of each of the parties involved in the particular project under construction. The insurer in this case is more likely to be more than one insurance company. Figure 6.2 shows the more important policies usually issued by each of the parties in connection with construction.

Other insurances are also required to provide cover for the ordinary day-today business and other activities but these are not dealt with here.

Property insurance

This insurance mainly provides protection to the works and any material, equipment and machinery connected with it. It is generally transacted through what has become

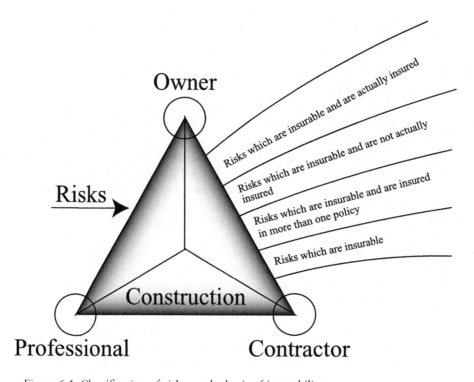

Figure 6.1 Classification of risks on the basis of insurability.

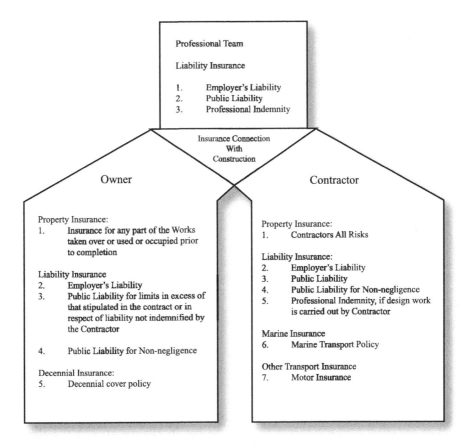

Figure 6.2 Insurances which may be required on a construction project.

known as Contractors' All Risks Insurance Policy or Erection All Risks Insurance Policy. Despite the All Risks tag on the policy, the insurance cover in either of these two policies excludes a number of risks and has a long list of conditions that must be met. In general, however, unless a risk is specifically excluded from the policy, it is considered to be included in the cover provided. It is in this negative approach that one can identify the extent of insurance cover provided.

Liability insurance

Liability insurance is intended to provide protection to the insured party against specific legal liabilities to which he may become exposed as a result of activities culminating in bodily injury and/or property damage.

The legal liabilities may be towards employees, in which case Employers' Liability Insurance would apply, or towards third parties who are not partly to the insurance contract, in which case Public Liability Insurance would apply.

In the case of the design professional, legal liabilities incurred in the course of his professional work are covered under Professional Indemnity Insurance.

Generally speaking, legal liabilities are incurred as a result of negligence and lack of care (see Chapter 5 above). However, in certain circumstances, insurance is required even when negligence has not been committed. Such insurance is also transacted within the liability type of insurance.

Non-negligence insurance

As stated earlier, and generally speaking, legal liabilities hinge on negligence but in fact two higher standards of liability exist in law. These are, in ascending order, strict liability and absolute liability, see page 149. In principles of law where strict liability applies, it is not sufficient for a party to absolve himself from liability by simply taking reasonable care. Strict liability would apply in construction, for example, due to statutory obligations or under the principle of the *Rylands v. Fletcher* in common law jurisdictions, see page 149.

In such circumstances, if the contractor is found to be non-negligent in respect of damage or loss but instead the owner is found to be responsible and is held strictly liable, then insurance would only apply if the non-negligence element of the risk is specifically included in the cover.

This type of risk to which the owner is exposed was highlighted in the case *of Gold v. Patman and Fotheringham,* in 1958.[30] The details of the case were as follows:

> By Clause 14 of the then current RIBA form of contract, the contractor undertook to indemnify the owner against claims in respect of damage to property arising out of the works, provided it was due to the negligence or default of the contractor or his subcontractors. By Clause 15, he was required without prejudice to this liability, to effect or cause any subcontractor to effect such insurances as might be specifically required by the Bills of Quantity. The bills contained a provision as follows:
>
> > 'The Contractor is to insure or make payment in connection with the following: (b) Insurance of adjoining properties against subsidence or collapse.'
>
> Without negligence on the part of the contractor or his subcontractors, bored piles sunk by specialist subcontractors damaged the adjoining property. The contractor, under a liability policy, had insured himself against this risk, but the adjoining owners not surprisingly preferred to sue the building owner, under the unqualified indemnity terms of party-wall awards previously made in their favour, and not the contractor. In an action against the contractor, the building owner contended that he had been in breach in failing to take out an insurance policy for the employer's benefit which would have covered the claim. Held, by the Court of Appeal, overruling Gorman J., that the contractor's obligation, on a proper interpretation of the provision of the bills, was only to insure himself, and not the building owner.

---

30 *Gold v. Patman and Fotheringham* [1958] 2 All ER 497, see also: *Hudson's Building and Engineering Contracts,* 11th edition, by I.N. Duncan Wallace, Sweet & Maxwell, 1995, 15–024 at page 1436.

Thus the owner found himself to be held liable by the courts in respect of damage against which he is neither insured directly nor indirectly through the contractor's insurances.

Confusion struck those involved in the RIBA Form of Contract and a gap in the insurance cover became obvious. To fill the gap, a clause incorporating a very wide cover was quickly drafted for inclusion in the 1963 RIBA Form of Contract. This clause took the insurance market in the United Kingdom by surprise and created some problems which necessitated a further revision of the insurance requirements of that document and was finally carried out with the help of the British Insurance Association.[31] The amended clause was incorporated in the 1968 edition of the RIBA Form of Contract requiring the contractor to take out and maintain an insurance policy in the joint names of the owner and the contractor providing the following cover:

A Any expense, liability, loss, claim, or proceedings which the owner may incur or sustain by reason of any damage to any property other than the works and as a result of carrying out the works and caused by the following risks:

   (a) Collapse;
   (b) Subsidence;
   (c) Vibration;
   (d) Weakening or removal of support;
   (e) Lowering of ground water, either temporary through pumping and any other constructional method, or permanent by installation of drainage or similar facilities.

B The sum insured is decided by the owner and is included in the contract documents.

C The following exceptions became acceptable and applied in the insurance policy transacted:

   (a) Damage caused by the negligence, omission or default of the contractor, his servants or agents or of any subcontractor, his servants or agents;
   (b) Damage attributable to errors or omissions in the designing of the works;
   (c) Damage which can reasonably be foreseen to be inevitable having regard to the nature of the work to be executed or the manner of its execution;[32]

31 'Insurance and Indemnity Clauses in Contracts and Standard Form of Agreements', Peter Madge, *Arbitration Journal*, July 1974.

32 Despite the use of difficult words such as 'foreseen' and 'inevitable' in the wording of this exception, the procedure, which seems to have become the normal practice in the issue of such a cover, leads to minimal potential for dispute between insurer and insured regarding their meaning. This procedure involves the insurer in arranging for an expert or experts to visit the site and to examine the drawings and specifications and report to the insurer as to the risks involved in the proposed construction. Any foreseen and inevitable damage, which can be identified by the insurer's experts, is then transmitted to the insured requiring him to take the necessary action to rectify the situation and in the alternative excluding such identified risks from the cover. Thus, the interpretation of 'foresee' and 'inevitable' becomes a foresight and not hindsight. This procedure may also be the reason why there have been very few claims made in respect of this cover.

(d) Damage which is the responsibility of the owner under the provisions of the contract;
(e) Damage covered under any other policy;
(f) Damage due to risks normally excluded from an insurance cover such as war, nuclear fuel, pressure waves, etc.;
(g) Personal injury.

The cover as detailed above was more acceptable to the insurance market than its predecessor as It was in line with the public liability cover already provided for the contractor. Now, the cover became extended to the employer/owner, an extension which was provided under a separate insurance policy. It is important therefore to emphasise that this cover is issued for the protection of the employer/owner and not the contractor and that it is required to be issued in the joint names of the contractor and the employer/owner for the following reasons:

1   It is preferable, if not essential, that the insurer of all the liability insurance policies, whether they include negligence or non-negligence, be the one and same insurer. If this is not the case, future events involving loss or damage may end up in a dispute between the two insurers. The insurer of the Public Liability Insurance (based on negligence) may repudiate the eventual claim on the basis that the cause of the loss is a non-negligent act and should, therefore, be covered by non-negligence policy. The situation may also be the reverse and in both cases a dispute may arise between the two insurers causing delay in settlement of the claim with all the disadvantages of such a situation. Even when one insurer is involved, there may be an agreement as to whether a loss is covered under one policy or the other in the case of there being a large excess imposed in the non-negligence policy. The impact of such an argument, however, is potentially less damaging and problematical than a dispute between insurers.

2   In order to properly assess the risks involved in non-negligence insurance, the insurer usually requires to study the method of construction to be adopted by the contractor.

3   The contractor who ought to be in control of the site and its activities should be a party to the insurance agreement in order that any conditions stipulated are abided by.

The situation changes if the employer/owner is the party responsible for taking out the insurance cover. Details of the insurance cover obtained and the conditions imposed by the insurer should then be made part of the tender documents. However, there is doubt in the minds of some eminent experts about the value of this cover.[33]

---

33 See *Hudson's Building and Engineering Contracts*, 11th edition, by I.N. Duncan-Wallace, Sweet & Maxwell, 1995, 15–024 at page 1437 and Mr. Duncan-Wallace's book entitled *Construction Contracts, Principles and Practice*, Sweet & Maxwell, 1986, par. 29–27 and 30–23.

Decennial insurance

Decennial insurance is generally transacted to cover the liability of those involved in construction for latent defects in the stability of the structure and for major defects in the weather shield for ten years. The ten-year cover matches the limitation period in respect of the stability and major defects in the structure or of an important part thereof in certain jurisdictions. Such liability is called decennial liability '*responsabilité décennale*', and exists mainly in jurisdictions where the civil code forms the basis of the legal system. The exact definition of the liability, whether contractual, or tortious, or both, and the parties to which it is attached differs from one jurisdiction to another and in some cases it is updated and refined as frequently as the changes in construction practices demand. For example, Article 1792 of the French Code Civil which deals with this topic was updated to its present format in January 1979, replacing the earlier version which itself came into force in January 1968.[34] The present Article 1792 states:

> Every constructor of a structure is legally responsible to the owner or those deriving title from him for any damage (including damage resulting from sub-soil conditions) which jeopardises the integrity of the structure or which by affecting one of its component elements or one of the equipment elements renders the structure unfit for its intended purpose.
>
> Such responsibility will not be imposed where the builder demonstrates that the said damage results from causation outside his authority and control.

The article classifies in its Sub-articles the definition of 'constructor', as follows:

1792–1. The following are deemed to be constructors of a structure:

(i) Any architect, contractor, technician or other person bound to the owner of the structure by a works contract
(ii) Any person who sells after completion a structure which that person has constructed or caused to be constructed
(iii) Any person who although active in the function of agent of the owner in fact performs a function which absorbs that of a person bound to the owner by a works contract.'

Sub-articles 1792–2 and 3 incorporate equipment in the liability net and sub-article 1792–4 deals with manufacturers, as follows:

1792–2. The presumption of responsibility established by Article 1792 extends equally to damage affecting the soundness of equipment components of a building, but only when the said components form an indissociable fixture of the service works, foundation works, or frame whether boundary enclosure or cladding.

---

34 'Liability in the French Construction Industry – Towards a Turnkey Approach', by G.L. McIlwaine, a paper presented at the IABSE Henderson Colloquium, Cambridge, 1984, and published in association with the Institution of Structural Engineers, under the title 'Liability', in 1985.

1792–3. The equipment elements not covered by 1792–2 above shall be the subject of a guarantee of good performance which shall be of a minimum duration of two years commencing from the date of reception of the structure.

1792–4. The manufacturer of a structure, or part of a structure or of an equipment component designed and produced when in service to meet precise and predetermined criteria, is severally responsible for the obligations laid down by Articles 1792, 1792–2 and 1792–3 which are otherwise borne by a person bound to the owner by a works contract who has installed without modification and in conformity with the specifications laid down by the manufacturer such part of the structure as is under consideration.

Articles 2270 and 2820 deal with the ten-year period of liability and the insurances required, respectively.

Needless to say, other jurisdictions have different wording but essentially they concur on the ten-year period. Examples are: the Iraqi Civil Code, Law No. 40 of 1951, amended by Law No. 48 of 1973, Article 870; the Egyptian Civil Code, Article 651; the Saudi Arabian Royal Decree M14 of 27 March 1977; the Italian Civil Code, Law No. 1086 of 5 November 1971; the Spanish Civil Code, Article 1591; the Belgian Civil Code Articles 1792 and 2270; the Dutch Civil Code, Article 1645; and the Venezuelan Civil Code, Article 1637.

The decennial insurance cover has been designed to be compatible with decennial liability and the word 'structure' is intended to include not only buildings but also large works –'*gros ouvrages*'- incorporated in construction activities.[35] The period of inception of decennial liability and therefore decennial insurance is defined as the date of the acceptance of the work by the client -'*reception*'. Such acceptance can be either the provisional acceptance or the final acceptance, after the completion of the Defects Notification Period, depending on the legal text but, in general, it begins from the latter date. The legal text also differs from one jurisdiction to another on the question of whether parties can agree to restrict that liability in any way. In Belgium, for example, it is not possible to restrict the amount or the period of decennial liability, as it is considered in the domain of public policy, whereas it is possible in Holland. It is also important to note that ordinary tortious liability involving the general duty of care may in some jurisdictions extend beyond the ten-year period such as is the case in Belgium where it extends to thirty years.

Therefore, where the decennial liability is neither restricted in amount nor in scope, the decennial insurance policy is of most benefit as it is issued in the name of the owner providing an immediate indemnity without awaiting either the allocation of responsibilities in respect of the damage or the outcome of any litigation which might ensue as a result. A subrogation clause is usually included, however, entitling the insurer to recover in the name of the insured from a negligent and liable party. It is claimed by some insurers that the subrogation clause can be omitted against parties nominated by the insured but this is an expensive additional cover.

---

35 'Memorandum on Professional Liability of Consulting Engineers in Belgium', by Vera Van Houtte, a paper prepared for FIDIC, April 1985.

The sum insured under the policy represents the full cost of rebuilding the premises or the project. It is sometimes possible to include for a certain indexation of the sum insured and it is understood that a cover for up to 10% compound rate per annum can be obtained. Additional sums may also be added against additional premium, such as professional fees, cost of removal of stock and machinery and cost of removal of debris.

The decennial insurance cover cannot, however, extend to wear and tear, or to any maintenance cost or minor defects and, in most policies, the principle of accidental damage is a prerequisite to indemnity. During the period of insurance, the benefits under the policy will transfer to any subsequent purchaser of the property or project.

Generally, there are two categories of exclusions attached to the policy which include risks that can be classified under the following headings:

- Uninsurable risks such as war, radioactivity, etc.
- Other risks, such as fire, which are normally covered by special insurance policies.

An important aspect of decennial insurance is the technical control service provided in association with the cover. The technical control service is provided by firms which specialise in this field with the task of enabling the insurer to mitigate the risks insured under the decennial insurance policy to an economically acceptable level for both the insured and the insurer.

The specialist firm is usually appointed by the owner, preferably before or during the design stage of the scheme and certainly prior to the constructions stage, since some of its functions start during that period. In most cases, it cooperates with the design team complementing their role through risk assessment and mitigation.

The technical control services offered by the specialist firm take the form of the following activities:

1 Checking the design assumptions, calculations, drawings, and specifications: this function extends to examination of the basic data and investigations carried out to ensure correct choice of foundations, fabric of the structure and external elements. The loading assumptions are examined with respect to safety, serviceability and durability of the scheme. Drawings, including workshop details, are checked and approved.
2 Site inspections: spot checks and site visits are made to ensure that execution of the works is carried out safely and in accordance with the drawings and specifications.
3 Material testing and quality control: testing of strength and performance of materials and components used in the construction is carried out with the assistance of testing laboratories.
4 Provision of reports during the execution of the works and finally a Certificate of Risk Assessment at the end of the construction period.

Needless to say, this type of insurance is an expensive protection which may cost up to 2% of the value of the project for the ten-year period of cover.

Overlaps and gaps

If the risks in a construction project are insured through the provision of a number of different policies issued for each of the parties involved including owner, design professionals, contractors and subcontractors, the number of policies issued would be extremely high. Besides the difficulty in checking these policies for errors and omissions, there remains the problem of the overlaps and repetition in parts of the insurance cover. These develop because of the use of standard insurance policies, each designed to give partial cover and protection against a few of the risks discussed in Chapter 3.

As well as overlaps, gaps in the insurance cover also emerge especially where none of the insurance policies cater for a specific risk. Figure 6.3 shows a diagrammatic display of the situation where multiplicity of policies leads to overlaps and gaps. The two main problems in having an overlap in an insurance cover are:

1   Where there is an overlap, a premium has been paid more than once.
2   There is potential for a dispute between the various insurers who have provided the cover in the case of a loss resulting in a demand from the insured for indemnity.

The problem with a gap in the insurance cover is of course that there is no insurance in respect of the risks represented by that gap. However, the situation becomes

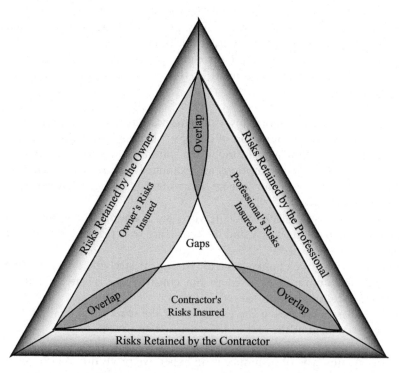

Figure 6.3 Retained and insured risks.

critical where the gap is not recognised by the insured parties when the insurance cover is taken out or when it is in operation.

A list of possible gaps which usually exist in the insurance arrangements as they are currently transacted is given in Chapter 14. However, a brief outlive of these gaps is given hereunder:

A   Gaps through uninsurable risks;
B   Gaps due to lack of cover, either in the insurance practice or through the wish of the insured;
C   Gaps due to the use of the conventional method of providing insurance.

Bonds and guarantees

Although bonds and guarantees are not insurance policies, they will be dealt with in this chapter to highlight the differences which exist between the two types of documents. These are fundamental differences, the most important of which is the fact that an insurance policy is a contract between two parties whereby the insured pays a premium to the insurer for the privilege of transferring the obligations, in respect of certain risks, to him. The bond is not a contract by itself since it rests on a latent contractual obligation between two parties. The surety acting for one party against a fee will, in the event of non-performance by that party, step in and fulfil the obligation. The responsibility passes to the surety only if the party in question is unable or unwilling to fulfil his responsibility and the surety may then pursue that party for recovery of his outlay so it is not an insurance for the party in default. The bond remains in force until the obligation of the contracting party in whose name the bond is issued is completely fulfilled. There is no cancellation clause and no termination date within the performance period. However, it is important to check that the form of bond issued does not include a clause permitting its cancellation in the case of variations to the contract between the owner and the contractor.

There are six main forms of bond and guarantee transacted within the construction industry. They are issued in the format that the issuer will pay an agreed sum when the beneficiary states that default had occurred. These are:

1   *Bid bonds:* Bid bonds are intended to assure the beneficiary that the bid or tender is a serious one and that, if it is accepted, the tenderer will proceed and effect the form of contract including whatever subsequent bonding arrangements he is required to provide.

2   *Performance bonds:* Performance or supply bonds are issued for the completion of the contract. Unless it is specifically stated, the bond is issued for the completion of the contract and not for its proper completion. This is the reason why keeping the bond beyond the contract period is of no value to the owner, unless the usual format of bond is changed to include the word 'proper', or such similar phraseology to incorporate the standard of performance into the bond. If it happens, however, that the contractor becomes insolvent during the construction period and it is then discovered that defective material and workmanship has been incorporated in the works, the cost of removing that work and replacing it with proper work comes within the responsibility of the insurer under the bond and subject to its limits.

3   *Advance payment guarantees:* These guarantees are issued to assure the beneficiary that any sums of money advanced will not be lost through default or poor performance by the party in receipt of the advance.

4   *Retention money bonds:* This type of bond is issued to allow the release of retention money usually held by the beneficiary.

5   *Maintenance bonds:* Maintenance bonds are issued to guarantee that as soon as the installation is completed, the contractor will fulfil his obligations throughout the commissioning and testing periods.

6   *Company suretyship:* This type of surety is transacted when a party acts as a guarantor for another in the lease purchase of equipment or similar. This transaction is extremely perilous for the surety since the beneficiary in this case is normally a finance company which has perfected a standard form of agreement in which the guarantee is made effective in respect of past and future transactions, irrespective of whether or not the surety is made aware of them. In a recent case in Ireland, it was held that such a transaction is a legally valid one but it was at the same time described as three blank cheques, one for the past, one for the present and another for the future.

7

THE INSURANCE SCENE

In a day and age when insurance has become an integral part of practically every business activity and when society has built into the legal system the benefits and effects of insurance, it is imperative for those responsible for decision-making to know and understand the structure of the insurance market. Furthermore, it is important for them to know and understand the relationship and interplay between the various authorities who control the insurance scene, how insurance is transacted and operated and who derives the benefits generated or the losses suffered.

As we are naturally more concerned here with construction-related insurance, it is important to identify those who are ultimately responsible and liable for the indemnity specified in the construction contract. It is also important to visualise the insurance scene and how it merges with the inherent risks in construction.

Structure of the insurance market

The structure of the insurance market has developed in most parts of the world in a way that three distinct layers of activity may be identified, each characterised by the function it performs.[1] At the base lies the function of handling and placement of risks within an insurance package which is performed by a broker, an agent or a sales department within an insurance company. The next layer is represented by the insurer, who performs the function of accepting and underwriting the contract of insurance.[2] The insurer, through underwriting insurance, exposes himself to concentrations of risk in both quantity and quality and seeks protection by ceding some part of the risks he has accepted through either co-insurance with other insurers or reinsurance with other companies. The latter function, forming the third layer of activity in the insurance market, is performed by reinsurance companies, which accept either a direct proportion of the risks accepted by the insurer or a certain identifiable section of these risks.

---

1 In some parts of the world, such as in Sweden, insurance brokers are not permitted to operate in handling and placing insurance.
2 The term 'underwriters' was originally given to insurers who signed their names at the foot of a policy showing the number of lines (shares) they had accepted out of the total sum insured.

The broker

The major portion of handling and placing of risks is carried out by insurance brokers, agents and insurance consultants. Insurers usually employ insurance agents, individually or collectively, and as such they represent one insurer or a number of insurers. The advice they give is therefore guided by the services offered by the companies they represent. Insurance brokers on the other hand act normally as agents of the insured to whom they offer their services as experts in the field of insurance generally or in certain fields in which they specialise. Their standing and qualifications differ from one country to another depending on whether control is exercised by the state (such as is the case for brokers in the United Kingdom, who are regulated under the Financial Services and Markets Act and similarly in Ireland under the Investment Intermediaries Act 2000), or by a trade association or not at all. Although brokers and agents draw remuneration for their services from the insurers through commission, they are expected to act on behalf of the insured and to the best of his interests in both functions of transacting insurance and handling any claims which may arise under the policy. A contract between the insured and the broker is made as soon as the insurance is placed through the broker enabling him to draw a commission, which forms the necessary consideration to complete the requisites of a contract. The commission paid is usually calculated as a percentage of the premium charged to the insured and, in general terms, it works out at around 10% to 25% covering expenses and profit. This commission is also paid if and when the policy is renewed. For this commission, the broker accepts the responsibility of, amongst others, payment of the full premium to the insurer, irrespective of whether or not he actually collects it. It is worthy of note that it is becoming more common for brokers to forgo the insurer's commission and negotiate a fee for their services with the insured.

It is important, however, to note that the broker does not take part in the indemnity contract transacted and does not accept any of the risks which form the subject matter of insurance.

In addition to the above responsibility, the broker undertakes, in principle, certain duties towards the insured, which can be summarised under the following four categories:

1 The right insurer and the appropriate policy

The broker is expected to know the insurance market and be familiar with the wide range of insurance companies active in the various branches of insurance. The broker should also be aware of the quality of the various insurers in view of the fact that the insurance contract is one of utmost good faith and total trust and because the quality of insurance can only be discovered when the insurer is called upon to pay a justifiable claim. It is then that the insured finds out whether or not the insurer is worthy of the trust placed in him. If an error had been made, the remedy can be lengthy and crowded with lawyers. It can also be divesting. In *Osman v. J. Ralph Moss Ltd.*, the Court of Appeal in the United Kingdom held that the brokers were negligent in, amongst other things, advising the plaintiff to insure with a company which was generally known to be in financial difficulty and which subsequently became insolvent.[3] The broker is further expected to be capable of dealing with the complexities of the insurance policy with all its sections, variations and fine print. In some circumstances, the most appropriate format for dealing with a particular situation may not be that of a standard policy which is usually used. The broker is, therefore, expected to understand the basic principles of insurance and to know the answers to questions appropriately identified in the following quotation from Rudyard Kipling:

> I keep six honest serving men
> (They taught me all I knew);
> Their names are What and Why and When
> And How and Where and Who.

In the selection of the insurer and the policy conditions and wording, the broker must act in the best interest of the insured having regard to the extent to which the insurer is prepared to go.

2 Assessment of risks and needs, and effecting insurance to suit

The assessment of risks and needs of the broker's client is a function of the broker who ought to be versed with its methods. The client may also need to be advised on the risks he should retain and those against which he should insure. But once the risks are identified, it is essential to ascertain that the insurance policy is issued in a way that accurately describes these risks and the subject matter of insurance. If this is not done, one may find when claims arise under the policy that indemnity is denied by the insurer because of a fault, or an error or an omission in the wording. Examples of this situation are plentiful, ranging from the trivial omission of a material fact to the blunderous misdescription of the insured's business and the operations he intends to carry out.

The plaintiff in *McNealy v. The Pennine Insurance Company* was a property repairer who worked also as a part-time musician.[4] He obtained a motor insurance policy through an insurance broker who filled the proposal form for him on the basis of information he had. When asked about his occupation, the plaintiff replied 'property repairer' and the broker did not enquire any further despite the fact that the particular policy chosen excluded a number of persons, one of which was 'whole or part-time musicians'. This fact was not discovered until an accident occurred involving the plaintiff whose claim for the loss was refused. The Court of Appeal in the United Kingdom held that the brokers had failed in their duty in that they had in their possession a leaflet setting out all the categories of excluded persons. The brokers should have gone through the leaflet with the plaintiff to ensure that he did not fall into any of the excluded categories.

The broker is furthermore expected to review the client's risks and requirements at renewal dates and, as they change, the insurance requirements change too.

In this respect, brokers have been innovative and have shown the way ahead in many instances such as pressing insurers to provide new forms of insurance (for

---

3 *Osman v. J. Ralph Moss Ltd.* [1970] 1 Lloyd's Rep. 313.
4 *McNealy v. The Pennine Insurance Company* [1978] 2 Lloyd's Rep. 18.

example, insurance to cover the loss of profits as a result of fire) and in premium calculations (reduction of premium to those who provide fire precautions). However, for these innovations to be a success, they must conform with the principles of insurance. If they do not, a disservice can result.

3 Sum insured and premium

The value of the items to be insured and the negotiation of the premium to be paid are other aspects on which the broker can advise. Whilst it is accepted that competition is generally beneficial and necessary for the progress of man, it can produce the opposite effect in the provision of services. Ruthless competition for the mere purpose of obtaining business can lead to failure and has to be viewed critically, particularly in the case of insurance. The insurer may become insolvent if his methods of premium calculation produce inadequate results. In construction, if the insurer is insolvent or if he does not honour a justifiable claim, the owner may end up with a silent site and/or the subject of unsatisfied judgments against him for personal injury or damage to third party property. Price alone should not be allowed to be the determining factor in the decision to accept an insurance cover.[5] See *Osman v. J. Ralph Moss Ltd.*, referred to above in note 3.

4 Claim settlement

The broker is also expected to help in the submission of claims when hazards eventuate and risks materialise. In carrying out his duties and responsibilities, the broker is expected to exercise due care, skill and diligence. If he fails in this respect, he may be held liable in contract or in tort or in both, in the same manner that a professional adviser may be held liable. It is worthwhile mentioning that, despite the fact that the broker is paid by the insurer, his first duty is towards the insured, and in the case of joint insurance the duty extends to all named as insured in the policy.

The insurance broker is also expected to know the legal interpretation of misrepresentation or concealment of a material fact by an insured when seeking insurance or completing a proposal form. He is expected to be able to judge whether or not certain facts are material to a risk against which indemnity is sought and to know that material facts must be disclosed to the insurer. To make such a judgment is not an easy task, especially when even judges differ, as happened in the case of the *Chariot Inns Limited v. Assicurazioni Generali S.P.A. and Coyle Hamilton Hamilton Phillips Ltd.*[6] The facts of this case were considered first by the High Court in Ireland when it was held that the insurance policy in question was valid, but when

---

5 The collapse of the Independent Insurance company in London in 2000 has given rise to uninsured losses the full extent of which will not be known for many years to come and will have significant impact on certain construction projects.
6 *Chariot Inns Limited v. Assicurazioni Generali S.P.A. and Coyle Hamilton Hamilton Phillips Ltd.* (1981).

the insurance company appealed to the Supreme Court, the appeal was allowed and the action against the insurer was dismissed. At the same time, the Supreme Court went on to hold that the brokers were liable to the plaintiff owners both in contract and in tort.

The facts of the case were that in January 1976, the plaintiffs, Chariot Inns Ltd., bought the licensed premises Chariot Innin Dublin. They then decided to extend the premises with the result that it was necessary to store some of the furniture in another premises, 82, Lower Leeson Street, partly owned by the plaintiffs. The insurance brokers for the plaintiffs were Coyle Hamilton Hamilton Phillips Limited, the second defendants, and their employee, Mr Hart, placed the fire insurance of the premises where the furniture was stored with the Sun Alliance Insurance Group. The policy was endorsed to cover the furniture brought in from the licensed premises.

A fire occurred in 82, Lower Leeson Street in April 1976 and a claim was made in respect of the damage to the premises and to the furniture.

When the premises of the Chariot Inn was bought, it was insured with the General Accident Insurance Company Ltd. but later, at renewal date when an increase in premium was sought, the plaintiffs instructed the brokers to place the insurance with the first defendants. Mr. Hart filled the proposal form on behalf of the insured. Having asked the questions stated in the form and filled the answers according to the information given, he did not feel that it was necessary to disclose the fire in 82, Lower Leeson Street in the proposal form. An insurance policy was issued in due course.

Fire occurred at the Chariot Inn in May 1978, causing extensive damage and a claim was lodged in June 1978, but the insurers repudiated liability on the basis of non-disclosure of the fire accident in 82, Lower Leeson Street. The owners then sued both the insurers and the brokers. At the High Court, the judge declared that the policy issued by the insurance company was valid and that the claim against the brokers could therefore be dismissed. He did, however, mention the potential liability of the brokers should his decision be reversed by the Supreme Court, in the event of an appeal. The insurance company appealed to the Supreme Court, and Mr Justice Kenny considered the question of what constitutes a material fact with respect to a risk against which an insurance cover is sought.

He came to the conclusion that it is not what the person seeking the insurance regards as material, nor is it what the insurance company regards as material. It is a matter or circumstance which would reasonably influence the judgment of a prudent insurer in deciding whether he would take the risk and, if so, in determining the premium which he would demand. The standard by which materiality is to be determined is objective, not subjective.

He thought that whilst it was not conclusive for the fire in 82, Lower Leeson Street and the damage to the plaintiffs' goods to be considered material facts, it was, however, a material fact that goods belonging to the Chariot Inn were damaged by fire in premises partly owned by the plaintiffs. He held that the broker owes a contractual duty to his client to possess the skill and knowledge which he then is required to exercise. He also held that the broker is liable in tort if he failed to exercise that skill and knowledge, and therefore Mr Hart should have known that the fire at Leeson Street, and the claim payment made, were material to the risk which was offered to the defendant-insurer to cover by insurance.

The courts in the United Kingdom seem to have always held in accordance with the principle that an insurance broker can be liable both in contract and in tort. One of the earlier legal cases where brokers were held to be 'liable to their clients both in contract and tort' was *Strong and Pearl v. Allison and Co. Ltd.* in 1926.[7]

This principle is also accepted in Canada and one of the important cases there was *Wilcox v. Norberg and Wiggins Insurance Agencies Ltd.* in 1979.[8] The view in Australia, however, seems to be different, as demonstrated by *Ogden & Co. Pty. Ltd. v. Reliance Fire Sprinkler Co. Pty. Ltd.* in 1975, where it was expressed that the existence of a contract between the parties precluded any liability under the tort of negligence.[9]

The liability of brokers extends also to the insurer if they fail to release relevant information held in their possession. It is unusual, however, to find that there is any other party who is owed a duty by the insurance broker but this cannot be totally excluded.

The liability in respect of the four aforementioned categories of duties owed by an insurance broker is divided into two levels: first, one which embodies the provision of a professional service with an implied requirement to exercise reasonable skill and care and, second, one which involves the clerical and routine services he performs imposing the higher level of strict liability. The latter function forms perhaps the major portion of the duties of an insurance broker.

The authority for such allocation of liability under common law goes as far back as the case of *Dickson & Co. v. Devitt,* in 1917, and perhaps even earlier.[10] The plaintiffs in that case instructed their insurance broker to arrange for the insurance against marine and war risks of some machinery to be shipped on the '*Suwa Maru* and/or steamers'. The broker's clerk inadvertently omitted the words 'and/or steamers' from the insurance slip relating to war risks. As it happened, the machinery was shipped on a steamer other than the '*Suwa Maru*' which was later lost in enemy action. It was held that the broker was liable on two grounds: first, failure to carry out the plaintiffs' specific instructions and second, (through his clerk) failure to exercise reasonable skill and care. Atkin J. said in that case:

> In my opinion, when a broker is employed to effect an insurance, especially when the broker employed is a person of repute and experience, the client is entitled to rely upon the broker carrying out his instructions, and is not bound to examine the documents drawn up in performance of those instructions and see whether his instructions, have, in fact, been carried out by the broker. In many cases the principal would not understand the matter, and would not know whether the document did in fact carry out his instructions. Business could not be carried on if, when a person has been employed to use skill and care with regard to a matter, the employer is bound to use his own care and skill to see whether the person employed has done what he was employed to do.

---

7 *Strong and Pearl v. Allison and Co. Ltd.* (1926),25 Ll. L. Rep. 504.
8 *Wilcox v. Norberg and Wiggins Insurance Agencies Ltd.* [1979] 1 WWR 414.
9 *Claude R. Ogden and Co. Pty. Ltd. v. Reliance Fire Sprinkler Co. Pty. Ltd.* [1975] 1 Lloyd's Rep. 52, 73, *per* MacFarlan J.
10 *Dickson & Co. v. Devitt* (1917), 86 LJKB 313.

However, where there is contributory negligence by the plaintiff, the damages awarded are reduced proportionately, as in the Canadian case of *Morash v. Lockhart and Ritchie Ltd.*[11] In that case, the brokers failed to send a renewal form to the insured which, in effect, would have informed him of the expiry of his fire insurance policy and would have invited him to renew the cover. This procedure is normally followed by insurance brokers a short time before the expiry of current insurances. The brokers in that case were held to be liable in negligence but, due to the fact that the fire, which caused the loss, occurred eighteen months after the expiry of the insurance cover, the damages awarded were reduced by 75% through contributory negligence.

Insurance brokers act not only as intermediaries between insurers and insureds but also between reinsurers and insurers providing the service of handling and placement of reinsurance transactions and drawing their remuneration from the reinsurers in the form of commission. Their duties, responsibilities and liabilities are similar to those when they are acting between the insurer and the insured.

The insurer

As stated earlier, the insurer obtains his insurance business through a broker, an agent, an insurance consultant, or through direct selling. Subject to local laws and traditions, the insurer may be regarded as the organiser of the risk distribution within society, or, on a smaller scale, within a community.[12] Such organisation can be in one of three forms, differing in a number of issues, but most importantly in the basic concept of who is to carry the risk of conducting the business of insurance. The three forms are as follows:

- Mutual insurance companies, which are owned by the insurance policy holders who, therefore, become liable for any losses made or beneficiaries if profits are made. Profits in this case would attract reductions in premiums and losses would require increases in premiums, retroactive if the capital available is not sufficient to cover the loss.
- Proprietary insurance companies, which have limited liability structure and are owned by shareholders with a guaranteed capital forming the limit of their liability. Profits and losses in such companies are shared by the shareholders, but where losses are concerned, a limit exists beyond which the insurer becomes insolvent. There have been a number of such events around the world in recent years.
- Lloyd's Underwriters of London, which is a unique organisation made up of underwriters who only accept insurance risks through brokers on behalf of

---

11 *Morash v. Lockhart and Ritchie Ltd.* (1979) 95 DLR (3d) 647.
12 In a number of countries where the applicable law stems from pure Islamic law, traditional concepts of insurance do not apply. This is due to the Islamic concept of law being the expression of 'Divine Will'. The terms of Divine Revelation were fixed at the time of Prophet Muhammad at whose death, in 632 AD, the Revelation ceased and the Divine Will was thus fixed for all time. Another example used to be the USSR, where motor liability insurance and tort liability insurance were not permitted, on the basis that they would, if available, allow the insured to escape the consequences of his negligence.

members who provide the back-up capital. The members carry unlimited liability and place a deposit with Lloyd's, the value of which determines the insurance they are entitled to underwrite.

Members usually join syndicates which specialise in a particular branch or branches of insurance in which they are interested to operate.

In the case of the last two forms of insurance organisations and where there is no possibility of retroactive increases in premium, the insurer must make his calculations with sufficient precision so that his premium income is not less than any payments he may be expected to make. For this purpose, premium calculations and future projections can be precisely made using statistical analysis of past performance of events and of known facts.

The reinsurer

When an insurer accepts and underwrites a number of risks, he will then face the same question that had faced the insured in the first place. Should he seek an indemnity in case of an event leading to a loss greater than what his economic potential could cover? Should he provide in his liquid funds a sufficiency to cover all the expected liabilities or should he take out insurance to cover some of the risks to which he is exposed and use his funds more productively elsewhere? Should he do as the insured did, i.e. substitute a possible variable expense (payment for damage or loss insured with him) with a fixed sum to be paid towards an insurance cover?

If the answer is in favour of spreading the risk, the insurer reinsures with a reinsurance company, thus transferring part of the risk he undertakes against payment of part of the premium originally charged. This business can be either transacted directly between the insurer, who is called the cedent, and the reinsurer or through a reinsurance broker. The reinsurance broker may be a broker dealing only with insurance business or a branch of a large broking firm dealing with reinsurance.

The reinsurance transaction can be either a one-off arrangement, known as facultative reinsurance, or an automatic arrangement with an agreed pattern for a specific branch of insurance under a contract between the insurer and reinsurers usually referred to as a reinsurance treaty. The treaty is usually an annual contract subject to renewal.[13]

Reinsurance transaction can also take place at a second level between one reinsurance company and another, which is then known as retrocession. The purpose of retrocession is to spread the risk further, mainly on a geographical basis, by involving reinsurers from other parts of the world. Thus one may find that a risk insured in one part of the world is then reinsured and reinsured again through retrocession to many tens of insurance and reinsurance companies. A schematic distribution of the risk, in what can be a typical example, where insurance is transacted through co-insurance, reinsurance and retrocession, is shown in Figure 7.1, adapted from the reference in note 13. Reference should also be made to Figure 7.2 which shows the

13 *Reinsurance, Principles and Practice*, Klaus Gerathewohl, Verlag Versicherungswirtschaft, e.V., Karlsruhe, 1980, Vol. 1, at page 366.

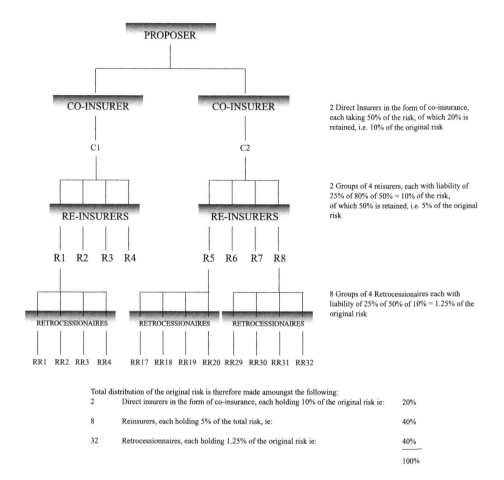

Figure 7.1 Distribution of risk between insurers and reinsurers.

distribution of not only the risk but also the premium in the more usual and direct way of insurance, reinsurance and retrocession, which normally takes place in construction insurance.

The fact that an insurance company buys reinsurance gives no rights to the insured so that if the insurer fails or refuses to perform his obligations to his insured, the insured has no right of action against the reinsurer.

Premium calculations

To calculate the necessary premium to cover a particular risk or a set of risks is a scientific process based on statistics and previous experience. As in most statistical calculations, there are assumptions in calculating the insurance premium which may transpire to be erroneous. To err in carrying out the necessary premium calculations is a risk of transacting insurance, but not to carry out such calculations at all can be disastrous and at the insurer's own peril.

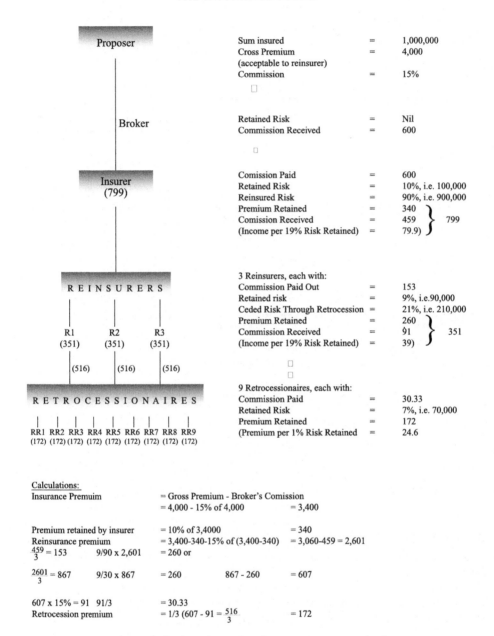

Figure 7.2 Distribution of risk and premium between insurers and reinsurers.

Statistics are available from different sources, the most important of which being that compiled by the insurance industry itself in respect of previous experience. To understand the importance of correct assessment of the insurance premium, one needs to know the financial distribution of the risk and the premium between the insurance matrix of broker, insurer and reinsurers.

Figure 7.2 shows the distribution of risk for a sum insured of one million units and a gross premium of 4,000 units. The broker's commission in this example is typically assumed to be 15% and is charged in respect of expenses incurred in the handling and placement of the risk plus profit. The remainder of 3,400 is the net premium in respect of insuring the risk. Assuming that the insurer calculates his preferred retention of the risk to be 100,000 units, the remaining portion is then reinsured, either facultatively or automatically under a treaty to three reinsurers, each taking 30% of the risk.

The insurer picks up 10% of the net premium, i.e. 340 units in respect of his 10% retention. This leaves 3,060 units of which he also retains, usually, a commission from the reinsurers in respect of his business expenses and placement profit which, if assumed at 15%, brings his income to 799 units (i.e. 340 plus 15% of 3,060=799, leaving a reinsurance premium of 2,601 units).

The reinsurers in turn are each assumed to retain a 9% share of the original risk, i.e. 90,000 units, and retrocede the remaining 21% to three retrocessionaires in equal parts. They each pick up 9/90 of the reinsurance premium (i.e. 9/90 of 2601=260 units leaving a retrocession premium of 1,821 units), plus a commission from the re-insurers for expenses and placement profit of the 21% share. Assuming this commission to be 15%, they each receive 91 units, bringing their income to 351 units (i.e. 15% of 1821 divided by 3=91, leaving a net retrocession premium of 1,548 units in the hands of three reinsurers, each with 516 units).

The retrocessionaires each pick up 7% of the original risk, i.e. 70,000 units, for a net premium of 172 units made up of 516 units divided into three.

Figure 7.2 also shows that the insurer, in order to be able to reinsure, must calculate the premium in such a manner that it is acceptable to reinsurers. If it is not and the reinsurer refuses to participate, the insurer is faced with either running the risk of over-exposure or insuring at a loss, i.e. at a premium larger than that received, and which at the same time eliminates his reinsurer's commission. Thus statistics and proper premium calculations have become an inseparable part of insurance.

It should be remembered, however, that construction insurance is attractive from the point of view of the large sums involved which reflect in the large premiums usually received at the commencement of an insurance period. Thus competition is extremely high to obtain this type of insurance business even when the indications are alarming. Unfortunately, most of the risks in construction fall into a category where it is accepted that statistics and their various methods of analysis produce only a guide rather than a precise method of premium calculation. Thus indications of alarm may be dampened by those who are optimistic by nature or who are anxious to obtain business.

There are many reasons for this feature of construction insurance which encompasses contractors' all risks, erection all risks, public liability, employer's liability and professional indemnity policies and where even large portfolios are non-homogeneous and subject to the risk of random fluctuation.[14] Some of these reasons stem from the following features of construction:

14 *Reinsurance, Principles and Practice*, op. cit., see note 13, vol. 1, at page 9.

- *Geographic location*: Construction is undertaken all over the world but, due to its very nature, it is mostly situated in high-risk areas and sometimes spread over large distances. To explain this feature more clearly, one has to think of projects which are particularly exposed to risk such as bridges and dams where river waters have to be dealt with or diverted. One may have to consider roads and highways through areas which are impassable until the roads are constructed.
- *Concentration of sums insured*: The cost of projects has escalated to such an extent that new terms have had to be invented to describe the new construction. Large projects gave way to jumbo projects which were then replaced by giant projects reaching in value over $1 billion. More recently the title pharaonic has been given to those projects greater than $1 billion in value.
- *Technical innovation*: The fast development in technical achievement that has taken place in recent years makes past experience and its statistics obsolete. An understanding of the behaviour of new materials and processes is gained through discovery rather than from accumulated knowledge. To be at the fore-front of technology is to be able to anticipate events and to have sufficient per-ception to imagine what can go wrong, thus replacing the ability to learn from past experience with forward prediction based on assumptions.
- *Period of insurance*: The time taken to complete a construction project involves a number of years during which events outside the control of the insured and insurer take place, affecting the very principles of insurance.

Documentation

There are four main documents which form the basis of the whole of the insurance transaction. These are:

- *First*, the 'Proposal Form' which is used in almost every type of insurance. It is essentially designed to provide the insurers with full details of the risks they are being asked to insure. It has been said that the best person to identify these risks is the proposer himself and thus when he completes a proposal form it becomes the basis on which the contract of insurance is based. The principle that an insurance contract is one of the utmost good faith is another feature of the proposal form, see page 185.

  The proposal form is also the means by which statistical information is collected by the insurer about the proposer. In almost all proposal forms, there are questions requesting information about the insurance history of the proposer in the branch of insurance where he is seeking cover. Thus, a record is estab-lished of any claims made, any losses sustained, any cancellation of previous insurance and any refusal to effect a renewal of an existing insurance policy or any rejection of an insurer of a proposal of insurance.

  The proposal form ends with the declaration that, to the best of the proposer's knowledge and belief, the answers given are true and complete. This declaration binds the information given by the proposer to the contract. It is imperative that all material facts be disclosed to enable the insurer decide whether or not to accept the proposal to insure, and if accepting, to determine the premium required. Even when there are no questions regarding certain facts relevant to the subject matter of insurance, the proposer must reveal these facts. If a broker is used, he is also under the same obligation to reveal any relevant information to the insurer. Finally, if in doubt about the relevance of certain facts, the proposer should disclose them in the proposal form.

The information collected from proposal form, together with the statistical information in the possession of the insurer, in relation to the whole matter of insurance in the respective branch, forms the basis of the premium calculation. A quotation is usually given to the proposer setting out the premium required.

- *The second* document is the insurance 'Policy'. If and when the insurer accepts the proposal and the quotation of a premium is accepted by the proposer, a contract is made. The terms of the contract are stated in a document referred to as the policy. Policy wordings differ from one insurer to another and from one country to another. They are nevertheless standard documents within the insurers' organisation. They are rarely altered in the transaction of simple insurance agreements. In construction, however proposed, the proposer and his technical, legal and insurance advisers must read policy wordings carefully and, if necessary, he should request the prospective insurer to alter them in accordance with the circumstances surrounding the particular construction project.

Construction insurance policies, regardless of the type of insurance, have certain sections in common. These are:

- A *recital clause* integrating the relevant information forming the basis of the insurance agreement.
- An *operative clause* describing the cover granted by the policy.
- A *schedule* setting out the details pertinent to the particular policy, such as the name and address of the insured, the insured matter, the sum insured or the limit of indemnity, the excess to be applied in each and every claim, the geographic limit, the period of insurance and the premium payable.
- A *list of exclusions*, which limits the cover in the negative sense, ensuring that whatever is not excluded is effectively covered, subject to the remaining sections of the policy.
- A *list of conditions*, which defines the rights and duties of the insured and the insurer. It usually includes a clause dealing with the procedure to be followed if a dispute arises between the parties to the insurance contract. It is of interest in this connection to mention that the established rule *contra proferentem* applies to insurance policies rendering an ambiguity to be construed in a sense unfavourable to the party responsible for drafting the document.

- *The third* document is the 'Endorsement' which may or may not be required in any particular insurance transaction. It is used to vary the terms of an existing policy and is usually read in conjunction with a policy and forms part of it. It is therefore vital, when one examines an existing policy, to remember to examine any endorsement attached thereto.
- *The fourth* document is the 'Claim Form'. If and when the dreaded moment, feared by the insured, arrives and a risk materialises causing a loss payable under the policy, the truth about the integrity and efficacy of the insurer would be revealed. A notification of the loss has to be made as soon as possible and

certainly within a specified period of time. This is important to the insurer who likes to be notified quickly so that inspections, investigations and financial arrangements can be made. A claim form must usually be completed by the insured, giving details of the events leading to the loss and other particulars associated with the claim.

The insurers' departmentation

Historically, insurance developed from four distinct branches, starting with Marine Insurance, then Fire, Life and finally Accident. Each of these branches went through a mitotic process, dividing itself into, or shedding away, specialist classes of insurance, some under the control of the same branch, others under a completely separate branch title. Specialisation within each of these branches dealt with an aspect of human endeavour so different from the others that the insurer himself had to be a specialist in order to deal successfully with the people he was insuring and the service he was providing. Professionals of different disciplines were needed in various departments in order to deal with the business related to their discipline. Thus engineers, doctors, lawyers, mathematicians, accountants and economists are employed for their expertise within the services offered and rendered by insurers and reinsurers.

This development took different routes in different countries. Thus, engineering insurance, as developed in the United Kingdom, encompassed all aspects of mechanical and electrical engineering but left the building and civil engineering under the control of the accident department. In Europe, engineering insurance grew to include all aspects of engineering, but recently, it seems to have been absorbed back by some sectors into the accident departmentation.[15]

The branches connected with construction may be categorised into nine subdivisions: Fire, Engineering, Accident dealing with other Accidental Damage to Property, Motor, Liability, Marine, Aviation, Life and Credit. In construction, one needs the services rendered by at least four of these branches and sometimes more, depending on the type and size of the project considered. The solution adopted in most cases is to cover the project under separate insurance policies issued by each of the different departments resulting in the policies identified in Figure 6.2. In some cases, the Contractor's Public Liability and Employer's Liability may be added to the main property damage policy (Contractors' All Risks) and one policy is issued to cover the risks encompassed by these three policies leading to a reduction in the total number of policies issued.

Whilst the situation may be satisfactory for small projects, where the gaps and overlaps produced by the above system are small enough to be acceptable, in large projects this solution presents more difficulties. Various alternatives, in the form of project insurance, decennial liability and others have been offered but none seem to present a perfect solution. So, it remains to be seen whether the acceptable answer emerges from a broker, an insurer, a reinsurer or from the construction-related disciplines.

---

15 For example, at the Munich Reinsurance Company, Munich, Germany.

8

THE INSURANCE CLAUSES IN
STANDARD FORMS OF CONTRACT

As mentioned earlier, standard forms of contract have been developed in construction by commercial organisations for the purpose of providing a balanced distribution of risk; for efficient administration of the contractual activities; for building on the experience gained from repeated use of these forms; but most of all for the optimum protection of one or both parties' interests.[1] They were mostly drafted and developed, however, by independent professional organisations, rather than by one or other of the parties to the contract, in order to establish or consolidate a fair and just contract. Knowledge accumulated through experience and recurrent use over a long period of time has brought about revisions and modifications in these forms with the aim either of achieving greater certainty in the intention of the wording or of providing a response to the needs of the parties and/or society.

In these construction contracts, except in turnkey projects where an owner initiates plans for the design and construction of a project without necessarily having a consulting engineer in place, the owner generally must execute at least two agreements, one with his design professional(s) and the other with a contractor.

On the international scene, where the owner, design professional and contractor would generally be of different nationalities, these agreements should preferably incorporate and reflect internationally accepted forms of contract and legal concepts. A number of standard forms of contract exist to assist the parties involved in a construction project. The principles involved are discussed in Chapter 1 and the insurance aspects of four different forms of contract are examined in this chapter. The first form concerns the owner and the design professional and the others concern the owner and the contractor. Except for the ICE Form of Contract, all other documents have been produced by the International Federation of Consulting Engineers.

The first standard form of contract is the 'Client/Consultant Model Services Agreement', which is more commonly known as the 'White Book'. It was first published in 1990 for general use in agreements of pre-investment and feasibility studies and in administration of construction and project management. It replaced the three documents previously used in formulating such agreements which were designated the 'International General Rules for Agreement, (IGRA 1979 D&S,

---

1   See Construction Contracts on page 7.

IGRA 1979 PI and IGRA 1980 PM)'.[2] The White Book is now in its 3rd edition, having been first revised in 1991 and subsequently in 1998. The second standard form of contract discussed here is between the owner and the contractor and is referred to as the 'FIDIC Conditions of Contract for Works of Civil Engineering Construction', which is currently in its 4th edition. It is more popularly known as the Red Book.[3] As is the case with many other similar forms of contract for civil engineering projects, this FIDIC document has its origin rooted in the ICE Conditions of Contract, whose insurance clauses are also discussed here.[4]

The fourth standard form of contract discussed here is also between the owner and the contractor and is another FIDIC Form of Contract for all types of construction works where the design is carried out by the employer/owner or on his behalf. It was first published in 1999 as one of four forms. It is more commonly known as the New Red Book.

Liability and insurance clauses of the client/consultant Model Services Agreement (The White Book) 3rd edition 1998

Clauses 16 to 20, inclusive, of the White Book deal with the liability and insurance aspects of the relationship between a consultant and his client, the owner/employer. The text of these clauses can be seen in Appendix O. The text is fairly simple and does not allow any misinterpretation. However, the following points are worthy of particular mention:

1   The liability of the consultant under this form of contract is limited in all of its four aspects of type, amount, duration and extent.

   Thus in respect of type, the consultant's liability is limited under Sub-clause 16.1 to payment of compensation as a result of an established breach of the stipulated duty under Sub-clause 5(i) of exercising reasonable skill, care and diligence in the performance of his obligations under the Agreement.

---

2   International Model Form of Agreement between Client and Consulting Engineer and International General Rules of Agreement between Client and Consulting Engineer for Pre-Investment Studies (I.G.R.A. 1979 P.I.) 3rd Edition, and for Design and Supervision of Construction of Works (I.G.R.A. 1979 D & S) 3rd Edition.
3   FIDIC's Red Book was first published in 1957. The second edition was produced in 1967, the third edition in 1977 and the fourth edition in 1987. A number of editorial amendments were introduced in 1988 and subsequently further more significant amendments were introduced in 1992. For more information on this form of contract, see Nael G. Bunni, 'The FIDIC Form of Contract – 4th Edition', second edition, Blackwell Science, 1997'.
4   Conditions of Contract in many parts of the world have been formulated adopting as a basis for their philosophy the ICE Conditions of Contract for Use in Connection with Works of Civil Engineering Construction produced by the Institution of Civil Engineers, London. The FIDIC document was adopted using as a model, for the insurance clauses, the first edition of the Conditions of Contract for Overseas Works mainly of Civil Engineering Construction published in August 1956. This latter document was prepared by The Association of Consulting Engineers in the U.K. jointly with the Export Group for the Constructional Industries and is approved by the Institution of Civil Engineers, London.

2 The amount of liability is limited under Sub-clauses 16.3 and 18.1. Sub-clause 16.3 provides as follows:

> If it is considered that either party is liable to the other, compensation shall be payable only on the following terms:
>
> (i) Such compensation shall be limited to the amount of reasonably fore-seeable loss and damage suffered as a result of such breach, but not otherwise.
> (ii) In any event, the amount of such compensation will be limited to the amount specified in Clause 18.1.
> (iii) If either party is considered to be liable jointly with third parties to the other, the proportion of compensation payable by him shall be limited to that proportion of liability which is attributable to his breach.'

3 The liability is also limited in duration as provided for in Clause 17, which provides as follows:

> Neither the Client nor the Consultant shall be considered liable for any loss or damage resulting from any occurrence unless a claim is formally made on him before the expiry of the relevant period stated in the Particular Conditions, or such earlier date as may be prescribed by law.

4 Whilst the limitation in amount of compensation in respect of liability is given under Clause 16, this limit is without prejudice to any agreed compensation specified under Clause 31(ii) in respect of amounts due to the consultant or otherwise imposed by the Agreement.

Furthermore, it is agreed under Sub-Clause 18.1 that each party agrees to waive all claims against the other in so far as the aggregate of compensation which might otherwise be payable exceeds the imposed limit. It is also agreed that if either party makes a claim for compensation against the other party and such a claim is not established, the claimant shall entirely reimburse the other for his costs, which would have been incurred as a result of the claim.

5 In respect of extent, the White Book requires the client under Sub-clause 18.2 to provide an indemnity to the consultant, in so far as the applicable law permits, against any adverse effect of all claims, including those by third parties, which might arise out of or in connection with the Agreement. However, there are some exceptions to this indemnity, which relate to claims:

- that are covered by the insurances arranged under the terms of Clause 19;
- made after the expiry of the period of liability referred to in Clause 17;
- arising from deliberate default or reckless misconduct; and
- relating to matters otherwise than in connection with the performance of obligations under the Agreement.

6 The extent of the liability is also limited to the scope of work as defined in Clause 3 of the Conditions and as set out in the Appendix.

7 The liability dealt with in Clauses 16 to 18 of the White Book may be covered by new insurance policies to be effected by the consultant, or by increasing existing insurances at the date of the client's first invitation to him for a proposal for the services, where such requirement is made in writing.

If so requested, the consultant shall make all reasonable efforts to effect such insurance or increase in insurance with an insurer and on terms acceptable to the Client. The cost of such insurance or increase in insurance is at the expense of the Client.

8   The White Book also provides that unless otherwise requested by the client in writing the consultant shall make all reasonable efforts to insure on terms acceptable to the client:

(i) against loss or damage to the property of the client supplied or paid for under Clause 6.

(ii) against liabilities arising out of the use of such property.

The cost of such insurance shall be at the expense of the client.

9   The insurance requirements are set out in Clauses 19 and 20. They include: first, a public liability insurance to a limit of indemnity and terms to be approved by the client; and second, a 'reasonable' professional liability insurance. The cost of the first cover, which is to include damage to any equipment purchased with funding from the client is borne by the Client. The cost of the second cover is however to be borne by the engineer.

The responsibility, liability, indemnity and insurance clauses of the FIDIC Form of Contract between Owner and Contractor (The Red Book) 4th edition 1992

Under this type of agreement between owner/employer and contractor, it is appropriate to start with the FIDIC Conditions of Contract, 4th edition, reprinted in 1992, which is better known as the Red Book. It is worthy of note that the previous editions of this form of contract followed closely the ICE Conditions of Contract for civil engineering works current at the time of publication and developed with the changes made to the various revisions of that document. Thus, the FIDIC Conditions of Contract for Civil Engineering Works are based on the English tradition of construction contracts and incorporate legal terms drawn from that tradition.[5] However, if we compare the 4th edition of the Red Book with its previous editions and also with the 5th edition of the ICE Form of Contract, which was current at the time of publication of the 4th edition, we find that it contains a number of important improvements, particularly where the insurance clauses are concerned. However, notwithstanding these improvements, it is relevant, if not essential, when discussing the insurance clauses for a form of contract between employer and contractor, to deal with and to examine the provisions of not only the FIDIC Red Book, but also the clauses of the latest edition of the equivalent ICE Form, which at this point in time is the 7th edition.

The clauses relevant to insurance are of course not only those that specifically refer to insurance but also those others that lead directly to it. Thus, the clauses dealing with responsibility, liability and indemnity of the parties involved in respect

---

5   *The FIDIC Form of Contract – The Fourth Edition of the Red Book*, 2nd edition, by Nael G. Bunni, 1997, Blackwell Science Ltd., Oxford.

of the risks that lead to loss and/or damage also form part of the insurance concept. Accordingly, these are Clauses 20 to 25 inclusive and Clause 65 of the FIDIC Red Book. Clauses 20 to 25 provide for the allocation and sharing of the normal risks of loss and/or damage between the contractor and owner/employer whereas Clause 65 deals with the special risks. (The owner is referred to as the employer in both documents and throughout the remaining part of this chapter.)

In the overall context of the insurance arrangements for a construction project as a whole, Clauses 20 to 25 together with Clause 65 complement those insurance clauses usually included in the agreement between the employer and his professional team, as discussed above. Combined, they serve to protect the parties involved in a construction project against the liabilities to which they are exposed if and when the risks inherent in construction eventuate. Thus, the insurance clauses provided in agreements between an employer and contractor and also between employer and professional team provide for allocation of the risks to the various parties concerned who then become responsible and liable for their effect. But liability may be incurred not only by one of the contracting parties towards another but also towards third parties; and therefore indemnity clauses are incorporated in the conditions to cater for the situation that may arise when one party is sued in respect of an occurrence for which another is either partly or totally liable. Indemnity and cross-indemnity clauses, however, can only provide a safeguard when the party providing the indemnity is solvent, and hence whereas insurance is needed in these cases to provide a protection to the solvent party, it is essential in all other cases where the liable party could become insolvent.

The insurance clauses are therefore drafted to be instrumental in shifting liability in respect of certain risks towards an insurer who, together with any reinsurer(s), is expected to be financially capable of providing a promise to pay for any loss or damage that may be caused as a result of these risks eventuating.

The FIDIC and ICE Conditions of Contract between employer and contractor are so drafted that the insurable risks identified can be insured in either one or more policies. However, these risks range over a number of classes of insurance and are, therefore, covered by separate policies that are generally handled by different people, different insurance departments or even different insurance companies. Such insurance arrangements may lead to complications resulting in gaps and overlaps in the insurance cover provided.

Although compatibility exists between the general practice in the insurance market and the manner in which the insurance clauses are structured in the agreements between employer and contractor on the one hand and between employer and professional team on the other, gaps and overlaps however, could exist in the insurance cover provided, see page 204. They occur mainly due to:

- Inadequacy of the standard insurance clauses, which are generally spread in any one project through at least two separate contracts and which are expected to apply to all projects in all locations and in all circumstances despite the uniqueness of the construction contract.[6]

---

6   In this connection, see later the section on project insurance on page 389.

- The legal language used by the draftsman of the first edition of the ICE and the FIDIC Forms of Contract was preserved in the subsequent revisions of these two documents, up to the 5th edition of the ICE Form, published in 1973; and up to the 3rd edition of the FIDIC Form, published in 1977. Such legal language is cumbrous and difficult to understand for many reasons, but mainly due to the length of sentences used; the number of ideas contained in each of them; and the convoluted method of expression with multi-layered negatives, see page 9.[7] When one is presented with a sentence of 166 words defining the Excepted Risks and again with a sentence of 174 words applying this definition to insurance in these earlier forms of contract, it is difficult to understand what is intended without subdividing the sentences into smaller units and re-reading the text a number of times. However, this subdivision is not sufficient by itself to clarify the meaning since the language is so interwoven in the sentence that its subdivision cannot be made without detailed knowledge of the original intention behind the sentence.

- As a result of the criticism made in the first edition of this book, the relevant insurance clauses in FIDIC's Red and Yellow Books, and to a certain extent in the ICE Form, were altered and modified to varying degrees in 1987 and 1991 respectively.[8] In all these forms of contract, the language was made easier to understand by introducing bullet points and sub-paragraphs, which reduced the number of words in one sentence. The draftsman of FIDIC's Yellow Book was more understanding of what was needed to convert the wording of these clauses to a clear text. Those responsible for the Red Book followed the example of the Yellow Book, but to a lesser extent. Despite these changes in the insurance clauses, a number of problems remained unresolved in all these forms of contract.

- In 1999, a new suite of contracts was published by FIDIC, which comprised four standard forms of contract: the new Red Book; the new Yellow Book; the Silver Book; and the Green Book. For more details of these forms of contract, see page 298 below. The insurance clauses of the first three of these forms of contract are very similar to each other in concept and in text despite the different specialist use intended for each of them. Thus, this similarity is unwelcome in clauses that ought to be inherently different and add to the conceptual problems which the draftsman failed to appreciate, as will be discussed below in Chapter 10 of this book.

---

7 Consider, for example, the first sentence of Clause 20 of the ICE Form quoted on page 234 below.

8 In 1987, the 4th edition of the Red Book and the Third Edition of the Yellow Book were published with these changes. In 1991, the 6th edition of the ICE Form was published incorporating only some of the changes introduced by FIDIC. In Ireland, the IEI Conditions of Contract for civil engineering works, which were also rooted in the ICE Form, followed FIDIC's changes rather than the ICE when their 1990 reprint of the 3rd edition was published. However, in the IEI Conditions, a number of the mistakes made in FIDIC's Red Book were eliminated and the insurance clauses followed the recommendations made in the 1st edition of this book, see note 16 on page 237 below. The 6th edition of the ICE Conditions, which was published in 1991, adopted most of these changes and was then followed by the 7th edition of the ICE Conditions in 1999. The 4th edition of the IEI Conditions followed the 6th edition when it was published in 1995.

- The manner in which insurance is taken out by each of the parties involved in the construction process makes it impossible to dovetail the cover and eliminate all possible gaps.
- The uninsurability of some of the risks to which a construction project is inevitably exposed result in gaps, see page 195.

Whilst it is essential for any gaps and overlaps to be reduced in number and extent, the aim should obviously be to have them eliminated, a task that could only be successful if these clauses were clearly understood by all concerned. Therefore, despite the fact that the employer, contractor and professional team are not expected to be experts in the fields of liability, indemnity and insurance, they nevertheless require a working knowledge of these subjects.

Two alternative approaches are proposed here to deal with gaps and overlaps. The first approach entails rectifying the current clauses without introducing any material change in their existing framework or composition. This alternative is dealt with in this chapter, on a clause-by-clause basis. The second alternative, dealt with in Chapter 9, is to abandon the present framework and composition of these clauses and to start from the beginning by adopting the elementary concepts which have been developed earlier in the previous chapters. This entails drafting a new set of clauses following a sequence based on the following logic:

- Identification and allocation of risks;
- Allocation of responsibility for risks based on control and influence of events resulting from risks, otherwise as described on page 47;
- Liability and indemnity;
- Insurance.

Starting with the first approach, the remaining part of this chapter will examine the current insurance clauses of the 4th edition of FIDIC's Red Book and the 7th Edition of the ICE Form and discuss how they could be modified to eliminate the gaps and overlaps contained in them. Chapter 11 will examine the insurance clauses of the new suite of contracts of FIDIC and address how they could be modified to eliminate some of the new problems they have created.

Alternative one:
the FIDIC and ICE insurance clauses modified

The insurance clauses in the agreement between employer and contractor, as currently provided for by the 4th edition of the FIDIC Conditions of Contract for works of civil engineering construction, the Red Book of 1987 as reprinted in 1992, are Clauses 20 to 25 inclusive and Clause 65. They however have a direct link with Clauses 39, 40, 48, 49, 50, 53, 66, 67 and 69 and an implied link with Clauses 8, 10, 11, 13, 14, 36, 53 and 67.

To modify the insurance clauses of FIDIC's Red Book without changing the intended meaning, it is necessary to simplify the transition from Responsibility to Insurance by using flow charts which provide a separate compartment for each of the relevant stages, see Figures 8.1 and 8.2.

RESPONSIBILITIES AND LIABILITIES OF THE CONTRACTING PARTIES AS DEFINED AND AGREED UNDER FIDIC RED BOOK, 4th Edition 1992 (for full text of these clauses, see below).

20.1 Care of the Works
The Contractor shall take full responsibility for the care of the Works and materials and Plant for incorporation therein from the Commencement Date until the date of issue of the Taking-Over Certificate for the whole of the Works, when the responsibility for the said care shall pass to the Employer. Provided that:
(a) if the Engineer issues a Taking-Over Certificate for any Section or part of the Permanent Works the Contractor shall cease to be liable for the care of that Section or part from the date of issue of the Taking-Over Certificate, when the responsibility for the care of that Section or part shall pass to the Employer, and
(b) the Contractor shall take full responsibility for the care of any outstanding Works and materials and Plant for incorporation therein which he undertakes to finish during the Defects Liability Period until such outstanding Works have been completed pursuant to Clause 49.

20.2 Responsibility to Rectify Loss or Damage
If any loss or damage happens to the Works, or any part thereof, or materials or Plant for incorporation therein, during the period for which the Contractor is responsible for the care thereof, from any cause whatsoever, other than the risks defined in Sub-Clause 20.4, the Contractor shall, at his own cost, rectify such loss or damage so that the Permanent Works conform in every respect with the provisions of the Contract to the satisfaction of the Engineer. The Contractor shall also be liable for any loss or damage to the Works occasioned by him in the course of any operations carried out by him for the purpose of complying with his obligations under Clauses 49 and 50.

20.3 Loss or Damage Due to Employer's Risks
In the event of any such loss or damage happening from any of the risks defined in Sub-Clause 20.4, or in combination with other risks, the Contractor shall, if and to the extent required by the Engineer, rectify the loss or damage and the Engineer shall determine an addition to the Contract Price in accordance with Clause 52 and shall notify the Contractor accordingly, with a copy to the Employer. In the case of a combination of risks causing loss or damage any such determination shall take into account the proportional responsibility of the Contractor and the Employer.

20.4 Employer's Risks
The Employer's risks are:
(a) war, hostilities (whether war be declared or not), invasion, act of foreign enemies,
(b) rebellion, revolution, insurrection, or military or usurped power, or civil war,
(c) ionising radiations, or contamination by radio-activity from any nuclear fuel, or from any nuclear waste from the combustion of nuclear fuel, radio-active toxic explosive or other hazardous properties of any explosive nuclear assembly or nuclear component thereof,
(d) pressure waves caused by aircraft or other aerial devices travelling at sonic or supersonic speeds,
(e) riot, commotion or disorder, unless solely restricted to employees of the Contractor or of his Subcontractors and arising from the conduct of the Works,
(f) loss or damage due to the use or occupation by the Employer of any Section or part of the Permanent Works, except as may be provided for in the Contract,
(g) loss or damage to the extent that it is due to the design of the Works, other than any part of the design provided by the Contractor or for which the Contractor is responsible, and
(h) any operation of the forces of nature against which an experienced contractor could not reasonably have been expected to take precautions.

22.1 Damage to Persons and Property
The Contractor shall, except if and so far as the Contract provides otherwise, indemnify the Employer against all losses and claims in respect of:
(a) death of or injury to any person, or
(b) loss of or damage to any property (other than the Works),
which may arise out of or in consequence of the execution and completion of the Works and the remedying of any defects therein, and against all claims, proceedings, damages, costs, charges and expenses whatsoever in respect thereof or in relation thereto, subject to the exceptions defined in Sub-Clause 22.2.

22.2 Exceptions
The 'exceptions' referred to in Sub-Clause 22.1 are:
(a) the permanent use or occupation of land by the Works, or any part thereof,
(b) the right of the Employer to execute the Works, or any part thereof, on, over, under, in or through any land,
(c) damage to property which is the unavoidable result of the execution and completion of the Works, or the remedying of any defects therein, in accordance with the Contract, and
(d) death of or injury to persons or loss of or damage to property resulting from any act or neglect of the Employer, his agents, servants or other contractors, not being employed by the Contractor, or in respect of any claims, proceedings, damages, costs, charges and expenses in respect thereof or in relation thereto or, where the injury or damage was contributed to by the Contractor, his servants or agents, such part of the said injury or damage as may be just and equitable having regard to the extent of the responsibility of the Employer, his servants or agents or other contractors for the injury or damage.

22.3 Indemnity by Employer
The Employer shall indemnify the Contractor against all claims, proceedings, damages, costs, charges and expenses in respect of the matters referred to in the exceptions defined in Sub-Clause 22.2.

24.1 Accident of Injury to Workmen
The Employer shall not be liable for or in respect of any damages or compensation payable to any workman or other person in the employment of the Contractor or any Subcontractor, other than death or injury resulting from any act or default of the Employer, his agents or servants. The Contractor shall indemnify and keep indemnified the Employer against all such damages and compensation, other than those for which the Employer is liable as aforesaid, and against all claims, proceedings, damages, costs, charges, and expenses whatsoever in respect thereof or in relation thereto.

Figure 8.1 Responsibilities and liabilities of the contracting parties as defined and agreed under FIDIC Red Book, 4th edition 1992 (for full text of these clauses, see below).

INSURANCE REQUIREMENTS UNDER THE FIDIC's RED BOOK, 4th EDITION 1992

↓ C.A.R.   **↓ T.P.L.**   **↓ E.L.**

21.1 Insurance of Works and Contractor's Equipment
The Contractor shall, without limiting his or the Employer's obligations and responsibilities under Clause 20, insure:
(a) the Works, together with materials and Plant for incorporation therein, to the full replacement cost (the term 'cost' in this context shall include profit),
(b) an additional sum of 15 per cent of such replacement cost, or as may be specified in Part II of these Conditions, to cover any additional costs of and incidental to the rectification of loss or damage including professional fees and the cost of demolishing and removing any part of the Works and of removing debris of whatsoever nature, and
(c) the Contractor's Equipment and other things brought onto the Site by the Contractor, for a sum sufficient to provide for their replacement at the Site.

21.2 Scope of Cover
The insurance in paragraphs (a) and (b) of Sub-Clause 21.1 shall be in the joint names of the Contractor and the Employer and shall cover:
(a) the Employer and the Contractor against all loss or damage from whatsoever cause arising, other than as provided in Sub-Clause 21.4, from the start of work at the Site until the date of issue of the relevant Taking-Over Certificate in respect of the Works or any Section or part thereof as the case may be, and
(b) the Contractor for his liability:
　(i) during the Defects Liability Period for loss or damage arising from a cause occurring prior to the commencement of the Defects Liability Period, and
　(ii) for loss or damage occasioned by the Contractor in the course of any operations carried out by him for the purpose of complying with his obligations under Clauses 49 and 50.

21.3 Responsibility for Amounts not Recovered
Any amounts not insured or not recovered from the insurers shall be borne by the Employer or the Contractor in accordance with their responsibilities under Clause 20.

21.4 Exclusions
There shall be no obligation for the insurances in Sub-Clause 21.1 to include loss or damage caused by:
(a) war, hostilities (whether war be declared or not), invasion, act of foreign enemies,
(b) rebellion, revolution, insurrection, or military or usurped power, or civil war,
(c) ionising radiations, or contamination by radio-activity from any nuclear fuel, or from any nuclear waste from the combustion of nuclear fuel, radio-active toxic explosive or other hazardous properties of any explosive nuclear assembly or nuclear component thereof, or
(d) pressure waves caused by aircraft or other aerial devices travelling at sonic or supersonic speeds.

23.1 Third Party Insurance (including Employer's Property)
The Contractor shall, without limiting his or the Employer's obligations and responsibilities under Clause 22, insure, in the joint names of the Contractor and the Employer, against liabilities for death of or injury to any person (other than as provided in Clause 24) or loss of or damage to any property (other than the Works) arising out of the performance of the Contract, other than the exceptions defined in paragraphs (a), (b) and (c) of Sub-Clause 22.2.

23.2 Minimum Amount of Insurance
Such insurance shall be for at least the amount stated in the Appendix to Tender.

23.3 Cross Liabilities
The insurance policy shall include a cross liability clause such that the insurance shall apply to the Contractor and to the Employer as separate insureds.

24.1 Accident of Injury to Workmen
The Employer shall not be liable for or in respect of any damages or compensation payable to any workman or other person in the employment of the Contractor or any Subcontractor, other than death or injury resulting from any act or default of the Employer, his agents or servants. The Contractor shall indemnify and keep indemnified the Employer against all such damages and compensation, other than those for which the Employer is liable as aforesaid, and against all claims, proceedings, damages, costs, charges, and expenses whatsoever in respect thereof or in relation thereto.

24.2 Insurance Against Accident to Workmen
The Contractor shall insure against such liability and shall continue such insurance during the whole of the time that any persons are employed by him on the Works. Provided that, in respect of any persons employed by any Subcontractor, the Contractor's obligations to insure as aforesaid under this Sub-Clause shall be satisfied if the Subcontractor shall have insured against the liability in respect of such persons in such manner that the Employer is indemnified under the policy, but the Contractor shall require such Subcontractor to produce to the Employer, when required, such policy of insurance and the receipt for the payment of the current premium.

25.1 Evidence and Terms of Insurance
The Contractor shall provide evidence to the Employer prior to the start of work at the Site that the insurances required under the Contract have been effected and shall, within 84 days of the Commencement Date, provide the insurance policies to the Employer. When providing such evidence and such policies to the Employer, the Contractor shall notify the Engineer of so doing. Such insurance policies shall be consistent with the general terms agreed prior to the issue of the Letter of Acceptance. The Contractor shall effect all insurances for which he is responsible with insurers and in terms approved by the Employer.

25.2 Adequacy of Insurances
The Contractor shall notify the insurers of changes in the nature, extent or programme for the execution of the Works and ensure the adequacy of the insurances at all times in accordance with the terms of the Contract and shall, when required, produce to the Employer the insurance policies in force and the receipts for payment of the current premiums

25.3 Remedy on Contractor's Failure to Insure
If the Contractor fails to effect and keep in force any of the insurances required under the Contract, or fails to provide the policies to the Employer within the period required by Sub-Clause 25.1, then and in any such case the Employer may effect and keep in force any such insurances and pay any premium as may be necessary for that purpose and from time to time deduct the amount so paid from any monies due or to become due to the Contractor, or recover the same as a debt due from the Contractor.

25.4 Compliance with Policy Conditions
In the event that the Contractor or the Employer fails to comply with conditions imposed by the insurance policies effected pursuant to the Contract, each shall indemnify the other against all losses and claims arising from such failure.

Figure 8.2 Insurance requirements under the FIDIC Red Book, 4th edition 1992
(for full text of these clauses, see later in this chapter).

Clause 20 – Care of the Works and Employer's Risks

Sub-Clause 20.1 – Care of the Works

With the publication of the 4th edition of FIDIC's Red Book, the format of the insurance clauses diverged from the ICE format and Clause 20 was divided into four sub-clauses. For clarity and ease of comparison with the ICE Conditions, these sub-clauses are set out below and the equivalent text of the two Forms of Contract is discussed together.

Responsibility and period

The FIDIC Conditions of Contract, in a manner similar to the ICE document, begin this section on responsibility, liability, indemnity and insurance with Clause 20, by setting out the responsibility for the works and their care. The first sentence complements the express duty of the contractor to complete the contract, as defined in Clause 8.1, with the duty of 'Care of the Works' setting out the period during which the contractor has to provide for that care, which ends with the date of issue of the 'Taking-Over Certificate' by the engineer.

In this connection, the legal precedent in common law goes as far back as 1867 where in the case of *Appleby v. Myers* it was held that the plaintiff contractor was not entitled to recover anything from the employer in respect of any portion of the machinery he had erected but which was destroyed in a fire prior to the completion of the work.[9] This case is quoted in *Hudson's Building and Engineering Contracts*, 11th edition, in connection with the following statement:[10]

> Indeed, by virtue of the express undertaking to complete (and in some contracts to maintain for a fixed period after completion) the contractor would be liable to carry out his work again free of charge in the event of some accidental damage occurring before completion even in the absence of any express provisions for protection of the work.

This sentence was in fact cited in the judgment made in the case of *Charon v. Singer Sewing Machines Ltd.*, in 1968. The defendants in this case employed the plaintiffs to convert a shop and living accommodation. The contract included the words 'allow for covering up and protecting the works during frosty and inclement weather or from damage from any other cause and reinstating any work so damaged'. Vandals broke into the shop, one day before completion of the work, and caused damage that had to be repaired by the plaintiff by doing work already carried out and paid for. The court held that the contractor had to bear the cost of repair.

---

9 *Appleby v. Myers* (1867) LR 1 CP 615, quoted in *Hudson's Building and Engineering Contracts*, 11th edition, by I.N. Duncan Wallace, Sweet & Maxwell, 1995, 4–251, page 645.

10 *Hudson's Building and Engineering Contracts*, op. cit., see note 9. The words quoted and the reference to the case of *Charon v. Singer Sewing Machines* (1968) 112 S.J. 536 appear on page 307 of the 10th edition and the case is referred to in 4–051 on page 508 of the 11th edition.

Thus, the duty of the contractor to take care of the works is connected with his duty to complete the contract.[11]

It is unfortunate that Clause 20 is not written in a logical and orderly manner in that it deals with the period of responsibility for the care of the works prior to defining the subject matter of that care. Furthermore, it does not refer to the relationship between the particular risks in respect of which loss and damage to the works might occur and the whole spectrum of risks to which the project is exposed. The allocation of the risks and the responsibility for them between the employer and the contractor are left in an ambiguous state even when the opportunity is given to the draftsman under Sub-clause 20.4 to define the Employer's Risks. This problem is inherited from the ICE Form, the source for the FIDIC Red Book and remains in the new suite of contract forms published by FIDIC.[12]

Clause 20, however, clearly states that the contractor is responsible for the works, together with materials and plant from the Commencement Date until the date of issue of the Taking-Over Certificate and remains so during the Defects Liability Period for any outstanding items until they have been completed. As defined in Sub-Clause 1.1(f)(i) of the Conditions, the works include the permanent as well as the temporary works. Therefore, if the Engineer issues a Taking-Over Certificate while there is still outstanding work to be completed during the Defects Liability Period, the contractor continues to be responsible for the care of such work until its completion. The wording of Sub-clause 20.1(b) indicates that the responsibility ceases for each item as and when it is completed pursuant to Clause 49. The definition of the extent of outstanding items in case of repairs to existing defective work is left to be established, in each individual case, by the engineer in his notification to the contractor in accordance with Sub-Clauses 48.1 and 49.2.[13] Thus, once items are identified as outstanding, they become the subject of the insurance required during the Defects Liability Period.

Clause 20 further provides in its paragraph (a) that if the engineer issues a Taking-Over Certificate in respect of a part of the permanent works, the contractor's liability in respect of that part ceases from the date of issue of the certificate and the responsibility for it passes from the contractor to the employer.

It is worthwhile noting that when the 6th edition of the ICE Conditions of Contract was introduced, the wording of Clause 20(1) followed its equivalent in the 4th edition of FIDIC's Red Book. In particular, the fourteen-day period after the date of the Certificate of Substantial Completion during which the contractor was responsible for the care of the works under the 5th edition of that Form was eliminated. That fourteen-day period was intended to allow the engineer time to inform the employer of the need to effect his own insurance arrangements and to allow the employer to effect such insurance. However, in practice, this caused more problems than it resolved since the contractor and his insurers found themselves responsible for the care of a project which had been handed over to the employer and thus had no control whatsoever over its destiny. The problem created was even more serious

---

11 See page 187 above in connection with the contractor's duty to complete.
12 See Chapter 10 below in this connection.
13 The relevant words in Sub-clause 48.1 are: 'The Engineer shall also notify the Contractor of any defects in the Works affecting substantial completion . . .' The relevant words in Sub-clause 49.2 are contained in paragraph (b) of that sub-clause.

as the fourteen days could in practice become extended to thirty-five days if the engineer delayed the issue of the certificate by the full permitted period of twenty-one days under Clause 48. The wording of Clause 20(1) remained unchanged in the 7th edition of the ICE Conditions of Contract.

Of course, both the engineer and the employer should realise the effect of issuing the Taking-Over Certificate under the 4th edition of FIDIC's Red Book (and the Certificate of Substantial Completion under the 7th edition of the ICE Conditions of Contract) on the insurance aspects of the completed project. They should ensure that any insurance covers required for the handed-over project are in fact in place before these certificates are issued and that such insurance would operate immediately thereafter.

Taking-over part of the works

Paragraphs (a) and (b) of Sub-Clause 20.1 of the 4th edition of FIDIC's Red Book are essentially the same as paragraphs (b) and (c) of Sub-clause 20(1) of the 7th edition of the ICE Conditions of Contract. They relate to the transfer of the responsibility for the care of the works from the contractor to the employer of any part of the works that had been taken over by the employer or in respect of which a Taking-Over Certificate or a Certificate of Completion has been issued. The insurance implication here is not obvious and the engineer is assumed to know that the insurers usually restrict their cover in respect of those parts of the work that have been taken over. This restriction is due to the difference between the insurance cover usually provided in a Contractors' All Risks policy during the construction period and that provided during the Defects Liability Period, see page 249. The Engineer is also expected to inform the employer of the necessity to make his own insurance arrangements for such part of the works and perhaps to find out whether the employer has in fact acted accordingly.

The effect of taking over part of the works on the insurance cover is completely ignored in both of these forms of contract, a fact that exposes such part of the works to the risk of damage or loss without the benefit of insurance. Accordingly, as in the case of the whole of the works, before issuing a Taking-Over Certificate or a Certificate of Completion in respect of a part of the works, the insurance arrangements for such part must be ascertained taking into consideration the subrogation rights of the insurers. The right of subrogation is the right of an insurer who has paid for any loss or damage to claim from another party who is responsible for part, or all, of the loss or damage and who is uninsured by that insurer, see page 181.

It is not unusual for this insurance gap, which occurs when a part of the works had been taken over, to result in a problem at the final stages of construction, as the parties involved are more likely to be concerned with technical engineering matters rather than with insurance details. The result may also be unpredictable, as was the legal decision in the case of *English Industrial Estates v. G. Wimpey.*[14] In that case,

---

14 *English Industrial Estates Corp. v. G. Wimpey & Co. Ltd.* [1973], 1 Lloyd's Report 118 (CA) Clause 16 of the JCT Contract stated: 'if at any time or times before practical completion of the Works the employer, with the consent of the contractor shall take possession of any part or parts of the same, then, notwithstanding anything expressed or implied elsewhere in the contract such parts shall, as from the date on which the employer shall have taken possession thereof be at the sole risk of the employer as regards any of the contingencies referred to in Clause 20(A).'

The FIDIC Red Book, 4th Edition, 1992 Reprint, Clause 20

20.1 Care of the Works

The Contractor shall take full responsibility for the care of the Works and materials and Plant for incorporation therein from the Commencement Date until the date of issue of the Taking-Over Certificate for the whole of the Works, when the responsibility for the said care shall pass to the Employer. Provided that:

(a) if the Engineer issues a Taking-Over Certificate for any Section or part of the Permanent Works the Contractor shall cease to be liable for the care of that Section or part from the date of issue of the Taking-Over Certificate, when the responsibility for the care of that Section or part shall pass to the Employer, and

(b) the Contractor shall take full responsibility for the care of any outstanding Works and materials and Plant for incorporation therein which he undertakes to finish during the Defects Liability Period until such outstanding Works have been completed pursuant to Clause 49.

20.2 Responsibility to Rectify Loss or Damage

If any loss or damage happens to the Works, or any part thereof, or materials or Plant for incorporation therein, during the period for which the Contractor is responsible for the care thereof, from any cause whatsoever, other than the risks defined in Sub-Clause 20.4, the Contractor shall, at his own cost, rectify such loss or damage so that the Permanent Works conform in every respect with the provisions of the Contract to the satisfaction of the Engineer. The Contractor shall also be liable for any loss or damage to the Works occasioned by him in the course of any operations carried out by him for the purpose of complying with his obligations under Clauses 49 and 50.

20.3 Loss or Damage Due to Employer's Risks

In the event of any such loss or damage happening from any of the risks defined in Sub-Clause 20.4, or in combination with other risks, the Contractor shall, if and to the extent required by the Engineer, rectify the loss or damage and the Engineer shall determine an addition to the Contract Price in accordance with Clause 52 and shall notify the Contractor accordingly, with a copy to the Employer. In the case of a combination of risks causing loss or damage any such determination shall take into account the proportional responsibility of the Contractor and the Employer.

20.4 Employer's Risks

The Employer's risks are:

(a) war, hostilities (whether war be declared or not), invasion, act of foreign enemies,

(b) rebellion, revolution, insurrection, or military or usurped power, or civil war,

(c) ionising radiations, or contamination by radio-activity from any nuclear fuel, or from any nuclear waste from the combustion of nuclear fuel, radio-active toxic explosive or other hazardous properties of any explosive nuclear assembly or nuclear component thereof,

(d) pressure waves caused by aircraft or other aerial devices travelling at sonic or supersonic speeds,

(e) riot, commotion or disorder, unless solely restricted to employees of the Contractor or of his Subcontractors and arising from the conduct of the Works,

(f) loss or damage due to the use or occupation by the Employer of any Section or part of the Permanent Works, except as may be provided for in the Contract,

(g) loss or damage to the extent that it is due to the design of the Works, other than any part of the design provided by the Contractor or for which the Contractor is responsible, and

(h) any operation of the forces of nature against which an experienced contractor could not reasonably have been expected to take precautions.

ICE Conditions, 7th Edition – Measurement Version – Clause 20

20 Care of the Works

(1)(a) The Contractor shall save as in paragraph (b) hereof and subject to sub-clause (2) of this Clause take full responsibility for the care of the Works and materials plant and equipment for incorporation therein from the Works Commencement Date until the date of issue of a Certificate of Substantial Completion for the whole of the Works when the responsibility for the said care shall pass to the Employer.

(b) If the Engineer issues a Certificate of Substantial Completion for any Section or part of the Permanent Works the Contractor shall cease to be responsible for the care of that Section or part from the date of issue of that Certificate of Substantial Completion when the responsibility for the care of that Section or part shall pass to the Employer.

(c) The Contractor shall take full responsibility for the care of any work and materials plant and equipment for incorporation therein which he undertakes during the Defects Correction Period until such work has been completed.

Excepted Risks

(2) The Excepted Risks for which the Contractor is not liable are loss or damage to the extent that it is due to

(a) the use or occupation by the Employer his agents servants or other contractors (not being employed by the Contractor) of any part of the Permanent Works

(b) any fault defect error or omission in the design of the Works (other than a design provided by the Contractor pursuant to his obligations under the Contract)

(c) riot war invasion act of foreign enemies or hostilities (whether war be declared or not)

(d) civil war rebellion revolution insurrection or military or usurped power

(e) ionizing radiations or contamination by radioactivity from any nuclear fuel or from any nuclear waste from the combustion of nuclear fuel radioactive toxic explosive or other hazardous properties of any explosive nuclear assembly or nuclear component thereof and

(f) pressure waves caused by aircraft or other aerial devices travelling at sonic or supersonic speeds.

Rectification of loss or damage

(3)(a) In the event of any loss or damage to

   (i) the Works or any Section or part thereof or

   (ii) materials plant or equipment for incorporation therein

while the Contractor is responsible for the care thereof (except as provided in sub-clause (2) of this Clause) the Contractor shall at his own cost rectify such loss or damage so that the Permanent Works conform in every respect with the provisions of the Contract and the Engineer's instructions. The Contractor shall also be liable for any loss or damage to the Works occasioned by him in the course of any operations carried out by him for the purpose of complying with his obligations under Clauses 49 and 50.

(b) Should any such loss or damage arise from any of the Excepted Risks defined in sub-clause (2) of this Clause the Contractor shall if and to the extent required by the Engineer rectify the loss or damage at the expense of the Employer.

(c) In the event of loss or damage arising from an Excepted Risk and a risk for which the Contractor is responsible under sub-clause (1)(a) of this Clause then the Engineer shall when determining the expense to be borne by the Employer under the Contract apportion the cost of rectification into that part caused by the Excepted Risk and that part which is the responsibility of the Contractor.

the contractor was engaged in 1969 to build a new extension to a factory belonging to the employer and leased to a certain lessee. The Conditions of Contract used were the Standard Form of Building Contract, JCT, which allocated to the employer, under Clause 16, the risk of fire in any part of the works that had been taken into possession. The employer allowed the lessee to install equipment and store material in a part of the newly constructed extension. Fire occurred in 1970 in the new extension causing considerable damage estimated then at £250,000. It was held that the contractors at the time of the fire had not in effect handed over to the employers, and that although the lessee was using part of the extension, it was the contractor's responsibility to insure it until actual hand-over. As a result, the contractor was held to be liable to the employer for the damage caused to the parts used by the employer.

It would be prudent therefore to include a term in this part of the clause clarifying the transfer of risk and the exact date when this formality is supposed to take place.

Paragraph (b) of Sub-Clause 20.1 also provides for the contractor to take full responsibility for the care of any outstanding work he undertakes to complete during the Defects Liability Period. As stated earlier, it is imperative that such work is identified clearly and its extent specified because the insurance requirements are based on such identification and specification.

Sub-Clauses 20.2 to 20.4 of FIDIC's Red Book and their equivalent Sub-clauses 20(3) and 20(3) of the ICE Conditions, are intended to identify and allocate the risks that lead to loss and/or damage between the employer and the contractor, if and when these risks eventuate. It is extremely important to realise that the provisions of these sub-clauses are not intended to deal with or cater for the risks that lead to economic loss or loss of time.

Although the wording of these sub-clauses shows a great improvement on the wording of the clauses in the previous editions of these two forms of contract, no attempt has been made to explain the difference in concept and in treatment under the contract of the matrix of risks of loss and/or damage and the matrix of economic and/or time loss. A lot is left to the imagination of the user of these conditions of contract as to where these risks of loss and/or damage fit into the general principles of risk allocation for the whole contract and how they interact with using insurance, where available, to lighten the burden of any probable liabilities as a result of one of these risks eventuating. In this connection, Clause 20 is worded in an ambiguous manner providing a good example of the piecemeal development of the principles of risk allocation without reference to what is intended by the provisions of this clause. To clarify this part of the contractual arrangements between the contracting parties, its intention must be set out in a precise manner.

The intention of Clause 20 is to divide the risks of loss and/or damage into two matrices, the first of which is allocated to the contractor and includes all risks 'whatsoever' other than those specifically mentioned in the second matrix. The second matrix of risks is allocated to the employer and is referred to in FIDIC's Red Book as the 'Employer's Risks' where each of these risks is specifically identified in Sub-clause 20.4. In the ICE Conditions, this second matrix of risks is referred to as the 'Excepted Risks' and dealt with in Sub-clause 20(2).[15] In FIDIC's Red Book, the second matrix of risks (i.e. the Employer's Risks) is further divided into two groups, the first of which is called the 'Special Risks', which includes the risks that are specifically identified in Sub-Clause 65.2 and the second group includes all other risks of the 'Employer's Risks' matrix. This subdivision is only mentioned in Clause 65, which provides that the contractor 'shall be under no liability whatsoever in consequence of any of the Special Risks referred to in Sub-clause 65.2 . . .'. As this group of risks is part of the Employer's Risks, which are dealt with in Clause 20 of the Conditions, the first edition of this book recommended that they should be

---

15 The term 'Excepted Risks' was used in the previous editions of FIDIC's Red Book, but was changed to 'Employer's Risks' in the 4th edition in order to reflect precisely what was intended by the matrix of risks, which it represented. Furthermore, the term 'Excepted' was confusing, firstly as it did not explain what it was excepting from, and secondly it sounded very like 'accepting', which has the opposite intention.

brought back to Clause 20.[16] There is no equivalent clause to the Special Risks clause in the ICE Conditions and Clause 65 dealt with the outbreak of war. In the 7th edition of the ICE Conditions, the war clause became part of Clause 63, which deals with 'Frustration'.

Sub-Clauses 20.2 to 20.4 of FIDIC's Red Book and their equivalent Sub-clauses 20(3) and 20(3) of the ICE Conditions are also intended to serve as the basis of liability for the cost of repair and making good any damage sustained by the works due to any of the risks that lead to loss or damage. They allocate the liability for the cost of such repair and making good to the contractor in respect of any of the risks in the first matrix referred to above. Any loss or damage which occurs as a result of the second matrix of risks, 'the Employer's Risks', is repaired and made good at the cost of the employer if it is not a Special Risk. If the damage is as a result of a Special Risk, the repair or making good is carried out in accordance with the provisions of Sub-Clause 65.3, as a variation to the contract necessitating an addition to the Contract Price valued in accordance with Clause 52: 'Valuation of Variations'.

---

16 Although this recommendation was not totally implemented, the reference to Clause 65 in the equivalent wording of the penultimate sentence of Sub-clause 20(1) of the third edition of FIDIC's Red Book has been deleted from the 4th edition. This indicates that all the special risks are indeed within Clause 20 of the 4th edition. It is also interesting to note that many of the recommendations made in the first edition of this book were heeded and that those that were not remain as a cause of a problem in both the Red Book and the ICE Conditions. The recommendations that were heeded are summarised below and those that were not are dealt with in the remaining part of this chapter:

- Simplification of the language and the introduction of shorter sentences and more bullet points;
- The liability for loss and/or damage occurring as a result of both Employer's and Contractor's Risks is now shared between the employer and the contractor;
- The new wording of Sub-clause 20.3 of the 4th edition of the Red Book allows flexibility in evaluating the cost of repair or making good as a result of an Employer's Risk;
- Tidying up some of the wording of the Employer's Risks, or the Excepted Risks in the ICE Conditions, and removing unnecessary differences between the two forms of contract. In particular, reference should be made to the risks of riot, the use or occupation by the employer of a part of the works, the design risk in connection with the words 'a cause solely due to' in the Red Book.
- Correcting the scope of the insurance of the Works to exclude only part of the Employer's Risks, to eliminate the necessity for a second Contractors' All Risks policy to be negotiated and obtained by the employer.
- Removing the phrase 'for which he (the Contractor) is responsible under the terms of the Contract' from the text of Clause 21 in both forms of contract considered here, thus clarifying the intent of this clause. Also, correcting the confusion caused by the words 'including for the purposes of this Clause any unfixed materials . . .', in the 5th edition of the ICE Conditions.
- The reference to excess or deductible in Clause 21, whether implied or expressly stated.
- The changes made to Clause 22, which simplify the language and make it more comprehensible.
- Re-introduction of joint insurance in Clause 23 and tidying up the wording of that clause, incuding the addition of the cross liability clause.
- Adding a number of general insurance requirements to Clause 25.

Therefore, where the contractor is concerned, the following implications can be construed from the operation of Sub-Clauses 20.2 and 20.3, in terms of the two matrices of risks and their sub-divided groups of risks:

1   The word 'if' in the phrase '. . ., *the Contractor shall, if and to the extent required by the Engineer, . . .*' in the first sentence of Sub-clause 20.3 can be interpreted to indicate that the contractor may in fact be relieved of his obligations to complete the items damaged due to an Employer's Risk, if the engineer does not require him to carry out any repair or making good after such damage.

2   The repair or making good to any damage caused to the works is paid for in two different ways:

   A   In the case of an Employer's Risk, as a variation under Clause 52 of the Conditions with a special provision for replacement of contractor's Equipment when it results as a consequence to a Special Risk; and

   B   At the cost of the contractor in the case of any risk other than an Employer's Risk.

The equivalent clause in the ICE Conditions is Sub-clause 20(3), which provides that loss or damage as a result of an Excepted Risk, if required to be rectified, should be paid for by the employer at his 'expense'. The word 'expense' is intended here to mean cost plus profit.

The last sentence of Sub-clause 20.2 of the FIDIC document deals with the situation where the contractor is engaged on work during the Defects Liability Period for the purpose of either completing any outstanding work or carrying out his obligations under Clauses 49 and 50. The responsibility for damage to the works caused by the contractor during this period is allocated to him. A similar liability provision is included in the ICE Conditions under the last sentence of paragraph (a) of Sub-clause 20(3).

Both documents also apportion liability for damage occurring as a result of risks eventuating from both matrices of risks, i.e. those within the definition of the Employer's Risks and others within the responsibility of the contractor, in accordance with their proportional responsibility and not in accordance with the dominant risk. See in this connection the last sentence of Sub-Clause 20.3 of FIDIC's Red Book; and paragraph (c) of Sub-clause 20(3) of the ICE Conditions.

Before leaving these two Sub-clauses, it is necessary to pay attention to the word 'whatsoever' in the phrase 'If any loss or damage happens to the Works, . . ., from any cause whatsoever, . . .' in FIDIC's Red Book, which is essentially an indemnity clause requiring legally an absolute clarity, precision and explicitness in its composition. The courts have interfered in the case of *Smith and Others v. South Wales Switchgear Ltd.*[17] and the subsequent case of *Photo Production Ltd. v. Securicor* in 1980,[18] as a result of such a clause.

---

17   *Smith v. South Wales Switchgear Ltd.,* [1978] 1 All ER 18.
18   *Photo Production Ltd. v. Securicor Transport Ltd.,* [1980] 1All ER 556.

The judgments made in the two cases referred to above included important statements in this connection. In order to understand these statements, the facts in each case are now narrated:

1 The first case concerned UBM Chrysler (Scotland) Ltd., motor manufacturers who had for some years engaged the South Wales Switchgear Co. Ltd. to carry out an annual overhaul of electrical equipment at their factory in Scotland. In 1970, UBM Chrysler once again requested this service, by their letter dated 18 March. The request was accepted and the contractors were sent a purchase note requesting the service specified in the note, which included the sentence 'subject to . . . our General Conditions of Contract 24001, obtainable on request'. The contractors did not request a copy of the General Conditions, although on 1 July 1970, a copy of the 1969 version was sent to them. On 16 July 1970, the contractors informed their client that instructions had been given for the work to be carried out.

It seems that three versions of the General Conditions existed: an original version; the 1969 version; and an amended one dated March 1970. All three versions had the number 24001 and included an indemnity clause which provided that

> in the event of the order involving the carrying out of work by the supplier and its subcontractors on land and/or premises of (client), the supplier will keep the (client) indemnified against . . . any liability, loss, claim or proceedings whatsoever under Statute or Common Law (i) in respect of personal injury, or death of, any person whomsoever in respect of any injury or damage whatsoever to any property, real or personal, arising out of or in the course of . . . the execution of the order.

On 25 July 1970, an employee of the contractors, Mr W. Smith who was engaged on an overhaul at the employer's factory, was seriously injured in an accident. Mr Smith brought an action for damages against the employer alleging negligence and breach of statutory duty. The Employer served a third-party notice on the contractors claiming indemnity under the indemnity clause of the General Conditions of Contract. The court held that the accident was wholly caused by the employer's negligence and breach of statutory duty and awarded Mr Smith damages against the employer. The court also held that the employer was entitled under the indemnity clause to be indemnified by the contract against the liability which they had incurred. On appeal, this decision was affirmed and the appellants appealed to the House of Lords contending that:

(i) since it was uncertain which of the three versions of the General Conditions of Contract would have been sent to the Contractor if the latter had requested a copy, it had not been proved that any of the versions had been incorporated into the contract, and

(ii) on its proper construction, the indemnity clause did not require the contractor to indemnify the employer against liability for the employer's own negligence or that of their employees.

Briefly, it was held:

(i) that the reference in the purchase order to the employer's 'General Conditions of Contract 24001, obtainable on request' was sufficient to incorporate into the contract the conditions as revised in March 1970.

(ii) that the principles applicable to a clause which purported to confer exemptions from liability on one party to a contract applied to a clause of indemnity. Accordingly, the party in whose favour the indemnity clause was made was entitled to indemnity against the consequences of his own negligence, or that of his servants, only where the indemnity clause contained an expressed provision to that effect, or where the words of the clause in their ordinary meaning were wide enough to cover negligence on the part of the party in whose favour the indemnity clause was made. The appeal, therefore, was allowed.

Viscount Dilhorne stated the following;

> While an indemnity clause may be regarded as the obverse of an exempting clause, when considering the meaning of such a clause one must, I think, regard it as even more inherently improbable that one party should agree to discharge the liability of the other party for acts for which he is responsible. In my opinion it is the case that the imposition by the proferens (the party in whose favour the indemnity clause was inserted) on the other party of liability to indemnify him against the consequences of his own negligence must be imposed by very clear words.

2   In the second case, Photo Production Ltd. employed Securicor Transport Ltd. to provide security services, including night patrols, at their factory. While visiting the factory one night to carry out his duties, an employee of Securicor deliberately started a fire which got out of control and a large part of the factory and stock were burnt down. Photo Production Ltd. sued Securicor for damages on the ground that they were liable for the act of their employee. Securicor pleaded an exception clause in their contract with Photo Production to the effect that under no circumstances were they to be

> responsible for any injurious act or default by any employee of the Company unless such act or default could have been foreseen and avoided by the exercise of due diligence on the part of the Company as his employer; nor, in any event, shall the Company be held responsible for: (a) Any loss suffered by the customer (Photo Production) through . . . fire or any other cause, except insofar as such loss is solely attributable to the negligence of the Company's employees acting within the course of their employment . . .

The trial judge held that Securicor were entitled to rely on the exception clause. The Court of Appeal reversed his judgment, holding that there was a fundamental breach of the contract by the defendants which precluded them from relying on the exception clause.[19] The defendants, Securicor Transport Ltd., appealed to the House of Lords, where the appeal was allowed on the basis that

(i) There was no rule of law by which an exception clause in a contract could be eliminated from a consideration of the parties' position when there was a breach of contract (whether fundamental or not) or by which an exception clause could be deprived of effect regardless of the terms of the contract. Lord Wilberforce stated that '. . . the question is one of construction, not merely of course of the exclusion clause alone, but of the whole contract'.

(ii) Although the defendants were in breach of their implied obligation to operate their service with due and proper regard to the safety and security of the plaintiff's premises, the exception clause was clear and unambiguous and protected the defendants from liability. Lord Wilberforce in his judgment made the following comments on the Unfair Contract Terms Act 1977 (UK) which is of extreme importance in the context of this section on insurance clauses:

> . . . in commercial matters generally, when the parties are not of unequal bargaining power, and when risks are normally borne by insurance, not only is the case for judicial intervention undemonstrated, but there is everything to be said and this seems to have been Parliament's intention for leaving the parties free to apportion the risks as they think fit and for respecting their decisions.
> . . .
> At the judicial stage there is still more to be said for leaving cases to be decided straightforwardly on what the parties have bargained for rather than on analysis, which becomes progressively more refined, of decision in other cases leading to inevitable appeals.

It follows that in order to avoid any risk of misinterpretation of the intention of Sub-clause 20.2, the lesson that might be learnt from the above two cases indicates that the wording of the first sentence of this part of the clause should perhaps be changed to the following:

> If any loss or damage shall happen to the Works, or to any part thereof, or materials or Plant for incorporation therein, during the period for which the Contractor is responsible for the care thereof, from any cause whatsoever including the negligence or default of the Employer his servants or agents, other than the risks defined in Sub-Clause 20.4, the Contractor shall at his own cost, rectify such loss . . . to the satisfaction of the Engineer.

Sub-Clause 20.4 – The Employer's Risks

As stated above, the Red Book identifies the risks of loss and /or damage as the Employer's Risks whereas the ICE Conditions still use the confusing title of the

---

19 A fundamental breach is one which entitles the party not in default to elect to terminate the contract.

Excepted Risks. As explained earlier, these risks are grouped in the Red Book into two groups, as follows:

Group 1 – Special Risks:
1.1 war, hostilities (whether war be declared or not), invasion, and acts of foreign enemies,
1.2 rebellion, revolution, insurrection, or military or usurped power, or civil war, insofar as these risks relate to the country in which the Works are to be executed,
1.3 ionising radiation or contamination by radio-activity from any nuclear fuel or from any nuclear waste from the combustion of nuclear fuel, radio-active toxic explosive or other hazardous properties of any explosive nuclear assembly or nuclear component thereof,
1.4 pressure waves caused by aircraft or other aerial devices travelling at sonic or supersonic speeds,
1.5 riot, commotion or disorder, unless solely restricted to employees of the Contractor or of his Subcontractors and arising from the Conduct of the Works.

Group 2 – Employer's Risks, which are not Special Risks:
2.1 loss or damage due to the use or occupation by the Employer of any Section or part of the Permanent Works, except as may be provided for in the Contract,
2.2 loss or damage to the extent that it is due to the design of the Works, other than any part of the design provided by the Contractor or for which the Contractor is responsible,
2.3 any operation of the forces of nature against which an experienced contractor could not reasonably have been expected to take precautions,
2.4 rebellion, revolution, insurrection, or military or usurped power, or civil war, insofar as these risks relate to a country outside that in which the Works are to be executed.

All the above risks are designated as risks belonging to the employer in standard forms of contract in order to produce harmony between such forms and standard insurance policies that are available in the insurance market. It is unclear, therefore, why there should be any significant difference between the ICE Conditions of Contract and one of its offspring, the Red Book, in an international matter such as insurance. However, differences other than simple wording arrangements exist and these are detailed below:

A   The Red Book includes as an Employer's Risk any operation of the forces of nature against which an experienced contractor could not reasonably have been expected to take precautions. Such a clause is an example of a good recipe for confusion and dispute. Words such as 'experienced contractor', 'reasonably', 'expected to take precautions', are extremely difficult to define or debate in precise manner.

B   There is a difference between the two sets of Conditions of Contract in connection with the design risk. The ICE Conditions refer to this risk as 'loss or damage to the extent that it is due to any fault defect error or omission in the design of the Works (other than a design provided by the Contractor pursuant to his obligations

242

under the Contract)'. The Red Book defines this risk as 'loss or damage to the extent that it is due to the design of the Works, other than any part of the design provided by the Contractor or for which the Contractor is responsible'. Whatever wording is used, it must be connected and read together with the duties of the designer in the agreement between him and his client, whether that client is the employer or the contractor. This is due to the fact that an accident due to the engineer's design may occur even if no negligence has been committed. Such an event could occur simply because of the state of art and knowledge in the field of engineering, as was the issue in question in the Australian case, *Manufacturers' Mutual Insurance v. The Queensland Government Railway*, where the decision of the court hinged on the definition of faulty design.[20]

The facts in this case were that an insurance policy covering loss or damage arising out of or in connection with a contract for the supply and erection of a railway bridge excluded 'loss or damage arising from faulty design'. The bridge, at Mirani, Pioneer River, Australia, was to replace one built in 1897, which was swept away by flood in 1956. The new bridge was to be constructed on the site of the old one. The insured sustained loss when, in an unprecedented flood, three concrete piers collapsed due to the inadequacy of their design to withstand the forces then experienced. The design of the piers complied with the standard to be expected from experienced professional engineers at that time. A dispute arose between the insured and the insurers as to whether such a claim in respect of this event was payable. An arbitrator was appointed and the question submitted to the arbitrator was: 'Are the insured entitled to be indemnified by the insurer in respect of the loss and damage sustained . . .?'

The arbitrator gave his reasons in a document that did not form part of the award he rendered. The award, however, referred to these reasons. The arbitrator, having stated that the insurance company accepted the onus of establishing that the exclusion referred to did apply, and having considered the question of the cause of the collapse of the piers, said:

> . . . the transverse forces which may operate upon piers in a stream are of a magnitude not previously realised or recognised by engineers, and that where piers will be subjected in a stream to such transverse forces, their shape is of great importance. It follows from what I have said that on the evidence which I heard, I have concluded that the prismatic piers as designed, and as they were being constructed were inadequate to withstand the transverse forces to which they could be subjected in the Pioneer River . . .
>
> . . . I am satisfied that the design of the piers in the sense of their prismatic shape resulted in loss, but the question which then follows for consideration is the meaning and effect of 'faulty design' as these words appear in the policy.

The arbitrator, holding *inter alia* that the design of the piers contained no element of personal failure, or non-compliance with the standards to be expected of the design engineers, awarded that the insured was entitled to be indemnified.

20 *Manufacturers' Mutual Insurance v. The Queensland Government Railway & Another,* 1968, Q.W.N. 12.

On 14 February 1967, the Supreme Court of Queensland ordered that the agreement to submit to arbitration be made a Rule of the Court. On 13 July 1967, the Supreme Court of Queensland dismissed an appeal by the insurers for a motion that the award be set aside or, alternatively, be remitted to the arbitrator for redetermination by him on the grounds that there was an error on the face of the award because the arbitrator misconstrued the meaning of the phrase 'faulty design'. The insurers appealed to the High Court of Australia, which allowed the appeal and reversed the judgment of the Supreme Court of Queensland. In allowing the appeal, the following important statement was made:

> Before this Court two points, not relied upon before the Full Court and now, it seems, taken principally to preserve whatever rights the respondent may have, were raised. First that the reasons for award did not form part of the award, and secondly that as the question of the construction of the policy had been referred to the arbitrator, an error of construction afforded no ground for interference by the Court. Were either point to be accepted, the award could not be impeached and that would be an end of the matter. In our opinion, however, neither point is valid . . .
>
> Accepting, of course, the learned arbitrator's findings of fact which have already been set out, we have come to the conclusion that upon the proper construction of the relevant exclusion, the loss which occurred did arise from faulty design. Let it be accepted, as the arbitrator found, that the piers, as designed, failed to withstand the water force to which they were subjected because they were designed in accordance with engineering knowledge and practice which was deficient, rather than because the designer failed to take advantage of such professional knowledge as there was nevertheless the loss was due to 'faulty design' and the arbitrator has done no more than explain how it happened that the design was faulty. We think it was an error to confine faulty design to 'the personal failure or non-compliance with standards which would be expected of designing engineers' responsible for the piers. To design something that would not work simply because at the time of its designing insufficient is known about the problems involved and their solution to achieve a successful outcome is a common enough instance of faulty design. The distinction which is relevant is that between 'faulty', i.e. defective design and design free from defect. We have not found sufficient ground for reading the exclusion in this policy as not covering loss from faulty design when, as here, the piers fell because their design was defective, although, according to the finding, not negligently so. The exclusion is not against loss from 'negligent designing'; it is against loss from 'faulty design', and the latter is more comprehensive than the former. For the foregoing reasons we consider that this appeal should succeed.

In the same judgment, J. Windeyer stated:

> . . . Doubtless a faulty design can be the product of fault on the part of the designer. But a man may use skill and care, and he may do all that in the circumstances could reasonably be expected of him, and yet produce

something which is faulty because it will not answer the purpose for which it was intended. His product may be faulty although he be free of blame . . .

This is of course a good example of a professional being responsible for an event but not liable for it, see page 144. Another interesting outcome from the above judgment is the following statement comparing faulty design and faulty workmanship:

. . . Faulty workmanship I take to be a reference to the manner in which something was done, a fault on the part of a workman or workmen. A faulty design, on the other hand, is a reference to a thing. If the words were 'faulty designing' the two phrases might perhaps be comparable: but the words are 'faulty design'.

The inclusion of the ICE Conditions of the words '. . . (other than a design provided by the Contractor pursuant to his obligations under the Contract)', or the similar wording in the Red Book, separates the responsibilities and liabilities for design carried out by the contractor in respect of his obligations under the contract from those resulting from design carried out by the employer or his agents.

Clause 21 – Insurance of the Works, etc.

This clause sets down the insurance requirements for which the contractor is responsible in respect of the works and any related items connected with their construction. It is worthy of note that the scope of the insurance cover is not directly related to the responsibilities and liabilities allocated to the parties under Clause 20. Specific items are provided in the Bill of Quantities for such insurance to be priced by the contractor when tendering. The employer pays for such insurance to safeguard against the possibility of severe losses occurring during the currency of the contract, which may render the contractor financially incapable of completing the project. Such a prospect is unacceptable not only to the contractor but also to the employer.

Clause 21 of the Red Book sets out the insurance requirements with respect to the works and the contractor's equipment. The contractor's equipment is a defined term in the Red Book and could in certain civil engineering projects involve large amount of money.[21] The 7th edition of the ICE Conditions refers only to the works and not to contractor's equipment. The scope and extent of the insurance cover required is based on the responsibilities and liabilities allocated not only to the contractor but also those allocated to the employer under Clause 20. The insurance gap, which existed in the 3rd edition of the Red Book and in the 5th edition of the ICE Conditions, where the contractor was required to insure the

---

21 Contractor's Equipment is defined in paragraph (v) of Sub-Clause 1.1(f) of the 4th edition of the Red Book as 'all appliances and things of whatsoever nature (other than the Temporary Works) required for the execution and completion of the Works and the remedying of any defects therein, but does not include Plant, materials or other things intended to form or forming part of the Permanent Works'.

works only against those risks of loss or damage for which he was responsible, has been eliminated in the 4th edition of the Red Book and in the 7th edition of the ICE Conditions.

Sub-Clause 21.1 provides that the works together with materials and plant for incorporation therein should be insured to the full replacement cost plus an additional sum of 15% of such replacement cost, or as may be specified in Part II of the Conditions. This additional percentage is required to cover any extra costs of and incidental to the rectification of loss or damage including professional fees and the cost of demolition and removal of any part of the works and of removal of debris of whatsoever nature. Furthermore, the term 'replacement cost' is qualified by defining cost as inclusive of profit within the context of this clause. This is essential if the contractor were to be able to include profit in any claim he makes to the insurer in respect of repair or replacement of a part of the works.[22]

In this connection, the additional percentage required in Sub-clause 21(1) of the ICE Conditions for the equivalent of this item is 10%.

Sub-Clause 21.1 also provides for insurance of the contractor's equipment and other things brought on to the site by the contractor, for a sum sufficient to provide for their replacement at the site. The replacement cost should therefore be reflected in the sum insured or the monetary value of the insurance policy to be provided for the relevant elements. From an insurance point of view, the sum insured is usually defined in the Schedule section of the insurance policy in a precise manner so as to establish the basis of claim settlement if a risk covered under the policy eventuates.

Besides determining the insurer's liability, the sum insured must be in precise terms because it forms the basis of calculating the insurance premium and the basis of any analysis of the performance of the insurance transaction. The calculation of the replacement cost is based on the contract price plus any subsequent adjustment made through variations, additions, omissions and normal inflation, such as, increases in the cost of material, plant, labour and machinery. Such normal inflation referred to as the primary inflation, should be covered by the Contractors' All Risks Policy in addition to two other elements of inflation in the event of loss or damage to any completed or partially completed part of the works. These are:

- The inflation between the time at which such part of the work is originally carried out and the time at which it is repaired or reinstated, which is referred to here as secondary inflation;
- The inflation that occurs during the delay in executing any uncompleted part of the works after such event of damage, which is referred to here as the transitional inflation.

22 It is an important insurance philosophy that its purpose is to place the insured in the same position he was in prior to the loss, and thus it is not accepted generally that an insured party would be permitted to make profit as a result of a claim in respect of an insured item and particularly so if such a claim is a result of his own negligence. However, if the loss or damage is not the result of negligence of the contractor or if an independent contractor is brought to carry out the repair, the question of entitlement to profit becomes less difficult to understand.

The FIDIC Red Book, 4th Edition, 1992 Reprint, Sub-Clauses 21.1 to 21.4

21.1 Insurance of Works and Contractor's Equipment

The Contractor shall, without limiting his or the Employer's obligations and responsibilities under Clause 20, insure:

(a) the Works, together with materials and Plant for incorporation therein, to the full replacement cost (the term 'cost' in this context shall include profit),

(b) an additional sum of 15 per cent of such replacement cost, or as may be specified in Part II of these Conditions, to cover any additional costs of and incidental to the rectification of loss or damage including professional fees and the cost of demolishing and removing any part of the Works and of removing debris of whatsoever nature, and

(c) the Contractor's Equipment and other things brought onto the Site by the Contractor, for a sum sufficient to provide for their replacement at the Site.

21.2 Scope of Cover

The insurance in paragraphs (a) and (b) of Sub-Clause 21.1 shall be in the joint names of the Contractor and the Employer and shall cover:

(a) the Employer and the Contractor against all loss or damage from whatsoever cause arising, other than as provided in Sub-Clause 21.4, from the start of work at the Site until the date of issue of the relevant Taking-Over Certificate in respect of the Works or any Section or part thereof as the case may be, and

(b) the Contractor for his liability:
  (i) during the Defects Liability Period for loss or damage arising from a cause occurring prior to the commencement of the Defects Liability Period, and
  (ii) for loss or damage occasioned by the Contractor in the course of any operations carried out by him for the purpose of complying with his obligations under Clauses 49 and 50.

21.3 Responsibility for Amounts not Recovered

Any amounts not insured or not recovered from the insurers shall be borne by the Employer or the Contractor in accordance with their responsibilities under Clause 20.

21.4 Exclusions

There shall be no obligation for the insurances in Sub-Clause 21.1 to include loss or damage caused by:

(a) war, hostilities (whether war be declared or not), invasion, act of foreign enemies,

(b) rebellion, revolution, insurrection, or military or usurped power, or civil war,

(c) ionising radiations, or contamination by radio-activity from any nuclear fuel, or from any nuclear waste from the combustion of nuclear fuel, radio-active toxic explosive or other hazardous properties of any explosive nuclear assembly or nuclear component thereof, or

(d) pressure waves caused by aircraft or other aerial devices travelling at sonic or supersonic speeds.

ICE Conditions, 7th Edition – Measurement Version – Clause 21

21 Insurance of Works etc.

(1) The Contractor shall without limiting his or the Employer's obligations and responsibilities under Clause 20 insure in the joint names of the Contractor and the Employer the Works together with materials plant and equipment for incorporation therein to the full replacement cost plus an additional 10% to cover any additional costs that may arise incidental to the rectification of any loss or damage including professional fees cost of demolition and removal of debris.

Extent of cover

(2)(a) The insurance required under sub-clause (1) of this Clause shall cover the Employer and the Contractor against all loss or damage from whatsoever cause arising other than the Excepted Risks defined in Clause 20(2) from the Works Commencement Date until the date of issue of the relevant Certificate of Substantial Completion.

(b) The insurance shall extend to cover any loss or damage arising during the Defects Correction Period from a cause occurring prior to the issue of any Certificate of Substantial Completion and any loss or damage occasioned by the Contractor in the course of any operation carried out by him for the purpose of complying with his obligations under Clauses 49, 50 and 51.

(c) Nothing in this Clause shall render the Contractor liable to insure against the necessity for the repair or reconstruction of any work constructed with materials or workmanship not in accordance with the requirements of the Contract unless the Bill of Quantities provides a special item for this insurance.

(d) Any amounts not insured or not recovered from insurers whether as excesses carried under the policy or otherwise shall be borne by the Contractor or the Employer in accordance with their respective responsibilities under Clause 20.

Finally, in international contracts, it would be useful to add a provision for any payments from the insurer to be made to the insured in foreign currency, and so it may be appropriate to add the following sentence to Sub-clause 21.1:

> The insurance in paragraphs (a) and (b) shall provide for compensation to be payable in the types and proportions of currencies required to rectify the loss or damage incurred.

This is an important provision, especially when the contractor is obliged to insure with a local insurance company which would arrange for settlement of any claims in local currency, despite the usual reinsurance arrangements which are made by insurers with international reinsurance companies.

Sub-Clause 21.2 defines the period of insurance and the extent of insurance cover required during the two constituents of this period, i.e. the Construction and the Defects Liability Periods. Under the ICE Conditions, Sub-clause 21(2), the Defects Liability Period is referred to as the Defects Correction Period, which is a more accurate terminology for what that period is intended to mean. Of course, the liability of the contractor continues after that period ends and its duration is dependent on the applicable law of the contract. For the purposes of this chapter, the term Defects Correction Period will be used.

During the construction period, insurance cover is required in the Red Book 'from the start of work at the Site until the date of issue of the relevant Taking-Over Certificate in respect of the Works or any Section or part thereof as the case may be'. This can create a gap in the insurance cover between the Commencement Date, which is a defined term, and the date of start of work at the site, and particularly so where the contract contains a substantial plant element.[23] It is difficult to understand why this restriction has been incorporated in the Red Book since it does not result in any saving in premium and only makes a complicated subject more complex. It is suggested in the Guide to the Red Book published by FIDIC that one way of avoiding this gap might be to alter Sub-Clause 21.2 through replacing the words 'start of work at the Site' by the defined term 'Commencement Date'.[24] However, a warning is given that such a change 'could lead to complications in arranging insurance' and that 'other wordings might be more appropriate in the context of a particular contract'. If this change is implemented in Sub-clause 21.2, then it is suggested in the Guide, and rightly so, that a similar change should be made in Sub-clause 25.1, in order to maintain consistency.

During the Defects Correction Period, the insurance cover required under Clause 21 is reduced to any damage due to: first, an event during the Construction Period but discovered in the Defects Correction Period; and, second, an event during the Defects Correction Period which is 'occasioned by the Contractor in the course of any operations carried out by him for the purpose of complying with his obligations under Clauses 49 and 50'. Clauses 49 and 50 are contained in a section of the Red Book entitled Defects Liability, and Sub-Clause 49.2 incorporates any work that a contractor undertakes under Clause 48, which deals with the Taking-Over Certificate. Similarly, these two clauses are contained in a section of the ICE Conditions entitled 'Outstanding Work and Defects' and Sub-clause 49(2) incorporates any work that a contractor undertakes under Clause 48, which deals with the Certificate of Substantial Completion. However, in the latter form of contract, Clause 51 is added to Clauses 49 and 50 and the contractor should provide cover for carrying his obligations in respect of ordered variations during the relevant period.

---

23 Commencement Date is defined in paragraph (i) of Sub-clause 1.1(c) of the 4th edition of the Red Book as 'the date upon which the Contractor receives the notice to commence issued by the Engineer pursuant to Clause 41'.

24 *Guide to Use of FIDIC Conditions of Contract for Works of Civil Engineering Construction, Fourth Edition*, Fédération Internationale des Ingénieurs-Conseils, Switzerland, 1989, page 18. This publication contains 173 pages of commentary and the text of the 4th edition of the Red Book, and is available from FIDIC, PO Box 86, CH 1000 Lausanne, 12–Chailly, Switzerland.

The insurance policy produced under this clause must be issued in the joint names of the employer and the contractor. It is not sufficient therefore to name the employer in the policy as a principal or to note his interest in its provisions. To be jointly insured, the policy must include in its schedule an explicit statement to that effect. Furthermore, joint insurance should extend to the Defects Correction Period and not as suggested on page 73 of the Guide to the use of FIDIC Red Book, referred to above, where it is stated that

> during the Defects Liability Period the insurance is only against that damage which the Contractor is required to repair under the terms of the Defects Liability Clause and so the Employer has no insurable interest in this part of the policy. This section of the insurance could, therefore be in the name of the Contractor alone.

As explained above, the required insurance cover extends beyond the repair of defects into loss or damage arising from a cause occurring prior to the commencement of the Defects Liability Period. It would, therefore, be wise and would cost no more to have the full cover in the joint names of the employer and the contractor.

Although it may be that, under certain legal concepts, it is immaterial whether or not joint insurance is implemented,[25] the practical implications are enormous. If the insured is only one of the parties to the contract, for example the contractor, then the following practical problems may arise:

1 Any alteration to the insurance cover provided by a policy issued in the name of one party is notified only to that party. The solution proposed by some to make it a contractual duty on the part of the insured party, in this example the contractor, to inform the other party, the employer, of any such alteration is of no use to the latter if the former defaults and an event occurs which results in an insolvent contractor and a silent site.

2 The right of negotiation with the insurers in the event of an occurrence covered under an insurance policy is reserved to the party named as insured.

3 The right of representation at arbitration in respect of a dispute under an insurance policy is limited to the party named as insured in that policy. If the employer is not named as joint insured, he will not be permitted to appear in the private proceedings of an arbitration process or have his views expressed therein.

4 The subrogation rights of the insurer may be exercised after an incident has caused a loss for which the insurer has paid.

5 Any payment made by the insurer in respect of damage or loss to the insured works is made to the insured party. If insurance is not jointly arranged, the party not named in the policy has no control over the sums paid and although it is generally accepted that the insurers will only pay when the repairs are carried out, the control over the manner in which they are carried out may be lost.

In any case, both forms of contract require the insurance of the works to be in the joint names of the employer and the contractor in respect of all loss or damage

---

25 *The ICE Conditions of Contract 5th Edition, A Commentary*, N. Duncan Wallace, Sweet & Maxwell, London, 1978, page 70.

arising from whatever cause other than the risks specifically excluded in Sub-Clause 21.4 of the Red Book or in Sub-clause 20(2) in the case of the ICE Conditions.

Clause 21 of the ICE Conditions differs in its text from its equivalent in the Red Book on a fundamental issue. It specifically recognises, in paragraph (c) of Sub-clause 21(2), that insurance is not required against the risk of incorporating in the works 'material and workmanship not in accordance with requirements of the Contract unless the Bill of Quantities provides a special item for this insurance'. Whilst the reasons for this provision are not stated, it is clear that this is a wise exclusion to be made by an employer on a contractor who is employed to carry out construction work in a correct manner and using appropriate material as specified.[26] This requirement is made irrespective of whether or not insurance is available in respect of such risks in the insurance market. Including the words 'unless the Bill of Quantities shall provide a special item for this insurance' caters for an exception to this rule. This last phrase is presumably meant for manufactured articles, which are usually manufactured outside the site and brought in for installation in the works. They are normally covered against the risks of defective material and defective workmanship, through Manufacturer's Risk insurance.

It is essential, however, to state at the outset that the relevant phrase in Clause 21 of the ICE Conditions is intended to specifically exclude from the insurance cover the necessity of insuring the defective parts themselves and not any other part which may become damaged or lost as a result of failure of a defective part.[27] Parts of the works that are not defective in themselves but sustain damage as a result of the failure of another part of the works, which is defective, are insured under the standard Contractors' All risks policy as their cost is included within the sum insured stated in the schedule of the policy. If it is intended that such an insurance policy exclude resultant damage, a special exclusion to that effect must be inserted in the policy. However, such a policy would not then be in accordance with the requirements of the ICE Conditions as the risk of resultant damage from defective materials or workmanship is not an Excepted Risk, which delineate the permitted exclusion from the insurance cover required under paragraph (a) of Sub-clause 21(2).

Similar analysis can be made in connection with faulty or defective design, which is an Excepted Risk, and therefore not the responsibility of the contractor. Insurance

---

26 See page 190 on why construction projects differ from manufactured articles.
27 A practical example of the difference between the damage to a defective part and resultant damage to a part that is not defective is clear from the following event: A high parapet wall was constructed at the top of a three-storey office block built using precast concrete elements. The wall, costing no more than £3,000 to build, collapsed under normal wind conditions and fell on to the roof of the office block. This resulted in the collapse of the precast concrete roof slabs that, in turn, caused eccentric distribution of forces on to the precast concrete frame, which collapsed in its entirety. The cost of the office block when it collapsed was in the region of £300,000. It was found that the wall was both defectively designed and defectively constructed. In this situation, if the insurance policy had been issued excluding from the cover any defective design and/or workmanship and any damage consequent upon such defects, the insurers would not be expected to pay anything. However, if the insurance cover excluded defective design and/or workmanship but included resultant damage, the insurers would be expected to pay £297,000 but not the cost of the defective wall of £3,000.

of such defective items is not required under Clause 21 since it follows the Excepted Risks. However, insurance of the parts that are not defective but become damaged as a result of failure of the defective part is required under Clause 21 as it is not an Excepted Risk. If an insurance policy is intended to exclude damage to parts which are not defective in themselves, the policy must state this exclusion clearly. If it does, it may be argued that the policy is not in accordance with the requirements of the ICE Conditions of Contract. One may ask why a contractor should be made responsible for damage resulting from defective design? The responsibility in this case emanates from Clause 20 where the contractor is made responsible for the loss or damage resultant from risks that do not come within the definition of the excepted risks under Sub-clause 20(2). This difficulty might be eliminated by separating the allocation of risks from the responsibility for providing insurance against the risks.

The Red Book, on the other hand, is silent on the topic of insuring defective material or defective workmanship. Under the Red Book, the contractor is expected to provide an insurance cover against all risks other than those excluded under Sub-Clause 21.4, which therefore include defective material and defective workmanship. This requirement, which must be intentional since it is one of the few problems that were not corrected following the publication of the first edition of this book, is unfortunate and detrimental to the very core of the construction contract. The insurance of defective material and workmanship in a construction contract is not insurance *per se* but a lottery, and an insurer following proper insurance principles would not consider insuring these defects, except in a few cases where he would be certain that the insured would implement procedures for eliminating such a risk. Otherwise, the matter in question is not whether damage would occur but rather when it would.

Furthermore, the use of substandard material and workmanship is much less costly than proper material and workmanship. Hence, where contracts are awarded on the basis of competitive tendering, it is only a matter of time before the lowest and perhaps successful tenders are based on the supply of the cheapest possible materials and workmanship, in many cases a standard lower than acceptable.[28] Lessons should be learnt from the experience of tendering procedures gained during times of recession. During such periods, competition for contracts is keenest and tenders are submitted at cost if not below, with contractors depending on claims made during the construction period to ensure a break-even or provide an insurance policy which not only excludes defective design but also resultant damage.

Besides defective material and defective workmanship, the contractor should be permitted to have in his Contractors' All Risks policy an exclusion of indirect and consequential losses which is a standard exclusion in this type of insurance policies. However, the liability for such losses remains the contractor's.

---

28 In this connection, some may argue that it is not essential to have both 'if' and 'when' unknown entities in the insurance equation and that the insurance prerequisites are satisfied if only one of those variables is known. They may give as an example life insurance where it is inevitable for a human to die and the only question is when. However, this comparison is erroneous since the question of benefit must be considered and whereas death does not provide financial gain to the person whose life is insured, the contrary is true in the case of defective material and defective workmanship.

Sub-Clause 21.3 of the Red Book provides for the situation where there is under-insurance of an item that is lost or damaged. In the case of under-insurance, the value of the claim is reduced in the proportion of the sum indicated in the policy to the sum representing the replacement cost of the damaged element. This insurance principle is referred to as 'the average clause' and is usually incorporated in most Contractors' All Risks policies. In such a case, Sub-Clause 21.3 allocates the responsibility for any amounts not insured or not recovered from the insurers in respect of the cost of the replacement or repair of the lost or damaged item to the party who is responsible for that item pursuant to Clause 20.

Sub-Clause 21.3 of the Red Book could be taken to refer indirectly to the excess that is usually imposed in all construction insurance policies on the insured in respect of each loss or damage claimed from the insurer. In some cases it is referred to as deductible and can be defined as a specified amount or percentage of each and every claim, which the insured has to bear in case of loss. An important character-istic of insurance is the relationship between the excess imposed and the premium charged: as the amount of excess increases, the premium charged in respect of the relevant insurance decreases. It is therefore important to specify an upper and lower limit for the excess allowed in the insurance requirements. A lower limit below which the contractor is not permitted to drop is usually specified in order to reduce the premium and also to stipulate and encourage effective protective measures to be implemented on site to reduce the probability of occurrence of damage. A higher limit beyond which the contractor is not allowed to go is usually specified in order to prevent the loss of benefit of insurance. Such limits should therefore be specified in the Form of Tender, which forms part of the tender documents.

It is for this reason that the Guide to the use of the FIDIC Red Book, referred to above, recommends the addition of the phrase 'and with deductible limits for the Employer's Risks not exceeding . . . (insert amounts)' to paragraph (a) of Sub-Clause 21.1. However, it is not clear why this provision for deductibles is not extended to the contractor's risks or why the provision was not incorporated in the text of Clause 21 when the Red Book was reprinted with amendments in 1992, which was after the publication of the Guide.

In the equivalent clause in the ICE Conditions, Sub-clause 21(2)(d), a specific reference is made to excesses, but more significantly as can be seen later, the excesses are dealt with in Sub-clause 25(2).

Clause 22 – Damage to Persons and Property – Indemnity

Sub-Clause 22.1 Damage to Persons and Property

Clause 22 of the FIDIC Conditions is the indemnity clause where the liability towards third parties emanating from the construction and completion of the works is identified, allocated and apportioned in accordance with planned criteria specified therein. It is therefore complementary to Clause 20 and is one of the most important but tortuous clauses in a construction contract. It deals with relationships beyond those between the contracting parties, i.e. engineering, construction, insurance and law, disparate fields which come together to define the outcome of possible events that may cause damage to property other than the works and injury to persons other than those involved in the project. For such a clause to succeed, it requires to be clearly drafted with little or no sacrifice of brevity, two contrasting demands resulting in a difficult task for the draftsman.

The difficulty in drafting can be appreciated by examining this clause not only in the Red Book but also in the ICE document. However, it can be said that a major improvement has been achieved in the latest editions of these two documents. Sub-clause 22(1) of the 5th edition of the ICE Conditions, which contained two sentences of 102 and 265 words, and its equivalent Sub-Clause 22.1 of the 3rd edition of the Red Book, are now much slimmer and much more comprehensible.

However, there still remains a number of matters that should be noted in Clause 22 in the abovementioned two documents and as the wording of these two clauses is very similar to each other, the provisions contained in them can be discussed together:

1   The indemnity specified in Sub-clause 22(1) is qualified in two respects. The first is by exceptions detailed under Sub-clause 22(2)(a) to (e) in the ICE document and Sub-clauses 22.2(a) to (d) in the Red Book. The second qualification is in respect of contributory negligence by either of the parties to the other who would be found liable in respect of a loss, damage or injury.

The manner in which the contributory negligence is dealt with in the two documents differs. In the ICE Conditions, the mechanism of contribution is dealt with in Sub-clause 22(4). It has been phrased in this manner as was necessary in view of the decision in the case of *A.M.F. International Ltd. v. Magnet Bowling & G.P. Trentham Ltd.* where it was found that an explicit wording was necessary in an indemnity clause if it were to operate legally in a situation of contributory tortious negligence. Sub-clause 22(4)(a) views the indemnity from the contractor to the employer and Sub-clause 22(4)(b) views it from the employer to the contractor.[29]

In the Red Book, contributory negligence is dealt with in Sub-Clause 22.2(d) which is in itself a qualification to the indemnity provided by the contractor in favour of the employer/owner. It only applies however in the situation described in that Sub-Clause. The wording of this Sub-Clause is the same as that in Sub-Clause 22(1)(d) of the 3rd edition of the Red Book on which Mr. Duncan-Wallace suggested that a similar remedy may apply through a claim for direct breach of contract on the *Mowbray v. Merryweather* principle but, obviously, at the risk of having to enter into a legal dispute on liability.[30]

2   There is no equivalent statement in the FIDIC Conditions to that expressed in Sub-clause 22(2)(a) of the ICE Conditions, which deals with damage to crops on the site.

29 *A.M.F. International Ltd. v. Magnet Bowling & G.P. Trentham Ltd.* [1968] 1 WLR 1028, as reported on in *Hudson's Building and Engineering Contracts*, 11th edition, by I.N. Duncan-Wallace, Sweet & Maxwell, (1995), 15–051, page 1451.

30 *Mowbray v. Merryweather* [1895] 2 QB 640, as reported on in *The ICE Conditions of Contract 5th Edition, A Commentary*, by IN. Duncan-Wallace, Sweet & Maxwell, (1978), page 76 and in *Hudson's Building and Engineering Contracts*, 11th edition, by I.N. Duncan-Wallace, Sweet & Maxwell, (1995), 15–049, page 1450.

The FIDIC Red Book, 4th Edition, 1992 Reprint, Sub-Clauses 22.1 to 22.3

22.1 Damage to Persons and Property
The Contractor shall, except if and so far as the Contract provides otherwise, indemnify the Employer against all losses and claims in respect of:
(a) death of or injury to any person, or
(b) loss of or damage to any property (other than the Works),
    which may arise out of or in consequence of the execution and completion of the Works and the remedying of any defects therein, and against all claims, proceedings, damages, costs, charges and expenses whatsoever in respect thereof or in relation thereto, subject to the exceptions defined in Sub-Clause 22.2.

22.2 Exceptions
The 'exceptions' referred to in Sub-Clause 22.1 are:
(a) the permanent use or occupation of land by the Works, or any part thereof,
(b) the right of the Employer to execute the Works, or any part thereof, on, over, under, in or through any land,
(c) damage to property which is the unavoidable result of the execution and completion of the Works, or the remedying of any defects therein, in accordance with the Contract, and
(d) death of or injury to persons or loss of or damage to property resulting from any act or neglect of the Employer, his agents, servants or other contractors, not being employed by the Contractor, or in respect of any claims, proceedings, damages, costs, charges and expenses in respect thereof or in relation thereto or, where the injury or damage was contributed to by the Contractor, his servants or agents, such part of the said injury or damage as may be just and equitable having regard to the extent of the responsibility of the Employer, his servants or agents or other contractors for the injury or damage.

22.3 Indemnity by Employer
The Employer shall indemnify the Contractor against all claims, proceedings, damages, costs, charges and expenses in respect of the matters referred to in the exceptions defined in Sub-Clause 22.2.

---

3   Damage to property which is the unavoidable result of the execution and completion of the works, or the remedying of any defects therein, in accordance with the contract is a condition which provides a release of the indemnity given by the contractor in favour of the employer/owner in Sub-Clause 22.2(c) of the Red Book. The ICE equivalent Sub-clause 22(2)(d) provides this release only to damage which is the unavoidable result of the construction of the works in accordance with the contract.

ICE Conditions, 7th Edition – Measurement Version – Clause 22

22 Damage to persons and property
(1) The Contractor shall except if and so far as the Contract provides otherwise and subject to the exceptions set out in sub-clause (2) of this Clause indemnify and keep indemnified the Employer against all losses and claims in respect of
  (a) death of or injury to any person or
  (b) loss of or damage to any property (other than the Works)
      which may arise out of or in consequence of the construction of the Works and the remedying of any defects therein and against all claims demands proceedings damages costs charges and expenses whatsoever in respect thereof or in relation thereto.

Exceptions
(2) The exceptions referred to in sub-clause (1) of this Clause which are the responsibility of the Employer are
  (a) damage to crops being on the Site (save in so far as possession has not been given to the Contractor)
  (b) the use or occupation of land provided by the Employer for the purposes of the Contract (including consequent losses of crops) or interference whether temporary or permanent with any right of way light air or water or other easement or quasi-easement which are the unavoidable result of the construction of the Works in accordance with the Contract
  (c) the right of the Employer to construct the Works or any part thereof on over under in or through any land
  (d) damage which is the unavoidable result of the construction of the Works in accordance with the Contract and
  (e) death of or injury to persons or loss of or damage to property resulting from any act neglect or breach of statutory duty done or committed by the Employer his agents servants or other contractors (not being employed by the Contractor) or for or in respect of any claims demands proceedings damages costs charges and expenses in respect thereof or in relation thereto.

Indemnity by Employer
(3) The Employer shall subject to sub-clause (4) of this Clause indemnify the Contractor against all claims demands proceedings damages costs charges and expenses in respect of the matters referred to in the exceptions defined in sub-clause (2) of this Clause.

Shared responsibility
(4)(a) The Contractor's liability to indemnify the Employer under sub-clause (1) of this Clause shall be reduced in proportion to the extent that the act or neglect of the Employer his agents servants or other contractors (not being employed by the Contractor) may have contributed to the said death injury loss or damage.
  (b) The Employer's liability to indemnify the Contractor under sub-clause (3) of this Clause in respect of matters referred to in sub-clause (2)(e) of this Clause shall be reduced in proportion to the extent that the act or neglect of the Contractor or his subcontractors servants or agents may have contributed to the said death injury loss or damage.

The limitation of relief to damage to property is attributed to the difficulty in obtaining insurance cover, under a Public Liability policy, for personal injury, which is the unavoidable result of the work.[31] Therefore, as the insurance requirements later expressed in Clause 23 are based on the indemnity Clause 22, it would not be worthwhile and might perhaps be futile to ask for an insurance cover not usually available in the market. The liability of the employer under the Red Book is, therefore, wider in that it includes any damage suffered which can be described as the unavoidable result of the completion of the works. However, both documents remain silent on the possibility of a loss being unavoidable due to the contractor's method of construction. Sub-Clause 22.2 should therefore continue to exclude this risk from the release in indemnity. This means that any damage which is the result of the contractor's method of construction would be the responsibility of the contractor.

4   Both documents omit to include in the exceptions from the indemnity required to be provided by the contractor to the employer those risks which are allocated to the employer under Clause 20 that might affect third parties, for example, defective design carried out by the employer or on his behalf. This indemnity clause should therefore have a further item added to its qualifying sub-clauses, explicitly referring to some of the Employer's Risks, and their effect.

5   Breach of statutory duty committed by the engineer or the employer is specifically excluded from the liability of the contractor in Sub-Clause 22(2)(e) of the ICE Conditions as a result of the decision in the case of *A.M.F. International v. Magnet Bowling,* discussed earlier in point 1 of this section.[32] This case also applies here since the contractor and the professional team normally share a breach of statutory duty through safety legislation. A similar clarification is needed in the equivalent Sub-clause 22.2(d). A cross reference to the Engineer's Professional Indemnity policy in respect of any breach of such duty must be made by the employer in the conditions of engagement of the professional team.

6   The indemnity expected from the contractor in favour of the employer in both the ICE Conditions and the Red Book is unlimited in amount. As this is unrealistic and could lead to misunderstanding, a false sense of security and inequality in tender evaluation, a sum should be fixed in the Form of Tender defining the limit set for this indemnity. The employer fixes this limit and it is then used for the operation of Clause 23, which deals with the insurance requirements emanating from the present clause. The proper place for the insertion of this limit is probably Clause 22(1).

Clauses 23 and 24 – Third Party Insurance – Injury to Workmen

These two clauses are dealt with simultaneously as they are concerned collectively with the question of insurance needed in response to the liabilities and indemnities provided under Clause 22.

---

31   *Engineering Law and the ICE Contracts,* Max W. Abrahamson, Applied Science Publishers Ltd., London, 4th ed., 1979, page 126.
32   *A.M.F. International Ltd. v. Magnet Bowling & G.P. Trentham Ltd.* [1968] 1 WLR 1028.

The FIDIC Red Book, 4th Edition, 1992 Reprint, Sub-Clauses 23.1 to 23.3

23.1 Third Party Insurance (including Employer's Property)
The Contractor shall, without limiting his or the Employer's obligations and responsibilities under Clause 22, insure, in the joint names of the Contractor and the Employer, against liabilities for death of or injury to any person (other than as provided in Clause 24) or loss of or damage to any property (other than the Works) arising out of the performance of the Contract, other than the exceptions defined in paragraphs (a), (b) and (c) of Sub-Clause 22.2.

23.2 Minimum Amount of Insurance
Such insurance shall be for at least the amount stated in the Appendix to Tender.

23.3 Cross Liabilities
The insurance policy shall include a cross liability clause such that the insurance shall apply to the Contractor and to the Employer as separate insureds.

24.1 Accident or Injury to Workmen
The Employer shall not be liable for or in respect of any damages or compensation payable to any workman or other person in the employment of the Contractor or any Subcontractor, other than death or injury resulting from any act or default of the Employer, his agents or servants. The Contractor shall indemnify and keep indemnified the Employer against all such damages and compensation, other than those for which the Employer is liable as aforesaid, and against all claims, proceedings, damages, costs, charges, and expenses whatsoever in respect thereof or in relation thereto.

24.2 Insurance Against Accident to Workmen
The Contractor shall insure against such liability and shall continue such insurance during the whole of the time that any persons are employed by him on the Works. Provided that, in respect of any persons employed by any Subcontractor, the Contractor's obligations to insure as aforesaid under this Sub-Clause shall be satisfied if the Subcontractor shall have insured against the liability in respect of such persons in such manner that the Employer is indemnified under the policy, but the Contractor shall require such Subcontractor to produce to the Employer, when required, such policy of insurance and the receipt for the payment of the current premium.

Under Sub-Clause 23.1, the contractor is required to insure, in the joint names of the contractor and the employer, against liabilities for death of or injury to any person, other than workmen referred to in Clause 24, or for loss of or damage to any property, other than the Works, arising out of the performance of the contract.

ICE Conditions, 7th Edition – Measurement Version – Clause 23

23 Third party insurance
(1) The Contractor shall without limiting his or the Employer's obligations and responsibilities under Clause 22 insure in the joint names of the Contractor and the Employer against liabilities for death of or injury to any person (other than any operative or other person in the employment of the Contractor or any of his subcontractors) or loss of or damage to any property (other than the Works) arising out of the performance of the Contract other than those liabilities arising out of the exceptions defined in Clause 22(2)(a) (b) (c) and (d).

Cross liability clause
(2) The insurance policy shall include a cross liability clause such that the insurance shall apply to the Contractor and to the Employer as separate insured.

Amount of insurance
(3) Such insurance shall be for at least the amount stated in the Appendix to the Form of Tender.

ICE Conditions, 7th Edition – Measurement Version – Clause 24

24 Accident or injury to operatives etc.
The Employer shall not be liable for or in respect of any damages or compensation payable at law in respect or in consequence of any accident or injury to any operative or other person in the employment of the Contractor or any of his subcontractors save and except to the extent that such accident or injury results from or is contributed to by any act or default of the Employer his agents or servants and the Contractor shall indemnify and keep indemnified the Employer against all such damages and compensation (save and except as aforesaid) and against all claims demands proceedings costs charges and expenses whatsoever in respect thereof or in relation thereto.

Excluded from the requirement to insure are the exceptions referred to in Sub-clause 22.2, but not those under paragraph (d) of the Red Book (or paragraph (e) of Sub-clause 22(2) in the case of the ICE Conditions).

Once again, a major improvement has been achieved in the wording of Clause 23 in both the 4th edition of the Red Book and in the ICE Conditions. The changes are very similar in both documents and include the re-introduction of joint insurance with the supplement of a cross-liability provision such that the insurance shall apply to the employer and the contractor as if they were separate insureds and accordingly each would be considered as a third party towards the other.

Once again, the fact that insurance is obtained does not alter the obligations and liabilities of the parties as defined in Clause 22. While these liabilities are not limited in amount, the required insurance is limited to an amount stated in the Appendix to Tender under Sub-Clause 23.2. The relevant provision under the ICE Conditions in this connection is Sub-Clause 23.3. The stated amount in respect of third party insurance is referred to as a minimum amount per occurrence with the number of occurrences being unlimited.

As stated above, the scope of insurance in the Red Book is also limited insofar as it may exclude the exceptions outlined in paragraphs (a), (b) and (c) of Sub-Clause 22.2, but not paragraph (d). This means that the insurance cover provided under this clause would have to cover any act or negligence of the employer, resulting in death of or injury to persons or loss of or damage to property, other than the works. Therefore, whilst the contractor is not required to indemnify the employer in case of a negligent act causing damage or injury to a third party, the insurance cover provided in compliance with Sub-Clause 23.1 should provide an indemnity to the limit specified under Sub-Clause 23.2. Similar arrangement is required under the ICE Conditions.

The period of insurance is indirectly specified by reference to the loss or damage arising out of the performance of the contract.

Turning to Clause 24, it stipulates that the employer shall not be liable for any damages or compensation payable to any workman or other person in the employment of the contractor or any subcontractor unless as a result of an act or default for which he is responsible. That liability includes acts or defaults of the employer's agents or servants. The clause also requires the contractor to indemnify the employer against all such damages and compensation unless they are the responsibility of the employer.

The liability in respect of accidents or injuries to employees of a contractor or a subcontractor is dealt with differently in various parts of the world. In many countries, there are specific statutory provisions applying to this type of liability and to the insurance transactions connected with it, making it difficult to formulate a single clause that could apply to all circumstances of a construction contract.

Sub-Clause 24.2 provides that the contractor must insure against his liability towards his workmen for the whole time at which he employs any person on the works. As the premium for this type of insurance is based on the pay-roll of the employer, Sub-Clause 24.2 provides that the contractor's obligation to insure would be satisfied if each subcontractor insures against his own liabilities towards his own workmen in a similar manner to that done by the contractor so that the employer is indemnified under the policy. In this connection, the contractor must require his subcontractors to present to the employer the policy of insurance and the receipt for the payment of the current premium. It should be noted that this policy is not required to be in the joint names of the contractor and the employer.

It should also be noted that there is no equivalent clause to Sub-Clause 24.2 in the ICE Conditions.

As can be gathered, therefore, many of the problems in Clauses 23 and 24 of the previous editions of the Red Book and the ICE Conditions have been resolved, except the following, which relate to Clause 24:

1   Due to the manner in which a premium is calculated in respect of Employer's Liability insurance, any person employed by the employer whose services are

loaned or made available to the contractor or sub-contractors will not be covered unless a term is specifically provided in Clause 24 of cover for such eventuality.

2   The liability in respect of accidents or injuries to employees of a contractor or subcontractor is dealt with differently in various parts of the world. In many countries, there are specific statutory provisions applying to this type of liability and to insurance transactions connected with it, making it necessary to consider the provisions of Clause 24 every time it is used in a particular jurisdiction.

Clause 25 – General insurance requirements

There are significant insurance requirements which are common to all three insurance policies required under Clauses 21, 23 and 24. These are set out in Clause 25 of both documents, the Red Book and the ICE Conditions. Sub-clause 25.1 of the Red Book stipulates that the contractor should provide to the employer evidence of compliance with the insurance requirements specified under the contract prior to the start of work at the site. If the insurances are required to be in force from the Commencement Date, which is the more appropriate start as used in the ICE document, and as discussed earlier,[33] then the wording of this sub-clause should be altered accordingly. In any event, the contractor is required to produce, within 84 days of the Commencement Date, the insurance policies required under the contract and to deliver them to the employer so that they can be checked and approved by him. The contractor is also required under the Red Book to notify the engineer when he has provided these policies and therefore the engineer is thus given a passive duty to act should matters not proceed in accordance with the provisions of the contract. It is not clear as to whether such duty exists under the ICE wording, but the matter ought to be resolved at the time of the first or second interim certificate when the contractor would presumably include the cost of the insurance in his monthly statement.

Sub-Clause 25.1 also requires the insurance policies to be in the general terms agreed prior to the issue of the Letter of Acceptance, which presumes that such terms would have been discussed and agreed at the relevant time.[34] Furthermore, the contractor is obliged under the provisions of Sub-Clause 25.2 to produce to the employer, when required to do so, the insurance policies in force and the receipts for payment of current premiums.

In the ICE Conditions, it is simply stated that the terms of the insurance shall be subject to the approval of the employer, provided that such approval shall not be unreasonably withheld. Some commentators have expressed doubt about the adequacy of the 84-day period for production of the insurance policies and especially so where complex international projects have to be insured through national insurance companies.[35] However, whilst this may have been the experience in the past, it is important to recognise that under Sub-Clause 25.1, the terms of the insurance policies must be discussed between the employer and the contractor before

---

33   See page 249 above.
34   There are a number of matters that are related to the Letter of Acceptance besides insurance, which are referred to in the following sub-clauses of the Red Book: 1.1(b)(i)(v); 1.1(b)(i)(vi); 1.1(e)(i); 5.2; 10.1; 14.1; 14.3; 25.1; 41.1; 57.2; and 66.1.
35   See the following note.

The FIDIC Red Book, 4th Edition, 1992 Reprint, Sub-Clauses 25.1 to 25.4

25.1 Evidence and Terms of Insurance

The Contractor shall provide evidence to the Employer prior to the start of work at the Site that the insurances required under the Contract have been effected and shall, within 84 days of the Commencement Date, provide the insurance policies to the Employer. When providing such evidence and such policies to the Employer, the Contractor shall notify the Engineer of so doing. Such insurance policies shall be consistent with the general terms agreed prior to the issue of the Letter of Acceptance. The Contractor shall effect all insurances for which he is responsible with insurers and in terms approved by the Employer.

25.2 Adequacy of Insurances

The Contractor shall notify the insurers of changes in the nature, extent or programme for the execution of the Works and ensure the adequacy of the insurances at all times in accordance with the terms of the Contract and shall, when required, produce to the Employer the insurance policies in force and the receipts for payment of the current premiums.

25.3 Remedy on Contractor's Failure to Insure

If the Contractor fails to effect and keep in force any of the insurances required under the Contract, or fails to provide the policies to the Employer within the period required by Sub-Clause 25.1, then and in any such case the Employer may effect and keep in force any such insurances and pay any premium as may be necessary for that purpose and from time to time deduct the amount so paid from any monies due or to become due to the Contractor, or recover the same as a debt due from the Contractor.

25.4 Compliance with Policy Conditions

In the event that the Contractor or the Employer fails to comply with conditions imposed by the insurance policies effected pursuant to the Contract, each shall indemnify the other against all losses and claims arising from such failure.

ICE Conditions, 7th Edition – Measurement Version – Clause 25

25 Evidence and terms of insurance

(1) The Contractor shall provide satisfactory evidence to the Employer prior to the Works Commencement Date that the insurances required under the Contract have been effected and shall if so required produce the insurance policies for inspection. The terms of all such insurances shall be subject to the approval of the Employer (which approval shall not unreasonably be withheld). The Contractor shall upon request produce to the Employer receipts for the payment of current insurance premiums.

Excesses
(2) Any excesses on the policies of insurance effected under Clauses 21 and 23 shall be no greater than those stated in the Appendix to the Form of Tender.

Remedy on Contractor's failure to Insure
(3) If the Contractor shall fail upon request to produce to the Employer satisfactory evidence that there is in force any of the insurances required under the Contract then the Employer may effect and keep in force any such insurance and pay such premium or premiums as may be necessary for that purpose and from time to time deduct the amount so paid from any monies due or which may become due to the Contractor or recover the same as a debt due from the Contractor.

Compliance with policy conditions
(4) Both the Employer and the Contractor shall comply with all conditions laid down in the insurance policies. Should the Contractor or the Employer fail to comply with any condition imposed by the insurance policies effected pursuant to the Contract each shall indemnify the other against all losses and claims arising from such failure.

---

the Letter of Acceptance is issued. It is at that time that the important features of these policies are determined, and all that would be necessary after the issue of the Letter of Acceptance is simply to have these policies processed. These features should include all the critical elements of the insurance cover, such as:

- The sum insured in the case of the Contractor's All Risks Policy and any required indexing to be applied on such sum during the contract period and which would be necessary to calculate the full replacement cost;
- The limit of indemnity in the case of the liability policies;
- The excesses to be applied in case of claims;
- The general and special exclusions from the insurance cover provided;
- The conditions attached to the policies; and
- Most importantly, perhaps, the mechanism of claim settlement and whether or not there is agreement on the appointment of a specifically appointed loss adjuster by the insurer and the insureds. Where such agreement is not made, a loss adjuster appointed by the insurer should be professionally qualified and should act in a recognised professional manner. If he does not, the whole process would lead to disputes and disrepute.

These provisions would entail comprehensive discussions and negotiations prior to the award of the contract but after submission of the tenders. Otherwise, it would be difficult, if not impossible, for a contractor to quote a definite price for the provision of these insurances. This may be the case especially if the critical features of the insurance policies are not precisely specified in the tender documents or if they are required to be altered during the pre-award negotiations.

More importantly, it would be a speculative task for the contractor to quote a definite price if insurance is not available for some of the risks required to be covered under the contract, such as the faulty material, workmanship or design cover required under the strict interpretation of the wording of Clauses 20 and 21 of the Red Book.

Sub-Clause 25.2 of the Red Book places an obligation on the contractor to notify the insurers of any change in the nature, extent or programme for the execution of the works. This provision is necessary as it is a usual condition of the Contractor's All Risks Policy that any such change must be notified to the insurer in case there is a material change in the risks covered. The insurance cover may be invalidated if the insurers are not made aware of such changes or, indeed, any other condition imposed under the terms of the policy. The contractor is also required to ensure the adequacy of the insurances at all times in accordance with the terms of the contract. This obligation is a continuing one and the contractor would be wise to clarify the position regarding any risks for which insurance is not available, at the earliest possible time, but in all circumstances not later than the pre-award negotiations. To this clause another requirement should be added, which would oblige the joint insureds to conform with the conditions of the insurance policies, as provided for in the ICE Conditions in Sub-clause 25(4), where the responsibility is placed on the shoulders of the joint insureds, the employer and the contractor, to comply with all the conditions laid down in the insurance policies.

If the contractor fails to effect and keep in force any of the insurances required under the contract or if he fails to provide the policies within the 84 days specified in Sub-Clause 25.1, then the employer may effect and keep in force such insurances. This provision is under Sub-Clause 25.3 (Sub-clause 25(3) in the ICE Conditions), which also entitles the employer to pay the premium necessary for effecting and keeping in force such insurances and then deducting the amounts paid from any monies due or to become due or as a debt due from the contractor.

The insurance contract being based on the principle of utmost good faith extends the application of its conditions from a period prior to the issue of the policy, the date of completion of the proposal form, to the date of completion of the period of insurance. The insured is thus required to abide by the following conditions:

(a) Conditions pertaining to the proposal form, which affect the validity of the whole of the insurance contract;
(b) Conditions stipulated in the insurance policy itself, which affect the validity of the cover from the date of any breach;
(c) Conditions related to claim settlement, the breach of which affects the particular claim.

Finally, Sub-Clause 25.4 of the Red Book requires that in the event that either the employer or the contractor fails to comply with the conditions of the insurance policy or policies, each shall indemnify the other against all losses or claims arising from such failure. Sub-clause 25(4) of the ICE Conditions provides for a similar condition.

Figure 8.3 shows a summary of the insurance arrangements as required in the 4th edition of the Red Book.

Clause 65 – Special Risks

Sub-Clause 65.1 – No Liability for War, etc. Risks

Clause 65 in the FIDIC document is very closely related to Clause 20 and a major portion of the Employer's Risks in Sub-Clause 20.4 is identical with what is termed as Special Risks under Clause 65, see page 242. This wording not only confuses the concept of Risk in the mind of the reader or the user but also confuses the very concept of Clause 65, which is to allocate the financial consequences when some of the risks eventuate.[36]

This situation may have arisen because of the fundamental change effected when the equivalent clause in the ICE document must have been considered. In the FIDIC

The FIDIC Red Book, 4th Edition, 1992 Reprint, Special Risks – Clause 65 Sub-Clauses 65.1 to 65.8

65.1 No Liability for Special Risks
The Contractor shall be under no liability whatsoever in consequence of any of the special risks referred to in Sub-Clause 65.2, whether by way of indemnity or otherwise, for or in respect of:

(a) destruction of or damage to the Works, save to work condemned under the provisions of Clause 39 prior to the occurrence of any of the said special risks,

(b) destruction of or damage to property, whether of the Employer or third parties, or

(c) injury or loss of life.

65.2 Special Risks
The special risks are:
(a) the risks defined under paragraphs (a), (c), (d) and (e) of Sub-Clause 20.4, and

(b) the risks defined under paragraph (b) of Sub-Clause 20.4 insofar as these relate to the country in which the Works are to be executed.

65.3 Damage to Works by Special Risks
If the Works or any materials or Plant on or near or in transit to the Site, or any of the Contractor's Equipment, sustain destruction or damage by reason of any of the said special risks, the Contractor shall be entitled to payment in accordance with the Contract for any Permanent Works duly executed and for any materials or Plant so destroyed or damaged and, so far as may be

---

36 'Recommendations for Review of the Conditions of Contract (International) for Works of Civil Engineering Construction', Dr Joachim E. Goedel [1985] ICLR 316.

required by the Engineer or as may be necessary for the completion of the Works, to payment for:

(a) rectifying any such destruction or damage to the Works, and

(b) replacing or rectifying such materials or Contractor's Equipment,

and the Engineer shall determine an addition to the Contract Price in accordance with Clause 52 (which shall in the case of the cost of replacement of Contractor's Equipment include the fair market value thereof as determined by the Engineer) and shall notify the Contractor accordingly, with a copy to the Employer.

65.4 Projectile, Missile

Destruction, damage, injury or loss of life caused by the explosion or impact, whenever and wherever occurring, of any mine, bomb, shell, grenade, or other projectile, missile, munition, or explosive of war, shall be deemed to be a consequence of the said special risks.

65.5 Increased Costs arising from Special Risks

Save to the extent that the Contractor is entitled to payment under any other provision of the Contract, the Employer shall repay to the Contractor any costs of the execution of the Works (other than such as may be attributable to the cost of reconstructing work condemned under the provisions of Clause 39 prior to the occurrence of any special risk) which are howsoever attributable to or consequent on or the result of or in any way whatsoever connected with the said special risks, subject however to the provisions in this Clause hereinafter contained in regard to outbreak of war, but the Contractor shall, as soon as any such cost comes to his knowledge, forthwith notify the Engineer thereof. The Engineer shall, after due consultation with the Employer and the Contractor, determine the amount of the Contractor's costs in respect thereof which shall be added to the Contract Price and shall notify the Contractor accordingly, with a copy to the Employer.

65.6 Outbreak of War

If, during the currency of the Contract, there is an outbreak of war, whether war is declared or not, in any part of the world which, whether financially or otherwise, materially affects the execution of the Works, the Contractor shall, unless and until the Contract is terminated under the provisions of this Clause, continue to use his best endeavours to complete the execution of the Works. Provided that the Employer shall be entitled, at any time after such outbreak of war, to terminate the Contract by giving notice to the Contractor and, upon such notice being given, the Contract shall, except as to the rights of the parties under this Clause and Clause 67, terminate, but without prejudice to the rights of either party in respect of any antecedent breach thereof.

65.7 Removal of Contractor's Equipment on Termination

If the Contract is terminated under the provisions of Sub-Clause 65.6, the Contractor shall, with all reasonable dispatch, remove from the Site all Contractor's Equipment and shall give similar facilities to his Subcontractors to do so.

65.8 Payment if Contract Terminated

If the Contract is terminated as aforesaid, the Contractor shall be paid by the Employer, insofar as such amounts or items have not already been covered by payments on account made to the Contractor, for all work executed prior to the date of termination at the rates and prices provided in the Contract and in addition:

(a) the amounts payable in respect of any preliminary items referred to in the Bill of Quantities, so far as the work or service comprised therein has been carried out or performed, and a proper proportion of any such items which have been partially carried out or performed;

(b) the cost of materials, Plant or goods reasonably ordered for the Works which have been delivered to the Contractor or of which the Contractor is legally liable to accept delivery, such materials, Plant or goods becoming the property of the Employer upon such payments being made by him;

(c) a sum being the amount of any expenditure reasonably incurred by the Contractor in the expectation of completing the whole of the Works insofar as such expenditure has not been covered by any other payments referred to in this Sub-Clause;

(d) any additional sum payable under the provisions of Sub-Clauses 65.3 and 65.5;

(e) such proportion of the cost as may be reasonable, taking into account payments made or to be made for work executed, of removal of Contractor's Equipment under Sub-Clause 65.7 and, if required by the Contractor, return thereof to the Contractor's main plant yard in his country of registration or to other destination, at no greater cost; and

(f) the reasonable cost of repatriation of all the Contractor's staff and workmen employed on or in connection with the Works at the time of such termination.

Provided that against any payment due from the Employer under this Sub-Clause, the Employer shall be entitled to be credited with any outstanding balances due from the Contractor for advances in respect of Contractor's Equipment, materials and Plant and any other sums which, at the date of termination, were recoverable by the Employer from the Contractor under the terms of the Contract. Any sums payable under this Sub-Clause shall, after due consultation with the Employer and the Contractor, be determined by the Engineer who shall notify the Contractor accordingly, with a copy to the Employer.

ICE Conditions, 7th Edition – Measurement Version – Clause 63

63 Frustration
(1) If any circumstance outside the control of both parties arises during the currency of the Contract which renders it impossible or illegal for either party to fulfil his contractual obligations the Works shall be deemed to be abandoned upon the service by one party upon the other of written notice to that effect.

War clause
(2) If during the currency of the Contract there is an outbreak of war (whether war is declared or not) in which Great Britain is engaged on a scale involving general mobilization of the armed forces of the Crown

(a) the Contractor shall for a period of 28 days reckoned from midnight on the date that the order for general mobilization is given continue so far as is physically possible to carry out the Works in accordance with the Contract and

(b) if substantial completion of the whole of the Works is not achieved before the said period of 28 days has expired the Works shall thereupon be deemed to be abandoned unless the parties otherwise agree.

Removal of Contractor's Equipment
(3) Upon abandonment of the Works pursuant to sub-clauses (1) or (2)(b) of this Clause the Contractor shall with all reasonable dispatch remove from the Site all Contractor's Equipment.

In the event of any failure so to do the Employer shall have like powers to those contained in clause 54(3) to dispose of any Contractor's Equipment.

Payment on abandonment
(4) Upon abandonment of the Works pursuant to sub-clauses (1) or (2)(b) of this Clause the Employer shall pay the Contractor (in so far as such amounts or items have not already been covered by payments on account made to the Contractor) the Contract value of all work carried out prior to the date of abandonment and in addition

(a) the amounts payable in respect of any preliminary items so far as the work or service comprised therein has been carried out or performed and a proper proportion of any such items which have been partially carried out or performed

(b) the cost of materials or goods reasonably ordered for the Works which have been delivered to the Contractor or of which the Contractor is legally liable to accept delivery (such materials or goods becoming the property of the Employer upon such payment being made to the Contractor)

(c) a sum being the amount of any expenditure reasonably incurred by the Contractor in the expectation of completing the whole of the Works insofar as such expenditure has not been recovered by any other payments referred to in this sub-clause and

(d) the reasonable cost of removal under sub-clause (3) of this Clause.

To this end and without prejudice to the provisions of sub-clause (5) of this Clause the provisions of Clause 60(4) shall apply to this sub-clause as if the date of abandonment was the date of issue of the Defects Correction Certificate.

Works substantially completed
(5) If upon abandonment of the Works any Section or part of the Works has been substantially completed in accordance with Clause 48 or is completed so far as to be usable then in connection therewith

(a) the Contractor may at his discretion and in lieu of his obligations under Clauses 49 and 50 allow against the sum due to him pursuant to sub-clause (4) of this Clause the cost (calculated as at the date of abandonment) of repair rectification and making good for which he would have been liable under the said Clauses had they continued to be applicable and

(b) the Employer shall not be entitled to withhold payment under Clause 60(6)(c) of the second half of the retention money or any part thereof except such sum as the Contractor may allow under the provisions of the last preceding paragraph.

Contract to continue in Force
(6) Save as aforesaid the Contract shall continue to have full force and effect.

Conditions, the contractor is neither liable nor expected to indemnify the employer in respect of any damage to the works or any other property or injury to anyone as a result of an event emanating from any of the Special Risks. In fact, it is the employer who is to indemnify the contractor in respect of such damage or any claims that may result therefrom. Only one qualification exists and that is in respect of defective work which is condemned under the provisions of Clause 39. In the 7th edition of the ICE document, the outbreak of war has been relegated to a sub-clause of Clause 63, which deals with frustration of the contract. Frustration is defined there as any circumstance outside the control of the parties, which renders it impossible or illegal for either of them to fulfil their contractual obligations. Thus, the equivalent ICE Sub-Clause to Sub-Clause 65.1 of the Red Book renders any direct comparison between the two documents inappropriate.

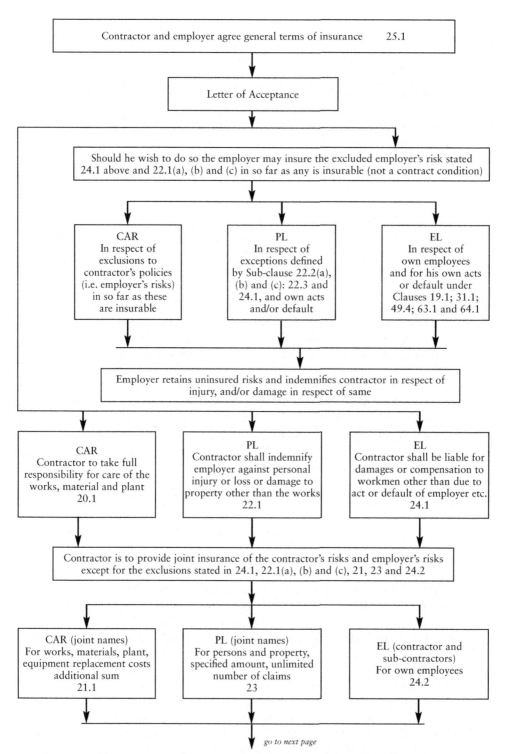

Figure 8.3 The Insurance Scheme as in the 4th edition of the Old Red Book.

from previous page

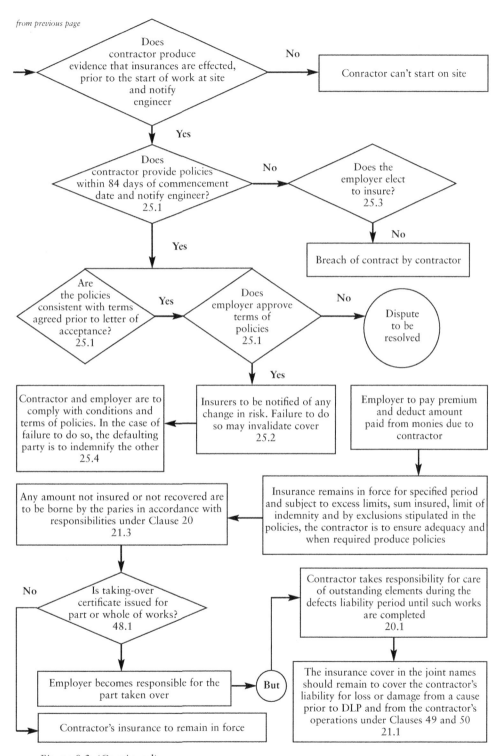

Figure 8.3 (Continued).

Two comments are necessary at this stage. These are:

(a) *The Risk of War, etc.:* Although the risk of war, which is an employer's risk under paragraph (a) of Sub-Clause 20.4, is expressly defined in Sub-Clause 65.6 as a special risk regardless of where it occurs in the world, by reference to Sub-Clause 65.2(b), there seems to be an intention that the combined risks of war, hostilities, invasion and acts of foreign enemies are also to be treated as Employer's Risks regardless of where they occur in the world. Similar intentions apply to the Employer's Risks as defined in paragraphs (c), (d) and (e) of Sub-Clause 20.4. However, it appears that this is intended to apply only in the context of the provisions of Sub-Clause 65.1 where destruction of or damage to the works and to other property and injury or loss of life arise due to the above risks. If that is the case, there is incompatibility with the provisions of Sub-Clause 65.2(b) where the definition of the risks of rebellion, revolution, insurrection, military or usurped power, and civil war are restricted to those relating to the country in which the works are to be executed. If destruction of an item covered by the description in Sub-Clause 65.1 happens due to war in the country where it is manufactured, it would be considered a special risk, why then would it not be a special risk under Sub-Clause 65.2(b) if it happens due to, for instance, civil war in that country?

(b) *The Risk of Pressure Waves:* The risk of pressure waves caused by aircraft or other aerial devices travelling at sonic or supersonic speeds is included as an Employer's Risk under paragraph (d) of Sub-Clause 20.4. The inclusion of this risk under the list of Employer's Risks originates from the ICE Form where in the UK, the Government undertook to pay compensation if damage resulted from the supersonic test flights originally made by Concorde. Consequently, insurers in the UK excluded this risk from their Contractors' All Risks Insurance Policy. Later when little or no damage was caused by supersonic test flights, the undertaking seems to have remained in respect of operational flights. In countries where the standard form of contract does not originate from the ICE Form, no specific reference is made to this risk. Therefore, there is no apparent reason for this risk to remain as one of the Employer's Risks and especially when insurance cover can be provided for it, if requested, under the Contractors' All Risks Insurance Policy.

Sub-Clause 65.2 – Special Risks

Sub-Clause 65.2 identifies the special risks by reference to the provisions of Sub-Clause 20.4. Thus, the special risks include the risks identified in paragraphs (a), (c), (d) and (e) of Sub-clause 20.4 and also those in paragraph (b) of that Sub-Clause in so far as the risks relate to the country in which the works are to be executed. Thus, a distinction is made between the risks of war, hostilities and acts of foreign enemies which are defined as special risks regardless of where they occur, and the risks of rebellion, revolution, insurrection, military or usurped power, and civil war which are recognised as special risks only if they occur in the country in which the works are to be executed.

In broad terms, the remaining part of Clause 65 deals with the consequences of the occurrence of a special risk. Figure 8.4 shows a flow chart of the consequences of these risks when they eventuate during the contract period.

Sub-Clause 65.3 – Damage to Works by Special Risks

Sub-clause 65.3 prescribes the contractual arrangements if the works, or any materials or plant on or near or in transit to the site, or any of the contractor's equipment, sustain destruction or damage by reason of any of the special risks. It provides that the contractor is entitled to payment in accordance with the contract for:

(a) Any permanent works duly executed;
(b) Any materials or plant destroyed or damaged; and
(c) In so far as the engineer requires or as may be necessary for the completion of the works, for
  (i) rectifying any destruction or damage to the works, and
  (ii) replacing or rectifying such materials or contractor's equipment.

With regard to (c) above, the phrase 'necessary for the completion of the Works' requires careful consideration. The intended meaning of this phrase is that the contractor is entitled to payment for materials and/or equipment under the pro-visions of Sub-clause 65.3 only if such material and/or equipment was present on the site for the purpose of completion of the works. Any materials or equipment present on the site for the convenience of the Contractors but not required for the comple-tion of the works at the time of occurrence of the Special Risk do not qualify for payment by the employer if it is destroyed or damaged as a consequence of the Special Risk. Furthermore, under Sub-Clause 65.3, payment for destroyed or damaged materials or Equipment is not dependent on the resumption of construction work or completion of the works after the occurrence of the Special Risk. This is because the basis of Clause 65 is the allocation of responsibility for these special risks to the employer.

Sub-Clause 65.3 continues by requiring the engineer to determine

> an addition to the Contract Price in accordance with Clause 52 (which shall in the case of the cost of replacement of Contractor's Equipment include the fair market value thereof as determined by the Engineer) and shall notify the Contractor accordingly, with a copy to the Employer.'

An important aspect of the wording in the above text is the use of the words 'the cost of replacement of the Contractor's Equipment [which] include the fair market value'. In this regard, the replacement cost or the fair market value must be reason-ably and fairly determined without any subjective influence. This aspect assumes great importance in international contracts and particularly where expensive items of machinery and equipment are used.

Any payment to the contractor under Clause 65 is determined by the engineer as an adjustment to the contract price in accordance with Clause 52 of the conditions of contract. As discussed earlier, this determination may not always be an addition.

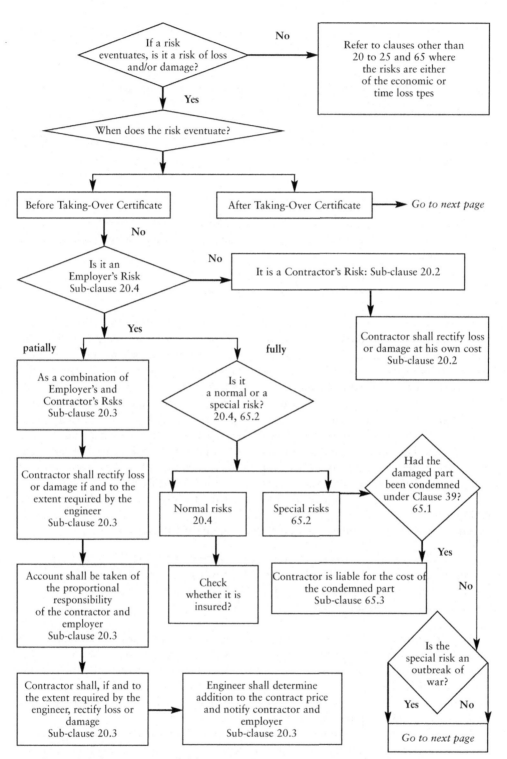

Figure 8.4 Consequences of risks eventuating.

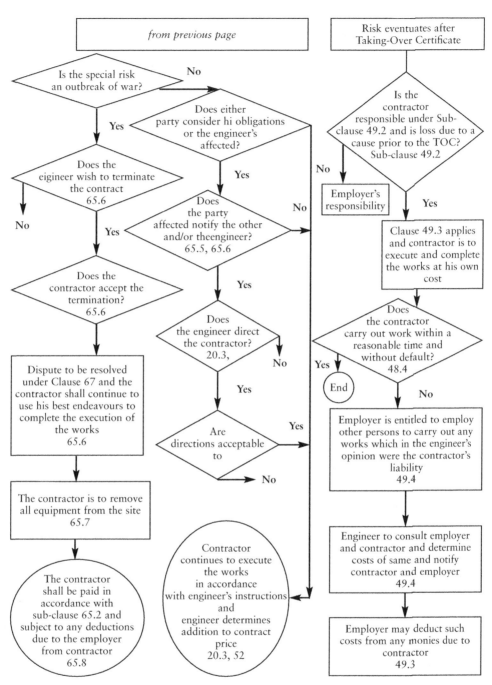

*Note 1: The contractor shall continue to use his best endeavours to complette the execution of the works 65.6
**Note 2: Contractor may invoke Clause 13 as being physically impossible to continue and/or Claue 67: dispute.

Figure 8.4 (Continued).

For example, a reduction may result if the part destroyed includes defective work, materials or plant and the contract is terminated under the provisions of Sub-Clause 65.6 without the engineer requiring the replacement of such defective work.

Sub-Clause 65.4 – Projectile, Missile

The definition of destruction, damage, injury or loss of life caused by the special risks is extended in Sub-Clause 65.4 to include any

> destruction, damage, injury or loss of life caused by the explosion or impact whenever and wherever occurring of any mine, bomb, shell, grenade, or other projectile, missile, munition, or explosive of war'.

Sub-Clause 65.5 – Increased costs arising from Special Risks

Sub-Clause 65.5 deals with increased costs arising from the Special Risks and allocates the liability for any increase in such cost of the works to the employer. In this regard, the wording of Sub-Clause 65.5 is extremely wide in that any increase in cost 'consequent on or the result of or in any way whatsoever connected with the said special risks . . .' is allocated to the employer. However, the contractor is required to notify the engineer forthwith of such increase as soon as it comes to his knowledge. Furthermore, the engineer is required to 'duly consult' with the employer and the contractor before determining the amount of the contractor's costs. The engineer is also required to notify the contractor in writing of the amount he has determined with a copy to the employer. Once again Clause 39 is the only qualification to this provision.

Sub-Clause 65.6 – Outbreak of War

The remaining parts of Clause 65 deal with the risk of war and the contractual arrangements, whether war is declared or not, and irrespective of where such risks eventuate in any part of the world. Sub-clause 65.6 provides that the employer is entitled to terminate the contract at any time after the outbreak of war by giving notice to the contractor and the contract becomes terminated upon such notice being given subject to the operation of Clause 67. Where the contractor is concerned, this sub-clause provides that if there is an outbreak of war, which materially affects the execution of the works, whether financially or otherwise, he is required to use his best endeavour to complete the works, unless the employer terminates the contract. This is an extremely onerous provision where the safety of employees is threatened and the employer delays his decision to terminate the contract.

If a contract is terminated under Sub-Clause 65.6, then the mechanism and consequences of such termination are as follows:

(a) The contract is terminated only when a notice is given by the employer to the contractor. Upon such notice being given, the contract terminates except as to the rights of the parties under Clause 65, 'Special Risks'; and Clause 67, 'Settlement of Disputes', but without prejudice to the rights of either party in respect of any antecedent breach of the contract. This means that once a contract is terminated under Sub-Clause 65.6, no provision of that contract survives other than the provisions of Clauses 65 and 67. This wording of Sub-clause 65.6 does not affect the rights and obligations of the parties to the contract in respect of a breach committed prior to the termination of the contract under that sub-clause.

(b) The rights of the parties following termination under Sub-clause 65.6 are dealt with under the following sub-clauses:

- Sub-Clause 65.3, 'Damage to Works by Special Risks', discussed above;
- Sub-Clause 65.7, 'Removal of Contractor's Equipment on Termination'; and
- Sub-Clause 65.8, 'Payment if Contract Terminated'.

In this regard, the provisions of Sub-Clause 65.3 have been discussed above, while the remainder of the present section considers Sub-Clauses 65.7 and 65.8.

Sub-Clause 65.7 – Removal of contractor's equipment on termination

As quoted above, Sub-Clause 65.7 provides that 'If the Contract is terminated under the provisions of Sub-Clause 65.6, the Contractor shall, with all reasonable dispatch, remove from the Site all Contractor's Equipment and shall give similar facilities to his Subcontractors to do so.'

Accordingly, Sub-Clause 65.7 provides for the right of the employer to require the Contractor, 'with all reasonable dispatch' to 'remove from the Site all the Contractor's Equipment'. This provision is intended to ensure that the effects of the occurrence of any of the Employer's Special Risks be reduced to a minimum by the removal from the site of such materials and equipment for which the employer would become liable if they remain on the site and sustain loss or damage.

Sub-Clause 65.8 – Payment if contract terminated

Sub-Clause 65.8 provides for the method and amount of payment to which the contractor is entitled if the contract is terminated under Sub-Clause 65.6. In such circumstances, the contractor is entitled to be paid by the employer for amounts not already paid in relation to all work executed prior to the date of termination at rates and prices provided in the contract. In addition, and as quoted earlier, the contractor is entitled to be paid the following amounts under Sub-Clauses 65.8(c) to (f):

(c) a sum being the amount of any expenditure reasonably incurred by the contractor in the expectation of completing the whole of the works insofar as such expenditure has not been covered by any other payments referred to in this Sub-Clause;

(d) any additional sum payable under the provisions of Sub-Clauses 65.3 and 65.5;

(e) such proportion of the cost as may be reasonable, taking into account payments made or to be made for work executed, of removal of Contractor's Equipment under Sub-Clause 65.7 and, if required by the Contractor, return thereof to the Contractor's main plant yard in his country of registration or to other destination, at no greater cost; and

(f) the reasonable cost of repatriation of all the Contractor's staff and workmen employed on or in connection with the Works at the time of such termination.'

There are various phrases used in the above provisions of Sub-Clause 65.8 which merit more detailed consideration. For example:

- The words '*expenditure* reasonably incurred' in (c) above must mean an amount of outlay that has already been incurred. Otherwise, the term 'cost' would have been used instead of 'expenditure', since 'cost' is defined in paragraph (i) of Sub-Clause 1.1 of the Red Book as 'expenditure . . . incurred or to be incurred . . .'.
- Therefore, the expression 'expenditure reasonably incurred . . . in the expectation of completing the whole of the Works' in paragraph (c) above, would mean an expenditure already incurred, which is essential for the project as a whole, such as, payment for the contractor's insurance policies pursuant to Clauses 21 and 23 of the general conditions of contract, or the expenditure incurred in obtaining the performance security required under Clause 10 of the general conditions of contract. It would not include an item not already incurred and neither would include consequential costs.
- The word 'proportion' in the expression 'such proportion of the cost as may be reasonable', which appears in paragraph (e) above, could mean anything from 0% to 100%. Thus, in certain circumstances, the word 'proportion' could mean zero if that were deemed to be the reasonable amount in the particular circumstances. Accordingly, when the latter meaning is incorporated in the remainder of Sub-Clause 65.8(c), the reasonable proportion of the cost of removal of the contractor's equipment may be very little if that equipment is simply removed and transported to a location nearer to the Site than the 'main plant yard' in his (the contractor's) country of registration.

9

THE INSURANCE CLAUSES IN FIDIC'S TRADITIONAL FORMS OF CONTRACT

A proposed redraft

The incredible increase in the number and size of international construction contracts witnessed during the 1970s and 1980s continued and is still on going in several countries around the world. Many contractors of different nationalities have become established in various foreign countries, some have formed joint ventures whilst others have forged long-term associations with other contractors and so their knowledge and experience have extended from their home industry to further fields.

By 1987, when the time came to revise FIDIC's Red and Yellow Books, it became clear that the insurance clauses in what were then the current editions of these forms of Contract had served their purpose and that it was time to reconsider their logic and composition. Priorities for such revision were proposed in the first edition of this book for the consideration of the drafting committees. These priorities could be summarised as follows:

1   A sequential transition from Risk to Responsibility to Liability to Indemnity to Insurance must form the basic logic of the new wording of these clauses in order to properly represent what happens in practice.
2   A clear language, readily understood by the practicing engineer from various nationalities, must be used. Where long sentences cannot be avoided, the reader must be helped by punctuation to break the sentence into smaller discrete units of thought.
3   The provisions intended for international use must not be based on a specific legal system. Terms that have different meanings in different jurisdictions should be avoided.
4   The insurance cover required under the Contract must dovetail into other insurance covers taken out by the parties involved in the construction contract. The option must remain open for either party to the contract to effect insurance cover with the minimum of gaps and overlaps.

A model set of insurance clauses was drafted following the above principles, but keeping as far as possible to the same content as was in the FIDIC Form, to deal with the various aspects of risk, responsibility, liability, indemnity and insurance for loss or damage during the construction and defects correction periods. It formed the main part of Chapter 8 of the first edition of this book, pages 234 to 241, and served as a basis from which any set of conditions of contract could draw the necessary wording with little or no modification.

These priorities were positively considered and taken into account by the revising committees of the Red and Yellow Books.[1]

As explained in note 1, the 4th edition of the Red Book, and to a greater extent the 3rd edition of the Yellow Book, took on board some of the proposals put forward in the previous chapter of the first edition of this book and also in the above model clauses. Therefore, it is worthwhile updating these model clauses and noting the philosophy behind them.

The proposed model insurance clauses, which deal with risks of loss and damage, the responsibility as allocated to the parties in respect of these risks, the consequent liability, indemnity and insurance are set out below.

Proposed Model Clauses for a Contract for Works of Civil Engineering Construction

1 Construction Risks, allocation and responsibility

The risks of loss or damage to property (including the Works) and those of personal injury or death which may arise from any cause whatsoever during the performance of the Contract shall be shared between the Employer and the Contractor, and allocation to them in accordance with Sub-clauses 1.1 and 1.2 of this Clause.

The Employer and the Contractor shall be responsible for the risks allocated to them which shall be referred to hereinafter as the Employer's Risks and the Contractor's Risks respectively. The Employer's Risks are divided into two categories, namely: Normal Risks and Special Risks, as provided in Sub-clause 1.1 of this Clause and Clause 3 herein. The allocation of risks and responsibility is based on the ability to exercise either control over the risk or influence over its consequence.

Employer's Risks of loss and/or damage

1.1 Employer's Risks of loss and/or damage:
For the purposes of the Contract, the Employer shall be deemed to have responsibility for the following risks:

---

1 The first edition of this book recorded the following acknowledgement: *'The clauses set out in this chapter incorporate suggestions made during discussions with two of FIDIC's Contract Committees: 1. FIDIC's Electrical and Mechanical Contracts Committee during its discussions with ORGALIME (Organisme de Liaison des Fabriques Metalliques Européenées), to consider the revision of the 2nd Edition of the Conditions of Contract (International) for Electrical and Mechanical Works; 2. FIDIC's Civil Engineering Contracts Committee during its discussions with F.I.E.C. (Fédération Internationale Européenées de la Construction), to consider the revision of the 3rd Edition of the Conditions of Contract (International) for Works of Civil Engineering Construction. The author is indebted to the participants in these discussions for their contribution towards what he considers to be a better form of these clauses.'*

- Employer's Normal Risks of loss and/or damage:

  (a) Damage to crops being on the Site (save in so far as possession has not been given to the Contractor).

  (b) The use or occupation of the Site by the Works, or any part thereof, or for the purpose of construcion, completing and maintaining the Works (including consequent losses of crops).

  (c) Any interference, whether temporary or permanent, with any right of way, light, air, water; or other easement or quasi-easement which are the unavoidable results of the construction of the Works in accordance with the Contract.

  (d) The right of the Employer to construct the Works or any part thereof on, over, under, in or through the Site.

  (e) Fault, error, defect or omission in the design of the Works (other than a design provided by the Contractor pursuant to his obligations under the Contract).

  (t) Pressure waves caused by aircraft or other aerial devices travelling at sonic or supersonic speeds.

  (g) Damage (other than that resulting from the Contractor's method of construction) which is the unavoidable result of the construction of the Works in accordance with the Contract.

  (h) Claims, demands, proceedings, damages, costs, charges and expenses in respect of any Act, or neglect, or breach of statutory duty, done or committed by the Engineer or the Employer, his agents, servants or other Contractors (not being employed by the Contractor).

  (i) Any risks, other than those allocated to the Contractor in the Appendix to these Conditions, which an experienced contractor could not have foreseen, or if deemed foreseeable could not reasonably have made provision against.

- Employer's Special Risks of loss and/or damage:

  (j) War, hostilities (whether war be declared or not), invasion, act of foreign enemies.

  (k) Ionising radiation, or contamination by radioactivity from any nuclear fuel, or from any nuclear waste from the combustion of nuclear fuel, radioactive, toxic, explosive, or other hazardous properties of any explosive, nuclear assembly, or nuclear component thereof.

  (l) Rebellion, revolution, insurrection, military or usurped power or civil war, insofar as it relates to the country in which the Works are being, or are to be executed or maintained.

  (m) Riot, strike, commotion or disorder, unless solely restricted to the employees of the Contractor or of his subcontractors and arising from the execution of the Contract.

  (n) Acts of Government.

Contractor's Risks of loss and/or damage

1.2 Contractor's Risks of loss and/or damage:
   For the purposes of the Contract, the Contractor shall be deemed to have responsibility for all the risks of loss and/or damage not stated under the

Employer's Risks including those risks which an experienced Contractor could have foreseen and could have reasonably made provision for in his tender.

2 Responsibility for Care of the Works and Passing of Risks

2.1 Unless the Contract is terminated in accordance with these Conditions, the Contractor shall take full responsibility for the care of the Works and any material for incorporation therein from the Commencement Date until the date when the Engineer shall have issued a Taking-Over Certificate for the whole of the Works pursuant to Clause 48.

The responsibility for the care of the Works shall pass on that date, from the Contractor to the Employer, which date shall be referred to as the Risks Transfer Date.

2.2 Provided that if, prior to the Risks Transfer Date, the Engineer shall issue a Taking-Over Certificate for any part of the Permanent Works, the Contractor shall cease to be responsible for the care of that specific part for which the Taking-Over Certificate was issued and the responsibility for the care thereof shall thereupon pass to the Employer.

2.3 It is also provided that the Contractor shall take full responsibility for the care of any outstanding work which he shall have undertaken to complete during the Defects Correction Period until all such outstanding work is completed.

2.4 If the Contract is terminated in accordance with the provisions of these Conditions, the Risks Transfer Date will be the date of expiry of the Notice of Termination.

3 Responsibility for Damage or Loss to the Works, before the Risks Transfer Date

If any of the Risks eventuate causing damage, loss or injury to any part of the Works, or any material for incorporation therein, before the Risks Transfer Date, the Contractor shall undertake the repair or restitution in accordance with Sub-clauses 3.1 to 3.5. Such repair or restitution shall be carried out so that, at completion, the Permanent Works shall be in conformity in every respect with the requirements of the Contract and the Engineer's instructions:

3.1 If the damage or loss arises from any of the Contractor's Risks, as defined in Sub-clause 1.2, then the Contractor shall be responsible for the repair and restitution at his own cost.

3.2 If the damage or loss arises from any of the Employer's Normal Risks, as defined in Sub-clauses 1.1(a) to 1.1(i), the Contractor, if required by the Engineer within 28 days of the risk eventuating, shall be responsible for the repair and restitution at the cost of the Employer.

3.3 If the damage or loss arises from any of the Employer's Special Risks, as defined in Sub-clauses 1.1(j) to 1.1(n), the Contractor, if required by the Engineer within 28 days of the risk eventuating, shall be responsible for the repair and restitution at the expense of the Employer.

It is furthermore agreed that the Contractor is entitled to payment for any of the following items destroyed, or damaged, due to any of the Special Risks:

(a) The Permanent Works or any material for incorporation therein;

(b) Material or other property of the Contractor used, or intended to be used, for the purposes of the Works.

3.4 If the cause of loss, damage or injury is attributable to a combination of Employer's and Contractor's Risks, the responsibility for repair and restitution shall be that of the Contractor. The Employer shall, however, reimburse the Contractor by an amount proportionate to the loss, damage or injury for which he is responsible and as detailed in Sub-Clauses 3.2 and 3.3 hereof.

3.5 If the subject of the damage or loss had previously been condemned by the Engineer under the provisions of Clause 39, then the Contractor, if required by the Engineer within 28 days of the risk eventuating, shall be responsible for the repair and restitution at his own cost.

4 Liability and Indemnity

Liability and Indemnity for Personal Injury to Third Party and Property Damage other than the Works

4.1 The Contractor shall be liable for and shall indemnify the Employer against all claims, demands, proceedings, costs, charges and expenses in respect of:

- personal injury and/or
- damage or loss to any property (other than the Works), arising out of any of the Contractor's Risks in the execution of the Works and before the Risks Transfer Date. After this date, and unless such injury or loss or damage is caused by the Gross Misconduct of the Contractor as defined in Sub-Clause 4.5, the Contractor's liability and indemnity shall be limited:
- in time, to the Defects Liability Period; and
- in extent, to any outstanding work which the Contractor shall have undertaken to complete under Clause 48 hereof and also to the extent of any defective material and/or defective workmanship and/or defective design carried out by him.

4.2 The Employer shall be liable for and shall indemnify the Contractor against all claims, demands, proceedings, costs, charges and expenses in respect of:

- personal injury and/or
- damage or loss to any property (other than the Works), arising out of the Employer's Risks in the execution of the Works throughout the Contract.

4.3 The Contractor's liability to indemnify the Employer, as aforesaid, shall be reduced proportionately to the extent that the Employer's Risks may have contributed to the said damage, loss or injury. Similarly, the Employer's liability to indemnify the Contractor, as aforesaid, shall be reduced proportionately to the extent that the Contractor's Risks may have contributed to the said damage, loss or injury.

Liability and Indemnity, for Accidents or Personal Injury to Workmen

4.4 The Employer shall not be liable for, or in respect of, any damages or compensation payable at law in connection with any accident or injury to any workman or other person in the employment of the Contractor or any subcontractor, or to any person employed by the Employer whose services may, for the time being, be loaned or made available to the Contractor or his subcontractors. The Contractor shall indemnify and keep indemnified the Employer against all such damages and compensation and against all claims, demands, proceedings, costs, charges and expenses whatsoever in respect thereof, or in relation thereto.

This agreement shall not apply, however, to any accident or injury, which results, or is contributed to, by any act or default of the Employer, his agents or his servants. The Employer shall be liable for such accident or injury to the extent of his act or default and shall indemnify the Contractor accordingly.

Gross Misconduct

4.5 'Gross Misconduct' shall mean an act or omission implying either a failure to pay due regard to serious consequences which a conscientious contractor would normally foresee as likely to ensue or a deliberate disregard of any consequences of such act or omission.

5 Insurance (Alternative A – Contractor to take out Insurance)

Insurance of the Works

5.1 Without limiting his obligations, responsibilities and liabilities, the Contractor shall insure, in the joint names of the Employer and the Contractor, the Permanent Works, the Temporary Works and any materials for incorporation therein against all loss or damage, from whatever cause, arising out of the Contractor's Risks, as defined in Sub-Clause 1.2. Furthermore, the insurance cover shall extend to include loss or damage to any part of the Works as a consequence upon failure of elements defectively designed or constructed with defective material or workmanship. The amount of insurance shall be to an adjusted full value and with deductible limits which shall not exceed the relevant figures stated in the Appendix to the Form of Tender. For the purposes of this Clause, the meaning of 'material' shall include material whether on the Site or otherwise allocated to the Contract in the Contractor's statements.

Insurance of Construction Plant

5.2 The Contractor shall also insure to its replacement value the plant he intends to use for the construction of the Works, with deductible limits which shall not exceed the relevant figures stated in the Appendix to the Form of Tender.

Third Party Liability and Employer's Liability Insurance

5.3 The Contractor shall, without limiting his or the Employer's obligations and responsibilities insure, in the joint names of the Contractor and the Employer, against liabilities for death of or injury to any person (other than as provided in Sub-Clause 4.4) or loss of or damage to any property (other than the Works) arising out of the performance of the Contract. The insurance policies shall be issued for at least the amount stated in the Appendix to Tender.

The insurance policy shall include a cross liability clause such that the insurance shall apply to the Contractor and to the Employer as separate insured.

The Contractor shall also insure against any liability arising under Sub-Clause 4.4 and shall continue such insurance during the whole of the time that any persons are employed by him on the Works. Provided that, in respect of any persons employed by any subcontractor, the Contractor's obligations to insure as aforesaid under this Sub-Clause shall be satisfied if the Subcontractor shall have insured against the liability in respect of such persons in such manner that the Employer is indemnified under the policy, but the Contractor shall require such Subcontractor to produce to the Employer, when required, such policy of insurance and the receipt for the payment of the current premium.

5.4 The insurance cover provided under Sub-Clauses 5.1, 5.2 and 5.3 of this Clause may include the permitted exclusions stated under Sub-Clause 5.5. The period of insurance shall be from the Commencement Date the Risks Transfer Date. The Insurance cover shall continue thereafter but only in respect of:

(a) any damage, loss or injury occasioned by the Contractor in the course of any operation carried out by him for the purpose of complying with his obligations under Clauses 49 and 50 and

(b) any damage, loss or injury arising during the Defects Liability Period, out of any of the Contractor's Risks occurring prior to the commencement of that period.

Insurance Exclusions

5.5 The Contractor may, without limiting his or the Employer's obligations and responsibilities, include in the insurance cover provided by him under this Clause any of the following exclusions, worded as specified in Sub-Clause 5.6 hereof.

(a) against the necessity for the repair, or reconstruction, of any work constructed with materials and workmanship not in accordance with the requirements of the Contract, unless the Bill of Quantities shall provide a special item for this insurance in respect of specified elements;

(b) for the Contractors' All Risks policy, the 'Employer's Risks' as defined in Sub-Clause 1.1;

(c) for the Third Party Liability policy, paragraphs (a) to (e) of Sub-Clause 1.1;

(d) in respect of consequential loss, including penalties for delay and non-completion; and

(e) wear and tear, shortages and pilferages.

5.6 The organisations in whose name these conditions are published shall, acting jointly, publish from time to time authorised wordings of the exclusions permitted by Sub-clause 5.5 hereof. Such authorised wordings shall take effect as if they are stated in Sub-Clause 5.5 at the date thirty days prior to the latest date for submission of tenders for the Works.

General Insurance Requirement

5.7 The Contractor shall, before the Commencement Date and whenever required, produce to the Employer for inspection any policy or policies of insurance required under these Conditions, together with the receipts in respect of premiums paid under such policy or policies. The Contractor shall also produce to the Employer any other insurance policy which he may be legally required to effect and keep in force, together with the receipts in respect of premiums paid.

5.8 All insurances shall be obtained from an Insurer and in terms subject to the approval of the Employer, which approval shall not be unreasonably withheld.

5.9 During the currency of the Contract any material alteration to such insurance made at the Contractor's request shall be immediately notified by the Contractor to the Employer and shall be subject to the approval of the Employer, which approval shall not be reasonably withheld. In the case of any material alteration made by the Insurer, the Contractor shall immediately provide written evidence to the Employer of such alteration. Such alteration shall not release the Contractor in any way from his obligations under the Contract.

Remedy on Contractor's Failure to Insure

5.10 If, upon request, the Contractor shall fail to produce to the Employer satisfactory evidence that there is in force the insurance referred to in Sub-Clauses 5.1, 5.2 and 5.3 or any other insurance which he may be required to effect under the terms of the Contract, then and in any such case the Employer may effect and keep in force such insurance and pay such premium or premiums as may be necessary for that purpose. Thereafter, from time to time, the amount so paid by the Employer, as aforesaid, shall be deducted from any monies which are, or may become due to the Contractor, or the same shall be recovered as a debt due from the Contractor.

Compliance with Insurance Conditions

5.11 The Contractor and the Employer shall comply with the conditions stipulated by the Insurers as given in the policies for insurance mentioned in Sub-Clauses 5.1, 5.2 and 5.3.

6 Insurance (Alternative B – Employer to take out Insurance)

General Insurance Requirements

6.1 The Employer may elect, prior to the Tender stage, to effect any of the insurances for which the Contractor is made responsible under Clause 5 hereof. In such a case, the Employer shall specify in the Tender Documents the extent, type and duration of insurance to be provided by him. The Contractor shall be given the opportunity to examine the details of the insurance cover provided by the Employer and shall be permitted to effect, at his own cost, any additional insurance cover he may require.

6.2 The Employer shall, before commencing the Works and whenever required, produce to the Contractor for inspection any policy or policies of insurance required under these Conditions, together with the receipts in respect of premiums paid under such policy or policies.

Notes on the proposed Model Clauses
Clause 1: Construction Risks, Allocation and Responsibility

Clauses 20 to 25 of FIDIC's Red Book Conditions of Contract, referred to generally as the insurance clauses, are numbered in this model as Clauses 1 to 6. They begin by identifying the risks to which the project is exposed and separating them into two parts or matrices. One matrix is allocated to the contractor and the second to the employer, as they are the two parties involved in the contract in question. As explained on page 236 of Chapter 8 above, the 4th edition of the Red Book, which followed the original proposals made in the first edition of this book, the employer is made responsible for the risks which belong to his agents, servants, or advisers and the contractor is made responsible for the risks which belong to subcontractors, suppliers and others who may be connected with his activities.

The allocation of risks is mainly based on the ability of either party or those others for whom he is responsible to either control events or influence their outcome. If this does not apply, the remaining part of the criteria put forward earlier in Chapter 2, page 47, Figure 2.4 is used as a basis. Clause 1.1 sets out clearly the division of risks into these two matrices, the first of which is allocated to the employer and is referred to as the Employer's Risks; the second being allocated to the contractor and are referred to as the Contractor's Risks. The first matrix is further divided into two groups to differentiate between the risks which can impede the execution of the contract and called 'Special Risks' and others which would have little effect on completion of the contract as a whole and are designated 'Normal Risks'. This differentiation has two consequences: the first is in respect of the method of payment to the contractor in case he is required to carry out a repair or make a restitution. This is later stipulated in Clauses 3.1, 3.2 and 3.3. The second consequence of this division in the Employer's Risks is in respect of the responsibility to insure, which is dealt with in Clauses 5 and 6.

The opening statements of Clause 1 make it clear that they only deal with the responsibility and the liability emanating from loss, damage, injury or combinations of them, which may occur during the construction period and the Defects Correction Period. They do not therefore cater for the responsibilities and liabilities that emanate from the remaining provisions of the contract, which result mainly in economic or time losses.

Sub-clauses 1.1 and 1.2 allocate the risks between the employer and the contractor adopting the same principles used in Contractors' All Risks and Erection All Risks insurance policies, i.e. by specifically stating what is excluded from the intended cover and thus maintaining cover on all other risks. Therefore, and in a similar manner, the Employer's Risks are specifically mentioned, leaving to the contractor all other risks with the exception of those risks that are allocated to him in the Appendix pursuant to Sub-Clause 1.1(i) and those others mentioned in Sub-clause 1.2.

The introduction of the Employer's Risks in Sub-clause 1.1(i), i.e. unforeseen risks, is necessary in order to leave the door slightly ajar to admit into the Employer's Risks events that could not be envisaged. They are not, however, intended to relate to the usual type of risk, which is known to exist to a possible or probable degree.

To eliminate any ambiguity that might exist in the minds of the contracting parties as to what is foreseen, it would be worthwhile to enumerate such risks in a separate document (for example Part II of the General Conditions of Contract).[2] Such enumeration would also be helpful in focusing the attention of the parties on mitigation of the risks and loss prevention. A chart such as that shown in Figure 9.1 could be compiled by the designer or contractor to look into the risks that lie ahead.[3] The chart simply incorporates on one axis the envisaged risks, on another axis the expected and programmed site activities and on a third axis, the allocation of risk to the parties involved. From such a chart one may see at a glance the site activities which could be affected if any of the risks eventuate and also to whom the responsibility for these risks is allocated.

Whilst it is unfortunate that precision cannot be maintained in the task of alloca-tion of risks, it is by definition impossible to achieve and especially so in a standard form which is expected to apply to all construction contracts anywhere in the world. Thus some flexibility is necessary for the sake of justice albeit at the expense of precision and at the risk of incorporating fertile ground for dispute as to what is 'foreseen' and what is not.

The group of the Employer's Normal Risks is basically compiled from those risks, which the contractor has either no control over, or no influence over their consequence. This criterion of allocation yields the following principles:

(a) *The employer's choice of site:* The employer's decision to construct the works on a particular site results in certain risks being attached to that decision. If the aforementioned criterion of allocation is used, the employer ends as being responsible for the risks attached to the site prior to the commencement of work and after taking over the completed project or any part thereof. Thus, damage to crops as in Sub-clause 1.1(a), or damage due to any unavoidable result of construction of the works as in Sub-clause 1.1(b), or a challenge to the right of the employer to construct the works on the site as in Sub-Clause 1.1(c), form risks allocated to the employer.

(b) *Unavoidable result of the construction of the works in accordance with the contract:* If the contractor carries out his duties and responsibilities in accordance with the contract, without negligence, and with proper materials and workmanship, any loss or damage caused by his actions is the unavoidable result of the works. Such risk of loss or damage is incorporated into the Employer's Risks.

---

2 'The Spectrum of Risks in Construction', FIDIC's Standing Committee on Professional Liability Report, 1985.

3 The proposal of this chart was the predecessor of what is now required under the Health and Safety legislations that came into effect in the 1990s in all the European Union countries. It is now a requirement in these countries for the designer of a project to have a file of the risks of injury or death that are identified in his project design, which file he would then hand to the contractor, as soon as appointed, who would complete it by adding the risks in respect of his methods of construction.

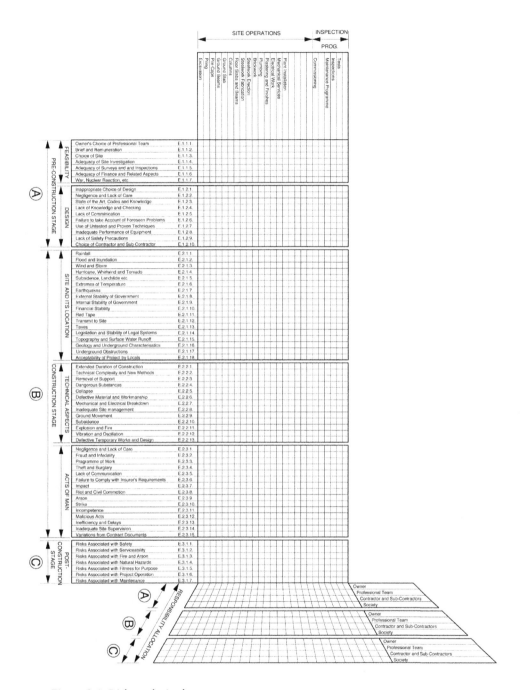

Figure 9.1 Risk analysis chart.

(c) *Act or neglect or breach of statutory duty by the employer or those for whom he is responsible:* Such acts are the responsibility of the employer and any risk of loss or damage emanating from them must therefore fall within the Employer's Risks.

The Employer's Special Risks are then identified and each should be defined in the contract to avoid any dispute arising as to precise meaning of these risks. The definitions given by a law dictionary are shown in Table 14 below.[4]

Clause 2: Responsibility for Care of the Works and Passing of Risks

Sub-clause 2.1 allocates the responsibility for care of the works to the contractor and thus establishes the principle of control over events and the influence over their consequences.

It further establishes the date at which such responsibility passes to the employer, referred to as the Risks Transfer Date, which is, basically, the date when the engineer shall have issued a Taking-Over Certificate for the whole of the works pursuant to Clause 48 of the Conditions of Contract.

Sub-clause 2.2 deals with the situation where one part of the works is taken over by the employer prior to the Risks Transfer Date. It stipulates that such action relieves the contractor of his responsibility for the care of that part of the works that had been taken over.

Sub-clause 2.3 stipulates that the contractor keeps the responsibility for the care of any outstanding work, which is not completed on the Risks Transfer Date. The care of such outstanding work remains with the contractor until the whole of the work is completed.

Sub-clause 2.4 defines the Risks Transfer Date if the contract is terminated in accordance with Clause 65.

Clause 3: Responsibility for Damage or Loss to the Works, before the Risks Transfer Date

Clause 3 places the responsibility for making the repair or the restitution in case of loss, damage or injury on the contractor, if these occur during the period in respect of which the contractor is responsible for the care of the works. It further specifies who is responsible for payment in respect of such repair or restitution, placing in Sub-Clause 3.1 the responsibility for the cost on the contractor if the risk causing the damage or loss is one of the Contractor's Risks. Sub-clause 3.1 goes on to state how such repair is to be carried out. If the risk causing the damage or loss is one of the Employer's Risks, the contractor shall be responsible for carrying out the repair or restitution only if the engineer so requires within twenty-eight days of the event. Such repair or restitution is to be carried out, however, at the cost of the employer if the risk is a Normal Risk and at the expense of the employer if it is a Special Risk.

---

4 The definitions of war, civil war, rebellion, insurrection, strike and riot are taken from the *Oxford Companion to Law* by David M. Walker referred to earlier note 2 to Chapter 5 and those of hostilities and military or usurped power from *Engineering Law and the ICE Contracts* by M.W. Abrahamson, referred to earlier note 31 to Chapter 8.

Table 9.1 Definitions of the Special Risks

Special Risk	Definition
War	A forcible contention between one state or group of states and another state or group of states through the application of armed force and other measures with the purpose of overpowering the other side and securing certain claims or demands.
Civil War	Large scale and sustained hostilities (as distinct from mere revolt or rising) between the organised armed forces of two or more factions within one state, or between the government and a rebel or insurgent group.
Hostilities (whether war be declared or not)	Hostile acts by persons acting as the agents of sovereign powers, or of such organised and considerable forces as are entitled to the dignified name of rebels as compared to mobs and rioters. This does not cover the act of a mere private individual acting entirely on his own initiative, however hostile his action may be.
Rebellion	Violent opposition by a substantial group of persons against the lawfully constituted authority in a state, so substantial as to amount to an attempt to overthrow that authority.
Insurrection	A term generally used as meaning an uprising against the constituted authority of a state lesser in scope and purpose than that which is designated a revolution or a rebellion.
Military or Usurped Power	Without using words of rigorous accuracy, military and usurped power suggests something more in the nature of war and civil war than riot and tumult . . . nor can it be properly contended that the words 'military power' do not refer to military power of a government lawfully exercised. The disjunctive 'or' is used to denote contrast between the words 'military' and 'usurped'. The words are not 'usurped, military power'.
Strike	A usual term given for a simultaneous and concerted cessation of work by an employer's employees, or a substantial group of them, normally in pursuance of an industrial dispute. . . . It is a breach of contract by each workman, unless due notice of intent to terminate employment be given by each, in which case it is considered a breach of contract, but is not illegal or criminal unless it involves committing criminal acts, nor is it an actionable conspiracy so long as the predominant motive is the furtherance of a legitimate interest, such as improvement in the conditions of work.
Riot	A tumultuous disturbance of the peace by three or more persons assembled together without lawful authority, with intent to assist each other, if necessary by force, against anyone who opposes them in the execution of a common unlawful purpose, and who execute or begin to execute that purpose in a violent manner so as to alarm at least one person of reasonable firmness and courage. The purpose is commonly to destroy property.
Commotion	Public disorder and physical disturbance.

The period of twenty-eight days is the period during which the employer has to make up his mind as to whether or not such repair is required to be carried out by the contractor. This period should be increased in the case of the Special Risks if it is felt that it is too short for the employer to make up his mind.

In all cases, the damaged or lost part of the works is assumed not to have been condemned by the engineer. If such condemnation has been made, then the contractor is expected to make the repair or restitution at his own cost. The repair is also expected to be carried out in such a manner that the permanent works will always be in conformity with the requirements of the contract documents and any instructions which may have been given by the engineer.

If loss or damage arises in consequence of a combination of more than one risk and if the risks are both Employer's and Contractor's Risks, then Sub-Clause 3.4 operates on the basis of proportionate responsibility.

Clause 4: Liability and Indemnity for Personal Injury to Third Party and Property Damage other than the Works

The liability of the contractor for any loss or damage or injury sustained by a third party, under the contract, flows essentially from any of the following:

- non-performance of his obligations under the contract;
- negligence in the execution of the contract works; and
- breach of any statutory requirements.

These are all encapsulated in the term 'Contractor's Risks' in the form of an exclusion from the general reference to an Employer's Risk in paragraphs (e), (g) and (h) of Sub-clause 1.1. Sub-clause 4.1, therefore, defines the liability accordingly and divides the timing into three periods based on the presence of the contractor on site. The first period is prior to the Risks Transfer Date when the contractor is continually present on site and any of his acts or omissions, which may fall within the aforementioned definition, may cause a resultant liability. The second period is after the Risks Transfer Date and during the Defects Correction Period when the contractor may be intermittently present on site whilst carrying out his obligations during that period. During the Defects Correction Period, the contractor may be liable if any of his acts or omissions falls within the aforementioned exceptions.

The third period begins after the end of the Defects Correction Period when the contractor can only be held liable in the case of a latent defect for which he is liable or for gross misconduct giving rise to gross liability. Gross misconduct is defined in Sub-Clause 4.5.

Sub-clause 4.2 defines the Employer's liability in respect of loss, damage or injury in terms of the Employer's Risks and Sub-Clause 4.3 deals with the situation where both the employer and the contractor are liable.

Sub-clauses 4.1, 4.2 and 4.3 provide for liability to be followed by indemnity and require the party held liable to indemnify the other party against all claims, demands, proceedings, costs, charges and expenses.

Sub-clause 4.4 defines the liability and indemnity for accidents or personal injury to workmen and the type of person covered by this clause of the contract. The liability of the contractor and the employer in respect of personal injury to workmen is based on the same principles, which are defined for third parties. As in Sub-Clause 4.1, the indemnity follows the liability.

Sub-clause 4.5 defines Gross Misconduct. The degree of negligence in a commission or an omission shifts from the normal to the gross when the commission or omission involves a moral issue or a wilful act or failure to pay regard to serious consequences. Thus a car driver is negligent when he drives at 50 km.p.h. in a 30 km.p.h. zone but he would be grossly negligent if he drove at 100 km.p.h. through a busy town at midday. Gross negligence emanates from a gross misconduct and attracts gross liability.

Clause 5: Insurance (Alternative A – Contractor to take out Insurance)

The indemnity clauses must be backed by an insurance plan to ascertain that funds will be available when needed to repair damage or to have reimbursement in case of a loss. The insurance plan must be such that either the contractor or the employer could implement it depending on the particular situation. Two alternatives are therefore prepared, one for the contractor to insure, Alternative A, and the second for the employer, Alternative B. In both alternatives, the policies required, as described in Chapter 12, are:

- A Contractors' All Risks policy with an extended cover to protect the employer and the contractor jointly against loss or damage to the works consequent upon failure due to defective design, workmanship and material;
- A Third Party or Public Liability insurance policy and
- An Employer's Liability insurance policy.

The required cover has three other features: the first is an adjustable sum insured as described in Chapter 12; the second is a limit for the deductible or excess generally applied in these policies; and the third is the provision of cover for materials to be incorporated in the works whether on site or allocated to the site. Sub-clause 5.1 is drafted incorporating these requirements.

Sub-clause 5.2 provides for the necessary insurance in respect of construction plant and incorporates a maximum limit for the deductible, which may be different from that imposed on claims in respect of the section dealing with the works. It is important to note that the idea behind setting a maximum limit for the deductible is to ensure that the scope of the insurance cover is not curtailed through the voluntary imposition of a very high excess limit, by the contractor. The value stated in the Form of Tender should therefore be much higher than that generally imposed by the insurer.

The indemnity provided in respect of third party liability and employer's liability has also to be backed by insurance and Sub-clause 5.3 provides for the requirement of such insurance. The third party liability cover should be in the joint names of the employer and the contractor, which necessitates the addition of what is called 'cross-liability clause'.

The period of insurance and the scope of cover provided during that period are detailed in Sub-clause 5.4. In recognition of the fact that there are risks which are either uninsurable or should not be insured, Sub-Clause 5.5 is drafted to enumerate and define the permitted exclusions from the insurance cover.

Sub-clause 5.6 defines further what is meant by permitted exclusions, as it is possible to phrase the wording of the exclusion in such a way that would not be acceptable. Standard wordings should therefore be issued by the authorities publishing the standard form of conditions, having first established that they are acceptable to the insurance market.

To ensure that the insurer and the conditions of the insurance policy are acceptable to the employer and that these conditions are observed by both joint insureds, it is necessary to have the contractor present the policy to the employer for approval.[5] It is also necessary to enumerate the important principles behind such conditions and this is done under Sub-Clauses 5.7, 5.8 and 5.9. These deal with production of evidence of payment of premium, Sub-Clause 5.10 deals with the remedy should the contractor fail to provide the required insurance policies and the stipulated cover.

Insurance (Alternative B – Employer to take out Insurance)

If the employer elects to provide the necessary insurances in respect of the contract, it would be wise to have such insurances negotiated and agreed with his insurers prior to completion of the contract documents. The insurers may wish to leave the question of premium calculation until the contractor is appointed and his methods of construction are known. The employer should reproduce in his tender documents a copy of the policies he has obtained and any other relevant information necessary for the provision and pricing of the cover. The contractor should be given the opportunity to consider and decide on any additional insurance covers that he might require.

The form and type of insurance cover that could be obtained by the employer could be quite different from that described in this chapter. It may take the form of any of the insurances described in Chapter 12 and may therefore include insurance policies other than those required by the aforementioned clauses.[6] Thus, the employer may choose to introduce within the Conditions of Contract with the contractor any of the following:

- Materials and plant for incorporation in the works whilst in storage or in transit to the site. This is already required under the Electrical and Mechanical Contracts and described later in this chapter.
- Machinery and hired plant used in the construction and/or the erection of the project;
- Professional indemnity insurance in respect of the design activities related to the project;
- Air-freight cover for urgent repairs that might be required in a project where some major elements are manufactured abroad;
- Unfair termination of contract;
- Non-negligence insurance cover;
- Project insurance in one of its many forms;
- Decennial insurance or latent defects cover;

5 See page 261 in connection with production of the insurance policies.
6 See 'Overlaps and gaps' in Chapter 6 and page 225 in Chapter 8.

- Credit risk insurance;
- Group personal accident, travel, medical and life assurance cover;
- Unfair call on any surety or bond with special attention to on-demand bond;
- Delay risk;
- Currency risks;
- Manufacturer's risks of defective material, defective workmanship and defective design;
- Confiscation of construction plant and machinery;
- Expropriation of overseas assets risk; and
- Difference-in-conditions insurance. These would include any of the above risks included in one insurance policy plus any specific risks either connected with the type of contract undertaken or with the locality of the contract such as war or earthquake or the risk of not being able to obtain construction materials or other matters necessary for the completion of the project following political events, etc.

Amendments required for an electrical and mechanical contract

Electrical and mechanical contracts differ from those of civil engineering works and certain amendments are needed in respect of the following matters as related to the clauses under discussion:

(a) The date of commencement of the works in relation to work on site: generally speaking, the work on site starts much earlier in civil engineering projects than that in electrical and mechanical projects. There is, therefore, a lag in E & M contracts between the date at which a contract is awarded and work commencement on site. During this time, electrical and mechanical equipment is usually being manufactured. During that stage, it is under the control and the responsibility of the manufacturer but, as it leaves his premises for shipment to site, it passes through the stages of rail or road transport to port, storage at port, transit on ship or aircraft, storage at port of destination, rail or road transport to site, storage on site for a period of time until it is required for installation on site by the contractor. During all these stages, insurance is required and is generally obtained through a combination of Marine Insurance, Goods in Transit Insurance and finally the Erection All Risks Insurance under the E & M contract. The problem in this system is, however, that the date of discovery of the damage to equipment, except in obvious circumstances, is not until the packing is removed and the equipment is taken out for installation. In certain circumstances, the damage is not discovered even then and not until commissioning of the equipment, or of the whole plant, is started. Months would have passed between the date of damage and the date of discovery and the insurers of the first stage of the transport process would obviously require a proof that they are responsible. Such a proof is not easily obtainable. It is therefore better to eliminate this gap by attaching the transport risks to the same insurer who provides the cover for the dominant risks, i.e. the Erection All Risks insurer in a policy called Marine Cum Erection Policy. If that is thought to be desirable, then a clause could be inserted between 5.1 and 5.2 stating the following:

> Without limiting his obligations, responsibilities and liabilities, the Contractor shall in the joint names of the Employer and the Contractor, insure the equipment for incorporation in the Works against all loss or damage from whatever cause whilst in transit or in storage, including loading and unloading risks, until delivery to site. Such insurance shall be effected with the same Insurer as for the insurance under Sub-Clause 5.1 hereof. The amount of insurance shall be to the full value of the equipment under the Contract but with deductible limits which shall not exceed the relevant figures stated in the Appendix to the Form of Tender.

If this clause is inserted, mention will have to be made to it in Sub-Clause 5.5 dealing with permitted exclusions as they apply to this type of insurance cover.

(b) Unlike civil engineering projects, the outcome of a project of an electrical and mechanical nature is generally a product which is destined to be marketed and sold to consumers. A prudent employer would obtain a Products Liability Insurance cover. In fact in some parts of the world, a statutory duty is imposed on manufacturers and others introducing products to the market to ensure that these products, when properly used, are safe and free from risk to health. Such an insurance cover can be taken either under an extension to Public Liability Insurance or as a separate policy.

It is conceivable that, in certain circumstances of loss, damage, or injury due to such products, a consumer who wishes to join all those in sight in his legal action may sue the contractor who erected the plant. It may therefore be prudent for the contractor, under these conditions of contract; to have a clause inserted making the employer responsible for keeping in force a Products Liability Insurance policy in the employer's name, during the productive lifetime of the project. Such a clause may be worded as follows:

> The Employer shall effect and keep in force a Products Liability Insurance during the productive lifetime of the project. Such insurance shall be effected against all sums, subject to the limit stated in the Appendix to the Form of Tender, for which the Employer shall be held legally liable to pay in respect of:
>
> (a) accidental bodily injury to or illness of any person;
> (b) accidental loss of or damage to property occurring anywhere in the world and caused by any defects in any goods sold, supplied, erected, installed, repaired, processed, manufactured or tested at or from the Works.

As in the previous point, if this clause is inserted, mention will have to be made to it in Sub-clause 5.5 dealing with permitted exclusions as they apply to this type of insurance cover.

(c) It is an obvious feature of the electrical and mechanical type of contract that testing and commissioning of individual pieces of equipment and the plant as a whole is part of the contractual agreement. Such testing and commissioning includes normally:

1 The mechanical or electrical testing of individual units or assemblies to establish that they were properly erected (referred to as cold testing) and

2 The operational commissioning where for the first time a machine is subjected to the load under normal operational conditions and sometimes under overload conditions. This type of testing is sometimes referred to as hot testing.

During this period, the works become a working unit exposed to a greater level of risk of breakdown, fire, explosion, mechanical or electrical derangement, etc. As it is unusual for insurers to cover such risks of testing in isolation (because many claims arising during this period originate from events in the erection period), the testing and commissioning period should be covered under the Erection All Risks policy and a slight adjustment in the wording of Sub-Clause 5.6 is necessary. Such adjustment can be made by adding the following sentence:

> The insurance required under this Clause shall be extended to include a period of four weeks of testing and commissioning of the Plant prior to the issue of the Taking-Over Certificate.

The period of four weeks indicated above is only for illustration purposes.

(d) For any contract where there is to be operation, consideration should be given to having a 'Machinery Breakdown' insurance cover and 'Loss Of Profits Following Machinery Breakdown' cover.

10

THE INSURANCE CLAUSES
OF THE NEW 1999 FIDIC FORMS
OF CONTRACT

In September 1999, the International Federation of Consulting Engineers, FIDIC, published a new set of standard forms of contract alongside those that were in use at that time. The new set is made up of the following four contract forms:

- *The Green Book:* The Short Form of Contract – Agreement, General Conditions, Rules for Adjudication and Notes for Guidance;
- *The New Red Book:* The Construction Contract (Conditions of Contract for Building and Engineering Works, Designed by the Employer) – General Conditions, Guidance for the Preparation of the Particular Conditions, Forms of Tender, Contract Agreement, and Dispute Adjudication Agreement;
- *The New Yellow Book:* The Plant and Design-Build Contract (Conditions of Contract for Electrical and Mechanical Plant, and for Building and Engineering Works, Designed by the Contractor) – General Conditions, Guidance for the Preparation of the Particular Conditions, Forms of Tender, Contract Agreement and Dispute Adjudication Agreement; and
- *The Silver Book:* The EPC and Turnkey Contract (Conditions of Contract for EPC Turnkey Projects) – General Conditions, Guidance for the Preparation of the Particular Conditions, Forms of Tender, Contract Agreement and Dispute Adjudication Agreement.

Unlike the standard forms of contract that were in use prior to September 1999, which were distinguished from each other on the basis of the type of project to which they applied, the new forms are distinguished on the basis of the allocation of the design function. It is because of this new distinguishing characteristic that the New Red and New Yellow Forms were not given a different colour, as FIDIC wished them to be identified by their respective function of design rather than by their colour.[1] In any case, it is important to remember that although the new Forms of

---

1 Despite FIDIC's wish, these documents are more easily identifiable by their colour with the added tag of 'old' and 'new' and the past few years since their publication date have proved that assertion. It is indeed a pity that these new Contract Forms were not given new colours, which would have eliminated the need for the added tag of 'old' and 'new'. The drafting Committee was advised to that effect by many correspondents, but chose to ignore the advice.

Contract have retained many of the principles and concepts of the old Forms, the differences between them are too numerous and too wide in format and in concept to consider the new set of documents as a revision of the old.

The New Red, New Yellow and the Silver Books have been drafted with the same format and to a large extent their text is similar in its wording. The draftsman, however, pursued this desire for similarity in wording too far in certain instances, and particularly so in the risk, responsibility and insurance provisions.[2] The format adopted for these new Forms of Contract is that of the Orange Book, which had been published in 1995 for use in Design-Build and Turnkey projects. The Orange Book is now obsolete, as it has been replaced by the New Yellow Book.

The Silver Book is totally new and to a large extent forms a departure from FIDIC's established position of providing forms of balanced risk allocation. The risks in the Silver Book are mostly allocated to the contractor. The Green Book is also a new venture for FIDIC in that it is intended for smaller contracts of less than US$0.5 million. Whilst the changes in format may not be sufficient to influence one's choice between the old and the new forms, the eighteen significant changes in concept would. These changes in concept will be discussed below and during the conference and the paper will also review the logic and rationale behind the major changes.

Accordingly, it is appropriate to consider first the New Green Book and then deal with the three new major Forms of Contract. From the insurance point of view, it is appropriate to consider first the relevant clauses of the New Red Book, as it is the closest to the traditional form of the old Red Book, and then consider the New Yellow Book, leaving the Silver Book to be discussed at the end.

The New Green Book

The Green Book, as stated in its Foreword, is intended to be used as a form of contract for engineering and building work of fairly simple or repetitive work of short duration with relatively small capital value,[3] but it may be suitable, subject to the type of work and circumstances, for contracts of greater value. The objective of the Green Book is for the Contract to express in clear and simple terms traditional procurement concepts.[4]

Furthermore, the form is drafted in a flexible format that includes all essential commercial provisions and a variety of administrative arrangements. Thus, it is envisaged that the employer may provide the design himself, by others on his behalf, or by the contractor in a design/build format. In the latter situation, tenderers would be required to submit a design with their tenders, which would be governed by the provisions of Clause 5 of the Green Book's Conditions, 'Design by Contractor'. It is also envisaged that in the Green Book there would be no traditional 'Engineer' or 'Employer's Representative' in the formal sense used by FIDIC in most of its other

---

2   For example, with the variations in the design function between the three new major Forms of Contract, there should be different insurance requirements set out independently for each of them. Furthermore, the requirements in turnkey projects demand as a prerequisite to its scope the provision of additional insurance policies.

3   As stated above, it is suggested that the intended capital value is around US$0.5 million.

4   See the first line of the Notes for Guidance, which forms the last section of the Green Book.

Clauses 13, 14 and 6 of the Green Book

13 Risk and Responsibility

Contractor's Care of the Works

13.1 The Contractor shall take full responsibility for the care of the Works from the Commencement Date until the date of the Employer's notice under Sub-Clause 8.2. Responsibility shall then pass to the Employer. If any loss or damage happens to the Works during the above period, the Contractor shall rectify such loss or damage so that the Works conform with the Contract.

Unless the loss or damage happens as a result of an Employer's Liability, the Contractor shall indemnify the Employer, the Employer's contractors, agents and employees against all loss or damage happening to the Works and against all claims or expense arising out of the Works caused by a breach of the Contract, by negligence or by other default of the Contractor, his agents or employees.

Force Majeure

13.2 If a Party is or will be prevented from performing any of its obligations by Force Majeure, the Party affected shall notify the other Party immediately. If necessary, the Contractor shall suspend the execution of the Works and, to the extent agreed with the Employer, demobilise the Contractor's Equipment.

If the event continues for a period of 84 days either Party may then give notice of termination which shall take effect 28 days after the giving of the notice.

After termination, the Contractor shall be entitled to payment of the unpaid balance of the value of the Works executed and of the Materials and Plant reasonably delivered to the Site, adjusted by the following:

(a) any sums to which the Contractor is entitled under Sub-Clause 10.4,
(b) the Cost of his suspension and demobilisation,
(c) any sums to which the Employer is entitled.

The net balance due shall be paid or repaid within 28 days of the notice of termination.

14 Insurance

Extent of Cover

14.1 The Contractor shall, prior to commencing the Works, effect and thereafter maintain insurances in the joint names of the Parties:

(a) for loss and damage to the Works, Materials, Plant and the Contractor's Equipment,

(b) for liability of both Parties for loss, damage, death or injury to third parties or their property arising out of the Contractor's performance of the Contract, including the Contractor's liability for damage to the Employer's property other than the Works, and

(c) for liability of both Parties and of any Employer's representative for death or injury to the Contractor's personnel except to the extent that liability arises from the negligence of the Employer, any Employer's representative or their employees.

Arrangements

14.2 All insurances shall conform with any requirements detailed in the Appendix. The policies shall be issued by insurers and in terms approved by the Employer. The Contractor shall provide the Employer with evidence that any required policy is in force and that the premiums have been paid.

All payments received from insurers relating to loss or damage to the Works shall be held jointly by the Parties and used for the repair of the loss or damage or as compensation for loss or damage that is not to be repaired.

Failure to Insure

14.3 If the Contractor fails to effect or keep in force any of the insurances referred to in the previous Sub-Clauses, or fails to provide satisfactory evidence, policies or receipts, the Employer may, without prejudice to any other right or remedy, effect insurance for the cover relevant to such default and pay the premiums due and recover the same as a deduction from any other monies due to the Contractor.

6 Employer's Liabilities

Employer's Liabilities

6.1 In this Contract, Employer's Liabilities mean:

(a) war, hostilities (whether war be declared or not), invasion, act of foreign enemies, within the Country,

(b) rebellion, terrorism, revolution, insurrection, military or usurped power, or civil war, within the Country,

(c) riot, commotion or disorder by persons other than the Contractor's personnel and other employees, affecting the Site and/or the Works,

(d) ionising radiations, or contamination by radio-activity from any nuclear fuel, or from any nuclear waste from the combustion of nuclear fuel, radio-active toxic explosive, or other hazardous properties of any explosive nuclear assembly or nuclear component of such an assembly, except to the extent to which the Contractor may be responsible for the use of any radio-active material,

(e) pressure waves caused by aircraft or other aerial devices travelling at sonic or supersonic speeds,

(f) use or occupation by the Employer of any part of the Works, except as may be specified in the Contract,

(g) design of any part of the Works by the Employer's personnel or by others for whom the Employer is responsible, and

(h) any operation of the forces of nature affecting the Site and/or the Works, which was unforeseeable or against which an experienced contractor could not reasonably have been expected to take precautions.

(i) Force Majeure,

(j) A suspension under Sub-Clause 2.3 unless it is attributable to the Contractor's failure,

(k) any failure of the Employer,

(l) physical obstructions or physical conditions other than climatic conditions encountered on the Site during the performance of the Works, which obstructions or conditions were not reasonably foreseeable by an experienced contractor and which the Contractor immediately notified to the Employer,

(m) any delay or disruption caused by any Variation,

(n) any change to the law of the Contract after the date of the Contractor's offer as stated in the Agreement,

(o) losses arising out of the Employer's right to have the permanent work executed on, over, under, in or through any land, and to occupy this land for the permanent works, and

(p) damage which is an unavoidable result of the Contractor's obligations to execute the Works and to remedy any defects.

forms of contract. Instead, the employer takes over the functions usually performed by the engineer or the employer's representative. However, the employer may appoint an independent engineer to act impartially by modifying Clause 3 of the Conditions 'Employer's Representative'.

These various options are explained at the end of the Green Book in a section entitled 'Notes for Guidance', which do not form part of the contract.[5] Accordingly, once the employer considers the options available to him under the Green Book, he is guided to select what he needs and deletes what he does not, ending in a contract form which crystallises his choice. The employer is then directed to complete an appendix, which incorporates the characteristics of his chosen contract, prior to inviting tenders. This appendix appears at the beginning of the Green Book as part of the Agreement, which will be eventually signed with the selected contractor.

5 Unlike FIDIC Contracts Guide for the major forms of contract in FIDIC's 1999 Suite, which was published separately during 2001, but was copyrighted in 2000, the Notes for Guidance of the Green Book were given in the last section of the book itself.

The flexibility of the document is a significant feature of the Green Book, particularly where insurance is concerned. This is due to the fact that the relevant clause, Clause 14, only specifies the general framework of the cover required to be obtained leaving the various details to be completed by the employer in the Appendix with extensive freedom to include any insurance requirement and in any detail he deems fit.

Analysis of Clauses 13 and 14 of the Green Book

As can be seen from its title, Clause 13 of the Green Book is the relevant clause to the topics of risk and responsibility. However, as we read the text of this clause, we find no mention of risk at all. In fact, other than in the title, the word 'Risk' does not appear anywhere in the Green Book. But, on close scrutiny, it becomes apparent that the reference to an 'Employer's Liability' in the second paragraph of Clause 13 is intended to lead the reader to Clause 6 of the form, where for some inexplicable reason the draftsman refers to risks as liabilities.[6] It is extremely peculiar that FIDIC which pioneered the adoption of the risk concepts in its various forms of contract[7] is now turning the clock back with its Green Book and confusing risk with liability. Even from a linguistic point of view, it is difficult to understand how one could confuse risk with liability. They are two terms which are entirely different etymologically, scientifically, legally and in every other sense. Risk is technically defined as 'A combination of the probability, or frequency, of occurrence of a defined hazard and the magnitude of the consequences of the occurrence'.[8] Liability, on the other hand is defined as 'the legal concept of one party being subject to the power of another, or to a rule of law requiring something to be done or not done. This requirement to do something or not to do it can be compelled by legal process at the other party's instance. It is sometimes called subjection.'[9] Liability may arise either from a voluntary act or by force of some rule of law. Thus, a person who enters into a contract becomes liable to perform what he has undertaken, or to pay for the counterpart performance, or otherwise to implement his part of the bargain.

Even if it was not wrong to use the term 'risk' and mean 'liability', and it is suggested here that it is wrong, the substitution of 'risks' with 'liabilities' is a detrimental step. The topics of risk and risk management are now part of a respected field of science and their principles should be strengthened and enhanced rather than diluted in any contract.

---

6   This part of the paper is based on an article by the author, published in [2002] ICLR 220.

7   The 3rd edition of the Yellow Book and to some extent the 4th edition of the Red Book, both of which were first published in 1987, were the first forms of contract that recognised the natural flow of Risk *to* Responsibility *to* Liability *to* Indemnity *to* Insurance. See in this connection, Nael G. Bunni, *The FIDIC Form of Contract – 4th Edition*, second edition, Blackwell Science, 1997.

8   British Standard No. 4778: Section 3.1 – Guide to concepts and related definitions: 1991. The British Standards Institution, Linford Wood, Milton Keynes, MK14 6LE, UK. In the same British Standard, the definition of 'hazard' is given as 'A situation that could occur during the lifetime of a product, system or plant that has the potential for human injury, damage to property, damage to the environment, or economic loss'. See also page 28.

9   David M. Walker, *The Oxford Companion to Law*, Clarendon Press, Oxford, 1980.

Moving on to the second paragraph of Sub-clause 13.1, the text presents us with an equally serious problem and that is in respect of the gap created by the division of risks (referred to as liabilities), between the employer and the contractor. The employer is allocated the risks described in Clause 6. The contractor is allocated the risks of all loss or damage happening to the works and of all claims or expense arising out of the works caused by 'a breach of the Contract, by negligence or by other default of the Contractor, his agents or employees'. To whom then are the other risks allocated? The risks referred to here are the risks that do not qualify within the meaning of an 'employer's risk' nor can they be described as 'a breach of the Contract' by the Contractor. This problem is of a similar nature to that created in the three major forms of contract published by FIDIC in 1999 through their Sub-Clause 17.1(b)(ii) where the basis of indemnity is negligence rather than legal liability.[10] This gap in risk allocation ultimately creates a gap in the insurance cover for the project, unless of course it is specifically dealt with in the Appendix.

Clause 13.2 deals with Force Majeure, which is a risk allocated to the employer in Sub-Clause 6.1(i), although it is referred to as 'a liability'. Force Majeure is defined in Sub-Clause 1.1.14 of the Green Book as

> an exceptional event or circumstance: which is beyond a Party's control, which such Party could not reasonably have provided against before entering into the Contract; which, having arisen, such Party could not reasonably have avoided or overcome; and, which is not substantially attributable to the other Party.

Whatever the merit, desirability or necessity for such a clause in a major form of contract such as the New Red, Yellow or the Silver Books, it is suggested here that there is none for a 'simple contract of short duration with relatively small capital value'. This is due to the complications it creates from the legal and insurance points of view. It is further suggested that the appropriate method of dealing with the risks as captured by the intended meaning of Sub-Clause 13.2 of the Green Book is to designate them as exceptional risks leading to specific remedies under the contract as is the case under Clauses 65 and 66 of the 4th edition of the old Red Book: 'Special Risks' and 'Release from Performance'.[11]

As to Clause 6, grouping in one clause all the Employer's Risks (provided they are properly identified as risks), including those that lead to pure economic and/or time loss together with the others that lead to loss and/or damage, is a good idea. This idea of grouping all the employer's risks in one clause is now adopted in the latest of the published FIDIC's forms, the Form of Contract for Dredging and Reclamation Works, published in 2001, which has fortunately resolved the problem of the confusion between 'risk' and 'liability'. Thus, for anyone who is intent on using a form for smaller contracts, it would be better to base it on this latest form of FIDIC.

Dealing with the insurance provisions of the Green Book, one may start from the beginning when a risk eventuates. The consequences might be either insurable or not. Whether or not it is required to be insured, Clause 6, which defines these risks

10   See note 36 to Chapter 1, but page 525, last paragraph.
11   See note 10 above, but page 528, where a similar proposal is made in connection with FIDIC's three major forms, but referring to the special risks as exceptional risks.

and refer to them as liabilities, is silent on how and when insurance, if available, is to be provided in respect of the liabilities or the risks specified. Thus, the link between liability, indemnity and insurance is lost. This situation may be due to the idea expressed in Sub-Clause 14.2 that the employer should set out his precise requirements relating to the required insurance cover in the Appendix, but the Appendix should then explain the relationship between Sub-Clauses 10.4 and 14.2.

Where insurance is concerned, it is the smaller employers and contractors that are usually not fully versed in the complexities of construction insurance and therefore, it is in these smaller contracts that they require specific standard conditions to assist them in providing a balanced arrangement and one that would work without conflict when events lead to accidents. Some of the questions that may escape the attention of those who are not used to this type of insurance include the following:

1 Clause 14.1(c) requires insurance cover to be provided 'for liability of both parties and of any employer's representative . . . except to the extent that liability arises from the negligence of the Employer, any Employer's representative or their Employees'. It is clear that there is no requirement to effect insurance against the negligence of the employer or the employer's representative. However, does this mean that an insurance cover is required for non-negligence of those named above?

2 The term 'Works' is defined in Sub-clause 1.1.19 of the conditions as meaning 'all the work and design (if any) to be performed by the Contractor including temporary work and any Variation'. As 'Works' would most probably include design carried out by the contractor, there is a standard requirement for professional indemnity insurance included in Sub-clause 14.1(a). Details of such cover must be included in the Appendix and therefore space must be allocated therein for such details.

3 What is the meaning of the phrase in Sub-clause 14.2 'evidence that any required policy is in force'? Is a letter from an insurance broker sufficient? The wordings of construction insurance policies differ greatly and it is meaningless to simply obtain as evidence anything other than the insurance policies themselves, including any endorsements issued and conditions attached.

The intricacies of construction insurance are many as can be seen from the previous chapters of this book.

The new major books: Red, Yellow and Silver

As stated earlier, the new three major books: the New Red Book; the New Yellow Book; and the Silver Book, are very similar in format. They each contain twenty clauses, seventeen of which have common titles and all of which have similar wording where the concepts match.[12]

---

12 The three clauses that carry different titles are: Clause 3 ('The Engineer' in the New Red and Yellow Books and 'The Employer's Administration' in the Silver Book); Clause 5 ('Nominated Subcontractors' in the New Red and 'Design' in the New Yellow and Silver Book Books); and Clause 12 ('Measurement and Evaluation' in the New Red and 'Tests after Completion' in the New Yellow and Silver Book Books).

However, although the format is the same in all three books, the Silver Book can be distinguished by the absence of the function of the 'Engineer' and by its allocation of the risks, which marks a shift from the employer to the contractor. The other two books, however, continue the tradition of having an 'Engineer' and remain as a whole within FIDIC's tradition of drafting standard forms of contract with a balanced allocation and sharing of risks between the employer and the contractor.

These distinguishing characteristics of the Silver Book should not be taken as a criticism of its concept and application. In particular, the Silver Book was conceived in response to the need created by those who favoured the use of private finance for infrastructure projects, and grew as a result of the demands associated with BOT or BOOT projects and with the new ideas of mixing together design, construction and operation. This entailed demanding a fixed, lump-sum contract price with least or no risk of an increase if and when unexpected events took place. Of course, privately financed projects require being financially viable with an assured return on the funds advanced. Therefore, although demanding a fixed, lump-sum contract price means that the employer would be paying a higher price for the construction of the project, he would not normally object to having to do so if he were assured of an acceptable return on his total investment.

The important point to recognise, however, as a result of these distinguishing characteristics of the Silver Book is that they require quite a different set of insurance arrangements to those normally associated with the Red Book. The New Yellow Book has the special feature of allocating the function of design to the contractor, which also requires a special insurance arrangement for that design element of the project. The problem is that despite these distinguishing features of the New Yellow and Silver Books, the insurance clauses of the three books are almost the same and differ on very few aspects, as can be seen later on in this chapter.

It is now appropriate to consider the new provisions in these three FIDIC forms of contract that might affect either the risks that lead to loss and/or damage or the insurance cover required to protect the parties in general terms.

New provisions that might affect the risks which lead to loss and/or damage, and insurance

The following provisions are considered here to be of importance from the point of view of the risks that lead to loss and/or damage, and also in general terms from the view of the insurance cover required to protect the parties should these risks eventuate:

1   A new provision has been added in both of the New Red and Yellow Books, which recognise the risk of the 'Employer's Financial Arrangements'. Clause 2.4 provides as follows:

Employer's Financial Arrangements

2.4 The Employer shall submit, within 28 days after receiving any request from the Contractor, reasonable evidence that financial arrangements have been made and are being maintained which will enable the Employer to pay the Contract Price (as estimated at that time) in accordance with Clause 14

[*Contract Price and Payment*]. If the Employer intends to make any material change to his financial arrangements, the Employer shall give notice to the Contractor with detailed particulars.

It is to be noted that if the employer fails to submit the evidence requested, the contractor is entitled under Sub-clause 16.1, after 21 days notice, to suspend the works, or reduce the rate of progress of the work. If the employer fails to submit the evidence requested within 42 days after the notice given under Sub-Clause 16.1, the contractor is entitled under Sub-clause 16.2, to terminate the contract as a final remedy.

2   The higher standard of performance of Fitness for Purpose applies whenever the contractor is required to design under the new Books. Thus, it is specified in the first paragraph of Sub-clause 4.1 of the New Yellow and Silver Books that 'When completed, the Works shall be fit for the purposes for which the Works are intended as defined in the Contract.' (See also Sub-clause 4.1(c) of the New Red Book and Sub-clause 11.3 in all three Books relating to Defects Notification Period.)

3   Whilst all the books recognise the matrix of pure economic risks, the wording of the provisions of Sub-clause 4.2 relating to performance security differs greatly from its equivalent under the traditional forms of contract published by FIDIC. So, for example:

  •   On-Demand Guarantee or Surety Bond is permitted under the new Books;
  •   The employer is not permitted to make a claim under the Performance Security except for amounts to which he is entitled in specified events;
  •   An indemnity is specified where the employer claims to the extent to which he is not entitled; and
  •   The employer is required to return the Performance Security within 21 days after receiving a copy of the Performance Certificate, see Sub-clause 11.9.

4   In the New Yellow and Silver Books where the design function is allocated to the contractor, there are many new provisions, the most important of which is the 'Employer's Requirements'. The first paragraph of Sub-clause 5.1 of the New Yellow Book provides as follows:

General Design Obligations

5.1 The Contractor shall carry out, and be responsible for, the design of the Works. Design shall be prepared by qualified designers who are engineers or other professionals who comply with the criteria (if any) stated in the Employer's Requirements. Unless otherwise stated in the Contract, the Contractor shall submit to the Engineer for consent the name and particulars of each proposed designer and design Subcontractor.

The term 'Employer's Requirements' is defined in Sub-clause 1.1.1.5 and its significant importance is perhaps reflected in the number of Sub-clauses in which it is referred to; twenty-five sub-clauses throughout the document. The drafting of these requirements is probably the main source of success or failure of the project and of the disputes that might arise in a project under this form of contract. Under the Silver Book, the contractor is responsible even for the accuracy of the Employer's Requirements.

5 Sub-clause 4.21 in the new books introduces the concept of 'Progress Reports' requiring that 'Unless otherwise stated in the Particular Conditions, monthly progress reports shall be prepared by the Contractor and submitted to the Engineer in six copies.' The Sub-clause continues to provide that these reports should be submitted monthly thereafter, each within seven days after the last day of the period to which it relates and gives in detail what they should contain.

6 There are significant changes in risk allocation in the new books, as can be seen for example in Clauses 15.5 and 19. Sub-clause 15.5 introduces the concept that the Employer is entitled to terminate the contract, at any time for his convenience, by giving notice of such termination to the Contractor. Clause 19 substitutes the concept of 'frustration' in the old Red Book with that of 'Force Majeure', which is defined in that clause as

an exceptional event or circumstance
(a) which is beyond a Party's control,
(b) which such Party could not reasonably have provided against before entering into the Contract,
(c) which, having arisen, such Party could not reasonably have avoided or overcome, and
(d) which is not substantially attributable to the other Party.

It may include, but is not limited to, exceptional events or circumstances of a number of events listed therein, so long as conditions (a) to (d) above are satisfied.

7 A new concept of *Limitation of Liability* is now introduced in the new Books. Sub-Clause 17.6 provides that the parties are not liable to each other

for loss of use of any Works, loss of profit, loss of any contract or for any indirect or consequential loss or damage which may be suffered by the other Party in connection with the Contract, other than under Sub-Clause 16.4 [*Payment on Termination*] and Sub-Clause 17.1 [*Indemnities*] . . .

The total liability of the contractor to the employer, under or in connection with the contract other than in limited circumstances is required to be stated in the Particular Conditions or (if a sum is not so stated) the Accepted Contract Amount. However, this Sub-clause is not intended to limit liability in any case of fraud, deliberate default or reckless misconduct by the defaulting Party.

8 Clause 18 'Insurance' in the new books is significantly different from its equivalent in the old Red Book. An analysis of this topic is available in print. [13]

The New Silver Book

As discussed above, the Silver Book is a new entrant into the field. To understand the philosophy and reasons behind its conception, it is best to quote from the

---

13 See note 10 above.

authoritative paper written by the Chairman of FIDIC's Contracts Committee and Leader of the Task Group who prepared the FIDIC 1999 Conditions of Contract.[14]

> Not only is it a fact of life that many employers have always demanded 'fixed, lump sum contract prices', and that FIDIC did not have a suitable standard form to cater for such demand, but in recent years the trend has been towards private financing (not only of private investment and speculative projects, but also of public infrastructure projects). The prerequisites for obtaining private finance for a project are vastly different from those of obtaining government or other public money. Private financing requires that the project is independently viable in financial terms, and that there will be, so far as possible, an assured return on the finance provided. The lenders on a BOT or similar project will do their calculations showing the outlay over the construction period and the income over the succeeding operation period. For the return to be reasonably assured, the bases for their calculations will have to be as firm as possible. If the construction work costs more than reckoned (inclusive of any contingency allowance), then the calculations will not hold. If the construction time is longer than planned, then the income will not begin to come in on time, and the calculations will not hold. Therefore, such lenders have to ensure that the risks of cost and time overruns of the construction contract are limited as far as humanly possible. Such lenders are aware that contractors will have to charge a premium for carrying the additional risks necessary to provide the required greater security of construction cost and time. The premium in certain cases may reasonably be large. However, they would rather accept such premium and include it in their calculations before embarking on the project, than discover later on that the project is no longer viable and that they are incurring an overall loss.

Thus, the Silver Book has been and is intended to be used for special projects and is not a 'contract for all seasons', as was practised in the use of the old Red Book. The Silver Book, if and when used, ought to be entered into with the utmost care, with all eyes open focusing on the risks that have been shifted, from a balanced contract between the parties to a contractor-borne risks. These risks are referred to in various sub-clauses of the contract, but the following are highlighted from an insurance point of view:

- *Sub-clause 3.5:* As stated above, there is no 'Engineer' in place to deal with the usual administrative matters and instead it is the employer who carries out such a role. Therefore, it is the employer who makes the determination when no agreement is reached between him and the contractor. Sub-clause 3.5 states in part the following: 'If agreement is not achieved, the Employer shall make a fair determination in accordance with the Contract, taking due regard of all relevant circumstances.'. The word 'fair' should be noted.

---

14 'The Silver Book: The Reality', by Christopher Wade [2001] ICLR 500.

- *Sub-clause 5.1:* This Sub-clause states, in part, the following:[15]

> The Employer shall not be responsible for any error, inaccuracy or omission of any kind in the Employer's Requirements as originally included in the Contract and shall not be deemed to have given any representation of accuracy or completeness of any data or information, except as stated below. Any data or information received by the Contractor, from the Employer or otherwise, shall not relieve the Contractor from his responsibility for the design and execution of the Works.

  Therefore, the contractor has to take on board not only strict liability for design, fitness for purpose standard of performance, but also be liable for 'any error, inaccuracy or omission of any kind in the Employer's Requirements as originally included in the Contract.'

- Sub-clause 5.8, which provides as follows:

  5.8 *Design Error*
  If errors, omissions, ambiguities, inconsistencies, inadequacies or other defects are found in the Contractor's Documents, they and the Works shall be corrected at the Contractor's cost, notwithstanding any consent or approval under this Clause.

- The consequence of any of the following risks is borne by the contractor, as these risks have been omitted from the list of Employer's Risks in the Silver Book

  (f) use or occupation by the Employer of any part of the Permanent Works, except as may be specified in the Contract,
  (g) design of any part of the Works by the Employer's Personnel or by others for whom the Employer is responsible, if any, and
  (h) any operation of the forces of nature which is Unforeseeable or against which an experienced contractor could not reasonably have been expected to have taken adequate preventative precautions.[16]

Differences between the three new FIDIC books in Clauses 17 to 19

As stated above, Clauses 17 to 19 of the three major books published by FIDIC in 1999 only differ on very few relevant aspects. These differences are set out below:

- Under Sub-clause 17.1, the contractor's indemnity to the employer extends in the Silver Book, to the whole design activity.
- Under Sub-clauses 17.3 and 17.4, the Employer's Risks (f), (g) and (h) of the

---

15 Second paragraph of Sub-clause 5.1.
16 A comparison between Sub-clause 17.3 'Employer's Risks', in the Silver Book and that in the other two books, would reveal that these risks have been shifted to the contractor.

New Red and Yellow Books and their consequences are shifted to the contractor under the Silver Book.[17]

- Under Sub-clause 17.5, the employer's indemnity to the contractor in respect of infringement of intellectual and industrial rights under the New Red Book is related to an unavoidable result of the contractor's compliance with the contract whereas under the Silver and Yellow Books it is related to an unavoidable result of the contractor's compliance with the employer's requirements.

- Under Sub-clause 17.5, the contractor's indemnity to the employer in respect of infringement of intellectual and industrial rights under the New Red Book is related to a claim which arises out of or in relation to (i) the manufacture, use, sale or import of any goods, or (ii) any design for which the contractor is responsible. However, under the Silver and Yellow Books, this indemnity arises out of or in relation to (i) the contractor's design, manufacture, construction or execution of the works, (ii) the use of contractor's equipment, or (iii) the proper use of the works.

- Under Sub-clause 17.6, which deals with the total liability of the contractor to the employer, the limit related to 'the Accepted Contract Amount' in the Red and Yellow Books is replaced under the Silver Book with 'the Contract Price stated in the Contract Agreement'.

- Under the second paragraph of Sub-Clause 18.1, the words 'before the date of the Letter of Acceptance', which refer to the timing of agreement by the parties on the terms of insurance in the Red and Yellow Books is replaced by 'before they signed the Contract Agreement' under the Silver Book.

- Under the seventh paragraph of Sub-clause 18.1, the words 'Whenever evidence or policies are submitted, the insuring Party shall also give notice to the Engineer' in the Red and Yellow Books is replaced with 'When each premium is paid, the insuring Party shall submit evidence of payment to the other Party' under the Silver Book.

- Under the second paragraph of Sub-clause 18.2, the words 'the Contractor in the course of any operations (including those under Clause 11 [Defects Liability])' in the Red Book, which relate to the insurance cover during the Defects Liability Period, are replaced with 'the Contractor in the course of any operations (including those under Clause 11 [Defects Liability] and Clause 12 [Tests after Completion])' in the Yellow Book; and are replaced with 'the Contractor or Subcontractor in the course of any other operations (including those under Clause 11 [Defects Liability] and Clause 12 [Tests after Completion]) in the Silver Book.

- Under paragraph (d) of Sub-clause 18.2, of the Red and Yellow Books, the insurance of the works is required to extend

---

17 These risks are identified in the New Red Book as follows:

'(f) use or occupation by the Employer of any part of the Permanent Works, except as may be specified in the Contract,

(g) design of any part of the Works by the Employer's Personnel or by others for whom the Employer is responsible, if any, and

(h) any operation of the forces of nature which is Unforeseeable or against which an experienced contractor could not reasonably have been expected to have adequate preventative precautions.'

to a part of the works which is attributable to the use or occupation by the Employer of another part of the Works, and loss or damage from the risks listed in sub-paragraphs (c), (g) and (h) of Sub-Clause 17.3 [*Employer's Risks*], excluding (in each case) risks which are not insurable at commercially reasonable terms, with deductibles per occurrence of not more than the amount stated in the Appendix to Tender (if an amount is not so stated, this sub-paragraph (d) shall not apply).

In the Silver Book, this paragraph is worded to extend the insurance to

cover loss or damage from the risks listed in sub-paragraph (c) of Sub-Clause 17.3 [*Employer's Risks*], with deductibles per occurrence of not more than the amount stated in the Particular Conditions (if an amount is not so stated, this sub-paragraph (d) shall not apply).

- Under the second paragraph of Sub-clause 18.3 of the Red and Yellow Books, the amount of the lower limit for third party liability insurance cover in respect of any loss, damage, death or bodily injury which may occur to any physical property or to any person per occurrence is stated in the 'Appendix to Tender'. In the Silver Book, this limit is stated in the Particular Conditions.
- Under the second paragraph of Sub-clause 18.4, the reference to indemnities to the 'Engineer' in the Red and Yellow Books is deleted in the Silver Book.
- Under Sub-clause 19.6 of the Red and Yellow Books, where termination takes effect, 'the Engineer shall determine the value of the work done and issue a Payment Certificate . . .'. This wording is replaced in the Silver Book with 'the Employer shall pay to the Contractor'.

It is now appropriate to analyse the provisions of these risk and insurance clauses to appreciate their effectiveness. The clauses for the new Red Book are used here for this analysis.

Analysis of Clauses 17 to 19 of the New Red Book

A   Clause 17 – Risk and Responsibility

Although this clause of FIDIC's new suite of contracts is entitled 'Risk and Responsibility', it encompasses other contractual provisions, including indemnities; limitation of liability; and the unrelated topic of intellectual and industrial property rights. In fact, Clause 17 starts from the wrong end by dealing first with 'Indemnities' and then it somehow back tracks to deal with 'Responsibility' and then takes a further leap backwards and returns to 'Risk' and finally marches forward on to 'Liability'. This illogical sequence hardly helps the non-lawyer professionals for whom these provisions are intended. The clause leaves even the expert in the field wondering about the purpose of this confused and baffling sequence.

The theory of risk has developed in the past twenty years or so to such an extent that it is now common knowledge that for a contract to be performed in an effective manner, the inherent risks must be allocated to the contracting parties on some logical basis, which should be made known to them. Thus, as has been said earlier, the main purpose of a contract is to identify the principles of allocating the risks facing the contracting parties. Once these principles are identified, the consequences flow in the natural pattern of Risk <u>to</u> Responsibility <u>to</u> Liability <u>to</u> Indemnity <u>to</u> Insurance.[18] The format of Clause 17 should, therefore, follow that same sequence, with the insurance provisions left to the next clause, i.e. Clause 18, if it is desired that they should be presented separately.

Accordingly, Clause 17 ought to start with the provisions for 'Risk' and not with 'Indemnities' and Sub-clause 17.3 should be 17.1. Furthermore, the wording of Sub-Clause 17.1 should then start by explaining that the risks included under Clause 17 of the conditions of contract are only those risks of loss and damage and not the whole spectrum of the risks to which the project is exposed. The term 'Employer's Risks' in the context of this clause should therefore be replaced by 'Employer's Risks of Loss and Damage', since these risks are confined to those which lead to some form of accidental loss or damage to physical property or personal injury, which in turn may lead to financial and/or time loss risks, directly or through the other clauses of the contract.

If this explanation is not given and the mistake of referring to the risks under Clause 17 as 'Employer's Risks' is not corrected, then there is serious danger that the reader, and of course the user, will conclude that having identified in Clause 17 the employer's risks, all the other risks are the contractor's risks, including the contractual economic and time risks in the remaining provisions of the contract. This problem can be highlighted by reference to Clause 17 of the Orange Book where the draftsman fell into that trap and stated expressly in Sub-clause 17.5 that 'The Contractor's risks are all risks other than the Employer's Risks listed in Sub-Clause 17.3'. This mistake has led to many instances of misunderstanding, conflict and at least one serious arbitral proceedings, where the employer pointed out that by Sub-clause 17.5 he bears no risks under the contract other than those specified in Sub-clause 17.3.

Accordingly, it is essential to understand that the employer's risks traditionally identified under Sub-clause 20.3 of the old Red Book and those identified under Sub-clause 17.3 of the new suite of contracts are only the amalgamation of risks that are beyond the control of the contractor alone or that of both the contractor and the employer. Furthermore, these risks might have an implied resultant loss or damage to physical property or cause bodily injury, all of which are insurable. In contrast, very few of the other risks to which the project is exposed are insurable.

There are other problems in Clause 17. The second problem is the allocation of the risks specified in sub-paragraph (h) of Sub-clause 17.3 to the employer.[19] Whilst this does not form a departure from the old Red Book, it was hoped that the new suite of contracts would be up to date with developments in this field. The origin of this sub-paragraph goes back to the ACE Form of Contract recognisable as the route for the FIDIC Red Book. Whilst it is true that the contractor has no control over the events identified in this sub-paragraph, he is in control over their consequences and can instigate protection measures. The contractor can also mitigate any losses that

---

18 *The FIDIC Form of Contract – The Fourth Edition of the Red Book*, 2nd edition, by Nael G. Bunni, 1997, Blackwell Science Ltd., Oxford.

19 The risks in paragraph (h) of Sub-clause 17.3 are: 'any operation of the forces of nature which is Unforeseeable or against which an experienced contractor could not reasonably have been expected to have taken adequate preventative precautions'.

Clauses 17 to 19 of the New Red Book – 1999 Edition

17 *Risk and Responsibility*

Indemnities

17.1 The Contractor shall indemnify and hold harmless the Employer, the Employer's Personnel, and their respective agents, against and from all claims, damages, losses and expenses (including legal fees and expenses) in respect of:

(a) bodily injury, sickness, disease or death of any person whatsoever arising out of or in the course of or by reason of the Contractor's design (if any), the execution and completion of the Works and the remedying of any defects, unless attributable to any negligence, wilful act or breach of the Contract by the Employer, the Employer's Personnel, or any of their respective agents, and

(b) damage to or loss of any property, real or personal (other than the Works), to the extent that such damage or loss:

(i) arises out of or in the course of or by reason of the Contractor's design (if any), the execution and completion of the Works and the remedying of any defects, and

(ii) is attributable to any negligence, wilful act or breach of the Contract by the Contractor, the Contractor's Personnel, their respective agents, or anyone directly or indirectly employed by any of them.

The Employer shall indemnify and hold harmless the Contractor, the Contractor's Personnel, and their respective agents, against and from all claims, damages, losses and expenses (including legal fees and expenses) in respect of (1) bodily injury, sickness, disease or death, which is attributable to any negligence, wilful act or breach of the Contract by the Employer, the Employer's Personnel, or any of their respective agents and (2) the matters for which liability may be excluded from insurance cover, as described in subparagraphs (d) (i), (ii) and (iii) of Sub-Clause 18.3 [Insurance Against Injury to Persons and Damage to Property].

Contractor's Care of the Works

17.2 The Contractor shall take full responsibility for the care of the Works and Goods from the Commencement Date until the Taking-Over Certificate is issued (or is deemed to be issued under Sub-Clause 10.1 [Taking Over of the Works and Sections]) for the Works, when responsibility for the care of the Works shall pass to the Employer. If a Taking-Over Certificate is issued (or is deemed to be issued) for any Section or part of the Works, responsibility for the care of the Section or part shall then pass to the Employer.

After responsibility has accordingly passed to the Employer, the Contractor shall take responsibility for the care of any work which is outstanding on the date stated in a Taking-Over Certificate, until this outstanding work has been completed.

If any loss or damage happens to the Works, Goods or Contractor's Documents during the period when the Contractor is responsible for their care, from any cause not listed in Sub-Clause 17.3 [Employer's Risks], the Contractor shall rectify the loss or damage at the Contractor's risk and cost, so that the Works, Goods and Contractor's Documents conform with the Contract.

The Contractor shall be liable for any loss or damage caused by any actions performed by the Contractor after a Taking-Over Certificate has been issued. The Contractor shall also be liable for any loss or damage which occurs after a Taking-Over Certificate has been issued and which arose from a previous event for which the Contractor was liable.

Employer's Risks

17.3 The risks referred to in Sub-clause 17.4 below are:

   (a) war, hostilities (whether war be declared or not), invasion, act of foreign enemies,

   (b) rebellion, terrorism, revolution, insurrection, military or usurped power, or civil war, within the Country,

   (c) riot, commotion or disorder within the Country by persons other than the Contractor's Personnel and other employees of the Contractor and Subcontractors,

   (d) munitions of war, explosive materials, ionising radiation or contamination by radio-activity, within the Country, except as may be attributable to the Contractor's use of such munitions, explosives, radiation or radio-activity,

   (e) pressure waves caused by aircraft or other aerial devices traveling at sonic or supersonic speeds,

   (f) use or occupation by the Employer of any part of the Permanent Works, except as may be specified in the Contract,

   (g) design of any part of the Works by the Employer's Personnel or by others for whom the Employer is responsible, and

   (h) any operation of the forces of nature which is Unforeseeable or against which an experienced contractor could not reasonably have been expected to have taken adequate preventative precautions.

Consequences of Employer's Risks

17.4 If and to the extent that any of the risks listed in Sub-Clause 17.3 above results in loss or damage to the Works, Goods or Contractor's Documents, the Contractor shall promptly give notice to the Engineer and shall rectify this loss or damage to the extent required by the Engineer.

If the Contractor suffers delay and/or incurs Cost from rectifying this loss or damage, the Contractor shall give a further notice to the Engineer and shall be entitled subject to Sub-Clause 20.1 [Contractor's Claims] to:

(a) an extension of time for any such delay, if completion is or will be delayed, under Sub-Clause 8.4 [Extension of Time for Completion], and

(b) payment of any such Cost, which shall be included in the Contract Price. In the case of sub-paragraphs (f) and (g) of Sub-Clause 17.3 [Employer's Risks], reasonable profit on the Cost shall also be included.

After receiving this further notice, the Engineer shall proceed in accordance with Sub-Clause 3.5 *[Determinations]* to agree or determine these matters.

Intellectual and Industrial Property Rights

17.5 In this Sub-Clause, 'infringement' means an infringement (or alleged infringement) of any patent, registered design, copyright, trade mark, trade name, trade secret or other intellectual or industrial property right relating to the Works; and 'claim' means a claim (or proceedings pursuing a claim) alleging an infringement.

Whenever a Party does not give notice to the other Party of any claim within 28 days of receiving the claim, the first Party shall be deemed to have waived any right to indemnity under this Sub-Clause.

The Employer shall indemnify and hold the Contractor harmless against and from any claim alleging an infringement which is or was:

(a) an unavoidable result of the Contractor's compliance with the Contract, or

(b) a result of any Works being used by the Employer:

    (i) for a purpose other than that indicated by, or reasonably to be inferred from, the Contract, or

    (ii) in conjunction with any thing not supplied by the Contractor, unless such use was disclosed to the Contractor prior to the Base Date or is stated in the Contract.

The Contractor shall indemnify and hold the Employer harmless against and from any other claim which arises out of or in relation to (i) the manufacture, use, sale or import of any Goods, or (ii) any design for which the Contractor is responsible.

If a Party is entitled to be indemnified under this Sub-Clause, the indemnifying Party may (at its cost) conduct negotiations for the settlement of the claim, and any litigation or arbitration which may arise from it. The other Party shall, at the request and cost of the indemnifying Party, assist in contesting the claim. This other Party (and its Personnel) shall not make any admission which might be prejudicial to the indemnifying Party, unless the indemnifying Party failed to take over the conduct of any negotiations, litigation or arbitration upon being requested to do so by such other Party.

Limitation of Liability

17.6 Neither Party shall be liable to the other Party for loss of use of any Works, loss of profit, loss of any contract or for any indirect or consequential loss or damage which may be suffered by the other Party in connection with the Contract, other than under Sub-Clause 16.4 [Payment on Termination] and Sub-Clause 17.1 [Indemnities].

The total liability of the Contractor to the Employer, under or in connection with the Contract other than under Sub-Clause 4.19 [Electricity, Water and Gas], Sub-Clause 4.20 [Employer's Equipment and Free-Issue Material], Sub-Clause 17.1 [Indemnities] and Sub-Clause 17.5 [Intellectual and Industrial Property Rights], shall not exceed the sum stated in the Particular Conditions or (if a sum is not so stated) the Accepted Contract Amount.

This Sub-Clause shall not limit liability in any case of fraud, deliberate default or reckless misconduct by the defaulting Party.

---

might occur should any of these risks eventuate. Perhaps, more importantly, all the risks identified in sub-paragraph (h) represent events that are insurable and are generally required to be insured under the terms of the contract. The employer ultimately pays for such insurance through the contract provisions leaving the contractor in charge of any necessary repair, its cost and any claim negotiations with the insurers following the filing of such claims. These risks are not included as employer's risks in the ICE domestic contract or the others rooted in it.[20]

The third problem in Clause 17 of the new suite of contracts is the newly introduced restriction in Sub-clause 17.1(b)(ii) of the contractor's indemnity to the employer for property damage. This indemnity is now based on negligence rather than on legal liability as was provided in Clause 22.1 of the old Red Book.[21] This change is a retrograde step and copied from standard forms of contract for building works in the United Kingdom without any benefit to either the contractor or the employer.[22] The only beneficiary as a result of this change is the insurance market, since to cover this gap a new policy is now needed, which is commonly referred to in the United Kingdom as the non-negligence insurance policy. It seems that in making

---

20 For example, the Form of Contract for civil engineering construction of the Institution of Engineers of Ireland.

21 Sub-clause 17.1(b)(ii) of the new suite of contracts provides that 'The Contractor shall indemnify and hold harmless the Employer, the Employer's Personnel, and their respective agents, against and from all claims, damages, losses and expenses (including legal fees and expenses) in respect of: . . . damage to or loss of any property, real or personal (other than the Works), to the extent that such damage or loss: . . . is attributable to any negligence, wilful act or breach of the Contract by the Contractor . . .'.

22 C. Seppala, 'FIDIC's New Standard Forms of Contract – Force Majeure, Claims, Disputes and Other Clauses', ICLR Vol. 17, Part 2, April 2000. Mr. Seppala explains on page 238 of his article that FIDIC adopted this change in line with the policy in the major UK and other standard forms.

this change, the draftsman of Clause 17 of the new suite of contracts took comfort from a footnote in *Hudson's Building and Engineering Contracts*, where reference is made to both the RIBA and the ICE forms of contract.[23] The reference to the ICE form of contract in that footnote is incorrect since civil engineering contracts do not distinguish between the indemnity required to be given by the contractor for property damage on one hand and that for bodily injury, disease or death of any person on the other. In fact, the standard forms of contract for civil engineering construction in the United Kingdom or elsewhere do not impose the restriction now introduced.[24]

The last major problem in Clause 17 relates to the allocation to the contractor of the risk of '*use or occupation by the employer of any part of the Permanent Works*' in the Silver Book. The reasoning for such allocation is extremely obscure since neither such use (or occupation by the employer of any part of the Permanent Works) nor its consequences could be within the control of the contractor. Consequently, it is not a risk that could be assessed or against which some preventative measure could be taken.

Finally, there are some minor problems of drafting in Clause 17, which should be addressed for the proper understanding of what is intended by such a clause. For example, Sub-Clause 17.2 is a 'Responsibility' clause; Sub-clause 17.5 is a 'Risk' clause; and accordingly they should be designated as such. Another example is the need to have a statement as to proportional apportionment of indemnities when both employer and contractor have contributed to damage, loss or bodily injury. This would be particularly important where an indemnity clause is strictly interpreted under the applicable law of contract.[25]

B *Clause 18 – Insurance*

The first major problem in this Clause is the fact that the 'Insuring Party', as defined in the contract, is not the same for all the insurance policies required under the contract and it may be either of the two parties, employer or contractor. This is a recipe for confusion, gaps and/or overlaps in the combined insurance package, which could cost the parties dearly. It could only be advantageous to those involved in the insurance market.

The second paragraph of Clause 18 assumes that there would be a meeting between the parties prior to the date of the Letter of Acceptance at which the whole insurance package would be discussed and agreement would be reached on a policy towards insurance, which would 'take precedence over the provisions of (Clause

---

23 *Hudson's Building and Engineering Contracts*, 11th edition, by I.N. Duncan-Wallace, Sweet & Maxwell (1995), Vol. II, page 1437.
24 Clause 22 of the ICE form, whether the 5th edition, which is referred to in the referenced footnote in *Hudson's Building and Engineering Contracts* (1995), or the 6th or the 7th edition, do not refer to negligence by the contractor and do not distinguish between the indemnity required for property damage as against that for bodily injury, disease or death of any person.
25 For example, under English law, indemnity clauses would be strictly construed if the indemnitee seeks to enforce the clause in spite of his own negligence or fault.

Clauses 17 to 19 of the New Red Book – 1999 Edition

18   *Insurance*

General Requirements for Insurances

18.1   In this Clause, 'insuring Party' means, for each type of insurance, the Party responsible for effecting and maintaining the insurance specified in the relevant Sub-Clause.

Wherever the Contractor is the insuring Party, each insurance shall be effected with insurers and in terms approved by the Employer. These terms shall be consistent with any terms agreed by both Parties before the date of the Letter of Acceptance. This agreement of terms shall take precedence over the provisions of this Clause.

Wherever the Employer is the insuring Party, each insurance shall be effected with insurers and in terms consistent with the details annexed to the Particular Conditions.

If a policy is required to indemnify joint insured, the cover shall apply separately to each insured as though a separate policy had been issued for each of the joint insured. If a policy indemnifies additional joint insured, namely in addition to the insured specified in this Clause, (i) the Contractor shall act under the policy on behalf of these additional joint insured except that the Employer shall act for Employer's Personnel, (ii) additional joint insured shall not be entitled to receive payments directly from the insurer or to have any other direct dealings with the insurer, and (iii) the insuring Party shall require all additional joint insured to comply with the conditions stipulated in the policy.

Each policy insuring against loss or damage shall provide for payments to be made in the currencies required to rectify the loss or damage. Payments received from insurers shall be used for the rectification of the loss or damage.

The relevant insuring Party shall, within the respective periods stated in the Appendix to Tender (calculated from the Commencement Date), submit to the other Party:

(a)   evidence that the insurances described in this Clause have been effected, and

(b)   copies of the policies for the insurances described in Sub-Clause 18.2 [Insurance for Works and Contractor's Equipment] and Sub-Clause 18.3 [Insurance against Injury to Persons and Damage to Property].

When each premium is paid, the insuring Party shall submit evidence of payment to the other Party. Whenever evidence or policies are submitted, the insuring Party shall also give notice to the Engineer.

Each Party shall comply with the conditions stipulated in each of the insurance policies. The insuring Party shall keep the insurers informed of any relevant changes to the execution of the Works and ensure that insurance is maintained in accordance with this Clause.

Neither Party shall make any material alteration to the terms of any insurance without the prior approval of the other Party. If an insurer makes (or attempts to make) any alteration, the Party first notified by the insurer shall promptly give notice to the other Party.

If the insuring Party fails to effect and keep in force any of the insurances it is required to effect and maintain under the Contract, or fails to provide satisfactory evidence and copies of policies in accordance with this Sub-Clause, the other Party may (at its option and without prejudice to any other right or remedy) effect insurance for the relevant coverage and pay the premiums due. The insuring Party shall pay the amount of these premiums to the other Party, and the Contract Price shall be adjusted accordingly.

Nothing in this Clause limits the obligations, liabilities or responsibilities of the Contractor or the Employer, under the other terms of the Contract or otherwise. Any amounts not insured or not recovered from the insurers shall be borne by the Contractor and/or the Employer in accordance with these obligations, liabilities or responsibilities. However, if the insuring Party fails to effect and keep in force an insurance which is available and which it is required to effect and maintain under the Contract, and the other Party neither approves the omission nor effects insurance for the coverage relevant to this default, any moneys which should have been recoverable under this insurance shall be paid by the insuring Party.

Payments by one Party to the other Party shall be subject to Sub-Clause 2.5 [Employer's Claims] or Sub-Clause 20.1 [Contractor's Claims] as applicable.

Insurance for Works and Contractor's Equipment

18.2 The insuring Party shall insure the Works, Plant, Materials and Contractor's Documents for not less than the full reinstatement cost including the costs of demolition, removal of debris and professional fees and profit. This insurance shall be effective from the date by which the evidence is to be submitted under sub-paragraph (a) of Sub-Clause 18.1 [General Requirements for Insurances], until the date of issue of the Taking-Over Certificate for the Works.

The insuring Party shall maintain this insurance to provide cover until the date of issue of the Performance Certificate, for loss or damage for which the Contractor is liable arising from a cause occurring prior to the issue of the Taking-Over Certificate, and for loss or damage caused by the Contractor in the course of any other operations (including those under Clause 11 [Defects Liability]).

The insuring Party shall insure the Contractor's Equipment for not less than the full replacement value, including delivery to Site. For each item of Contractor's Equipment, the insurance shall be effective while it is being transported to the Site and until it is no longer required as Contractor's Equipment.

Unless otherwise stated in the Particular Conditions, insurances under this Sub-Clause:

(a) shall be effected and maintained by the Contractor as insuring Party,

(b) shall be in the joint names of the Parties, who shall be jointly entitled to receive payments from the insurers, payments being held or allocated between the Parties for the sole purpose of rectifying the loss or damage,

(c) shall cover all loss and damage from any cause not listed in Sub-Clause 17.3 [Employer's Risks],

(d) shall also cover loss or damage to a part of the Works which is attributable to the use or occupation by the Employer of another part of the Works, and loss or damage from the risks listed in sub-paragraphs (c), (g) and (h) of Sub-Clause 17.3 [Employer's Risks], excluding (in each case) risks which are not insurable at commercially reasonable terms, with deductibles per occurrence of not more than the amount stated in the Appendix to Tender (if an amount is not so stated, this sub-paragraph (d) shall not apply), and

(e) may however exclude loss of, damage to, and reinstatement of:

   (i) a part of the Works which is in a defective condition due to a defect in its design, materials or workmanship (but cover shall include any other parts which are lost or damaged as a direct result of this defective condition and not as described in sub-paragraph (ii) below),

   (ii) a part of the Works which is lost or damaged in order to reinstate any other part of the Works if this other part is in a defective condition due to a defect in its design, materials or workmanship,

   (iii) a part of the Works which has been taken over by the Employer, except to the extent that the Contractor is liable for the loss or damage, and

   (iv) Goods while they are not in the Country, subject to Sub-Clause 14.5 [Plant and Materials intended for the Works].

If, more than one year after the Base Date, the cover described in sub-paragraph (d) above ceases to be available at commercially reasonable terms, the Contractor shall (as insuring Party) give notice to the Employer, with supporting particulars. The Employer shall then (i) be entitled subject to Sub-Clause 2.5 [Employer's Claims] to payment of an amount equivalent to such commercially reasonable terms as the Contractor should have expected to have paid for such cover, and (ii) be deemed, unless he obtains the cover at commercially reasonable terms, to have approved the omission under Sub-Clause 18.1 [General Requirements for Insurances].

Insurance against Injury to Persons and Damage to Property

18.3 The insuring Party shall insure against each Party's liability for any loss, damage, death or bodily injury which may occur to any physical property (except things insured under Sub-Clause 18.2 [Insurance for Works and Contractor's Equipment]) or to any person (except persons insured under Sub-Clause 18.4 [Insurance for Contractor's Personnel]), which may arise out of the Contractor's performance of the Contract and occurring before the issue of the Performance Certificate.

This insurance shall be for a limit per occurrence of not less than the amount stated in the Appendix to Tender, with no limit on the number of occurrences. If an amount is not stated in the Appendix to Tender, this Sub-Clause shall not apply.

Unless otherwise stated in the Particular Conditions, the insurances specified in this Sub-Clause:

(a) shall be effected and maintained by the Contractor as insuring Party,

(b) shall be in the joint names of the Parties,

(c) shall be extended to cover liability for all loss and damage to the Employer's property except things insured under Sub-Clause 18.2 arising out of the Contractor's performance of the Contract, and

(d) may however exclude liability to the extent that it arises from:

    (i) the Employer's right to have the Permanent Works executed on, over, under, in or through any land, and to occupy this land for the Permanent Works,

    (ii) damage which is an unavoidable result of the Contractor's obligations to execute the Works and remedy any defects, and

    (iii) a cause listed in Sub-Clause 17.3 [Employer's Risks], except to the extent that cover is available at commercially reasonable terms.

Insurance for Contractor's Personnel

18.4 The Contractor shall effect and maintain insurance against liability for claims, damages, losses and expenses (including legal fees and expenses) arising from injury, sickness, disease or death to any person employed by the Contractor or any other of the Contractor's Personnel.

The Employer and the Engineer shall also be indemnified under the policy of insurance, except that this insurance may exclude losses and claims to the extent that they arise from any act or neglect of the Employer or of the Employer's Personnel.

The insurance shall be maintained in full force and effect during the whole time that these personnel are assisting in the execution of the Works. For a Subcontractor's employees, the insurance may be effected by the Subcontractor, but the Contractor shall be responsible for compliance with this Clause.

18)'. It remains to be seen as to how this provision would operate in practice and the effect it would have.

There are many drafting ambiguities in this clause, which should be clarified if the contract is to be operated successfully. Examples are:

- Sub-clause 18.1 provides that 'Wherever the Employer is the insuring Party, each insurance shall be effected with insurers and in terms consistent with the details annexed to the Particular Conditions.' [26] What is intended by the term 'details'? If, as stated, these details are expected to furnish the terms of the insurances supplied by the employer, then surely this must mean that nothing less explicit than the policies of insurance themselves have to be annexed.

Clauses 17 to 19 of the New RED Book – 1999 Edition

19   Force Majeure

Definition of Force Majeure

19.1   In this Clause, 'Force Majeure' means an exceptional event or circumstance:

(a) which is beyond a Party's control,

(b) which such Party could not reasonably have provided against before entering into the Contract,

(c) which, having arisen, such Party could not reasonably have avoided or overcome, and

(d) which is not substantially attributable to the other Party.

Force Majeure may include, but is not limited to, exceptional events or circumstances of the kind listed below, so long as conditions (a) to (d) above are satisfied:

(i) war, hostilities (whether war be declared or not), invasion, act of foreign enemies,

(ii) rebellion, terrorism, revolution, insurrection, military or usurped power, or civil war,

(iii) riot, commotion, disorder, strike or lockout by persons other than the Contractor's Personnel and other employees of the Contractor and Subcontractors,

(iv) munitions of war, explosive materials, ionising radiation or contamination by radio-activity, except as may be attributable to the Contractor's use of such munitions, explosives, radiation or radio-activity, and

(v) natural catastrophes such as earthquake, hurricane, typhoon or volcanic activity.

---

26   See the second line of the third paragraph of Clause 18.

Notice of Force Majeure

19.2 If a Party is or will be prevented from performing any of its obligations under the Contract by Force Majeure, then it shall give notice to the other Party of the event or circumstances constituting the Force Majeure and shall specify the obligations, the performance of which is or will be prevented. The notice shall be given within 14 days after the Party became aware, (or should have become aware), of the relevant event or circumstance constituting Force Majeure.

The Party shall, having given notice, be excused performance of such obligations for so long as such Force Majeure prevents it from performing them.

Notwithstanding any other provision of this Clause, Force Majeure shall not apply to obligations of either Party to make payments to the other Party under the Contract.

Duty to Minimise Delay

19.3 Each Party shall at all times use all reasonable endeavours to minimise any delay in the performance of the Contract as a result of Force Majeure.

A Party shall give notice to the other Party when it ceases to be affected by the Force Majeure.

Consequences of Force Majeure

19.4 If the Contractor is prevented from performing any of his obligations under the Contract by Force Majeure of which notice has been given under Sub-Clause 19.2 [*Notice of Force Majeure*], and suffers delay and/or incurs Cost by reason of such Force Majeure, the Contractor shall be entitled subject to Sub-Clause 20.1 [*Contractor's Claims*] to:

(a) an extension of time for any such delay, if completion is or will be delayed, under Sub-Clause 8.4 [*Extension of Time for Completion*], and

(b) if the event or circumstance is of the kind described in sub-paragraphs (i) to (iv) or Sub-Clause 19.1 [*Definition of Force Majeure*], and in the case of sub-paragraphs (ii) to (iv), occurs in the Country, payment of any such Cost.

After receiving this notice, the Engineer shall proceed in accordance with Sub-Clause 3.5 [*Determinations*] to agree or determine these matters.

Force Majeure Affecting Subcontractor

19.5 If any Subcontractor is entitled under any contract or agreement relating to the Works to relief from force majeure on terms additional to or broader than those specified in this Clause, such additional or broader force majeure events or circumstances shall not excuse the Contractor's non-performance or entitle him to relief under this Clause.

Optional Termination, Payment and Release

19.6 If the execution of substantially all the Works in progress is prevented for a continuous period of 84 days by reason of Force Majeure of which notice has been given under Sub-Clause 19.2 [*Notice of Force Majeure*], or for multiple periods which total more than 140 days due to the same notified Force Majeure, then either Party may give to the other Party a notice of termination of the Contract. In this event, the termination shall take effect 7 days after the notice is given, and the Contractor shall proceed in accordance with Sub-Clause 16.3 [*Cessation of Work and Removal of Contractor's Equipment*].

Upon such termination, the Engineer shall determine the value of the work done and issue a Payment Certificate which shall include:

(a) the amounts payable for any work carried out for which a price is stated in the Contract;

(b) the Cost of Plant and Materials ordered for the Works which have been delivered to the Contractor, or of which the Contractor is liable to accept delivery: this Plant and Materials shall become the property of (and be at the risk of) the Employer when paid for by the Employer, and the Contractor shall place the same at the Employer's disposal;

(c) any other Cost or liability which in the circumstances was reasonably incurred by the Contractor in the expectation of completing the Works;

(d) the Cost of removal of Temporary Works and Contractor's Equipment from the Site and the return of these items to the Contractor's works in his country (or to any other destination at no greater cost); and

(e) the Cost of repatriation of the Contractor's staff and labour employed wholly in connection with the Works at the date of termination.

Release from Performance under the Law

19.7 Notwithstanding any other provision of this Clause, if any event or circumstance outside the control of the Parties (including, but not limited to, Force Majeure) arises which makes it impossible or unlawful for either or both Parties to fulfil its or their contractual obligations or which, under the law governing the Contract, entitles the Parties to be released from further performance of the Contract, then upon notice by either Party to the other Party of such event or circumstance:

(a) the Parties shall be discharged from further performance, without prejudice to the rights of either Party in respect of any previous breach of the Contract, and

(b) the sum payable by the Employer to the Contractor shall be the same as would have been payable under Sub-Clause 19.6 [*Optional Termination, Payment and Release*] if the Contract had been terminated under Sub-Clause 19.6.

- Sub-clause 18.1 provides that '*When each premium is paid, . . . the insuring Party shall submit evidence of payment to the other Party. . . .*'[27] This wording does not provide the intended meaning. Payment of each insurance premium should be made to initiate or maintain the insurance cover and evidence should be provided whenever required.
- Sub-clause 18.2(d) specifies the deductibles to be applied to the insurance cover for some of the Employer's risks. Should the insurance cover for the Contractor's risks be subject to no deductibles?
- What is the meaning of 'insurable at commercially reasonable terms' in Sub-Clause 18.2(d); in the last paragraph of Sub-clause 18.2; and in Sub-clause 18.3(d)(iii)?

C Clause 19 – Force Majeure

A Force Majeure clause is an increasingly common feature of international contracts. It is the fashion, but is it necessary or even desirable? For FIDIC, one might suspect that importing Force Majeure from the old Yellow and Orange Books into the new suite of contracts was a desire to show a closer position to the civil law concepts and a move away from the common law principles. As one might suspect, this is a similar development to that of changing the title of Clause 66 of the 3rd edition of the old Red Book, 'Frustration', to 'Release from Performance' in the 4th edition.

Whilst these changes are outside the scope of this book, it is perhaps worth exploring the difference between the two doctrines of frustration and Force Majeure briefly, with particular reference to construction and construction insurance.

With the exception of the White Book, the FIDIC construction contracts in their various forms and titles have always been based on the premise that liability for non-performance of contractual obligations is a strict one. Failure to perform these required duties under the relevant contract would give rise to a claim for damages. Where FIDIC's White Book is concerned, which is intended for professional services, liability is based on the requirement of exercising reasonable skill and care in the performance of the duties under the contract. The rationale for the above rule in the FIDIC construction contract may lie in its common law origin, but in any case, except for specified events in the contract, the contractor is obliged to complete the contract.[28]

Where strict liability applies, why a party failed to fulfil its obligation is immaterial, and it is no defence for that party to plead that it has done its best.[29] As a party enters into contractual obligations freely, it accepts certain risks that are allocated to it and promises to bear these risks if and when they eventuate. In this way, the contracting parties are able to plan ahead with calculable certainty their schemes and arrange their business affairs. There are, however, specific risks that are beyond the capacity of a party to accept. In such circumstances, it would be better to name

---

27 The wording chosen is 'When each premium is paid . . .', which is not sufficiently explicit. It is not a question of when, since there is usually no insurance cover unless the premium *has already been* paid.

28 See page 187 above and also for example Clause 13 in the 4th edition of the old Red Book and Clause 19.7 of the New Red Book.

29 *Raineri v. Miles* [1981] AC 1050, 1086.

these risks and specify the method of dealing with and managing them. As construction contracts grow in size and complexity, such unacceptable risks become harder to identify and define in an explicit manner in the contract, hence the need for a doctrine of frustration or Force Majeure to excuse non-performance of promises. Frustration occurs whenever the law recognises that, without default of either party, a contractual obligation has become incapable of being performed because the circumstances in which performance is called for would render it a thing radically different from that which was undertaken by the contract: It was not this that I promised to do.

As argued by those who advocate the use of a Force Majeure clause, the advantage of such a clause is that it offers to the parties, should they wish to avail themselves of it, the opportunity to escape from the narrowness of the doctrine of frustration by including within their clause an event which would not be sufficient to frustrate the contract. However, such a clause does give the court power to review each word of the whole of the clause.[30] It is understood that, in certain jurisdictions, it is argued that conflict as to the interpretation of a Force Majeure clause becomes a matter for litigation rather than arbitration.

It is said that the doctrine of frustration is much narrower than the doctrine of Force Majeure and that uncertainty is inherent in the former, but that such uncertainty might be eliminated to a large extent by the incorporation into a contract of a suitably drafted Force Majeure clause. Then, the enquiry of the court can be limited and focused on the terms of the clause rather than the whole general notion of what is reasonable and fair under the doctrine of frustration.[31]

Is it not much more sensible and less likely to produce conflict in the first place, if neither is stated in the contract conditions, leaving the matter to the provisions of the contract law in the relevant jurisdiction?

However, such a move in the context of adopting Force Majeure is neither necessary nor desirable because:

- First, incorporating a clause such as Clause 19 into a contract not only duplicates what is usually provided for in the civil code of a civil law jurisdiction, but also enlarges the scope of the meaning and application of Force Majeure. This could result in the parties getting into a muddle and a contradictory situation;
- Second, the original concept of the Special Risks in Clause 65 of the old Red Book is all the protection that the contractor needs rather than Force Majeure;
- Third, most of the risks which now come under the FIDIC definition of Force Majeure are insurable and required to be insured. Therefore, no real benefit accrues to the contractor from being protected by such a clause without having to slip into uncharted waters.

Therefore, whilst it must be agreed that the treatment of the risks specified in Clause 19 should be a special one, it is erroneous to swing to the extreme end of the scale

---

30 *Force Majeure and Frustration of Contract*, edited by E. McKendrick, Lloyd's of London Press, 2nd edition, 1995, page 43.
31 As in note 30 above, but on page 39.

and designate them in the category of Force Majeure, particularly when that term has legal implications in certain jurisdictions. The answer for the purposes of these conditions of contract should be to designate them as what they are, i.e. an exceptional set of risks, which require different treatment to that given to the normal set of risks to which the project is exposed.

11

THE INSURANCE CLAUSES OF
THE NEW 1999 FIDIC FORMS
OF CONTRACT
A proposed redraft

It is unwise to criticise without offering a reasonable alternative. Therefore, based on the discussion presented in the previous chapter, the clauses set out below are put forward as a proposal for the replacement of Clauses 17 to 19 of the new Red Book of FIDIC. The New Yellow Book and the Silver Book require some modification to suit the risks shifted from the employer to the contractor and in particular the design risk. These are dealt with at the end of this chapter.

The proposed replacement of Clauses 17 to 19 of the New Red Book[1, 2]

17 Risk and Responsibility

Employer's Risks of Loss & Damage

17.1 The risks of loss and damage to the Works, Goods or Contractor's Documents for which the Contractor is not liable are:

   (a) Employer's Exceptional Risks of Loss and Damage, which are:

      i.   war, hostilities (whether war be declared or not), invasion, act of foreign enemies,

      ii.   rebellion, terrorism, revolution, insurrection, military or usurped power, or civil war, within the Country,

      iii.   riot, commotion or disorder within the Country by persons other than the Contractor's Personnel and other employees of the Contractor and Subcontractors, and

---

1   My grateful thanks are due to Mr Eamonn Conlon of Masons and Mr Anthony Harkness, insurance consultant, for the time they have given to studying and refining the many drafts of these alternative clauses.

2   Any one who is attracted to using part or all of these clauses should carefully consider their effect on the conditions of contract incorporating them and should take full responsibility and liability for such use.

iv. munitions of war, explosive materials, ionising radiation or contamination by radio-activity, within the Country, except as may be
attributable to the Contractor's use of such munitions, explosives,
radiation or radio-activity.

(b) Employer's Normal Risks of Loss and Damage, which are:
  i. pressure waves caused by aircraft or other aerial devices travelling
  at sonic or supersonic speeds,
  ii. use or occupation by the Employer of any part of the Permanent
  Works, except as may be specified in the Contract, and
  iii. design of any part of the Works by the Employer's Personnel or by
  others for whom the Employer is responsible, if any.

Responsibility for Care of the Works

17.2 The Contractor shall take full responsibility for the care of the Works and
Goods from the Commencement Date until the Taking-Over Certificate is
issued (or is deemed to be issued under Sub-Clause 10.1 [Taking Over of the
Works and Sections]) for the Works, when responsibility for the care of the
Works shall pass to the Employer. If a Taking-Over Certificate is issued (or is
deemed to be issued) for any Section or part of the Works, responsibility for
the care of the Section or part shall then pass to the Employer.

After responsibility has accordingly passed to the Employer, the Contractor
shall take responsibility for the care of any work which is outstanding on the
date stated in a Taking-Over Certificate, until this outstanding work has been
completed.

If any loss or damage happens to the Works, Goods or Contractor's Documents during the period when the Contractor is responsible for their care,
from any cause not listed in Sub-Clause 17.1 [Employer's Risks of Loss &
Damage], the Contractor shall rectify the loss or damage at the Contractor's
risk and cost, so that the Works, Goods and Contractor's Documents conform
with the Contract.

The Contractor shall be liable for any loss or damage caused by any actions
performed by the Contractor after a Taking-Over Certificate has been issued.
The Contractor shall also be liable for any loss or damage, which occurs after
a Taking-Over Certificate has been issued and which arose from a previous
event for which the Contractor was liable.

Consequences of Employer's Risks of Loss and Damage

17.3 If any of the risks listed in Sub-Clause 17.1(a) above occur, the parties' rights
and obligations are set out in Clause 19 below.

If and to the extent that any of the risks listed in Sub-Clause 17.1(b) above
result in loss or damage to the Works, Goods or Contractor's Documents, the
Contractor shall promptly give notice to the Engineer and shall rectify this loss
or damage to the extent required by the Engineer. If the Contractor suffers
delay and/or incurs Cost from rectifying this loss or damage, the Contractor
shall give a further notice to the Engineer and shall be entitled subject to Sub-
Clause 20.1 [Contractor's Claims] to:

(a) an extension of time for any such delay, if completion is or will be delayed, under Sub-Clause 8.4 [*Extension of Time for Completion*], *and*

(b) payment of any such Cost, which shall be included in the Contract Price. In the case of sub-paragraphs ii and iii of Sub-Clause 17.1(b) [*Employer's Normal Risks of Loss & Damage*], reasonable profit on the Cost shall also be included.

After receiving this further notice, the Engineer shall proceed in accordance with Sub-Clause 3.5 [Determinations] to agree or determine these matters.

Risk of infringement of Intellectual and Industrial Property Rights

17.4 In this Sub-Clause, 'infringement' means an infringement (or alleged infringement) of any patent, registered design, copyright, trade mark, trade name, trade secret or other intellectual or industrial property right relating to the Works; and 'claim' means a claim (or proceedings pursuing a claim) alleging an infringement.

Whenever a Party does not give notice to the other Party of any claim within 28 days of receiving the claim, the first Party shall be deemed to have waived any right to indemnity under this Sub-Clause.

The Employer shall indemnify and hold the Contractor harmless against and from any claim alleging an infringement which is or was:

(a) an unavoidable result of the Contractor's compliance with the Contract, or
(b) a result of any Works being used by the Employer:

    (i) for a purpose other than that indicated by, or reasonably to be inferred from, the Contract, or
    (ii) in conjunction with any thing not supplied by the Contractor, unless such use was disclosed to the Contractor prior to the Base Date or is stated in the Contract.

The Contractor shall indemnify and hold the Employer harmless against and from any other claim which arises out of or in relation to (i) the manufacture, use, sale or import of any Goods, or (ii) any design for which the Contractor is responsible.

If a Party is entitled to be indemnified under this Sub-Clause, the indemnifying Party may (at its cost) conduct negotiations for the settlement of the claim, and any litigation or arbitration which may arise from it. The other Party shall, at the request and cost of the indemnifying Party, assist in contesting the claim. This other Party (and its Personnel) shall not make any admission which might be prejudicial to the indemnifying Party, unless the indemnifying Party failed to take over the conduct of any negotiations, litigation or arbitration upon being requested to do so by such other Party.

Limitation of Liability

17.5 Neither Party shall be liable to the other Party for loss of use of any Works, loss of profit, loss of any contract or for any indirect loss or damage which may be suffered by the other Party in connection with the Contract, other than under Sub-Clause 16.4 [Payment on Termination], Sub-Clause 17.6 [Indemnities by the Contractor], and Sub-Clause 17.7 [Indemnities by the Employer].

Indirect loss shall include, but not limited to, for the purpose of this clause loss of profits, loss of use, loss of production, loss of business or loss of business opportunity.

The total liability of the Contractor to the Employer, under or in connection with the Contract other than under Sub-Clause 4.19 [Electricity, Water and Gas], Sub-Clause 4.20 [Employer's Equipment and Free-Issue Material], and Sub-Clause 17.4 [Risk of Infringement of Intellectual and Industrial Property Rights] shall not exceed the sum stated in the Particular Conditions or (if a sum is not so stated) the Accepted Contract Amount.

This Sub-Clause shall not limit liability in any case of fraud, deliberate default or reckless misconduct by the defaulting Party.

Indemnities by the Contractor

17.6 The Contractor shall indemnify and hold harmless the Employer, the Employer's Personnel, and their respective agents, against and from all claims, damages, losses and expenses (including legal fees and expenses) in respect of:

(a) bodily injury, sickness, disease or death of any person whatsoever employed on or in connection with the Works; and

(b) damage to or loss of any property real or personal (other than the Works),

arising out of or in the course of or by reason of the Contractor's design (if any), the execution and completion of the Works and the remedying of any defects, unless attributable to any negligence, wilful act or breach of the Contract by the Employer, the Employer's Personnel, or any of their respective agents.

Indemnities by the Employer

17.7 The Employer shall indemnify and hold harmless the Contractor, the Contractor's Personnel and their respective agents against and from all claims, damages, losses and expenses (including legal fees and expenses) in respect of:

(a) bodily injury, sickness, disease or death, which is attributable to any negligence, wilful act or breach of the Contract by the Employer, the Employer's Personnel, or any of their respective agents;

(b) Damage to crops being on the Site (save in so far as possession has not been given to the Contractor);

(c) The use or occupation of land (provided by the Employer) by the Works or any part thereof or for the purpose of the construction and completion of the Works (including Consequential Losses of Crops) or interference whether temporary or permanent with any right of way light air or water or other easement or quasi-easement which are the unavoidable result of construction of the Works in accordance with the Contract;

(d) The right of the Employer to construct the works or any part thereof on over under in or through any land;

(e) Damage which is the unavoidable result of the Contractor's obligations to execute the Works and remedy any defects in accordance with the Contract; and

(f) The Employers' Risks as set out in Sub-Clause 17.1 above.

The indemnities provided pursuant to Sub-Clause 17.6 and this Sub-Clause by the parties towards each other shall be proportionally reduced, if any act or neglect by either Party to the other Party contributed to the said bodily injury, sickness, disease, death, damage or loss.

18 *Insurance*

General Requirements for Insurances

18.1 If a policy indemnifies additional joint insured, namely in addition to the Employer and the Contractor (i) the Contractor shall act under the policy on behalf of these additional joint insured except that the Employer shall act for Employer's Personnel, (ii) additional joint insured shall not be entitled to receive payments directly from the insurer or to have any other direct dealings with the insurer, and (iii) the insuring Party shall require all additional joint insured to comply with the conditions stipulated in the policy.

Each policy insuring against loss or damage shall provide for payments to be made in the currencies required to rectify the loss or damage. Payments received from insurers shall be used for the rectification of the loss or damage.

The Contractor shall, within 28 days of the date of the Letter of Acceptance, or as otherwise agreed, submit to the Employer:

(a) evidence that the insurances described in this Clause have been effected, and

(b) copies of the policies for the insurances described in Sub-Clause 18.2 [Insurance for Works and Contractor's Equipment] and Sub-Clause 18.3 [Insurance against Injury to Persons and Damage to Property].

(c) evidence of payment of each insurance premium.

Whenever evidence or policies are submitted to the Employer, the Contractor shall also notify the Engineer of such submission.

The Employer and the Contractor shall comply with the conditions stipulated in each of the insurance policies. In the event that the Contractor or the Employer fails to comply with any condition imposed by the insurance policies effected pursuant to the Contract each shall indemnify the other against all losses and claims arising from such failure. The Contractor shall keep the insurers informed of any relevant changes to the execution of the Works and ensure that insurance is maintained in accordance with this Clause.

Neither Party shall make any alteration to the terms of any insurance without the prior approval of the other Party. If an insurer makes (or attempts to make) any alteration, the Party notified by the insurer shall promptly give notice to the other Party.

If the Contractor fails to effect and keep in force any of the insurances it is required to effect and maintain under the Contract, or fails to provide satisfactory evidence and copies of policies in accordance with this Sub-Clause, the Employer may (at its option and without prejudice to any other right or remedy) effect and keep in force any such insurance for the relevant coverage and pay the premiums due. The Employer may from time to time deduct the amount of these premiums so paid from any monies due or which may become due to the Contractor or recover the same as a debt due from the Contractor, and the Contract Price shall be adjusted accordingly.

Nothing in this Clause limits the obligations, liabilities or responsibilities of the Contractor or the Employer, under the other terms of the Contract or otherwise. Any amounts not insured or not recovered from the insurers shall be borne by the Contractor and/or the Employer in accordance with these obligations, liabilities or responsibilities. However, if the insuring Party fails to effect and keep in force an insurance which is available and which it is required to effect and maintain under the Contract, and the other Party neither approves the omission nor effects insurance for the coverage relevant to this default, any moneys which should have been recoverable under this insurance shall be at the cost of the Contractor.

Payments by one Party to the other Party shall be subject to Sub-Clause 2.5 [Employer's Claims] or Sub-Clause 20.1 [Contractor's Claims] as applicable.

Insurance for Works and Contractor's Equipment

18.2 The Contractor shall insure the Works, Plant, Goods and Materials (including any unfixed Materials and Plant or other things whether on the Site or otherwise intended for the Works) and Contractor's Documents for not less than the full reinstatement cost plus an additional 10% to cover any additional costs that may arise incidental to the rectification of any loss or damage including cost of demolition, removal of debris, professional fees and profit. This insurance shall be effective from the Commencement Date until the date of issue of the relevant Taking-Over Certificate for the Works.

The Contractor shall maintain this insurance to provide cover until the date of issue of the Performance Certificate, for loss or damage for which the Contractor is liable arising from a cause occurring prior to the issue of the Taking-Over Certificate, and for loss or damage caused by the Contractor in the course of any other operations (including those under Clause 11 [Defects Liability]).

The Contractor shall insure the Contractor's Equipment for not less than the full replacement value, including delivery to Site. For each item of Contractor's Equipment, the insurance shall be effective while it is being transported to the Site and until it is no longer required as Contractor's Equipment.

The insurance policies under this Sub-Clause:

(a) shall be in the joint names of the Employer and the Contractor, who shall be jointly entitled to receive payments from the insurers, payments being held or allocated between the parties for the sole purpose of rectifying the loss or damage,

(b)   shall cover all loss and damage from any cause not listed in Sub-Clause 17.1 [Employer's Risks of Loss and Damage], with deductibles per occurrence of not more than the amount stated in the Appendix to Tender.

(c)   shall also cover loss or damage to a part of the Works not in the occupation of the Employer which is attributable to the use or occupation by the Employer of another part of the Works; and

(d)   may however exclude loss of, damage to, and reinstatement of:

(i)   a part of the Works which is in a defective condition due to a defect in its design, materials or workmanship (but cover shall include any other parts which are lost or damaged as a direct result of this defective condition and not as described in sub-paragraph (ii) below),

(ii)   a part of the Works which is lost or damaged in order to reinstate any other part of the Works if this other part is in a defective condition due to a defect in its design, materials or workmanship, and

(iii)   a part of the Works which has been taken over by the Employer, except to the extent that the Contractor is liable for the loss or damage.

Insurance against Injury to Persons and Damage to Property

18.3   The Contractor shall insure against each Party's liability for any loss, damage, death or bodily injury which may occur to any physical property (except things insured under Sub-Clause 18.2 [Insurance for Works and Contractor's Equipment]) or to any person (except persons insured under Sub-Clause 18.4 [Insurance for Contractor's Personnel]), which may arise out of the Contractor's performance of the Contract and occurring before the issue of the Performance Certificate.

This insurance shall be for a limit per occurrence of not less than the amount stated in the Appendix to Tender, with no limit on the number of occurrences, and with deductibles per occurrence of not more than the amount(s) stated in the Appendix to Tender. If an amount is not stated in the Appendix to Tender, this Sub-Clause shall not apply.

The insurance policies specified in this Sub-Clause:

(a)   shall be in the joint names of the parties defined in the Appendix to Tender and shall contain a cross-liabilities clause such that the cover shall apply separately to each Insured as though a separate policy had been issued for each of them.

(b)   may however exclude liability to the extent that it arises from a cause listed in Sub-Clause 17.7 [Indemnity by the Employer].

Insurance for Contractor's Personnel

18.4   The Contractor shall effect and maintain insurance against liability for claims, damages, losses expenses (including legal fees and expenses) arising from injury, sickness, disease or death to any person employed by the Contractor or any other of the Contractor's Personnel.

The Employer and the Engineer shall also be indemnified under the policy of insurance, except that this insurance may exclude losses and claims to the extent that they arise from any act or neglect of the Employer or of the Engineer or their respective servants or agents or any other contractor (not being employed by the Contractor).

The insurance shall be maintained in full force and effect during the whole time that these personnel are assisting in the execution of the Works. For a Subcontractor's employees, the Subcontractor may effect the insurance, but the Contractor shall be responsible for compliance with this Clause.

19 Employer's Exceptional Risks of Loss and Damage

Notice of an Employer's Exceptional Risk of Loss and Damage

19.1 If a Party is or will be prevented from performing any of its obligations under the Contract by an Employer's Exceptional Risk of Loss and Damage, then it shall give notice to the other Party of the event or circumstances constituting such a risk and shall specify the obligations, the performance of which is or will be prevented. The notice shall be given within 14 days after the Party became aware, (or should have become aware), of the relevant event or circumstance constituting such risk.

The Party shall, having given notice, be excused performance of such obligations for so long as such risk prevents that Party from the performance thereof.

Notwithstanding any other provision of this Clause, an Employer's Exceptional Risk shall not relieve either Party from making payments to the other Party under the Contract.

Duty to Minimise Delay

19.2 Each Party shall at all times use all reasonable endeavours to minimise any delay in the performance of the Contract as a result of an Employer's Exceptional Risk.

A Party shall give notice to the other Party when it ceases to be affected by the Employer's Exceptional Risk.

Consequences of an Employer's Exceptional Risk

19.3 If the Contractor is prevented from performing any of his obligations under the Contract by an Employer's Exceptional Risk of which notice has been given under Sub-Clause 19.1 [Notice of an Employer's Exceptional Risk of Loss and Damage], and suffers delay and/or incurs Cost by reason of such an Employer's Exceptional Risk, the Contractor shall be entitled subject to Sub-Clause 20.1 [Contractor's Claims] to:

(a) an extension of time for any such delay, if completion is or will be delayed, under Sub-Clause 8.4 [Extension of Time for Completion], and

(b) if the event or circumstance is of the kind described in sub-paragraph (a) of Sub-Clause 17.1 [Employer's Risks of Loss and Damage] and, in the case of sub-paragraphs 17.1(a) (ii) to (iv), occurs in the Country, payment of any such Cost.

After receiving this notice, the Engineer shall proceed in accordance with Sub-Clause 3.5 *[Determinations]* to agree or determine these matters.

Optional Termination, Payment and Release

19.4 If the execution of substantially all the Works in progress is prevented for a continuous period of 84 days by reason of Exceptional Risks of which notice has been given under Sub-Clause 19.1 [Notice of an Employer's Exceptional Risk of Loss & Damage], or for multiple periods which total more than 140 days due to the same notified Exceptional Risks, then either Party may give to the other Party a notice of termination of the Contract. In this event, the termination shall take effect 7 days after the notice is given, and the Contractor shall proceed in accordance with Sub-Clause 16.3 [Cessation of Work and Removal of Contractor's Equipment].

Upon such termination, the Engineer shall determine the value of the work done and issue a Payment Certificate which shall include:

(a) the amounts payable for any work carried out for which a price is stated in the Contract;

(b) the Cost of Plant and Materials ordered for the Works which have been delivered to the Contractor, or of which the Contractor is liable to accept delivery: this Plant and Materials shall become the property of (and be at the risk of) the Employer when paid for by the Employer, and the Contractor shall place the same at the Employer's disposal;

(c) any other Cost or liability which in the circumstances was reasonably incurred by the Contractor in the expectation of completing the Works;

(d) the Cost of removal of Temporary Works and Contractor's Equipment from the Site and the return of these items to the Contractor's works in his country (or to any other destination at no greater cost); and

(e) the Cost of repatriation of the Contractor's staff and labour employed wholly in connection with the Works at the date of termination.

Release from Performance under the Law

19.5 Notwithstanding any other provision of this Clause, if any event or circumstance outside the control of the parties (including, but not limited to the Exceptional Risks) arises after the date of the Letter of Tender which makes it impossible or unlawful for either or both parties to fulfil its or their contractual obligations or which, under the law governing the Contract, entitles the parties to be released from further performance of the Contract, then upon notice by either Party to the other Party of such event or circumstance:

(a) the parties shall be discharged from further performance, without prejudice to the rights of either Party in respect of any previous breach of the Contract, and

(b) the sum payable by the Employer to the Contractor shall be the same as would have been payable under Sub-Clause 19.4 [Optional Termination,

Payment and Release] if the Contract had been terminated under Sub-Clause 19.4.

Notes

1 Add to Clause 1 (Definitions) the definition of Employer's Exceptional Risks and Normal Risks as defined in 17.1.
2 Delete the definition of Force Majeure.

The proposed replacement of Clauses 17 to 19 of the New Yellow and Silver Books

The insurance clauses of the Yellow and Silver Books need to deal with the design risk and also with any operational risks that would attach to a contract of the BOT or BOOT type. Accordingly, in addition to the above clauses proposed as a replacement to Clauses 17 to 19, there should be specifically written clauses to deal with the following insurances, which are only optional for the New Red Book type project, but necessary insurance requirements for a project of the New Yellow or Silver Books:

- Materials and plant for incorporation in the works whilst in storage or in transit to the Site.
- Machinery and hired plant used in the construction and/or the erection of the project;
- Professional indemnity insurance in respect of the design activities related to the project;
- Air-freight cover for urgent repairs that might be required in a project where some major elements are manufactured abroad;
- Manufacturer's risks of defective material, defective workmanship and defective design;

The following insurance covers remain as optional addition:

- Unfair termination of contract;
- Expropriation of overseas assets risk; and
- Confiscation of construction plant and machinery;
- Non-negligence insurance cover;
- Project Insurance in one of its many forms;
- Decennial insurance or latent defects cover;
- Credit risk insurance;
- Group personal accident, travel, medical and life assurance cover;
- Unfair call on any surety or bond with special attention to on-demand bond;
- Delay risk;
- Currency risks;
- Difference-in-conditions insurance. These would include any of the above risks included in one insurance policy plus any specific risks either connected with the type of contract undertaken or with the locality of the contract such as war or earthquake or the risk of not being able to obtain construction materials or other matters necessary for the completion of the project following political events, etc.

12

INSURANCES REQUIRED UNDER THE
FIDIC AGREEMENTS

A list of the insurance policies required by each member of the construction trinity, owner/employer, professional and contractor is shown in Figure 6.2 on page 197. Most of these policies are in fact required to be issued under the terms of the two standard forms of contract normally agreed between the owner/employer and the design professional on the one hand and between the owner/employer and the contractor on the other. In the international field, the standard form of contract normally agreed between the owner/employer and the design professional is FIDIC's White Book.[1] It requires no provision of insurance other than professional indemnity insurance with compatible public liability insurance cover, as discussed in Chapter 13 below. The contract normally agreed between the owner/employer and the contractor in the international field is one of a number of standard forms issued by FIDIC, which are referred to usually by the colour of their cover: The old Red Book, the old Yellow Book, the Orange Book, the New Red Book, the New Yellow Book, the Silver Book and the Green Book.[2] They all require that either the contractor or the employer should obtain the following three insurance covers:

---

1 'Client/Consultant Model Services Agreement', 1st edition, published first by FIDIC, Switzerland, in 1990, replacing the previously used three documents, namely: IGRA 1979 D&S (The International Model Form of Agreement between Client and Consulting Engineer for Design and Supervision), IGRA 1979 PI, IGRA 1980 PM. The White Book is now in its 2nd edition since 1998.
2 The precise titles of the FIDIC standard forms of contract are as follows:
   - **The old Red Book:** Conditions of Contract for Works of Civil Engineering Construction, 4th Edition 1987, Part I – General Conditions with forms of tender and agreement and Part II – Conditions of particular application with guidelines for preparation of Part II Clauses, reprinted 1988 with editorial amendments and later reprinted in 1992 with further amendments;
   - **The Yellow Book:** Conditions of Contract for Electrical and Mechanical Works including erection on site with forms of tender and agreement, 3rd edition 1987, Part I – General Conditions and Part II – Special Conditions, reprinted 1988 with editorial amendments;
   - **The Orange Book:** Conditions of Contract for Design-Build and Turnkey, 1st Edition 1995, Part I – General Conditions and Part II – Guidance for the preparation of conditions of particular application, Forms of tender and agreement;
   - **The new Red Book:** Conditions of Contract for Construction, 1st edition 1999 (for Building and Engineering Works, Designed by the Employer, General Conditions,

1   Property insurance (own property) which mainly includes the works to be constructed and other property on the site, see page 245, is covered normally by a Contractors' All Risks insurance policy (CAR);

2   Liability insurance to protect the owner/employer and the contractor against their legal liability in respect of bodily injury and disease of the contractor's employees working on the project, and arising as a result of the contractor's work on the project, is covered normally by the Employer's Liability insurance policy (EL);

3   Liability insurance to protect the owner/employer and the contractor against their legal liability, other than in item 2 above, within a specified limit of indemnity in respect of bodily injury of third parties (other than the employees) and damage to their property including that belonging to employees, and arising as a result of the contractor's work on the project, is covered normally by a Public Liability insurance policy (PL).

These insurance covers can be underwritten either for a particular project or annually for the contractor in respect of work on all projects he is executing. The insurance cover itself can be in the form of either three individual policies or a single composite policy combining three different sections dealing with own property damage, public liability and employer's liability. Perhaps the most popular form is the composite policy which ensures that all three risks are covered by the same insurer, thus reducing the number of insurance contracts to one and consequently minimising any gaps which might otherwise exist between the three insurance covers. A single insurer also means that there will be no dispute as to which insurer is responsible for a particular claim.

A composite insurance policy normally carries the name of the property policy and thus it is referred to as Contractors' All Risks Composite insurance policy, except in a project where the civil engineering content as compared with that of mechanical and electrical installation is small, in which case the policy is called Erection All Risks insurance policy. The two policies are essentially the same and only differ on minor detail, see page 295.

Contractors' All Risks insurance policy

As mentioned in Chapter 1, this type of insurance developed in the first half of this century from a fire extended cover which was originally required for construction

Guidance for the Preparation of the Particular Conditions, Forms of Tender, Contract Agreement, and Dispute Adjudication Agreement;
*   **The new Yellow Book:** Conditions of Contract for Plant and Design-Build for Electrical and Mechanical Plant, and for Building and Engineering Works, Designed by the Contractor, 1st edition 1999, General Conditions, Guidance for the Preparation of the Particular Conditions, Forms of Tender, Contract Agreement and Dispute Adjudication Agreement;
*   **The Silver Book:** Conditions of Contract for EPC Turnkey Projects, 1st edition 1999, General Conditions, Guidance for the Preparation of the Particular Conditions, Forms of Tender, Contract Agreement and Dispute Adjudication Agreement; and
*   **The Green Book:** Short Form of Contract 1st edition 1999, Agreement, General Conditions, Rules for Adjudication and Notes for Guidance.

work. This policy took shape as the extended cover was enlarged to include an increasing number of risks associated with construction. The 'All Risks' designation was added to the original description of this policy as a contractors' insurance policy to indicate the numerous risks covered, but those familiar with this policy always understood that there were a number of exclusions. However, the format of the Contractors' All Risks (CAR) insurance was influenced by another development, which was taking place in parallel at that time, namely the standardisation of the General Conditions of Contract. It is unclear as to which had the greater influence on the other, but when the Standard General Conditions of Contract were finally introduced after the Second World War their insurance clauses fitted the wording of the CAR policy.

Essentially, the purpose of the CAR policy was, and still is, to provide the answer to the risk of property damage connected with the construction of a project. The basic properties of the insurance cover can be best explained by dissecting the policy to its basic constituents, and dealing with each section individually.

Policy wording

There are many CAR policy wordings in circulation around the world but only few can be referred to as a standard policy wording. However, those that are recognised as having standard wordings fall in the category of a 'Standard Form' of agreement, see page 8, which means that the wording represents the intentions of the insurer, tailored to fit the insurance concept as understood by him. A specimen of one such standard policy is appended in Appendix B. It is a composite policy comprising two parts, which provide in the first a cover against own property damage and in the second a cover against public liability. As in most, if not all, Contractors' All Risks policies, the specimen in Appendix B incorporates the following sections:

- Recital Clause
- Operative Clause
- Definitions
- Schedule which defines, amongst other, the following terms:
  - The Insured
  - The Project
  - Property Insured
  - Sum Insured or Limit of Indemnity
  - Premium and Excess
  - Period of Insurance
- General Exclusions
- Special Exclusions
- Conditions
- Memoranda
- Signature Clause
- Endorsements

These sections may appear in the above-mentioned sequence or in some other arrangement which when analysed would give the same meaning. Therefore, in order to provide a clear understanding of the C.A.R. policy, the wording must be analysed and understood.

Recital clause

The Recital Clause sets out the outline of the insurance agreement embodied in the policy and describe the supporting documents and the principles upon which the policy is based. It appears at the beginning of the policy and refers to the Proposal Form completed by the insured, to the details provided in the schedule of the policy and to the other clauses incorporated therein.

Operative clause

The Operative Clause defines the type and extent of the insurance cover provided by the agreement, and appears in individual policies after the Recital Clause. In composite policies, however, there is an Operative Clause at the beginning of each part incorporated in the agreement, describing the cover provided by it. The insurance cover as described in this clause is generally restricted to 'unforeseen' or 'accidental' events, putting the onus of proof in case of a claim on the insured to prove that the events in question were unforeseen or accidental. Should this restriction be omitted from the Operative Clause and included under the 'Exceptions' part of the policy, the onus of proof shifts then from the insured to the insurer (see note 15, page 368).

Definitions

This section of the policy is used to define any special terms incorporated in the 'Schedule', serving as a back-up section for detailed information. Thus, for example, in the specimen policy included in Appendix B, the period of cover is defined and the definition expanded to indicate the relationship between the period of cover and extent of the insurance provided. The date of commencement of the Defects Notification period is used in that definition to indicate the point in time after which the cover provided in the Operative Clause is restricted to:

> Loss or damage caused by the Contractor in the course of any operations he may carry out for the purpose of complying with the obligations under the Defects Notification clause of the contract.

In fact, this wording is not compatible with the requirements of either the ICE or the FIDIC Conditions of Contract, which require the cover during the Defects Notification Period to extend to cover:

> Loss or damage arising from a cause occurring prior to the commencement of the Defects Notification period.

Thus, for example, the cover is to include a situation where a cast iron pipe fitting is cracked in an accident during the construction period but the damage is not discovered until later during the Defects Notification period, when the fitting is subjected to operational water pressure.

The restriction in cover takes place as soon as the Defects Notification period is reached, which means that the insurer must be informed of any extension in the period of construction due to delays of whatever kind. This does not usually apply in the case of an annual policy which is issued to a contractor who is constantly involved in one type of building operation. However, as delays are common in civil engineering projects, one must ensure that the policy is extended to cover the project up to the new date of commencement of the Defects Notification Period. Neglecting to do so may have catastrophic results if an accident occurs and the insurer declines liability. Under this heading, it should also be noted that, unless otherwise specifically stated in the policy, the insurer's liability expires in respect of a completed part that has been taken over by the employer before the completion date specified in the policy. The terminology used in the Operative Clause of the specimen policy in Appendix B, '. . . during the period of insurance stated in the Schedule, or during any further period of extension thereof, . . .', is useful in this respect.

Another definition given usually in the Contractors' All Risks policy concerns the sum insured, and relates to under-insurance. Under-insurance is a term used to describe the situation which occurs when the sum insured as indicated in the policy is less than the real value of the item in question. Since premium calculation is based primarily on the value of the sum insured, the insurer penalises the insured when a claim arises in under-insurance circumstances by reducing the amount recoverable from him in the same proportion as the sum indicated in the policy bears to the real value of the item insured.

This condition is referred to as 'the average clause' and is usually included in most CAR policies. To avoid the problem of under-insurance, the insurer usually agrees with the insured to adjust the various elements of the sum insured at the end of the construction period in accordance with the final account agreed between the contractor and the professional team under the main contract.

Schedule

The Schedule is basically a summary of the main features of the insurance agreement where such terms as the insured, the property or the project insured, the sum insured or the limit of indemnity and the excess to be applied are identified. Most of these terms are self-explanatory, but a few do require a closer examination:

The Insured

The individual contractors' All Risks insurance policy is normally issued in the joint names of the owner/employer and the contractor. When it is issued in its composite format, the part related to property damage should be treated as joint insurance whereas the section related to public liability may be issued either in the joint names of the owner/employer and the contractor or only in the name of the contractor with indemnity extended to the owner/employer through what is known as the Principal Clause, see page 250.

The Project

The project is identified in the Schedule through its title and its site location both of which are necessary for the purpose of assessment of premium and excess. The geographic location of the site is also referred to in the Operative Clause of the policy as part of the description of the property insured.

It is important to check that the geographic location of the project is not excluded from the cover of the policy, as this would render the insurance agreement worthless. In some cases, the geographical limits are defined in the Operative Clause in which case the limits should be checked to ensure that they include the geographic location of the site.

The Property Insured

The property insured under a Contractors' All Risks insurance policy includes the contract works plus either or all of the items listed under B, C and D below:

A. The contract works, whether permanent or temporary to a certain sums as detailed under the heading of Sum Insured.
B. Construction machinery and equipment to the limits stated in the Schedule. A list of construction machinery with individual values and specific details may have to be incorporated in the policy for a proper assessment of premium. The difference between construction machinery and equipment within the terms of the insurance policy is dependent on the ability of the former to move directionally under its own motive power, whereas the latter remains stationary during its operation. Therefore, bulldozers, excavators, cranes, graders, etc. are considered to be construction machinery whereas pumps, concrete mixing plants, compressors, etc. are referred to as construction equipment. This distinction is necessary because the risk of damage or injury resulting from the use of construction machinery is much higher than that in the case of construction equipment, thus necessitating a different system and level of rating when the insurance premium is calculated. Construction machinery and equipment are collectively referred to in the General Conditions of Contract as plant.

The specimen policy on the other hand mixes the meaning of plant and equipment and incorporates them under one item. Such terms should be precisely and similarly defined in both documents. The word 'plant' as used in the specimen policy denotes such material that is used for temporary works and has no mechanical implications.
C. Material brought to the site for incorporation in the works or allocated to the contract, but is in transit or stored elsewhere.
D. Personal effects of the insured's employees working on the site provided that they are not insured under a separate policy.

Sum Insured or Limit of Indemnity

The phrase 'sum insured' is an insurance term applied to property insurance whereas 'limit of indemnity' is the equivalent term applied to liability insurance, both of which are used to indicate the monetary value of the insurance contract provided.

344

The sum insured in respect of the items listed under the heading 'Insured Property' is usually defined in the Schedule in a precise manner, so as not to involve the 'average clause', see page 343. It is important to establish the correct sum insured because it forms the basis of determining the insurer's liability, the basis of calculating the amount of premium, and the basis of statistical analyses of the results of conducting insurance business. The sum insured may include any or all of the following elements:

A. The contract works which may be insured either to its 'full value' as required in Clause 21(a) of the ICE Conditions of Contract or to 'an estimated current value' as specified in the FIDIC counterpart. Both values are intended to mean the original tender sum plus any subsequent adjustment made through variations, additions, omissions and increases in cost of material, plant and labour. It is therefore clear in both documents that the normal inflation, referred to usually as primary inflation, which is caused by increases in such costs is to be covered under the Contractors' All Risks insurance policy. However, what is not so clear is whether any other element of inflation is required to be covered by the CAR policy. There are two resultant elements of inflation that should be considered in the event of damage to any completed or partially completed part of the works, which are:

- The inflation between the time at which such part of the work is originally carried out and that at which it is repaired or reinstated. Such inflation is usually referred to as secondary inflation.
- The inflation that occurs due to the delay in executing any uncompleted part of the works after such event of damage. This element is usually referred to as the transitional inflation.

In the 1980s, inflation played an important role in escalating the cost of construction work and although the construction industry as well as the insurance market felt its effect, precise definitions of the sum insured did not emerge. Thus, one might encounter any of the following terms used to describe the sum insured in contract conditions or in CAR policies: Contract Price; Full Value; Estimated Current Value; Total Contract Sum; Estimated Total Completed Cost; Replacement Value, and so on. However, whilst it is easy to establish what an insured may wish to be indemnified for, it is not so easy to establish a method of calculating the sum insured. If the insured wishes to be indemnified for a damaged item to its full value on the day of repair, the sum insured would have to be based on such full value irrespective of the original contract price or the original construction cost incurred. But how can such a value be estimated at the inception of insurance and precisely calculated at the end of construction? What seems to be the only acceptable method of dealing with this problem is to include in the conditions of contract a statement requiring the contractor to adjust the sum insured at regular and specified intervals during construction in respect of the primary inflation and to specify a percentage increase in respect of secondary and transitional inflation.

B. Construction machinery and equipment are required to be insured to their 'full value', under the ICE Conditions of Contract. The FIDIC Conditions, however, require these items to be insured to their 'reinstatement value' and thus recognise the possibility of primary and secondary inflation.

C. Material brought to the site for incorporation in the works or allocated to the contract is treated in the same manner as construction machinery and equipment in both documents.

D. The personal effects are normally included under an item with a predetermined sum insured.

E. There are other items which can be included in the cover and these are:

- Professional fees which may have to be paid in respect of services associated with any reinstatement or repair work can be included as a predetermined percentage addition to the value of the work. Such services are covered when they become necessary as a result of an accident leading to a claim covered under the policy. However, any fees incurred in the preparation and submission of the claim by the insured to the insurer are not covered under the policy.

- The cost of demolition and removal of debris from the site can be added to the sum insured to a specified limit defined in the Schedule. Such cost becomes part of any claim payable under the policy arising out of an accident causing irreparable damage or collapse, subject to any maximum limit that would generally be specified in the Schedule.

- The cost of any temporary emergency works executed after the occurrence of a loss to safeguard the works and to prevent any further damage from taking place is automatically covered under the policy.

- Additional working costs associated with the acceleration of repair or reinstatement work after a loss has taken place may be covered if specifically included. Thus, extra charges for overtime, night work, work on public holidays, express freight of parts required for repair might all be covered if previously agreed upon (see Memo 3 of the specimen policy in Appendix B). In some cases, airfreight of spare parts required for the repair or reinstatement of any damaged parts is also covered but, usually, there is a maximum monetary limit imposed on the amount recoverable under this heading.

- Additional costs due to a different method of construction being adopted in the repair or reinstatement of any damaged part to that used originally is a controversial subject and should be discussed between the insured and the insurer at the negotiation stage of the insurance contract. It is a well-established principle that the cost of any betterment in the quality of the insured item, after repair, is not insurable, see page 185. In some cases, however, there is no alternative but to follow a reconstruction procedure that would render the repaired item being of better quality than it originally was prior to the damage. In such a case, it is arguable as to whether or not such costs are recoverable in the case of a claim.

- Local taxes such as value added tax and its applicability to the cost of the works.

Where public liability is concerned, the limit of indemnity is defined in the Schedule under three headings:

1 Limit of indemnity for bodily injury;
2 Limit of indemnity for property damage;
3 Total limit of indemnity under the insurance policy inclusive of both bodily injury and property damage.

In this connection, the limit has to be clearly defined as to whether it is in respect of any one accident or a series of accidents arising out of one event. Furthermore, the meaning of 'event' has to be also given in precise terms since it would have significant effect on the amounts recoverable in respect of both the limit of indemnity and the excess to be applied where a claim is made. It is also important to clarify whether any limits apply in respect of individual claims.

Premium and Excess

The underwriting of Contractors' All Risks insurance is not an exact science and the appropriate premium applicable to the insurance of a particular project can only be calculated if an exhaustive investigation is carried out into the risks attached to its construction. As such a task is very laborious and involves a detailed study of the two imprecise elements of risk, i.e. event and probability, see page 28, not many insurers give the subject its due consideration. Many are content to apply past experience to this task and choose a premium similar to one that had proved to be adequate when applied in similar circumstances. Others apply a similar approach and choose figures from tables and nomograms already prepared for this purpose. This latter approach can be extremely inadequate even when the project is a simple one and without any complications. But in order to establish the type of project proposed for insurance, the least that the insurer must do is to examine the contract documents prepared by the design team and the method of construction proposed by the contractor.

Prudent insurers realise that no two construction projects are exactly alike, even for similar work. The factors which have to be taken into consideration are described in detail in Chapter 3 but briefly they include underground conditions, climatic conditions, geographic location, quality of design, ability of the design team and contractor, period of construction, etc.

Another factor which influences the premium is the total value at risk expressed in terms of the various elements of the sum insured, see page 344, and the amount of excess. Excess is an amount borne by the insured in respect of each and every loss or occurrence leading to a claim covered under the policy. Etymologically, the word 'excess', as used in this context, is confusing and a better term to convey what is intended would probably be 'deductible' or even better still 'insured's retained liability'. The latter term reflects the fact that the insured is sharing with the insurer the exposure to risk-making it easier for one to recognise the effect of increasing the insured's share of risk on the amount of premium he must pay.

The amount of excess is usually determined by the insured during the stage of premium negotiation, but there is always a minimum amount in respect of material damage, specified by the insurer. No such limit, or very low limit, is imposed on public liability claims. Besides decreasing the premiums, the excess is applied for the following reasons:

- To reduce the administration costs of processing claims;
- To involve the insured in the retention of some liability and sharing of risks and therefore to encourage him to take more care to avoid loss or damage; and
- To reduce risk assumed by the insured to a limit which he can bear.

Thus it may be necessary or advisable to apply different excesses to different circumstances and types of loss resulting in a number of excesses, for example:

- 'Acts of God' where losses are usually high;
- Testing and commissioning of mechanical and electrical plant; and
- Special risks or risks of Force Majeure.

In some insurance policies, the excess is treated as an exclusion from the cover and therefore appears as an item within the exclusion section of the policy.

Period of Insurance

As referred to under heading of 'Definitions', the period of insurance is divided into two parts, coinciding with the construction period and the period of defects notification of the project. As many aspects and features of this type of insurance, including in specific terms the extent of the applicable insurance cover, depend on whether the project is within the construction period or the Defects Notification Period, it is important to pay attention to clearly defining the relevant dates of start and end of these two periods.

Where the work undertaken by a contractor is of repetitive nature and where it is in building construction rather than civil engineering, an annual Contractors' All Risks insurance policy is sometimes arranged. The annual policy differs from the single policy per project in that it insures the contractor against all activities in a particular year rather than a particular project. The premium charged in respect of an annual policy is based on the Contractor's annual turnover. In annual policies, the distinction between the two parts of the period of insurance becomes less important.

General Exclusions

Despite the 'All Risks' tag in a CAR insurance policy, there are by necessity, by choice or by preference, a number of exclusions which restrict the cover granted by the policy.[3] It is through this negative approach that one can identify as to whether or not any particular risk is covered by the policy. Hence if a risk is not excluded, it is deemed to be included in the cover. The General Exclusions apply to all parts of the policy excluding from its cover any loss, damage or liability directly or indirectly caused by:

1   Inevitable or foreseen losses. In this connection, it must be stated that the word 'inevitable' is intended to mean 'certain to occur' and not 'likely to occur'. The latter wording, if used, results in a wider exclusion.

---

3   By necessity, there must be exclusions due to the fact that some risks are uninsurable (see page 195). By choice, there are exclusions based on the agreement between the insurer and the insured conferring no benefits or disadvantages to either. By preference, there are exclusions which can be deleted if an additional premium is charged.

2 War, invasion, act of foreign enemy, hostilities (whether war be declared or not), civil war, rebellion, revolution, insurrection, mutiny etc.

3 Nuclear reaction, nuclear radiation or radioactive contamination. In the case of a nuclear reactor, the cover of the 'hot part' usually ends when fuel is introduced and that of the 'cold part' ends when the owner takes over.

4 Sonic waves caused by aircraft or other aerial devices travelling at sonic or supersonic speeds. This exclusion emanates from the United Kingdom where the government undertook to pay compensation if damage resulted from the supersonic test flights made by Concorde. Later, when little or no damage materialised from the supersonic test flights, the undertaking seems to have remained in respect of operational flights. In some countries, this exclusion is not a necessary one in standard policies.

5 Confiscation, commandeering, requisition or destruction of or damage to property by order of any lawfully constituted authority.

6 Wilful act or wilful negligence of the insured. The 'Insured' in this clause is defined as anyone who has the legal right to represent the insured officially and therefore the insured's employees do not come within that definition and their wilful act or wilful negligence is not excluded from the insurance cover of the policy.

7 Cessation of work whether total or partial. Insurers do not like to keep cover on a project where work has ceased for one reason or another and no one is left behind on site to fulfil the duty of 'Care of the Works'. This exclusion applies from the date of cessation of work. In some cases, however, a specified period of time is permitted before the cover ceases.

Special Exclusions

These exclusions apply only to the relevant part of a composite policy and represent risks which, if covered, would either:

- Require a considerable increase in premium; or
- When they eventuate, cause adverse effects to the main contract insured under the policy; or
- interfere with other types of insurance dealt with under other separate insurance policies.

The Special Exclusions section includes the following:

1 *Consequential losses, including loss of use, penalties, fines, loss of contracts, and loss arising from delay in completing or negotiating contracts:* The phrase 'consequential loss' has no clear and precise definition or meaning in construction insurance. It has different meanings to different people and so the engineer may understand it as any loss which follows as a result of a logical progression from another event. The general insurance market may define consequential loss as a loss which is not tangible or physical and therefore leaves the property insurer to interpret such a definition as meaning an

interruption loss or loss of profits.[4] The liability insurer may interpret it as an economic or financial loss. The lawyer will define it from the legal precedent in *Croudace Construction Ltd. v. Cawoods Concrete Products* (1978) as loss or damage which does not result directly and naturally from the alleged breach of duty.[5]

Although there are specific insurance policies to cover this type of risk, the Contractors' All Risks insurance policy may be extended to cover specified types of consequential losses against an additional premium. This, however, is not the normal practice and is only done in few special circumstances.

2 *Mechanical and/or electrical breakdown or derangement of construction plant:* This type of exclusion normally applies to construction machinery, which could be very expensive to repair, maintain and/or replace. A special machinery breakdown insurance policy is available to cover this risk, which may be added to the ordinary Contractors' All Risks insurance policy upon payment of an additional premium, usually an expensive addition.

In projects involving the erection of industrial plant and machinery and where the content of construction machinery is comparatively low, this exclusion is eliminated, at least in respect of the works being erected. Similarly, in these projects, commissioning normally would have to be added to the cover granted due to the fact that it includes the risk of breakdown. A certain period is normally designated for commissioning, during which the type of testing programme to be allowed is indicated.

3 *Wear and tear, corrosion, oxidation, deterioration due to lack of use and normal atmospheric conditions:* This exclusion is normally confined to the part which is corroded or oxidised and does not extend to include the resultant damage caused to other property or items. The insured should insist on this point being clarified by stating in the wording of the exclusion that resultant damage to other items or property is included in the insurance cover.

A statement such as 'Loss or damage due to wear and tear, rust or gradual deterioration is excluded' means that all damage is excluded no matter how it is caused. An additional premium may be required to cover this resultant damage.

4 *Defective material and workmanship:* Defects in material and workmanship can cause extensive damage not only to elements in which the defective part has been incorporated but also to surrounding elements. Such damage can be either immediate or latent, and in the latter case the intensity and extent of the effect on the surrounding elements are much greater, and correspondingly more expensive to repair. Furthermore, should the damage occur after completion of a construction contract, the loss might extend to the suspension of the use for which the project was designed, causing consequential financial losses. The exclusion of loss or damage due to defective material and workmanship can be

---

4 Clause 22 in the 4th edition of the FIDIC Conditions of Contract for Works of Civil Engineering Construction, the old Red Book, requires the Contractor to 'indemnify the Employer against . . . material or physical damage to any property . . .', which in effect excludes consequential losses from such indemnity. The ICE equivalent clause does not restrict damage to physical damage.

5 *Croudace Construction Ltd. v. Cawoods Concrete Products Ltd.* (1978) 8 BLR 20.

either total or limited to the defect itself, in which case resultant damage to surrounding elements with sound material and workmanship is covered under the insurance policy. It is important to have the exclusion precisely worded in order that there is no dispute later in the case of a claim as to what was intended to be excluded.

A prudent insurer would be very careful in granting total insurance cover in respect of defective material and workmanship, since the result would almost certainly mean providing insurance cover for an event that would definitely produce adverse consequences. The owner/employer, the professional team and the competent contractor should reject such a cover due to the potential abuse that such insurance might encourage.

There is a close relationship between material, workmanship and design in any construction project. Consequently, a defect in any of these three essential requisites of a properly constructed project may cast a shadow over the other two. Furthermore, there is usually overlap between the functions of specification, design, supply, methods of work, methods of construction and supervision, which could confuse the parties involved in the project as to where each starts and where it ends. This aspect becomes more important as attempts made at codifying what is expected from the design professional, contractor and supplier are not totally clear and successful.

The Canadian case of *Pentagon Construction (1969) Co. Ltd. v. United States Fidelity and Guarantee Co.* (1978), showed the confusion of the judges in respect of the meaning of faulty or improper material, workmanship or design.[6] The three judges in the Appeal Court reached different opinions as to the meaning of design but agreed on the meaning of workmanship. It was a case where the contractor Pentagon Construction Co. Ltd. was engaged to construct a sewage treatment works which included, amongst other items, the construction of a circular concrete tank. Structural steel members were to be erected across the top of the tank, spanning between the walls of the tank and welded at their ends to a steel plate embedded in the concrete wall. These steel members were to give lateral support to the tank walls and also to provide support to some equipment suspended into the tank.

The contract was insured under a Contractors' All Risks insurance policy which excluded: 'a) Loss or damage caused by: (i) faulty or improper material or (ii) faulty or improper workmanship or (iii) faulty or improper design.'

The specification required the contractor to test the tank by filling it with water, which test was carried out after the completion of the concrete work, but unfortunately prior to the welding of the steel members to the end plate embedded in the wall. The tank consequently bulged and the contractor submitted a claim to the insurers in respect of the resultant loss. The insurers repudiated liability for the claim, basing their argument on the quoted exclusion.

The dispute was heard first by the court of first instance where it was held that neither the design nor the workmanship was to blame. The insurers appealed and argued that the design was faulty because it omitted to specify that the tank should

---

6 *Pentagon Construction (1969) Co. Ltd. v. United States Fidelity and Guarantee Co.* [1978], 1 Lloyd's Rep. 93.

not be tested before completion of the welding of the steel lateral supports to the wall. At the Appeal Court, the three judges agreed that the tank failed because of the contractor's failure to weld the steel members prior to testing, which amounted to improper workmanship. It was further agreed that:

Workmanship is not limited to the work or result produced by a worker. It includes the combination or conglomeration of all the skills necessary to complete the contract including, in this case, the particular sequence necessary to achieve the performance of the contract. Failure to follow that sequence could constitute faulty or improper workmanship and in this case did so. One judge decided that he did not have to consider the question of faulty or improper design. The other two judges differed about the meaning of faulty or improper design.

5 *Defective design:* Defective design is an excluded risk from the insurance cover provided normally by the Contractors' All Risks policy. However, as in the previous exclusion, this clause may either exclude only such items that are defectively designed or it may be worded in such a way that all damage or loss arising from defective design is excluded from the cover, including resultant damage. The exclusion could also be worded to exclude cover in respect of consequential losses resulting from such damage or loss. Unlike defective material and workmanship, defective design is a risk that would not have any adverse effect if it were included in the CAR policy as is sometimes done in project insurance policies taken out by the employer.[7] Due to the numerous ways of expressing this exclusion, it is advisable to use clear and precise wording and to avoid using such vocabulary as properly, part, reasonable, proper, etc.

6 *Partial possession or handing over of the project:* Due to the limited cover provided by the CAR policy during the Defects Notification Period, any section handed over to or possessed by the owner/employer passes automatically into that period and thus into the restricted type of cover mentioned earlier.

7 *Loss of or damage to aircraft, vessels, watercraft, or plant mounted on such vessels:* Loss of or damage to such vessels is the subject of insurance of a different type to that dealing with construction, and cover is granted through special policies issued by insurers in the marine, aviation or other insurance fields. Usually, therefore the cover under this heading is excluded from the Contractors' All Risks insurance policy.

8 *Mechanically propelled vehicles which are licensed for public road use:* These vehicles are subject to Statutory Law under the Road Traffic Acts of the relevant jurisdiction and are therefore subject to the insurance requirements stipulated in these acts.

9 *Loss of or damage to files, drawings, accounts, bills, currency, stamps, deeds, evidence of debt, notes, securities or cheques:* These losses are again subject to a different type of insurance and require an additional cover to be provided.

10 *Loss discovered only at the time of taking an inventory.*

---

7 'Lessons to be Learnt', 1983 Report of FIDIC's Standing Committee on Professional Liability, page 10.

Conditions

This section of the policy forms an administration guide of the insurance contract and provides a set of conditions to be observed by the insured. Non-compliance with these conditions could nullify the insurance cover and this is sometimes stated as the first condition. A number of the conditions have a close relationship with some aspects of the main contract and the manner in which the contractor performs on site. The important conditions usually included in a Contractor's All Risks insurance policy are discussed below:

1   *Changes in the risks:* Once the policy is issued, any change in the risk from that originally rated by the insurer must be notified to him in writing. The onus is placed on the insured to have this notification made and recorded as soon as possible after the change. The unfortunate part of this condition is the use of the word 'change' or sometimes 'material change' since the definition of these words is very loose, especially when applied in construction, where changes are very often effected during the construction process and considered to be a natural consequence of the process and therefore no reference to the insurer is made. Disputes in respect of whether insurance cover is available might then arise should such 'changes' cause accidental damage that would result in claims against the insurer.

2   *All reasonable precautions to be taken:* This condition stipulates that the contractor must take all reasonable precautions to ensure the safety of the works and to prevent any loss or damage from taking place.

Once again the use of the word 'reasonable' can cause a lot of controversy. This condition, however, can be found in many formats, some more restrictive than others. For example, one condition in a policy denied any liability by the insured if the loss or damage was 'caused by, or in connection with contracts imperfectly, inefficiently, or improperly fulfilled'. If this condition is precisely applied, the insurer could repudiate almost all claims which might be submitted to him on the basis that the work must have been imperfectly, inefficiently or improperly fulfilled for the loss or damage to have arisen.

3   *The insured is to minimise the loss, once an event has taken place:* This condition is normally included in order to oblige the contractor to mitigate the consequences and minimise the amount of a loss.

4   *How to make a claim:* The procedure to be followed and the timing within which claims must be reported are generally described in one of the conditions of a policy. The condition normally includes a statement to the effect that the insurer is allowed access to the site and to any documents relevant to the claim.

The method of settlement of claims is usually stipulated under this section together with directions regarding reinstatement of sums insured after the insurer makes payments.

5   *Settlement of disputes:* Settlement of disputes between the insurer and the insured is generally based on arbitration and a condition is included to cover this eventuality and to set out the procedure to be followed.

6   *Cancellation, jurisdiction, fraud, etc.:* Other conditions would probably be included in a policy to cover the captioned items and the insured is very strongly advised to examine these carefully to establish their relevance. One example of

such a condition in a CAR policy was drafted by a well-known insurance company for a competent and experienced contractor in respect of a major civil engineering project in the Republic of Ireland. The condition stated that the insurance cover was only valid for a site within the United Kingdom. The consulting engineer fortunately discovered this error after he received the policy from the contractor's insurance broker and before he forwarded it to the employer for approval. No one will ever know what the attitude of the insurer might have been in the event of a major claim, had that error not been corrected in time.

Memoranda

This section of a Contractors' All Risks insurance policy incorporates any extensions to or details of the basic cover provided by the other sections of the standard policy. In the specimen policy appended in Appendix B, there are three memoranda: the first deals with details of the sum insured, as referred to on page 344 above. The second memorandum explains the basis of settlement of claims in the case of a loss covered by the policy, see page 355; and the third memorandum reminds the insured that the cost of additional working after a loss is not covered by the policy unless previously agreed upon and endorsed accordingly in the Endorsements section.

Signature Clause

The insurance contract is one that must be signed in order to establish clearly and beyond any doubt its scope and all the other features of the insurance agreement made between the insured and the insurer. The signature must also be that of an authorised person to effect such an agreement.

Endorsements

Endorsements are needed to change the cover provided by an existing insurance policy. Either party to the insurance contract may require such change, and so for example, the insurer may insist on special limits in the cover due to the nature of the project or due to the nature of the site and its location. On the other hand, the insured may request and the insurer may agree to include additional risks within the cover provided.

It is, therefore, important when examining any policy of insurance to look for the endorsements issued, if any, and to consider their implications.

Settlement of claims

The basis of settlement of claims could appear in any one of the sections already mentioned above, in one form or another. Due to its importance, the basis of claim settlement is discussed here separately, which usually takes the form of one of the following:

1   If the damaged works can be repaired, then the settlement of the claim is based on the cost of the necessary repairs to restore the works to their condition, immediately before the accident, less any salvage value. The exception to this

would be in the case where the cost of repair exceeds the total value of the damaged item, in which case the replacement value forms the basis of settlement.

2   If the accident causes total loss, then the value of the claim is calculated on the basis of the actual value of the loss or damaged property, immediately before the accident, less any salvage.

Profit in respect of carrying out the work of reconstruction or repair is not usually payable to the insured since an insurance policy is a contract of indemnity and the purpose of the policy is to place the insured as closely as possible in the same financial state after a loss as that immediately prior to the occurrence giving rise to the claim. The contractor, therefore, unless otherwise agreed, makes his profit only once through his original contract with the employer. This principle can be viewed as an additional liability borne by the insured over and above the excess specified in the policy. It operates as a reminder that all efforts must be made to prevent accidents and to implement loss prevention measures and programmes.

In many cases, the contractor is obliged to repair the damage temporarily until a final decision is made on the course of action to be taken to carry out the final repairs. In such a case, if the temporary repair is part of the final repairs, thus not affecting the final cost of repair, the cost of temporary repairs is not paid for by the insurer unless agreed upon prior to their implementation or forming part of a programme to prevent the occurrence of further damage and greater loss, see page 000. In this connection, most insurers, if not all, will pay for the cost of any work done after an occurrence to minimise the eventual loss or damage, be it temporary work or permanent. The costs of any additions, alterations and/or improvements are also not payable to the insured unless they are within the value of the original design.

Payment of claims is generally made only after the insured has carried out the repairs or the replacement of the loss or damaged part except where the loss is of a large magnitude, in which case a number of payments on account may be negotiated with the insurer, to be paid in stages subject to certification by the professional team usually the design team.

Insured perils

Although the principle of the insurance cover under the Contractors' All Risks insurance policy is based on the expression 'what is not excluded, is therefore included', the most important perils covered are named below; see Chapter 3 for more information on these perils:

- Fire;
- Windstorm;
- Rainfall, flood and inundation;
- Subsidence;
- Landslide, rockslide and avalanches;
- Negligence, carelessness and lack of skill;
- Consequence of defective design, material and workmanship;
- Collapse;
- Earthquake; and
- Theft and burglary.

Public liability insurance policy

The wording of a public liability insurance policy differs from one part of the world to another and from one insurer to another, whether it forms part of the Contractors' All Risks insurance policy, as in the specimen provided in Appendix B, or as a contract on its own. The same basic rules that govern other insurance contracts apply in the case of public liability but there are some differences in matters of detail, which are discussed under this section.

- Under liability insurance, the Operative Clause states that the insurer will indemnify the insured against liability to pay compensation or damages for accidental bodily injury to any person or for accidental loss of or damage to any property which occurs during the period of insurance arising out of the performance of the contract, subject to certain limitations and exclusions. This wording appears to be sufficiently wide to govern liabilities under the law of torts in respect of negligence, nuisance, trespass, strict liability and statute, subject of course to the limitations and the exclusions referred to in the policy.
- Costs and expenses recoverable by the insured or incurred by him with the written permission of the insurer are recoverable under the terms of the policy.
- The definition of the term 'bodily injury' to which the insurance cover refers includes nervous shock, disease and illness. This term is used in preference to 'personal injury' to eliminate the possible misinterpretation that injury to reputation or mental sensitivity is also included in the cover. The latter term is usually used in professional indemnity insurance.
- A clause is included in the liability section to specify whether the limit of indemnity fixed in the Schedule is for any one occurrence (the aggregate of occurrences being unlimited) or it is the limit for all occurrences within the period of insurance. The latter type is a more restricted cover, having a real limit of liability that is potentially much lower than the former type. Consequently, it is also much cheaper.
- Exclusions from this type of insurance usually include the following:

  (a) Injury to the insured's employees. This is covered under the employer's liability insurance;
  (b) Property in the insured's ownership, custody or control as this is the subject of another insurance cover;
  (c) Property insured in the CAR policy;
  (d) Inevitable loss or damage;
  (e) Liability accepted by agreement. This exclusion must be worded carefully in order not to contradict the agreements made by the insured under the General Conditions of Contract or other contracts and agreements made by the insured during construction such as contracts with subcontractors, plant hirers, suppliers, etc. Strictly viewed, such agreements may invalidate the cover provided by the public liability insurance policy; and
  (f) Other risks contrary to the principles of insurance as already described in Chapter 6.
- Where the policy is issued in the joint names of two or more insureds, a cross liability clause is necessary in view of the exclusions referred to above to indicate

that for the purposes of the agreement in question each of the insured must be treated as a separate entity as if a separate policy had been issued in the name of each of the parties named as insureds, i.e. each of the insured is to be treated as a third party with respect to the other.

Employer's liability insurance policy

This liability arises out of the rule of vicarious liability, as discussed in Chapter 5 above, which could be defined as the liability of one person for the conduct of another.

The insurance policy indemnifies the employer against his legal liability for bodily injury or disease sustained by any employee under a contract of service or apprenticeship to him where such bodily injury or disease is caused during the period of insurance and arises out of and in the course of the employee's employment. Liability assumed under contract is usually included in the cover of this policy. There is no limit to the indemnity covered in this policy since the limit imposed by statute is usually extremely high.

The property of employees is covered under the public liability policy by a specific extension to or endorsement of the basic form of that policy.

13

PROFESSIONAL INDEMNITY INSURANCE

The design professional and other professionals are nowadays exposed to floodgates of 'liability in an indeterminate amount for an indeterminate time to an indeterminate class' of people.[1] In a way, the availability of indemnity insurance has encouraged the courts to increase the exposure beyond the reasonable limits of the notions of 'reasonable care and skill' and the 'neighbourhood relationship'. This exposure may lead to one or a combination of any of the following hazards, as classified in Chapter 2:

1   Personal injury including bodily injury and death;[2]
2   Physical damage to or loss of property;
3   Economic, time and intangible losses.

Professional liability claims may arise from any of the risks involved in such hazards as a consequence of:

(a) The standard that society expects from the professions;[3]
(b) The personality of the design professional and his relationship with the owner and the contractor;
(c) The unique features of the construction contract, see page 190;
(d) Lack of communication and other sources of mistakes, errors, omissions, breaches of professional duty and other such reasons which lie behind litigation in construction.

If the design professional is liable, it is extremely doubtful that he would be financially capable of providing indemnity from his own resources. In fact, it is even

---

1   See Chapter 5 and in particular page 178, where the quotation given is mentioned in the decision from the legal case of *Ultramares Corporation v. Touche* (1931).
2   The difference between 'bodily injury' and 'personal injury' is that the former term includes nervous shock, disease and illness but not injury to reputation or mental sensitivity, which the latter term includes.
3   'Design Professionals and Conflict', a chapter in a collection of information on Design Professionals and Professional Liability Exposures published under the title of *Professional Liability Loss Prevention Manual*, 1974, Design Professionals Insurance Company, USA.

doubtful if the design professional can finance the high cost of any necessary investigation to establish the cause of the event and whatever basis for defence, if any, might exist. Even if he could, the experience would most probably leave him penniless as well as adversely affected by the inevitable reactions concerning professional reputation, morale and judgments from colleagues. It must also be remembered that in the case of most, if not all, professional organisations, a member who becomes bankrupt faces immediate expulsion. Protection against pecuniary considerations may however come first and the proposition of placing all one's worldly possessions in the name of one's spouse is not always one of the practical propositions to be considered or the answers to be given. As one attorney advised his American clients, it is foolish to follow that route since statistically the incidence of divorce is more probable than that of professional negligence. The other option available to the design professional in providing protection against financial disaster is to maintain a programme of self-insurance by setting aside a sum of money every year in a reserve account for use in the case of a successful claim against him. However, this is a very expensive method of protection as such sums are not tax deductible. Alternatively, this method may be used by a number of professionals on a a collective scheme of self-insurance.

Indemnity through insurance

Insurance is perhaps the ultimate option to which a design professional may resort to protect himself, his firm and his clients against hazards that might eventuate as a result of errors, omissions or breaches of professional duty.

Professional negligence can be covered by a specially designed insurance policy called a Professional Indemnity Policy. It protects the insured against his liability to pay damages in respect of personal injury, loss or material damage due to his negligence or that of his own employees in the course of professional conduct of his business. The policy is of the liability type and thus a very fine but definite line separates it from the Public Liability Policy normally issued to protect the insured against his liability for his normal daily conduct. The two must not be confused with each other since, for the professional person, the cover afforded by one complements that given by the other, and does not overlap with it.

In comparison with other insurance covers, this type of insurance is of recent origin and it is still developing in its attempt to match theory with practice. Practice is dictated by considerations such as the 'changing standards of practice, constantly expanding judicial interpretations of duties owed, dramatic increases in claim frequency and severity, and vastly increased expectations on the part of design professionals' clients'.[4]

Two recent judgments of the House of Lords in England may have far reaching implications on the liability and thus on the professional indemnity insurance of

---

4　The words quoted are taken from a statement made for the Record by Paul L. Genecki, Senior Vice President, Victor O. Schinnerer & Co. Inc. for the Subcommittee on Commerce, Transportation and Tourism Committee on Energy and Commerce, United States House of Representatives, April, 1985.

architects and engineers. These are: *Royal Brompton Hospital National Health Service Trust v. Hammond and Others (Appellants) and Taylor Woodrow Construction (Holdings) Limited (Respondents)* [2002] UKHL 14; and *Co-Operative Retail Services Limited Young Partnership and Others (Appellants)* [2002] UKHL 17.[5]

The contract in both cases was the JCT form, used for building works, but these two judgments may affect other standard forms of the construction contract through analysis and interpetation. Furthermore, both cases revolved around the English Civil Liability (Contribution) Act 1978 and the meaning of the term 'the same damage' contained therein. It was emphasised in these two cases that 'damage', which means harm, should be differentiated from 'damages', meaning restitution, citing in this context *Birse Construction Ltd v. Haiste Ltd.*[6]

The first case arose out of certification by the architect that the contractor was entitled to payment in respect of a claim for prolongation and disruption. The second case arose out of the loss suffered by an employer as a result of a fire in the works, which not only caused economic loss, but also delay. The insurers under a contractors' all risks policy issued in the joint names of the employer and the contractor in compliance with the contract conditions, paid for the cost of the repair, but not for the delay. The employer sued the architect and the engineer, alleging that they were partly negligent. They were unsuccessful in claiming contribution from the contractor towards their eventual liability to the employer.

Policy wording

Due to the recent and continued development of professional indemnity Insurance, a standard policy form is not available and each of the few insurers underwriting this special type of insurgence has a different wording. However, a minimum basic wording still applies and, as in other insurance policies, the task of understanding the exact meaning of the contract of insurance, as expressed in the policy, is not an easy one. The best method of understanding the policy is to dissect it into separate sections, which must be analysed by the insured or a specialist in this field on his behalf, to discover whether or not he is purchasing the protection most suitable to his own practice. It must be realised that due to the specialist nature of this insurance, not all insurance brokers are qualified to give advice and to evaluate the cover in this field. Without such knowledge and expertise, one could be operating under a considerable handicap.

The policy wording offered is normally dependent on its origin and the type of profession to be covered. Additions or deletions, however, could be implemented through endorsements issued with the policy and the premium paid in respect of the final cover will depend on the extent of cover purchased.

---

5 *Royal Brompton Hospital National Health Service Trust v. Hammond and Others (Appellants) and Taylor Woodrow Construction (Holdings) Limited (Respondents)* [2002] UKHL 14; and *Co-Operative Retail Services Limited Young Partnership and Others (Appellant)* [2002] UKHL 17.

6 *Birse Construction Ltd v. Haiste Ltd* [1996] 1 WLR 675, 682 per Roch LJ.

The policy wording will most probably contain the following basic elements:

- Insuring Clause;
- Schedule defining, amongst other terms, the following:
  - The insured;
  - The Insured's professional activity which is normally defined through the information sought in the proposal form by the Insurer;
  - Limit of indemnity;
  - Period of insurance;
  - Premium and excess;
- Exceptions (sometimes referred to as exclusions);
- Conditions;
- Memoranda which include any extensions beyond the basic insurance cover provided by the insuring clause; and
- Signature clause on behalf of the insurer.

To understand the professional indemnity policy and how this type of insurance is underwritten, one must study each of these elements separately and relate them to each other and to the impact they have on the activity of a design professional.

The Insuring Clause

The Insuring Clause sets out the basis on which the insurer will indemnify the insured and clearly qualifies that indemnity through the presence of legal liability. The clause thus reflects the quality of the insurance cover provided by the policy and how it operates. It is, therefore, imperative that it is studied carefully and related to the legal system under which the professional is operating.

The scope of cover varies from one policy wording to another. Some policies refer in their cover to liability arising from errors or omissions or negligent acts. Others include the preceding phrase: 'for any sum or sums for which the insured may become legally liable', which widens the cover to any legal judgment made under the operative legal system and within the professional activity accepted by the insurer. A wider cover can also be obtained to include specifically for a breach of any statutory duty imposed by legislation. For design professionals working outside their own country, it is wiser to add the indemnity in respect of any duty imposed under any law or custom existing in the country of the project. Thus, one may arrive at the following wording:

> The Underwriters hereby agree to indemnify the Insureds . . . for any sum or sums which the Insureds may become legally liable to pay arising from any claim or claims made against them during the period stated in the Schedule in consequence of any act of neglect, error or omission or for breach of statutory duty or any law or custom existing in the country as a result of a breach of a Professional duty in the professional conduct of their business as stated in the Schedule. . . .

The Insuring Clause also refers to the other elements of the policy through specific mention of the insured, the sum insured, period of insurance and the business activity of the insured. The scope for error in the policy, or misunderstanding between the insurer and insured is perhaps highest in the definition of the business activity, which should be described carefully and scrupulously in the proposal form, later in the policy and ultimately in the renewal forms. Such description should not be so restrictive as to necessitate an endorsement to cover a slight change in business description.

The Schedule

This section contains the particular features of the insurance contract which identify at a glance the insured and his address, the insured's professional activity, the limit of indemnity, the period of insurance, the applicable premium and amount of excess. These terms are now examined individually:

The insured

Generally, the Insured is the person named in the policy and any partner, director, officer, or employee of the named insured while performing his duties on behalf of the named insured. In the present system of 'liability beyond infinity' in most parts of the world, see page 176, the definition of the insured should extend to 'heirs and successors'. If the insured is an organisation with a board of directors and executive managers, the definition should extend to include the organisation, its board, its management and the employees. The policy can be extended to include any principal who retires, for his liability in respect of professional activities prior to his retirement. Alternatively, a separate policy may be issued to cover the risk of retired principals if such professional indemnity cover had originally been in force for a number of years prior to retirement. Such a retirement policy, which is generally issued for a period of around five years, has the advantage that the retired principal can be assured of his protection irrespective of whether or not the firm's cover is kept in force.

The insured's professional activity

As stated earlier in Chapter 6, see page 185, an insurance policy is a contract of utmost good faith based on trust placed by both sides in each other before and after the agreement is made. The insurer places his trust in the insured to give all material facts with respect to the risk to be insured, and the insured places his trust in the insurer that the promise to pay will be honoured. As already defined, see page 209, a material fact is 'a matter or circumstance which would reasonably influence the judgment of a prudent insurer in deciding whether he would cover the risk and, if so, in determining the premium which he would demand'.

In the case of professional indemnity insurance, the type of activity performed by the professional is of material importance on two counts. In the first, it is a material fact in deciding whether or not the risk is acceptable to the insurer, for example the risk in design work involving toxic waste disposal. In the second, it is a material fact in determining the premium which would apply, and therefore where other matters are of equal importance, the premiums charged in the case of various professions differ considerably. Based on figures released in the USA, the highest rates apply to surgeons, followed by physicians (81% of the rates applicable to surgeons), followed by architects and engineers (53%), followed by lawyers (30%) and followed by dentists (22%).[7] This pattern and the relative percentage figures have probably changed since the date at which they were issued and the indemnity net has certainly caught a larger number of professions including surveyors, insurance brokers and accountants. In the eighteen months prior to April 1985, the accountants' basic auditing business was 'on the wrong end of a flood of multi-million-dollar writs, filed by what the profession likes to dismiss as an unholy alliance of fleet-footed lawyers and embittered shareholders and creditors to bankrupt companies'.[8] The situation for accountants seems to persist but with ever increasing sums being sought in compensation.[9] Within the construction industry, the highest rates apply to structural engineers followed by mechanical and electrical engineers followed by geotechnical engineers followed by civil engineering professionals and finally by architectural/engineering firms. This can be seen from the table of insurance cost as a percentage of gross fees released by the American Consulting Engineers Council in their 1984 Professional Liability Statistical Report; see Table 13.1.

Other details of the insured's professional activity are also usually sought in the proposal form such as: type of clientele; geographic distribution of activity; associations with other organisations, and a detailed breakdown of the type of projects undertaken within the discipline practised.[10] The information given in the

Table 13.1 Professional indemnity insurance cost as a percentage of gross fees[11]

Construction field	1981	1982	1983	1984
Structural	3.31%	2.96%	2.93%	3.26%
Other	2.47%	2.52%	2.73%	2.37%
Mechanical/Electrical	2.24%	4.36%	2.36%	2.54%
Geotechnical	2.24%	4.29%	3.27%	2.37%
Civil	2.23%	2.75%	2.14%	2.36%
Architecture/Engineering	1.82%	1.67%	1.97%	2.28%

7 *Guidelines for Improving Practice*, op. cit., see Chapter 3, note 4, Vol. IV, No. 6.
8 *The Financial Times*, London, 10 April 1985.
9 *The Financial Times*, London, 1 March 2002, see article entitled 'Attempts at a speedy solution look doomed', by Adrian Michaels, relating to the firm Andersen, an accountancy firm described as 'keen to clear up quickly at least one piece of the Enron turmoil and has offered about $750m to Enron shareholders to settle suits seeking compensation for the energy trader's demise. However, people close to the discussions admit that shareholders are seeking more than $3bn and are sceptical that Andersen's offer comes close to what the firm can afford.'
10 'Client Selection and Limitations of Liability', by Claude Y. Mercier, a paper published in *Loss Control Bulletin* No. 61, September, 1982, prepared by National Program Administrator Inc. in cooperation with Simcoe & Erie General Insurance Company, Canada. Amongst the recommendations in this paper are the following: 'Ascertain that your client has the funds and the financing in place to bring the project to completion without undue hardship; be wary of speculative developers; whenever possible, use recognised standard forms of client/designer agreements . . .'
11 This table is taken from the American Consulting Engineers Council Professional Liability Statistical Report for 1984.

proposal form becomes part of the contract agreement when the insurance policy is issued.[12]

Limit of indemnity

The professional indemnity policy is designed to indemnify the insured in respect of any liability arising out of any claim during a period of insurance up to a certain limit plus all costs and expenses, if incurred, with the written consent of the insurer in the defence or settlement of the claim.[13]

Very few policies exclude indemnity in respect of legal costs and expenses. Some, however, totally restrict payment of these costs to cases where the amount exceeds the limit of indemnity, and the amount paid in respect of the costs and expenses is limited to the ratio of the limit of indemnity to the total value of the settlement.

It is important to realise that the limit of indemnity in most cases is an aggregate sum in any one year of insurance and therefore any payment made in respect of a claim during a particular year would reduce the limit of indemnity for any outstanding or forthcoming claims. When claims are made and settled within the insurance year, the insured may request and the insurer may agree to reinstate the limit of indemnity to that originally agreed. However, in most cases, a claim is made but not settled before a considerable period of time has elapsed, which denies the insured the possibility of such reinstatement. Because professional indemnity insurance is of the 'claims made' or retrospective liability type (see 'Period of insurance'), it would therefore be better to try to obtain a policy with a limit of indemnity which would apply to each individual claim, despite the additional premium which would normally have to be paid.

It is always difficult to assess the appropriate limit of indemnity that should be selected by a professional when taking out a professional indemnity policy. This difficulty arises because one would have to estimate in advance the maximum probable loss which is likely to arise as a result of professional conduct within a period of twelve months. In the case of construction projects, this quandary is even more difficult to analyse as the value of the maximum probable loss depends on the value of each of the projects designed, whereas the fee income, on which the premium is based, is a percentage of the total value of all projects. Furthermore, the profit margin earned forms a small proportion of fee income.

There are, however, various methods of selecting the limit of indemnity appropriate to a particular firm. These are:

- A 'rule of thumb' method where one applies a multiplier of between 1.5 to 4 to the gross annual fee income, i.e. a firm with £1.0 million gross annual fee income should have a cover in the region of £2.5 million if the multiplier is 2.5. This assessment leads generally to high limits of indemnity.
- The Maximum Probable Loss (M.P.L.) is a more scientific but laborious method which entails analysing the probable maximum loss of the largest projects undertaken. The use of the word 'probable' and not the word 'possible' should

12 See Appendix E – Specimen of a Proposal Form for Professional Indemnity Insurance.
13 See Appendix F – Specimen of a Professional Indemnity Insurance Policy.

be noted. The probability of these losses occurring simultaneously is then assessed and the limit of indemnity is next based on the figures available.

- Statistical surveys of limits of indemnity carried out by various organisations are sometimes published giving an indication of the normal level of indemnity cover taken out by others practising in the same field of activity. Such a survey is published annually, for example, by the American Consulting Engineers Council and the information presented in Table 13.2 is taken from their 1983 and 1984 Reports.
- Certain professional organisations require all or some of their members to provide a professional indemnity insurance with a specified minimum amount of indemnity. Such figures may provide guidance but are generally on the low side.

Period of insurance

The professional indemnity insurance policy is usually an annual policy renewable every year on the completion of a special form by the insured giving the up-to-date information necessary for the insurer to assess the risk and premium. The most important aspect of the period of insurance of this type of policy is what is known as 'claims-made' basis of indemnity. This term means that the cover provided is only extended to claims made during the period of insurance irrespective of the date of the negligent act, or fault, or omission that is giving rise to the claim. In some cases, it is referred to as a 'Retrospective Liability' policy. Thus, if a negligent act was committed but not discovered for a period of three years, any claim arising out of that act is dealt with under the policy in force three years after the date of that negligent act, if such policy were in existence at the time. Thus, when the cover

Table 13.2 Distribution of limit of indemnity by size of firm[14]

Number of Employees	Under $250,000	$250,000 to $499,999	$500,000 $999,999	$1m to $2.4m	$2.5m to $9-99m	Over $10m
1–5	27%	29%	20%	21%	3%	0
6 -10	14%	20%	28%	37%	1%	0
11 – 25	9%	16%	29%	42%	3%	1%
26 -100	2%	4%	19%	68%	7%	0
101 – 500	0	0	6%	42%	38%	14%
Over 500	0	10%	0	0	40%	50%

Year	Percentage number of all firms insuring					
1981	20%	20%	25%	33%	–	2%
1982	16%	18%	23%	40%	–	2%
1983	12%	16%	23%	42%	5%	2%
1984	9%	17%	23%	49%	6%	2%

14   This table is taken from the American Consulting Engineers Council Statistical Report for the years 1983 and 1984.

lapses on the policy in force at the date of the negligent act the insurer is released from any unclaimed indemnity in respect of that policy. The insurer has the advantage, in this type of insurance policy, of knowing on an annual basis the extent of premium he has earned and the amount of liability to which he is exposed in respect of any one year's insurance.

For this reason, it is important to ensure that all previous business is included in any cover provided either when such a policy is first issued or when a new insurer becomes involved. In the latter case, when the insured changes his insurer, it is essential that any situation with a potential claim be disclosed to the current insurer in order for the insured to remain covered if a claim is made later.

Premium and Excess (or deductible)

In Professional Indemnity insurance, premiums and excesses are significantly higher than those in other classes of insurance. The amount of each of these two elements is related and sensitive to the other in such a way that many insureds choose to increase their excess (referred to sometimes as 'deductible', which is the amount borne by an insured in respect of each and every occurrence or claim as the case may be) as a method of reducing the premium. It is not therefore unusual to find that the excess in professional indemnity Insurance represents one twentieth of the limit of indemnity. Premium calculations, however, depend not only on the size of the excess but also on the type of activity practised, the concentration of hazard and amount of risk in staging that activity, the limit of indemnity, the claim experience of an insured and that of the discipline in which he is practising, the extent and type of cover provided by the policy, the size of the firm and type of client for whom work is carried out.

Table 13.1, referred to above, shows the variation in premium with respect to the discipline within the construction industry. The effect of variations in the limit of indemnity on the premium can be approximately assessed by applying an increase of 20% to 30% to the premium for every time the limit of indemnity is doubled. Similarly, the effect of increasing the level of excess on the premium can be approximately assessed by applying a decrease of 8% to 15% to the premium for every time the excess is doubled.

The actual amount of premium is calculated by applying the agreed percentage rate to the estimated fees for the coming year and later making an adjustment at the end of the period of insurance using the audited accounts to calculate the actual value of fees. The term 'fees' should therefore be accurately defined when agreement on the premium rate is made between the insurer and the insured. Such definition should indicate whether the calculation should include any of the following:

1   Taxes applicable to the actual fee content of billings;
2   Billings and fees charged in respect of sale of documents, printing, expenses incurred, etc.;
3   Fees charged in respect of acting as expert witness in a court or arbitration proceedings;
4   Fees invoiced but not received;
5   Fees charged in respect of abandoned reports and designs; and

6   Fees charged in respect of design work delayed through slow and extended staging of pre-construction phase, ending in construction many years later. If these fees are not included in the calculation of premium until construction work is proceeded with, a cash flow problem may be created for the insured.

The excess is required on each and every professional indemnity policy mainly to keep the administration cost of such insurance to a minimum. Most insurers encourage a high excess to be borne by the insured and thus grant the appropriate reduction in the premium. The insured should also prefer higher excess limits not only because of the reduction in the premium but also to maintain the freedom to deal with claims below the excess in the manner most suitable to him, see QC clause on page 369. When deciding on the amount of excess, the insured must consider the probability of having more than one claim against him in any one year of insurance. Care must be taken to establish a clear definition of the excess and whether it is to be deducted from each and every claim, or alternatively from each claim or series of claims arising out of one loss. The difference between these two alternatives can be quite considerable if the excess is large and a number of claims arise as a result of one accident.

Although the excess is usually a fixed sum, there are situations where it is agreed to calculate it as a percentage of the claim subject to a maximum and a minimum limit. An example of such an agreement is a co-insurance contract by two insurers where the first accepts a percentage of the loss. It is also worthwhile mentioning that where the excess is a fixed sum, arrangements can be made, against an additional premium, to have an upper limit applied to the amounts deducted in respect of the excess in any single year of insurance, irrespective of the number of losses or claims experienced.

The exceptions/exclusions

This section of the policy restricts the cover of the insurance policy either by excluding indemnity in respect of liability, which is not intended to be covered under the policy, or liability, which can be covered, but in fact is not.

Excluding indemnity in respect of liability not intended to be covered under professional indemnity insurance can appear in the form of any of the following exclusions:

- liability arising out of bodily injury, sickness, disease or death sustained by any person as a result of and in the course of his employment by the insured under a contract of service or apprenticeship with the insured. This liability is normally covered under employer's liability insurance;
- liability arising out of the ownership, use, occupation or leasing of property mobile and/or immobile by, to or on behalf of the insured (this liability is normally covered under public liability insurance); and
- other exclusions as set out in Appendix F below.

Excluding indemnity in respect of liability, which can be covered but in fact is not can appear in the form of any of the following exclusions:

- liability arising out of any claim made against the insured as a result of any dishonest, malicious or illegal acts of any present or previous employee of the insured (this liability can be covered under professional indemnity insurance for an additional premium);
- liability arising out of patent or copyright infringement (this liability can also be insured under this policy for an additional premium);
- liability arising in certain geographical locations; and
- other exclusions as set out in Appendix F below.[15]

It is important to try to evaluate the relevance of each one of the exclusions included under this section of the policy and establish whether deleting it is desirable or worthwhile. Of course, the deletion of an exclusion means a wider insurance cover and thus higher premium. It follows that before one can evaluate the comparative cost of different insurance policies, the extent of the cover provided in them must first be established precisely.

In this connection, it is essential to note that if a term appears in the Insuring Clause of a policy, the onus of proof where a dispute arises relating to its interpretation is on the insured but if the term appeared in the 'Exceptions', then the onus of proof shifts to the insurer.[16]

Conditions

Professional indemnity insurance policies usually contain a number of conditions which govern the manner in which they operate. In the specimen provided in Appendix F, there are eight conditions setting out the features of the insurance cover provided. Some of these features are common to other types of insurance and therefore familiar. Others are either peculiar to this type of policy or have a specific relevance to professional indemnity and are therefore discussed below:

- *Notification of claims:* Every policy contains a paragraph specifying when and how the insurer is to be notified of claims. A precise number of days from the date of the event giving rise to the claim is usually specified as a condition in the policy within which a claim must be notified to the insurer. Alternatively, it

---

15 See the Exclusions section of the policy in Appendix F below, which sets out a specimen of a Professional indemnity Policy.
16 F.N. Eaglestone in his book *Insurance under the JCT Forms*, 2nd edition, Blackwell Science, 1995, explains 'the significance of the words "accidental" and "accident" appearing in the operative clause on the one hand, and an exception concerning inevitable or deliberate injury or damage on the other lies in the fact that in the former case the onus of proving an accident or an accidental occurrence is upon the insured, whereas in the latter case it rests upon the insurer to prove the exception applies'. He states that the authorities on this statement can be found in the legal cases of *Munro Brice & Co. v. War Risks Association* [1918], 2 KB 78, which decided that the burden of proof is upon the insured to prove an accident when these words appear in the operative clause of the policy; and in *Bond Air Services v. Hill* [1955], 2 QB 417, which decided that the onus is upon the insurer to prove the application of an exclusion.

would be required that the claim is notified as soon as possible after the event. If a time period is specified, it is then mandatory and so is any specific requirement as to how such notice is made.

The wording of this condition can be contentious in that some policies stipulate that notice of claim must be given as soon as the insured becomes aware of any occurrence, which may subsequently give rise to a claim. The definition of the words 'becomes aware' is extremely vague and can be interpreted in different ways by the insured or the insurer giving rise to a dispute or providing ground to void indemnity, as occurred in the case of *Williamson and Vellmer Engineering v. Sequoia Insurance Company,* see page 105. A less contentious wording might be set out in the following terms: 'immediate notice shall be given in writing of any claim or intimation of a possible claim made against the insured which may give rise to a claim under this policy.'

- *Co-operation:* Each policy contains a condition requiring the co-operation of the insured in providing documents, information, etc. for the purpose of conducting an investigation into the reasons behind the claim, should the insurer so wish. This is normally done when the cause of the failure or the event giving rise to the claim is not clear and the insurer wishes to exercise his option to defend the claim. All professional indemnity policies have a clause entitling the insurer to take over the conduct of the defence or the settlement of a claim, normally subject to what is known in practice as the QC clause (see condition 2 of the specimen). The QC clause is the answer provided by the professional indemnity insurance policy for the difficult situation which might arise when the insured finds himself faced with on the one hand a claim from his client and on the other a difference of opinion between him and the insurer as to whether or not the claim should be resisted.

This difference of opinion between the insured and the insured could arise in two different situations. First, the insured might wish to settle his client's claim and bring the matter to an end whereas the insurer desires to contest it. Second, the insured might wish to contest his client's claim in order to assert a point of principle whereas the insurer may desire to have the claim settled on the basis of economic expediency. In such situations the QC clause provides that a claim may only be contested if there is a reasonable probability of success.

Memoranda

This section of the professional indemnity insurance policy incorporates any extensions to the basic cover stated in the Insuring Clause together with any explanatory notes in relation to any of the Conditions, Exceptions or any other part of the policy. The extensions may include any of the following:

Partners' previous business

This extension, which is referred to as the Anteriority Cover, is necessary in the case of the insured either effecting professional indemnity insurance for the first time or changing his insurer. Its necessity emanates from the claims-made characteristic of this policy, see page 365.

Former partners and retiring partners

Once partners leave a firm, or retire, or terminate their services for one reason or another, they lose their control over whether or not the professional indemnity insurance policy is renewed from year to year. Even if it is, they have no control over the relevant terms under which it is renewed, for example lower limit of indemnity, higher excess, etc. If a retiring partner wishes to have protection against his potential liability during the retirement years, an extension of the basic insurance cover becomes necessary. Some experts in the field consider that it is the duty of the remaining partners in the professional firm to maintain this extension to the cover even after death of the departing partner since his heirs and successors need to be protected if the relevant applicable law permits a potential liability, either in contract or in tort. The extension is sometimes referred to as the 'Posteriority Cover' and is usually obtained through two alternative methods. The first method is to continue to name the partner as insured under the policy with the restriction that the cover only applies to those professional services performed prior to the date of termination of service. Under this method, if a claim is subsequently made, it may have an effect upon the firm's record resulting possibly in a higher insurance premium in respect of future renewals.

The second method is to acquire a separate professional indemnity insurance policy to cover the particular partner who is retiring or leaving the firm. The only requirement that most insurers make is that the firm or the professional should have been continuously insured for a number of years, usually three years, prior to the termination of activity. The terms of the policy are generally the same as those included in the most recent wording of the policy.

There is usually no problem in obtaining this extension if the firm itself is continuing its activity and renewing its insurance cover, but a problem would arise if the whole firm ceases to operate. To provide a condition in the policy, which stipulates that Posteriority Cover will be automatically granted if the firm ceases work, is a difficult task. However, if such a condition is not possible, a Posteriority Cover should be discussed at the time of negotiating the wording of the whole policy and an agreement in principle should be reached in advance and laid down in the policy.

Infidelity and dishonesty

Indemnity in respect of liability arising out of any dishonest, malicious or illegal acts of any employee of the insured is normally excluded from the basic cover. By adding this extension, the exclusion must be deleted.

Loss of documents

This extension covers the insured against any legal liability that may be incurred as a result of loss, damage or destruction of any document belonging to the insured, entrusted to him or in his custody. The costs and expenses incurred by the insured in replacing or restoring such document can also be covered under this extension.

Libel and slander

This extension indemnifies the insured against liability at law for damages, costs and expenses associated with claims for libel and slander.

Right of subrogation against employees

In the case of a claim covered by the insurance policy and where the insurer has paid for the loss, the right of subrogation permits the insurer to step into the position of the insured and avail himself of all the rights and remedies, permitted by that position, against any blameworthy party. This extension is therefore designed and worded in a way to prevent the insurer exercising the right of subrogation against a specifically named party or parties. In the case of an insured organisation where employees can be identified as responsible for acts resulting in claims against the insured, it is wise to obtain this extension in order to extend the insurance cover and protect the employees.

Recovery of legal costs incurred in the process of claiming professional fees

It is recognised that one of the sources of claims against professionals is the reluctant client who does not wish to pay for the professional service rendered. When a final note of fees is submitted, the client in such a case refuses to pay claiming for damages in respect of part of the service rendered by the professional. This extension of cover permits the insured to pursue the client legally by paying any legal costs incurred in the proceedings. It is usual, however, that a condition is imposed on such an extension which requires the insured to inform the insurer of the intention to institute such proceedings and that a legal adviser has specified that such legal action would meet with a reasonable probability of success.

Liability to third parties in accordance with the Hedley Byrne case

If the basic professional indemnity policy does not cover claims made by third parties against the insured under the principle of law established by the case of *Hedley Byrne & Co. Ltd. v. Heller and Partners Ltd.* in 1963, see page 171, such an extension can be obtained to give the required cover.

Environmental impairment

This extension, if provided by an insurer, would cover legal liability to pay for personal injury, damage to property, diminution or impairment of property, and interference with any right or amenity protected by law, caused or contributed to by: the emission, discharge, dispersal, disposal, seepage, release or escape of any liquid, solid or gas, or the generation of smells, vibrations, light, electricity, radiation, changes in temperature or any other sensory phenomena which directly or indirectly cause irritation, contamination and/or pollution of the environment and for the cost of litigation, or for expenses incurred in moving, nullifying or clearing up harmful substances. This extension is extremely wide and covers environmental hazards and risks, which could lead to extremely costly consequences. Events have shown this result and have caused the insurance market to retract, giving more thought to such extensions, less cover and higher assessment of premium rates.[17]

17 'Who is Really to Blame for the Current Insurance Crisis?', by G. Buddy Nichols and Kenneth W. Smith, *Risk Management*, July, 1985; and see also 'Naked Came the Insurance Buyer', by Carol J. Loomis, *Fortune*, 10 June 1985, Time Inc. All Rights Reserved.

The following quotation is taken from the Annual Report of the Munich Reinsurance Company for 1984/1985:

> Environmental pollution damage, which is often covered by Industrial Liability policies, can occur suddenly as a result of industrial accidents or gradually by slow impairment of the surroundings. The late claims potential in this risk sector is so unpredictable that insurers may have difficulty in coming to grips with it. The circumstances, as well as the dimensions of the damage involved, gives rise to the question whether the environmental risk is insurable at all.

The future is uncertain as to cost of this extension to cover and whether or not it could be granted when the legal theories in the field of professional liability are expanding.

Signature Clause

Due to the fact that the courts cannot form an opinion as to the terms upon which an insurance policy would have been concluded, the insurance contract must be in writing and this clause is required on all insurance policies, the professional indemnity insurance policy being no exception. The signature of a peson authorised to issue the policy must be placed on the insurance policy for it to be effective. This is not usually witnessed by the insured if an insurance broker is employed and, therefore, reliance is placed on him to see to it that the policy is correctly issued and signed.

Other aspects of the professional indemnity insurance cover

There are many aspects which are of specific importance with respect to professional indemnity insurance and these should be considered carefully when negotiating the wording or renewing of a policy. They include the following:

Breach of warranty

Breach of warranty is not normally covered by the usual professional indemnity policy since intrinsically the design professional is not expected to give a warranty or a guarantee of performance. Therefore, guarantees in respect of price, time, performance and other specific serviceability aspects are beyond the present legal liability attached to the design professional and if these were contractually agreed, the insurance cover must be investigated first.

Cancellation of cover

The wording of many professional indemnity insurance policies includes a term which reserves the right of the insurer to cancel the policy after giving notice of a specified number of days. Such notice could be normally as short as seven days and as long as six months. However, short notices could pose a real problem for the insured, especially if this right of cancellation is exercised after a major claim is notified. Whilst there is no coverage problem by the insurer in respect of the notified particular claim, there could be a problem in respect of other claims flowing from the same event or defect. The new insurer, if one is found, will exclude cover in respect of the defect that had given rise to the first claim and the insured will be facing future ramifications without an insurance cover. The new insurer will be providing Anteriority Cover only in respect of claims arising out of acts committed prior to the date of issue of the policy, provided the insured can prove that he was not aware of any circumstances which might lead to a claim.

Waiver of right to cancel cover

In some professional indemnity insurance policies, the wording includes a term which entitles the insurer to waive his right to cancel the cover in the case of discovery of misrepresentation or non-disclosure. This waiver is useful to both insurer and insured in cases where the insurer elects to continue the insurance cover with the exclusion of the indemnity in respect of the claim which has arisen or which might arise in relation to the circumstances which ought to have been disclosed originally; see condition 7 of Appendix F below. The condition, as stated, provides the mechanism through which an innocent mistake is not allowed to escalate into the avoidance of the whole insurance cover for the remaining period of insurance or preventing a future renewal of the cover if both parties are desirous of the continuation of the insurance arrangement.

Settlement of claims

As stated earlier the term 'claim' has a particular significance in relation to professional indemnity insurance in that all policies require the notice of claim to be submitted to the insurer within a specified short period of time or as soon as practicable after the professional has become aware of a claim or a potential claim. It certainly is important that claims, potential claims or suspected claims are notified to the insurer even when doubt exists, as failure to do so might nullify the cover. This is doubly important if the insured were changing his insurer, in which case any occurrence that might later give rise to a claim against the insured should be notified. Any claims arising from occurrences, when notified, are deemed to have been made during the currency of the policy. The text of the relevant condition dealing with notification of claims should be worded accordingly.

When a notice of claim is given, the insurer might take over the defence of that claim or might investigate the claim to find out whether or not it is covered under the terms of the insurance policy. The process of investigating and establishing the cause or causes that lie behind professional indemnity claims and losses is an expensive and complex one. In many cases, the cost of defending a claim is higher than the value of any repair necessary to rectify the damage caused by the alleged act of negligence, error or omission. Furthermore, it is a feature of professional indemnity insurance that the period of time usually taken for a claim to be settled is a long one. This feature combined with the high premiums associated with this type of policy makes professional liability insurance a very enticing class of business, especially when bank interest rates are high. To illustrate the aspect of long delay in settlement of claims, Table 13.3 shows the total amount of claims paid and outstanding at the end of each year for twelve years and also at the end of that period, 31 December 1981, as taken from the results of the National Program Administrator Inc. and Simcoe and Erie General Insurance Company in Canada.[18] For this purpose the results represent both

18 'An Overview of the Programs' Underwriting Results', by Claude Y. Mercier, a paper published in *Loss Control Bulletin* No. 60, July 1982, prepared by National Program Administrator Inc. in cooperation with Simcoe & Erie General Insurance Company, Canada.

Table 13.3 Combined loss development, Canadian architects and engineers[19]

Year	A	B	C	D	E	F
1970	72	1	99	414,793	678,878	64
1971	159	5	97	889,025	2,349,525	164
1972	132	10	93	789,740	1,604,886	103
1973	281	15	95	1,113,006	4,344,224	290
1974	300	15	95	955,495	3,483,225	265
1975	354	42	88	1,933,039	9,868,383	411
1976	532	61	89	2,814,172	6,227,449	121
1977	508	80	85	3,138,057	8,923,324	184
1978	616	107	83	3,480,322	7,668,710	120
1979	680	190	72	4,815,045	8,933,280	86
1980	630	347	45	3,097,050	7,103,070	129
1981	701	537	24	7,938,198	7,938,198	–
Total	4,965	1,410	–	31,377,942	69,123,152	120

A: No. of claims as at year end;
B: No. of claims unresolved as of 31st December 1981;
C: % of claims settled by 31st December 1981;
D: Total claims value as at year-end;
E: Total claims value as of 31st December 1981;
F: % growth in claim value as to 31st December 1981.

the architecture and engineering disciplines. The average number of years taken to settle a typical claim against a design professional is 6.17 years.

Two other parameters of claims in the field of professional indemnity insurance can be observed. These are frequency and severity of claims. The frequency is measured in terms of number of claims per 100 firms per year. The severity is measured in terms of the average cost per claim which is derived by dividing the insurer's total incurred losses, through claims paid, by the total number of claims. Frequency figures taken from the experience of Victor O. Schinnerer & Co. Inc. in the United States of America show that the frequency was 12.5 claims per 100 per year in 1960 and this had risen to 36.3 in 1978 and 44.8 in 1980.[20] The severity increased from $5,481 per claim in 1960 to $43,659 in 1978 to $66,250 in 1982 to an estimated figure of $70,000 in 1985. As mentioned above, these figures represent the payment made by the insurer and do not take into account the excess which is borne by the insured.

Professional indemnity insurance group schemes

Professional indemnity insurance is usually underwritten for professionals in their individual capacity or as members of a professional firm, either in the form of a partnership or as a company. Individuals or firms of similar professional discipline may however get together into a group with the objective of seeking professional

19 Taken from the previous reference.
20 Statement for the Record made by Paul L. Genecki, Senior Vice President, Victor O. Schinnerer & Co. Inc. for the Committee on Environment and Public Works, United States Senate, April 1985, Washington DC, USA.

indemnity insurance under one master policy or scheme. This latter method has many advantages, if it is correctly established and if an enthusiastic and competent insurance broker is found to put the threads together into a working cover. The advantages include the following:

1 Availability of cover

Professional indemnity insurance features as one type of insurance where a long-term relationship between the insurer and the insured can be extremely important and beneficial to both. The time lag that usually occurs between insurance, claim notification and claim settlement obscures the usual interaction of the insurance market forces.[21] These forces can be briefly identified as inflation, investment climate and competition. As inflation rises, the cost of claims when they mature into payments, after many years, increases, thus resulting in higher premium rates. Investment climate is a cyclic force, which affects insurance in general, but professional indemnity in particular through the high premiums generated. The third force is competition, which can cause deterioration in the quality of service performed if that quality cannot be easily measured and assessed. The quality of insurance, in terms of the promise made to pay in case of a loss, cannot be easily measured or ascertained if the promise is not put to the test, an event that does not often happen until a long time has elapsed.

These forces combined with the claims-made nature of this insurance produce a volatile market which may withdraw from providing cover to an individual professional or a firm of professionals, a situation which can be at best inconvenient but at worst may be disastrous for that firm.[22] However, in the case of a group scheme, the availability of insurance cover would be more certain even in years of high loss ratios, low investment incomes and low inflation.

2 Bargaining power

By uniting, the group stands a better chance than an individual can of achieving a better policy wording, wider insurance cover, more just and equitable settlement of claims, and above all a more dependable insurance promise.

Under this heading, one can include as an advantage the possibility of attracting a better quality insurance broking service, which can improve the quality of other insurances required by the design professional. One of these insurance covers needed is a specially worded public liability insurance policy to protect the design professional in his activities on site which might cause a claim not covered under the professional indemnity insurance policy. Such claims might arise in respect of any physical act by the professional person; so for example if, instead of instructing the contractor to carry out a certain function, the design professional performs that act

---

21 See page 374, settlement of claims of where it is stated that the average number of years taken to settle a typical claim against a design professional is 6.17 years.

22 'Current Conditions and Trends of the Insurance Market', by A. Brough, Vice President, Frank B. Hall & Co. Inc., a paper delivered to the American Consulting Engineers Council, 8 May 1985; and 'Professional Liability Insurance', by Paul L. Genecki, Senior Vice President, Victor O. Schinnerer & Co. Inc., a report to PEPP Board of Governors, 16 July 1985, where, in both articles, the shrinking American insurance market for professional indemnity insurance is described.

personally and if the act were to result in damage or injury, the appropriate cover under which indemnity is provided is the public liability and not the professional liability cover. The public liability insurance policy issued for a design professional must therefore be worded in such a way that it matches the Professional Indemnity insurance and complements its cover. The standard public liability insurance policy covers the professional usually in his normal day-to-day activities.

Group schemes with retention of control: 'Funded Group Scheme'

A group of professionals of the same discipline might go one step further than the ordinary professional indemnity insurance scheme by retaining some control over its handling and administration. This control can only be exercised if the members of the group, collectively, retain a share of the risk, which is otherwise offered to the insurer. The share of the risk retained by the group is usually in respect of a layer of indemnity of a specific sum above the excess applied by the insured in the policy. In view of this risk-sharing arrangement, the premium charged to the insured is reduced by a certain amount or percentage depending on the size of the share of the risk allocated to him. The resultant reduction in premium (i.e. the difference between what is normally charged for the professional indemnity insurance with the excess applied by the insured and that charged for the higher excess) might be viewed as a premium for the group scheme, hereinafter referred to as the Funded Group Scheme. A design professional may look upon such a scheme as providing an insurance cover with higher imposed excess than that he would usually apply in the case of an ordinary professional indemnity insurance policy, but with a difference. The difference being that in the case of the usual insurance policy, the insured bears the excess in respect of claims made against him whereas in the case of the Funded Group Scheme the insured is asked to bear two excesses, a first-layer excess in respect of a claim made against him and a second-layer excess, above the first, in respect of claims made against other participants in the group. The first-layer excess means a reduction in the premium he is charged by the insurers whereas the second-layer excess means a shared premium he obtains with other participants against the shared risk of paying that excess if and when a claim is made against other professionals in the group.

Besides the two advantages mentioned above for group schemes (i.e. availability of cover and bargaining power), the Funded Group Scheme has also the following advantages:

Knowledge

It has been said that 'Doctors bury their mistakes, architects cover them with ivy, and engineers write long reports which never see the light of day'.[23] The 1983 Report of FIDIC's Standing Committee on Professional Liability states:[24]

---

23 'Lessons from Failures of Concrete Structures', by Jacob Feld, American Concrete Institute Monographs No. 1, Detroit, Michigan, 1964, page 1.
24 'Lessons to be Learnt', Report of the Standing Committee on Professional Liability of the International Federation of Consulting Engineers, 1983, Lausanne, Switzerland, at page 11.

One of the recognised basic methods of acquiring knowledge is by learning to avoid mistakes, made by others. Therefore, the problems which result from mistakes should not be covered by a cloak of secrecy but brought out into the open, provided that complete confidentiality of such matters which may cause harm to others can be maintained.

It is unfortunate that when a failure occurs, there is no mechanism by which lessons can be drawn and distributed through the profession so that others do not repeat the mistakes made. Of course, confidentiality must be preserved together with a scientific approach beyond any motive other than the acquisition of knowledge. This is only possible in the case of a Funded Group Scheme.

Loss prevention programmes

When professionals are organised into such a group, then funds become available to the group through premiums received in respect of the second-layer excess and knowledge accumulates through alerts from the group members; it is possible then to organise loss prevention programmes. These programmes are extremely valuable in reducing the risks of error and omission to which the design professional is exposed, as experienced by the Association of Soil and Foundation Engineers in the United States[25] and the Association of Consulting Engineers of New Zealand,[26] two pioneers in this field. An interesting analysis from the experience of the latter organisation is given in Table 13.4, which illustrates the important areas of failure in performance. Information such as this can be obtained, disseminated and distributed through loss prevention studies organised by the profession. Source number 2 in Table 13.4 was further analysed and the results are shown in Table 13.5.

Through loss prevention programmes and other efforts made by the Association of Soil and Foundation Engineers in the United States, it was possible to reduce the number of claims made against their members from being the worst amongst other engineers in 1968 to the best in 1982.

Assessment of premium

With the information available to the managers of the Funded Group Scheme, it is possible to establish the appropriate level that should apply in any particular year to the design professional and also to the different disciplines within the profession. The purpose of this benefit is to establish stability of the insurance transaction for the design professional, the fund and the insurer. It is of course important in this respect to establish correctly the level of premium required for each layer of the three-layered insurance transaction as detailed in Figure 13.1.

25 'The ASFE Guide to In-Home Loss Prevention Programs', published by the Association of Soil and Foundation Engineers, 8811 Colesville Road/Suite 225, Silver Springs, 1983, Maryland, USA.
26 Alerts or claims received by the New Zealand Architects Co-operative Society Ltd., NZACS and the Consulting Engineers Advancement Society Inc., CEAS, were published in the proceedings of a seminar on Good Communications, July 1981.

Table 13.4 Distribution of claims and alerts as analysed by the New Zealand Architects Co-operative Society Ltd. and the Consulting Engineers Advancement Society Inc.

Source of claims and alerts based on the experience of:	NZACS 8 years (%)	CEAS 3 years (%)
1 Technical error	30	21
2 Failure to communicate with client, contractor or other consultant	12	32
3 The claim is regarded as a 'try on'	8	17
4 Design professional has been brought into an area of litigation through being named as a third party	17	4
5 Designer did not supervise his design or was otherwise out of touch with the work	6	7
6 The claim arose from a dispute over fees	7	3
7 Alerts which have not yet been confirmed as claims	20	16
Total	100	100

Table 13.5 With whom does communication fail?

With whom does communication fail?	NZACS data (%)	CEAS data (%)
1 Client	31	53
2 Contractor	7	26
3 Technical data; i.e. information from product suppliers, standard specifications, etc.	37	11
4 Other consultants	22	6
5 Consultant's own staff	3	4
Total	100	100

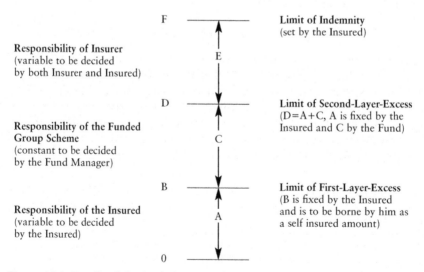

Figure 13.1 Details of the funded group scheme.

378

14

ALTERNATIVE METHODS
OF INSURING

Since its inception, construction insurance has been transacted on the basis of each of the parties involved in construction providing insurance cover compatible with the responsibilities allocated to it and the liabilities to which it is exposed. Its purpose has been and still is to ensure that the completion of the project is not hindered by financial problems emanating from loss or damage due to hazards and risks eventuating during the period of construction. Although the owner/employer and the designer must obviously be conscious of the lifetime hazards and risks related to the location of the project, construction insurance is usually only expected to deal with the project's exposure to risk during its construction and the Defects Notification Period that follows.

The conventional method

Besides the usual insurance cover required by any business activity, four insurance policies became associated with construction: Contractors' All Risks, Public Liability, Employer's Liability and Professional Indemnity; see Chapters 12 and 13 above. The Contractors' All Risks insurance policy was specifically developed for the purpose of covering, in one document, a project under construction against any physical or tangible loss or damage due to all insurable risks. The liability policies were adopted for use in construction and the four policies, combined, became firmly established as the appropriate method of providing insurance protection when, in the 1950s, standard forms of conditions of contract required them to be issued. This method became conventional and accepted in many parts of the world, helped by the international nature of the insurance business; see Chapter 7 above. It remained essentially the same, despite the inevitable evolution of different concepts based on local custom, national traditions and domestic law.

The conventional method of providing insurance protection has functioned, in the main, well enough with few problems and disputes, although as time passed overlaps and gaps in the cover usually provided became evident.[1] Overlaps and gaps in the insurance cover played an important role in large projects and their effects increased proportionately with the increase in size of a construction project.

---

1   'FIDIC's Project Insurance Seminar', *Consulting Engineer*, August 1978.

Overlaps and gaps

Overlaps occurred generally when the same insurance cover was obtained by more than one of the parties involved in construction. The effect of overlaps is essentially two-fold; the first effect is cost, and is reflected in higher premiums paid initially by the contractor, reimbursed in part or in total by the owner/employer and finally by the consumer or the community at large. The second effect is a complicated insurance arrangement, which is reflected in multiplicity of insurance policies, possibly of different wordings, and from different insurers, each with a promise that might never be kept.

Duplication in the insurance cover takes place mainly due to any one or a combination of the following:

- In projects where more than one main contractor is employed on an individual construction project, the number of policies required multiply in proportion to the number of main contractors employed. The obvious overlap in cover occurs with respect to public liability insurance required from each of the contractors employed on the project.
- One of the problems created by the employment of nominated subcontractors in accordance with the earlier editions of the FIDIC and the ICE Conditions of Contract was related to the multiple productions of insurance policies. Clause 59(1)(a) of these conditions was interpreted by some contractors to mean that all nominated subcontractors must provide their own Contractors' All Risks and Public Liability policies. The overlaps, which resulted from this duplication of cover must not only have been very expensive but also intolerable. However, that clause was fortunately revised in the latest editions of these conditions of contract to read as follows:

> . . . the nominated Subcontractor will undertake towards the Contractor such obligations and liabilities as will enable the Contractor to discharge his own obligations and liabilities towards the Employer under the terms of the Contract . . .

The interpretation of this revision was taken to mean that the nominated subcontractor does not have to produce his own Contractors' All Risks and Public Liability policies but that he would be insured under the main contractor's policies.[2] However, despite that revision, some contractors continued to demand from nominated subcontractors similar insurances to those he is required to provide, resulting in the overlap.

It is worthwhile noting that should there be a single set of insurance policies, their wording should be adhered to by nominated subcontractors covered by such policies, as any ordinary subcontractor would have to do, in order to enable

---

2 *Engineering Law and the ICE Contracts*, 4th edition, Max W. Abrahamson, Applied Science Publishers Ltd., London. Mr Abrahamson argues on page 209 that 'the mere fact that the employer requires the main contractor to supply a bond does not automatically make it reasonable for the main contractor to require a bond from the subcontractor . . .' A similar analogy may be drawn in respect of the insurance policies.

the main contractor to discharge his own obligations and liabilities towards the owner/employer.

- Although it is more expensive in the overall context for each subcontractor to obtain his own insurances and therefore to duplicate the cover, it nevertheless might be worthwhile for the main contractor to shed part of his obligations under the contract to subcontractors. This trend, which became established through the employment of nominated subcontractors, was also required by some contractors from domestic subcontractors. The insurance overlap is then even more serious and more costly.

- In some international contracts, the contractor is obliged to insure locally with a national insurance company. Certain conditions may sometimes be imposed by the national insurance company which might cause the contractor to be concerned that, when claims arise, he is either unprotected or he is not in a flexible position to do what is needed to be done. In such a situation, the contractor is obliged to take out a second insurance cover outside the country in which he is working. For example, where strict currency controls exist in a certain jurisdiction, the contractor might have to obtain a second insurance policy outside the jurisdiction, should he require to purchase replacement parts manufactured abroad.

Gaps on the other hand are more numerous than overlaps when the insurance cover is executed in accordance with the conventional method. They also have a more adverse effect, sometimes a ruinous one, if not taken care of. They can be divided into three categories, as follows:

A Gaps created through the existence of uninsurable risks;
B Gaps created due to lack of insurance cover either through the manner in which the business is transacted; or due to the insured's lack of knowledge and awareness of the risks; or due to the Insured's wish to self-insure;
C Gaps due to the use of the conventional method of providing insurance protection.

A Gaps through uninsurable risks

These gaps must exist no matter what method is adopted to provide insurance as they are due to the very nature, concept and legal principles of insurance. They are:

1 *Fortuity*: A risk is uninsurable if its probability of occurrence within a specified period of insurance is 100%, i.e. the risk is certain to occur. Fortuity is therefore a prerequisite to insurance, see page 195. Unfortunately, however, it is not easy to define words such as fortuitous, foreseen or accidental in advance of an agreement, thus leaving an amount of uncertainty about the quality of the insurance contract. It is ironic for such uncertainty to surround an area of which an insured wants to be most certain, i.e. that he is in fact insured.

2 *Causation*: Insurance policies are written with conditions, exclusions and memoranda which define the cover and its limitation as far as possible. When a loss or damage occurs resulting in a claim, it is necessary to establish the cause in order to find out whether any of those conditions, exclusions and memoranda

apply. Depending on such application, the insurance cover may be valid or it may not. However, a gap occurs in cases where it is not possible to establish readily the cause with sufficient precision to satisfy an assessor of the validity of the insurance claim. The gap is created by the delay in settlement of the claim, especially if the delay is protracted and the claim is a large one.

3   *Disclosure of all relevant information and compliance with conditions*: As discussed in Chapters 6 and 7 above, the insurance agreement is a contract of utmost good faith and as the insured is in possession of all relevant facts when he makes his insurance proposal, the onus is on him to divulge all relevant information connected with the insurance matter, as it may influence the insurer either in accepting to insure or in assessing the applicable premium. If this is not done adequately, the insurer may deny the cover when a claim is made.

   Similarly, the insured must comply with the policy conditions during the currency of the insurance period as these are the conditions of his agreement with the insurer. Non-compliance with a relevant clause may leave the insured in breach of contract and thus without cover.

4   *Consequential loss*: Although some consequential losses are insurable, there are others that are not. To separate the insurable type from the uninsurable, one has to consider the possibility of assessing the amount of loss suffered. If the amount of loss can be assessed, such as is the case in loss of profits following delay in completion, the risk is insurable. Otherwise, it is not, see page 349 for definitions of consequential loss.

5   *Tangible loss or damage must occur*: The insurance policy covers any tangible loss or damage due to certain risks which form the basis of the insurance contract. However, if such a risk eventuates and is discovered prior to the occurrence of loss or damage, the onus is on the insured to take all necessary steps to rectify the situation at his own cost. For example, if a structural defect is discovered in a foundation prior to its causing collapse of the supported structure, no claim is entertained by the insurer for rectifying the defect even if that defect or its consequence is an insured risk. But, if the defect is not discovered and the structure collapses, the insured is covered in respect of the loss or damage sustained by the insured element. This principle forms one of the major obstacles in providing a meaningful cover against latent defects in construction projects.

6   *Defective workmanship and material*: As discussed in Chapter 12, insurance is not, and should not be, extended to cover a part which is defective in its workmanship or in the material used in its construction. It must be noted however that this limitation does not apply to consequent damage of parts properly constructed or to manufactured parts in respect of which a manufacturer's guarantee can apply.

7   *Political risks and risks cover by government*: In some countries, certain risks are covered by the state and are therefore not the subject of insurance. For example, the risk of nuclear reaction is covered in the United Kingdom by its government. In the strict sense, this type of risk does not form a gap. However, where such risks are not covered by government or where government cannot provide such cover, as is the case in respect of political risks, a gap would exist.

B   Gaps due to lack of cover, through either insurance practice or insured's wish (if not lack of knowledge)

The gaps classified under this category are insurable in principle but are created through lack of insurance cover due to:

- The insurance practice in a particular part of the world;
- The insured's wish not to insure; and
- The insured's lack of awareness of the presence of the gap or of the insurability of the risks it represents.

The gaps, therefore, vary from one part of the world to another and from one insured to another. They can, however, be summarised as follows:

1   *The Wording of the policy and its exclusions*: As described earlier in Chapter 12, the wording of an insurance policy transcribes the exact nature of the cover provided and its extent. The exclusion section of the policy defines the hazards and the risks that are not insured. Meticulous checking of the policy is therefore required if one is to ensure that an unintended gap is not incorporated in the insurance cover.

2   *The Sum Insured*: The sum insured provides an upper limit of indemnity in respect of a particular insurance policy. As discussed on page 344 it has many possible meanings and values. A gap can therefore be easily created if the wrong sum insured is adopted when the insurance policy is negotiated. The difference between choosing the appropriate sum insured and another can be the difference between a completed contract and a frustrated one.

3   *Limit of indemnity for third party liability*: The liability of the owner and the contractor towards third parties is unlimited in value. Insurers, however, when providing indemnity policies insist on placing a limit in respect of their own liability under these policies. A gap is therefore created once that limit is set. The magnitude of the gap is obviously related to the amount of cover provided under these policies.

4   *Owner's/employer's risks*: In all construction contracts, there are hazards and risks which must be allocated to the owner/employer as he is the party who initiated the project, provided the site and decided the brief. These risks can relate to the project itself, to the other parties involved in construction or to third parties. For example, the excepted risks in the General Conditions of Contract form the owner's risks in terms of the project, and so do the limitations stated in the liability clause of the Conditions of Contract which relate to the risk of personal injury of other parties and loss or damage to their property. Some of these risks are insurable but in most situations they are not, in which case a gap is created and the owner will find that in the event of loss or damage he must pay for the cost of repair.

5   *Consequential losses*: Consequential losses are normally excluded from the insurance cover provided by the usual forms of Contractors' All Risks policy. Such losses are mainly economic in nature and include loss of revenue as a result of non-completion and higher completion costs due to loss or damage causing delay in construction, etc. Consequential losses, however, may be insurable under

a special type of policy, if required. However, if these risks are not covered, a gap would be created.

6   *Contractor's design*: If the contractor is expected to design any part of the project or the temporary works associated with its construction, a design insurance cover is required to supplement the normal Contractors' All Risks cover. Usually, such an insurance is transacted through a professional indemnity type of policy which becomes essential for the contractor if a gap is to be avoided.

7   *The design cover*: As described in Chapter 13, the design cover is usually obtained through a professional indemnity insurance policy which has developed with certain peculiarities. It is, for example, a claims-made insurance policy, see page 365, and therefore if the cover is not maintained, an insurance gap would be created. The cover is also restricted in its limit of indemnity to sums much below the exposure usually attached to the designer. To make matters even more critical, claim settlement usually takes a long period of time during which other claims related to the particular year of insurance might arise, thus reducing the available sum insured or exhausting it. All these aspects of the professional indemnity insurance policy and others, more comprehensively discussed in Chapter 13, create their own gaps. But, perhaps the most important gap is the difference between the cover provided by the professional indemnity insurance policy (which is damage due to faulty design) and that required to cover damage due to defective design. The latter term, defective design, is used to express inadequate but not necessarily faulty design, and therefore, it includes lack of knowledge due to the state of the art of engineering and other related fields. The fact that innovation plays an extremely important role in construction, with new materials and technology appearing in a perpetual surge, increases the gap even further.

8   *Indeterminate phraseology*: The use of words or phrases, either in contract conditions or in insurance policies, which are indeterminate in meaning could be conducive to disputes with insurers when they are called upon to fulfil their promise to pay in case of a claim. Such words as 'foreseen' or 'experienced' or 'proper' are sometimes necessary for drafting a general statement, which is intended to be adaptable to a number of situations and circumstances, but they could result in a gap if more than one meaning were to be construed from them. Similarly, the use of words or terms that portray a condition rarely attainable in practice or in reality might only provide fertile ground for dispute and possibilities for denial of liability.

One might also come across words of similar meanings used incorrectly such as faulty instead of defective, defect as a cause against defect as effect, maintenance against repair, error instead of mistake, supervise instead of inspect, cost instead of expense and responsibility instead of liability. Perhaps what is needed is a recognised definition for all of these terms and the subtle difference between them.

9   *Some political risks*: A number of political risks such as riot and strike can be insured in certain parts of the world, especially if political stability exists at the time of negotiating the insurance cover. However, other political risks could be extremely difficult to insure, as the consequences could be disastrous for both insured and insurer. A terrible example of such risks is terrorism, which has developed in recent times to affect every human being. The incorrect use of that term is another risk.

C Gaps due to the use of the conventional method of providing insurance protection

Two main gaps are created by the use of the conventional method of providing insurance cover and these are:

1 *The design gap*: The conventional method of insuring the design risk under a separate policy with a different Insurer to that providing the cover for the Contractors' All Risks insurance results in a gap in the following circumstances:

- where both insurers disclaim liability in respect of an event expected to be covered by one or the other of the insurance policies;
- where the extent of cover provided by the professional indemnity insurance policy is restricted to lack of care, error or omission which is narrower than the exposure to which the designer is subjected;
- where the limit of indemnity under the professional indemnity insurance policy is lower than the cost of repair or reinstatement;
- where the professional indemnity insurer refuses to pay a claim by an owner/employer against an insured design professional and uses tactics against him in such a way as to make him accept a much lower sum than that to which he is entitled. An example of these tactical moves is to deny the validity of the professional indemnity policy. alleging that the insured had invalidated his policy by not adhering to one of the legal principles of insurance, such as the disclosure of all material facts in the proposal form. At the same time, the insurer offers the claimant owner/employer an *ex gratia* payment much lower than that to which he is entitled. As the owner/employer is not in a position to ascertain the validity or otherwise of the policy, he is faced with the situation of fighting his case against the design professional who in turn has to join the insurer in the proceedings, which might prove unsuccessful, or accept the *ex gratia* payment.

2 *Period of insurance*: The cover in respect of the Contractors' All Risks insurance policy ends when the owner/employer takes over the project at the commencement of the Defects Notification Period and other than the restricted cover available during that period (when and if the contractor returns to the site or remains there to comply with the terms of the agreement), the insurance cover ceases. The exposure to loss or damage remains.

For public liability, the conventional method only requires the contractor to insure during the construction period and thus one needs to check that the gap created during the Defects Notification Period, when and if the contractor returns to site, is covered by another policy.

The cover for design risks, as discussed earlier, is maintained through a claims-made policy and therefore can only be maintained if the designer continues to renew the insurance policy annually after completion of the project. It is clear, however, that the owner has no control over the validity of such a policy or the terms under which it is renewed.

Recent developments in construction, insurance and law

Recent developments in the field of construction, as well as in other fields related to construction, have produced unlimited opportunities for disagreement and controversy between the parties involved in a construction contract. These developments have presented their own peculiar hazards and risks on construction. They stem from the following factors:

- A breakdown in the relationship of trust between the owner/employer, the professional adviser and the contractor due mainly to financial considerations. This breakdown is encouraged by the recent litigious tendencies of society combined with a number of legal pressures to which the construction industry is exposed and which proved to be insensitive to its special and peculiar characteristics.
- Society's wish for an increase in performance by the construction industry requiring it to match that exhibited by pure science or by technology.
- The continued lack of action by the construction industry to clarify its role and the role of each of the parties; to define their respective responsibilities toward each other and towards others outside the particular project or scheme in which they are involved. This situation has led to a recent move towards a new idea of 'partnering' between the parties to a contract. In a legal context, partnering may constitute no more than an expression of intent between two parties to behave in accordance with a set of agreed principles.[3] There are a number of objectives that may be achieved by the parties involved in a construction contract through partnering. These could be briefly summarised as follows and should perhaps be the subject of any partnering agreement executed on a construction project in the future:

    - Management and co-ordination;
    - Early warning/joint problem-solving;
    - Dealing with variations;
    - Monitoring of progress;
    - Snagging and repair work; and
    - Incentives.

- A haphazard approach to the identification, mitigation and allocation of the increasing number of risks to which the construction industry has become exposed.
- Society's appetite for a decreasing level of risk and an increasing level of indemnity from any source that happens to be available. In some instances, the basic principles of insurance have been sacrificed in order to provide a protective cover to an otherwise uninsurable situation.
- The willingness of some owners and decision-makers to accept the involvement of inexpert practitioners in plans and schemes of operation, which may initially be cheaper to execute but ultimately, having unknowingly sacrificed quality, more expensive in terms of life-time cost.

---

3 'The Contractual Basis for Partnering and Alliancing', an information sheet published by Freshfields Bruckhaus Deringer, International lawyers, November 2001.

The performance of the conventional method of insuring construction projects has become clouded with these factors. When it failed to provide the required performance, an alternative was sought.

Alternative methods of insuring

As explained in earlier chapters, the conventional method of providing insurance protection to construction projects, as stated in the standard forms of contract, require that such insurances be taken out by the contractor in the form of three insurance policies. These are:

- Contractors' All Risks or Erection All Risks insurance policy,
- Third party liability insurance policy; and
- Employer's liability insurance policy.

These policies mainly cover the contractors' risks but leave the owner/employer free to take out his own insurance for whatever risks he considers should be insured. However, although this method has worked well in practice, many gaps and sometimes overlaps exist because of:

- First, due to lack of either knowledge or appreciation of the consequences, the owner/employer does not always examine the necessity of insuring the risks to which he is exposed;
- Second, although it is generally required in the standard forms that the owner/employer is named as joint insured in the insurance policies, the cover is not intended to extend to his exclusive risks;
- Third, these insurance policies leave untapped a large number of available covers (which are normally required by a contractor but which are either not essential or not capable of being retained by the contractor without any detrimental effect on the project). An exhaustive but not complete list of these insurance covers is given in the following section below.
- Fourth, as the size, the technical or the legal and administrative complexity of a construction project increases, the matrix of risks attached to its execution becomes larger and interdependent. Further, as the matrix of risk becomes larger, so do the insurance requirements and covers for such projects.

Thus, gaps and insurance overlaps develop. As explained earlier, the problem with gaps, if and when they occur, is that the insurance cover would not operate when certain events and risks materialise, in which case, the owner/employer and the contractor could face disastrous consequences. An example of such consequences is what occurred on the Tarbela Dam project in the mid 1970s.[4] The problem with

---

4 Tarbela Dam, which is on the Indus River in Pakistan, suffered damage during its construction due to an accident, the cause of which was contested by the insurers. The cost of repair was estimated at the time, 1978, as being in the region of $50 million. For more details, see 'Insuring Civil Engineering Projects – Some Notes and reference texts for the FIDIC Seminar on Project Insurance', prepared by Charles T. Bright and Eric de Saventhem, London, 19-20 June 1978; and 'Insurance for Large Civil Engineering Projects', published by FIDIC, 1981 updated in 1997.

overlaps is the duplication in cover and therefore the additional unnecessary cost of the insurance transacted. Furthermore, for the larger type of project with considerable investments and where usually there are many subcontractors and suppliers, there is multiplicity of insurance policies. Therefore, the search for a better method of providing insurance that would generate a more certain cover should risks eventuate is continually being pursued and only a few solutions have been proposed. Of course, these solutions have their own problems, advantages and disadvantages and one can be sure that, until they are tried and tested, the list will not be an exhaustive one. The alternative methods of insuring that have been proposed can be enumerated as follows:

- The Difference-in-Conditions additional insurance;
- Project insurance, sometimes referred to as 'Principal-controlled insurance' or 'Wrap-up insurance';
- National schemes for the provision of insurance.

The Difference-in-Conditions additional insurance, 'DIC'

This process basically supplements the conventional method of providing insurance by simply extending the standard insurance cover provided, firstly in the three main insurance policies (CAR, PL, EL) required in the contract between the owner/employer and contractor; and secondly, in the contract between the owner/employer and professional, (PI). This extension to the standard insurance policies is generally done to pick up any shortfall in insurance and to close the gaps already mentioned earlier in this chapter. In order to carry out this task successfully, the services of a professional person or organisation experienced in this field of insurance, or preferably in the three related disciplines of construction, insurance, and law should be secured. This person or organisation acts as a risk manager to:

- Analyse the particular project and identify the risks peculiar to it;
- Draw up plans for mitigating the risks and allocating those remaining to the various parties involved on a sound and acceptable basis;
- Allocate the responsibilities and liabilities which emanate from these risks, if and when they eventuate;
- Investigate and devise a plan of insurance for the remaining risks where most, if not all, the gaps and overlaps are dealt with in an acceptable manner. The acceptable manner is not necessarily through insurance. Self-retention of the exposure to some risks should be considered as an alternative, either because it would be cheaper in the end or better from an overall point of view.

In general, the Difference-in-Conditions insurance should result in the same insurers being involved in both the conventional policies and the extended coverage, including any additional policies that might need to be issued: otherwise similar or other gaps would persist. It is also a method whereby the conventional policies are looked upon as a base layer with the limits of indemnity provided therein being used as an excess in the supplementary policies. This produces what is known in the insurance market as an excess-of-loss policy. For example, if the contractor is in possession of an annual public liability policy with a limit of indemnity of $1 million and the required limit of indemnity for the public liability cover on a particular project is $5 million, then the additional policy required would be issued for a limit of indemnity of $5 million with an excess of $1 million.

The difference-in-conditions method of insuring might produce a number of endorsements or additional policies and insurance arrangements, such as:

- Public liability insurance to higher limits than stipulated in the contract;
- Professional indemnity insurance cover for contractors' design work;
- Professional indemnity insurance for all professional services on an individual project to a specified limit of indemnity and for a period beyond the completion of the project;
- Non-negligent liability and damage insurance which extends the traditional cover based on negligent acts to non-negligent acts which might cause damage to the project, to adjacent property and/or injury to third parties;
- Marine and transport insurance;
- Air freight;
- Group personal accident, travel, medical, etc.;
- Credit risk;
- Unfair termination of contract;
- Unfair call of bond;
- The employer's risk of delay;
- Currency risk;
- Manufacturer's risks;
- Confiscation of construction plant;
- Expropriation of overseas assets risk;
- Special risks;
- Insurances after substantial completion;
- Decennial liability insurance cover in respect of latent defects in the individual project;
- Loss of profits insurance to cover economic losses;
- Interruption insurance, where the project includes operation; and
- Products liability insurance, where the project includes a manufacturing process.

Project insurance, Principal-controlled insurance or Wrap-up insurance

The title 'Project insurance', or as it is sometimes called 'Principal-controlled insurance' or 'Wrap-up insurance', conveys different meanings to different insurance markets and specialists. Some confer that title on the conventional method of providing insurance when the owner/employer arranges it. Others are inclined to reserve it for insurance arrangements by the owner/employer that include some or all of the Difference-in-Conditions insurance covers referred to above. In some insurance markets, it simply means a single professional indemnity insurance policy taken out by the owner/employer to cover all those providing professional service on an individual project for the duration of construction, testing, commissioning and subsequently the Defects Notification Period. In some cases, the cover is extended to a period of up to five years after the completion of the project, in which case it is more likely to be referred to as a policy. Certainly, the all-inclusive policy is more correctly given the title of wrap-up rather than project insurance.

Various forms of this type of insurance cover have been issued in the Netherlands, the United States and Canada as far back as 1960; for example the James Bay Project in 1962 in Canada. More recently, however, it has been transacted for major projects world-wide, such as the Channel Tunnel between England and France;[5] the Thames Barrier Scheme in England; the Oresund Link between Denmark and Sweden; the Heathrow Link and the Jubilee Underground Line in London; the ITAIPU-HEP dam in Brazil and Paraguay; oil field development and pipelines in Colombia; sewage outfall tunnels in Australia; and the Port Tunnel in Ireland. However, the details of the cover provided differed greatly from one project to another.

It is therefore important to begin this section by defining in precise terms the insurance cover being discussed and the intended meaning of Project Insurance, Principal-controlled insurance or Wrap-up insurance. In principle, such insurance is a combined insurance policy incorporating the following:

- The cover provided by the two conventional insurance policies normally issued to the contractor, Contractors' All Risks and Public Liability leaving out the Employer's Liability which is as a rule legally compulsory and dependent to a large extent on the contractor's payroll.
- A cover in respect of any of the insurable risks, which the owner/employer might wish to have insured, and which remain uninsured if the traditional method of insuring is chosen. This might include any of the risks mentioned in the Difference-in-Conditions, as described earlier in this chapter (it is to be noted again that if professional indemnity insurance is added, the policy more often than not would be referred to as the wrap-up type).
- The provision that the policy is issued in the joint names of the owner/employer, those providing professional services, all contractors and subcontractors involved on a particular project. The cover should include all temporary works, equipment and ancillary buildings used by the contractor in the construction process.
- A clause entitling either the owner/employer or the contractor to assume control over and to take out the insurance cover. It is thus not essential for a project insurance policy to have the owner/employer in control of its management. However, if he is, it is essential that he includes, in the invitation to tender, a specimen of the policy he has negotiated with the insurer so as to enable the contractor to include in his tender for any additional insurances he may wish to have. In particular, it is the contractor who knows best the risks against which he should insure and the contractor's insurance policy would then be tailored to his needs and the activities that he undertakes.

It is also assumed, however, that the usual professional indemnity insurance policy taken out by the design professional is kept in force alongside project insurance,

---

5 Details of the policy issued for the Channel Tunnel is given in *Insurance Under the JCT Forms*, by Frank Eaglestone, 2nd edition, Blackwell Science Ltd., 1996, Appendix 7.

but for a modest limit of indemnity. This is due to the fact that the professional indemnity cover must remain in force to protect the professional's other work for which project insurance is not arranged. Furthermore, as such cover is of the claims-made type, it would be necessary to maintain it in case of claims that might be made after the completion of the project. In some cases, the insurer of project insurance retains the right of subrogation against the professionals involved but this right is restricted to the limit of indemnity under the Professional Indemnity cover he possesses.

Frequently, the owner/employer undertaking a one-off construction project would lack the know-how in the insurance market, is not experienced in the purchase of insurance cover and would not be equipped to deal with the various intricacies of investigating the insurance market, placing the insurance business and maintaining an expert overview during the currency of insurance. The contractor, on the other hand, is generally in constant touch with the market and would have sufficient detailed knowledge of such factors as the method of construction to be employed and the temporary works to be used. Thus, he would be in a better position to obtain competitive rates in respect of the risks to be insured. In such a situation, the contractor should be entrusted with arranging the project insurance, unless the owner/employer feels justified in taking over this task.

In this connection, it must be remembered that the contractor's insurance brokers do not like to see the insurance transactions slipping through their hands into the brokers employed by the owner/employer. As a result, it is claimed that usually the contractor has a long-standing relationship with his insurers and with the loss adjusters used by them, and thus it would be more appropriate for him to arrange the necessary cover.

Advocates of project or wrap-up insurance give it credit for a number of advantages, which can in reality be attained even in the conventional method of providing insurance, if dealt with correctly. For example, it is claimed that the owner/employer has greater control over the operation of the insurance policy, its cancellation, renewal, effecting changes in its conditions, or in settlement of claims. However, such control can easily be exercised in the conventional method through joint insurance. It is also claimed that the owner/employer has the choice of deciding which risks are to be insured and which are to be retained. The decision in respect of allocation of risks can be made by the owner/employer at the pre-tender stage no matter what insurance method is adopted. In fact, if the owner/employer takes the matter of insurance seriously at the pre-tender stage, he can make a decision in respect of all features of insurance, irrespective of who takes out the insurance. The features include the value of the sum insured; how it is to be adjusted during the period of construction; the limit of indemnity for public liability; the maximum level of excess, which should be applied; and the period of insurance in respect of various parts of the project.

On the other hand, the use of project or wrap-up insurance is questioned for reasons that do not necessarily pose a problem. For example, in the case of the owner/employer taking out insurance, there are some who claim that because the identity of the contractor is not known at the time of negotiating the insurance cover, the premium rating cannot be calculated accurately. This is true, but there is nothing to stop the owner/employer and the insurer agreeing on the particulars of the insurance cover leaving the adjustment of rating to be carried out after the contract is awarded. Similarly, there is the criticism that in owner-controlled project insurance, the contractor is restricted to the terms of the policy negotiated by the owner and therefore cannot be responsible for any inadequacy which may be found in its terms. To resolve this issue, the contractor is given the chance to price, in his tender, for any additional insurance cover he may require provided he is informed of the terms negotiated by the owner. In pricing such additional insurance, the contractor is expected to give details of what he is seeking to insure.

Despite all the points discussed above, there is agreement that there are both advantages and disadvantages attached to project insurance. These are:

A   *Advantages:*

- The principal advantage is the inclusion of all the risks required to be covered under one single policy, including the risk of defective design in the case of a wrap-up policy. A single policy would mean a single point of indemnity and therefore, when risks eventuate, there should not be difficulties or waste of time in apportioning blame between the various parties involved. The inclusion of the design cover, in particular, overcomes the problem of delay in settlement of claims: such delay normally occurs when the cause of loss or damage and the allocation of blame has to be established. It also prevents any arguments between the parties involved in the project as to who is responsible and liable for the loss or damage, if the subrogation rights of the insurer are waived. The cover against defective design can be either a complete one to a certain maximum monetary limit or a partial cover. The latter refers to a defective design cover that excludes the defect itself and thus leaves the defect itself to be covered by the professional indemnity insurance policy, which would then have to be issued separately by each professional involved in the project.

- If the design cover is included in the cover, the design professional would not be isolated from his client, the owner/employer, by the insurer taking over the conduct of the claim, a procedure that is usually applied when the insurer is notified of a claim being made under the policy. In project insurance, the whole team is insured.

- In large projects where there is more than one main contractor, often of different disciplines, the vetting and matching of the various insurance policies produced from different insurers and reinsurers, possibly of different nationalities could cause delay at a time when all concerned want the work on site to commence as soon as possible after the tenders are awarded. In such a situation, where damage or loss occurs involving the different insurers, the settlement of claims can be extremely complicated, particularly if insurers do not agree on a joint assessment of the claim.

- It is claimed by some specialists that there would be a saving in premium if a single project insurance policy was issued for the whole project and for all the contractors working on its construction. The basis for such a claim is the established fact that a large proportion of the premium paid is in respect of the administrative costs and overheads of insurance brokers, insurers and reinsurers.

However, it should be noted that the savings involved cannot usually be estimated or ascertained, since to do so would entail going through the process of both methods of insuring: the traditional method with all the added benefits and also the project insurance method. To the author's knowledge, at least in one such situation, it was found that project insurance was too expensive to obtain and the owner/employer reverted back to the traditional method.

- In project insurance, the gap(s) created by some insurers in separating the owner/employer's risks from the contractor's risks could be eliminated without having to obtain a second set of policies.

- Where wrap-up or project insurance is transacted, it would be possible to establish a standard claim settlement procedure using a single loss adjuster, approved and trusted, by both the insured and the insurer. Furthermore, such an arrangement could be developed into a safety and loss prevention programme for the whole project.

B *Disadvantages:*

- The main disadvantage in project insurance is that the cover ceases once the project is completed, unless the policy duration is extended beyond the date of completion of the project. If the insurance cover is not extended beyond the completion date, one might find that when hazards and risks eventuate and failures occur after the date of completion, the project is not protected. For this reason, amongst others, professional indemnity insurance is usually required by a design professional whether project insurance is taken out or not, so that at least defective design could be covered. However, it must be noted that after the project is completed, there is usually no control over whether or not the professional is maintaining his professional indemnity cover and, if he is, whether he is maintaining it to the required level.

  If the insurer of both the wrap-up and the professional indemnity policies is the same, some return of premium might be arranged; otherwise a cost overlap occurs unless of course the wrap-up insurance in respect of design is obtained for amounts in excess of the limit of indemnity provided by the professional indemnity cover.

- There could also be some duplication of professional indemnity insurance cover, especially if the insurer's subrogation rights are waived.

- In order to include the design risk in the insurance cover provided by project insurance, the insurer usually requires that a full disclosure of the designer's claim record be made. This might lead to the insurance market taking over control of the design profession and be able to control who might or might not be employed within the construction industry. Similarly, at the time of arranging the insurance cover in owner-controlled project insurance, the identity of the contractor(s) is not known and, therefore, premium rating would present difficulties. Some insurers might over-load the premium in anticipation of all classes of contractor being employed whilst others would require control over the rating once the identity of the contractor(s) becomes known.

- It is claimed, that in project insurance, both the owner/employer and the contractor become more claim-conscious and willing to claim against the designer.

This of course would naturally result in the deterioration of the relationship between the three parties involved in the contract.

- In owner-controlled project insurance policies, if the claim experience gained by the insurer is recorded against the owner/employer, then the contractor becomes less concerned with preserving a clean record.
- In owner-controlled project insurance policies, the owner/employer should carry out a thorough analysis of the hazards and the risks to which the project is exposed. For this purpose, he must employ a specialist professional who is knowledgeable in both design and construction in the type of project concerned. Although this is an advantage by itself, there is a cost involved, which could be considerable for large projects, and this additional cost element might be considered as a disadvantage.
- Where international contracts are concerned and in some developing countries where it is required that insurance is placed locally with national insurance companies, and very often sufficient experience in project insurance is unavailable.
- There could be some duplication of cover, as contractors might need to continue part of their traditionally held insurance covers for latent defects and for third party claims.
- Where insurers insist on vetting the design, there could be a move towards conservative design.

Procurement of the wrap-up insurance policy

The owner/employer or the contractor may take out this type of policy policy. If there were more than one contractor, a managing contractor would have to be appointed. Such a contractor can be either a general contractor or employed for the purpose of managing the construction of the project. If the wrap-up insurance policy is taken out by the owner/employer, it would be necessary for a specialist in construction insurance to draft, in broad terms, the provisions of the wrap-up insurance policy, having first carried out a proper risk management exercise, including identification, assessment and evaluation of the hazards and risks inherent in the particular project and their probable consequences. Such analysis should provide an informed basis upon which the owner/employer can decide on the sharing and apportionment of responsibility for the risks between him and the contractor(s) and the extent to which he or the contractor(s) should be insured in respect of these risks under the provisions of the relevant contract(s).

The precise terms of the wrap-up insurance policy can then be determined through a procurement procedure on a case-by-case basis. When the terms of the wrap-up insurance policy are determined, they should be incorporated in the bidding documents for the tenderers' attention and information. The tenderers should be asked to study the terms of the wrap-up policy being provided by the owner/employer and to indicate, in their tenders, whether or not any additional insurance policies are required by them for the proper construction, completion and operation of the project. If an additional cover is required, then they should indicate the cost of such proposed insurances.

It would be advantageous to set up a risk management programme for the whole project to consider the effect of one contract on the others. Similarly, there would be an advantage in setting up a claims-management programme with a single loss-adjusting organisation that would act with an independent and impartial attitude towards both insured and insurer. The wrap-up insurance policy should be in the joint names of both owner/employer and contractor(s), and the risk management programme should be implemented prior to completion of the terms of the wrap-up insurance policy. Furthermore, as management of insurance claims is one of the main advantages of a wrap-up policy, claims procedure and claims prevention programmes should be implemented by the single loss-adjusting organisation selected.

Procurement of a wrap-up insurance policy by an owner/employer necessitates the adoption of a careful procedure, which would depend on the type and complexity of the project, perhaps on the following lines:

The employer should:

- Employ an experienced and knowledgeable insurance and risk management expert or firm,[6] to prepare, with the help of the design team(s), hazard and risk analysis and a detailed specification for the wrap-up insurance policy. The specification for the wrap-up insurance policy should be worded in broad terms so as to cover the minimum number of insurable inherent risks in the project. The cost of this work could be financed as part of the cost of the works.

- Invite suitable firms with insurance and risk management expertise to pre-qualify for the position of a Risk Management Firm by making presentations to the employer. The Presentation should include, amongst other things, the following:

  - The extent of past experience in this type of work;
  - The method in which it is proposed to carry out the work;
  - The personnel to be involved in the work;
  - The expected method of payment;
  - The staging of the work; and
  - The reasons as to why they should be appointed to act.

- Select one such firm to act as risk management firm, based on the quality of the service intended to be provided. It should be made clear to the risk management firm that their remuneration should only be by way of payment from the employer and that no commission is to be charged or accepted by them from any source whatsoever. In this connection, attention is drawn to the fact that this is different from the traditional method of remuneration adopted by insurance brokers who place the required insurances in the market and earn their remuneration through a commission from the insurers. It is also important to ensure that the broker will not act, on the same project, for any of the contractors who have been or will be employed for the execution of the project.

---

6   The expert could be either an individual or a firm of satisfactory expertise.

- Seek bids through the risk management firm, once appointed, from a short list of insurers, which should be approved by the employer, for the provision of the wrap-up insurance cover.
- Ask the risk management firm to make a recommendation for the terms and provisions of a wrap-up policy best suited for the particular project on the basis of the bids received. Once the terms and provisions of the final wrap-up policy are completed, return to the pre-selected firms for a competitive quotation of the required premium and the stages of payment for the finally approved wrap-up policy. Once approved, the risk management firm can proceed with the placement of the wrap-up policy and liaise with the expert and with the design team(s) in all of its tasks. The risk management firm should also be asked to arrange for quarterly reports to be submitted throughout the duration of the construction of the project, and until its completion. After completion, an agreed procedure should be attempted for the submission of reports and for maintaining control on the events during the Defects Notification Period and beyond.

It should be noted that the bidding documents for all the main contractors should include a copy of the hazard and risk analysis and the procedure outlined above in the procurement of the insurance cover together with the policy, as finally obtained.

National insurance schemes

Although the owner pays at the beginning for taking out the policies, the cost of insurance must, in the end, be borne by the consumer or society in general since insurers are in a trading market where they increase their premiums when they face losses and reduce them when profits are lucrative and when the business is aggressively sought. In construction insurance, the losses and profits are not easily established due to the time lag between the negotiation of the terms of the policy and premium assessment at one end and, at the other, the date at which the project is completed and the insured risks cease to be the concern of the insurer. At that stage, the premium charged plus investment profits, less administrative costs and overheads, less claims paid, becomes the earned income of the insurer. If this is in the negative on an overall basis, the insurer knows that he must raise his future premium rates if he is to survive and remain in business. The picture is clear in some branches of insurance, such as motor insurance, where losses can be easily and quickly assessed and where a decision is taken annually on the level of premium that should be charged to avoid incurring loss. In construction insurance, where a long period of time is required for projects to be completed and for losses to be assessed and settled, the picture is more difficult to grasp and the requirement for premium adjustment takes a long time to become evident.

In view of this, and if the consumer ultimately pays, a better alternative for insuring construction projects may be through national insurance schemes, either compulsorily through legislative enactments, or voluntarily through the construction industry itself.

If it is arranged in a compulsory fashion, it requires the necessary legal enactments and would be run in such a way that a premium is paid to a fund by the owner/employer when a contract is awarded to a contractor. Similarly, if it were arranged voluntarily, the premium would be paid into a special fund held by an organisation formed by the construction industry. These funds would provide cover for liabilities in negligence of any of the parties involved in the design and construction during the construction period and for a predetermined period of time after completion. The main attraction of such a scheme is to protect the consumer against lengthy and expensive legal actions, as some figures suggest that for every pound spent in rectifying a defect, four pounds is spent on legal fees. Registration of design professionals and contractors would be necessary under such a scheme where the track record of all involved assumes major importance for future work.

Appendix A

SPECIMEN

Policy No............................

QUESTIONNAIRE AND PROPOSAL FOR CONTRACTORS' ALL RISKS INSURANCE

The Insurers undertake to deal with this information in strict confidence

1. (a) Name and address of Principal (b) Name(s) and addresses of Contractor(s) (c) Name(s) and addresses of Sub-contractor(s) (d) Name and address of Consulting Engineer.	
2. Location of Site.	
3. Title and description of contract (if project consists of several sections, specify section(s) to be insured). Please attach necessary informative documents and plans.	
4. (a) Contract value, including materials supplied by principal (please attach Specification and Schedule of Prices). (b) Replacement value of construction equipment, e.g. scaffolding, auxiliary bridges, timbering and casing, tools and tackle, power generating sets, water supply and sewage installations, temporary buildings, fuel, etc. (c) Replacement value of construction machinery (please enclose list of the various items), NB: To be answered only if damage to construction machinery is to be covered.	
5. Work to be carried out by sub-contractors (if to be included).	

6. (a) Estimated construction period .... months from.................... to....................	
(b) Subsequent maintenance period .... months from.................... to.................... (To be answered only if to be included).	
7. Please give full details (as far as applicable) regarding (a) earthquake hazard (b) geological conditions, including subsoil (c) ground water level (d) name of, and distance to, nearest river, lake, sea, etc. (e) levels of such river, lake or sea (i) low water (ii) mean water (iii) highest level ever recorded (f) level of deepest excavation	
8. Are any existing buildings affected by the work to be carried out under the contract, e.g. by extensions, changes, underpinning, etc. (please forward details)	
9. Are extra changes for overtime, night- work, work on public holidays, express freight, etc. to be included?	
10. Is Public Liability to be included? If so, what Limits of Indemnity are required (any one event, aggregate limit for all claims)?	

We hereby declare that the statements made by us in this Questionnaire are true to the best of our knowledge and belief and we hereby agree that this Questionnaire shall form the basis and be part of the Policy.

Dated at .... this.................day of .... 20.............

Signature ....

399

Appendix B

SPECIMEN

Policy No......................

CONTRACTORS' ALL RISKS POLICY

Whereas the Insured named in the Schedule hereto has made to the

(hereinafter called 'the Insurers') a written proposal by completing a Questionnaire which together with any other statements made in writing by the Insured for the purpose of this Policy is deemed to be incorporated herein,

Now this Policy of Insurance witnesseth that in consideration of the Insured having paid to the Insurers the premium mentioned in the Schedule and subject to the exclusions, provisions and conditions contained herein or endorsed hereon the Insurers will indemnify the Insured in the manner and to the extent hereinafter provided.

GENERAL EXCLUSIONS

The Insurers will not indemnify the Insured in respect of loss, damage or liability directly or indirectly caused by or arising out of

(a) war, invasion, act of foreign enemy, hostilities (whether war be declared or not), civil war, rebellion, revolution, insurrection, mutiny, riot, strike, lock-out, civil commotion, military or usurped power, or malicious persons acting on behalf of or in connection with any political organisation, confiscation, commandeering, requisition or destruction of or damage to property by order of the government de jure or de facto or by any public authority;

(b) nuclear reaction, nuclear radiation or radioactive contamination;

(c) wilful act or wilful negligence of the Insured

(d) cessation of work whether total or partial.

In any action, suit or other proceeding where the Insurers allege that by reason of the provisions of Exclusion (a) above any loss, destruction, damage or liability is not covered by this insurance the burden of proving that such loss, destruction, damage or liability is covered shall be upon the Insured.

PERIOD OF COVER

(a) **Construction Period**
     The liability of the Insurers shall commence, notwithstanding any date to the contrary specified in the Schedule, with the unloading of the property specified in the Schedule at the Contract Site and shall expire on the data specified in the Schedule.

     The Insurer's liability expires also for parts of the insured contract works taken over or put into service by the Principal prior to the expiry date specified in the Policy whichever shall be earlier.

(b) **Maintenance Period**
     If a maintenance period is specified in the Schedule, the liability of the Insurers during this period shall be limited to any loss or damage caused by the insured Contractor(s) in the course of the operations carried out for the purpose of complying with the obligations under the Maintenance Clause of the contract.

GENERAL CONDITIONS

1. The due observance and fulfilment of the Terms of this Policy in so far as they relate to anything to be done or complied with by the Insured shall be a condition precedent to any liability of the Insurers to make any payment under this Policy.

2. The Schedule and the Section(s) shall be deemed to be incorporated in and form part of this Policy and the expression 'this Policy' wherever used in this contract shall be read as including the Schedule and Section(s). Any word or expression to which a specific meaning has been attached in any part of this Policy or of the Schedule or of the Section(s) shall bear such meaning wherever it may appear.

3. The Insured shall take all reasonable precautions to prevent loss, damage or liability and to comply with sound engineering practice, statutory requirements and manufacturer's recommendations and maintain in efficient condition all contract works, construction plant, equipment and construction machinery insured by this Policy.

4. The Insured shall immediately notify the Insurers in writing of any material change in the risk insured hereunder; the scope of cover and/or the premium shall, if necessary, be adjusted accordingly.

5. Representatives of the Insurers shall at any reasonable time have access to the site or premises and to all pertinent data, documents, drawings, etc. and shall have the right to inspect any property insured.

6. In the event of any occurrence which might give rise to a claim under this Policy, the Insured shall:
   (a) immediately notify the Insurers by telephone or telegram as well as in writing;
   (b) take all steps within his power to minimise the extent of the loss or damage;
   (c) preserve the damaged parts and make them available for inspection by a representative or surveyor of the Insurers;
   (d) furnish all such information and documentary evidence as the Insurers may require;
   (e) inform the police authorities in case of loss or damage due to theft or burglary.

   The Insurers shall not in any case be liable for loss, damage or liability of which no notice has been received by the Insurers within 14 days of its occurrence.

   Upon notification being given to the Insurers under this condition, the Insured may carry out the repairs or replacement of any minor damage; in all other cases a representative of the Insurers shall have the opportunity of inspecting the loss or damage before any repairs or alterations are effected. Nothing herein shall prevent the Insured from taking such steps as are absolutely necessary for the security and continuation of the contract work. If a representative of the Insurers does not carry out the inspection within a period of time which could be considered as adequate under the circumstances the Insured is entitled to proceed with the repairs or replacement.

7. The insured shall at the expense of the Insurers do and concur in doing and permit to be done all such acts and things as may be necessary or required by the Insurers in the interest of any rights or remedies, or of obtaining relief or indemnity from parties (other than those insured under this policy) to which the Insurers shall be or would become entitled or subrogated upon their paying for or making good any loss or damage under this Policy, whether such acts and things shall be or become necessary or required before or after the Insured's indemnification by the Insurers.

8. All differences arising out of this Policy shall be referred to the decision of an Arbitrator to be appointed in writing by the parties in difference or if they cannot agree upon a single Arbitrator to the decision of two Arbitrators, one to be appointed in writing by each of the parties, within one calendar month after having been required in writing so to do by either of the parties, or, in case the Arbitrators do not agree, of an Umpire to be appointed in writing by the Arbitrators before entering upon the reference. The Umpire shall sit with the Arbitrators and preside at their meetings. The Arbitrators and the Umpire shall be qualified Engineers. The making of an award shall be a condition precedent to any right of action against the Insurers.

GENERAL CONDITIONS (Cont/d ...)

9. If a claim is in any respect fraudulent, or if any false declaration is made or used in support thereof, or if any fraudulent means or devices are used by the Insured or anyone acting on his behalf to obtain any benefit under this Policy, or if a claim is made and rejected and no action or suit is commenced within three months after such rejection or, in case of arbitration taking place as provided herein, within three months after the Arbitrator or Arbitrators or Umpire have made their award, all benefit under this Policy shall be forfeited.

10. If at the time any claim arises under the Policy there be any other insurance covering the same loss, damage or liability the Insurers shall not be liable to pay or contribute more than their rateable proportion of any claim for such loss or damage.

SECTION I OF POLICY NO............................

PROPERTY INSURED
(Material Damage)

The Insurers hereby agree with the Insured that if at any time during the period of insurance stated in the Schedule, or during any further period of extension thereof, the property or any part thereof described in the Schedule shall suffer any unforeseen loss or damage from any cause, other than those specifically excluded, in a manner necessitating repair or replacement, the Insurers will pay or make good all such loss or damage up to an amount not exceeding in respect of each of the items specified in the Schedule the sum set opposite thereto and not exceeding in all the total sum expressed in the said Schedule as insured hereby.

The Insurers will also reimburse the Insured for the cost of clearance of debris following upon any event giving rise to a claim under this Policy but not exceeding in all the sum set opposite thereto in the Schedule.

EXCLUSIONS TO SECTION I

The insurers shall not, however, be liable for:
(a) The deductibles stated in the Schedule to be borne by the Insured in any one occurrence other than fire, lightning or explosion;
(b) consequential loss of any kind or description whatsoever including penalties, losses due to delay, lack of performance, loss of contract;
(c) loss or damage due to faulty design;
(d) cost of replacement or rectification of defective material and/or workmanship, but this exclusion shall be limited to the items immediately affected and shall not be deemed to exclude loss or damage resulting from an accident due to such defective material and/or workmanship;
(e) wear and tear, corrosion, oxidation, deterioration due to lack of use and normal atmospheric conditions;
(f) mechanical and/or electrical breakdown or derangement of construction plant, equipment and construction machinery;
(g) loss of or damage to vehicles, licensed for general road use or waterborne vessels or aircraft;
(h) loss of or damage to files, drawings, accounts, bills, currency, stamps, deeds, evidences of debt, notes, securities or cheques;
(i) loss discovered only at the time of taking an inventory.

PROVISIONS APPLYING TO SECTION I

Memo 1. Sum Insured: It is a requirement of this insurance that the amounts of insurance stated in the Schedule shall not be less than

for item 1: the full value of the contract works at the completion of the construction, inclusive of materials, wages, freight, customs duties, dues and materials or items supplied by the Principal;

for items 2 and 3: the replacement value of construction plant, equipment and construction machinery.

The Insured undertakes to notify the Insurers of any facts effecting a material increase or decrease of the sums insured provided always that such increase or decrease shall take effect only after the same has been recorded on the Policy by the Insurers, before the occurrence of any claim hereunder.

If, in the event of loss or damage, it is found that the sum insured is less than the amount required to be insured, then the amount recoverable by the Insurer under this Policy shall be reduced in such proportion as the sum insured bears to the amount required to be insured.

Memo 2. Basis of Loss Settlement: In the event of any loss or damage the basis of any settlement under this Policy shall be
(a) in the case of any damage which can be repaired – the cost of repairs necessary to restore the property to its condition immediately before the occurrence of the damage less salvage, or
(b) in the case of a total loss – the actual value of the property immediately before the occurrence of the loss less salvage,
provided always that the provisions and conditions have been compiled with.

The Insurers will make payments only after being satisfied by production of the necessary bills and documents that the repairs have been effected or replacement has taken place, as the case may be.

All damage which can be repaired shall be repaired, but if the cost of repairing any damage equals or exceeds the value of the property immediately before the occurrence of the damage, the settlement shall be made on the basis provided for in (b) above.

The cost of any provisional repairs will be borne by the Insurers if such repairs constitute part of the final repairs and do not increase the total repair expenses.

The cost of any alterations, additions and/or improvements shall not be recoverable under this Policy.

Memo 3. Extension of Cover: Extra charges for overtime, nightwork, work on public holidays, express freight, etc. are covered by this Insurance only if previously and specially agreed upon.

SECTION II OF POLICY NO.....................

THIRD PARTY LIABILITY

The Insurers will indemnify the Insured against all sums which the Insured shall become legally liable to pay as damages consequent upon
(a)   accidental bodily injury or illness (whether fatal or not) to third parties
(b)   accidental loss or damage to property belonging to third parties
occurring in direct connection with the performance of the contract insured by this Policy and happening on or in the immediate vicinity of the Contract Site during the Period of Insurance.

In respect of a claim for compensation to which the indemnity provided herein applies, the Insurers will in addition indemnify the Insured against
(a)   all costs and expenses of litigation recovered by any claimant from the Insured, and
(b)   all costs and expenses incurred with the written consent of the Insurers.

The Liability of the Insurers under this section shall not exceed the limits of indemnity stated in the Schedule (Section II).

EXCLUSIONS TO SECTION II

The Insurers will not indemnify the Insured in respect of

1.   expenditure incurred in doing or redoing or making good or repairing or replacing any work or property covered or coverable under Section I of this Policy;
2.   damage to any property or land or building caused by vibration or by the removal or weakening of support or injury or damage to any person or property occasioned by or resulting from any such damage (unless especially agreed upon Endorsement);
3.   liability consequent upon
     (a)   bodily injury to or illness of employees or workmen of the Contractor(s) or the Principal or any other firm connected with the contract work or members of their families;
     (b)   loss of or damage to property belonging to or held in care, custody or control of the Contractor(s), the Principal or any other firm connected with the contract work or an employee or workman of one of the aforesaid;
     (c)   any accident caused by vehicles licensed for general road use or by waterborne vessels or aircraft;
     (d)   any agreement by the Insured to pay any sum by way of indemnity or otherwise unless such liability would have attached in the absence of such agreement.

SPECIAL CONDITIONS APPLYING TO SECTION II

1.   No admission, offer, promise, payment or indemnity shall be made or given by or on behalf of the Insured without the written consent of the Insurers who shall be entitled if they so desire to take over and conduct in the name of the Insured the defence or settlement of any claim or to prosecute for their own benefit in the name of the Insured any claim for indemnity or damage or otherwise and shall have full discretion in the conduct of any proceedings or in the settlement of any claim and the Insured shall give all such information and assistance as the Insurers may require.
2.   The Insurers may so far as any accident is concerned pay to the Insured the limit of indemnity for any one accident (but deducting therefrom in such case any sum or sums already paid as compensation in respect thereof) or any lesser sum for which the claim or claims arising from such accident can be settled and the Insurers shall thereafter be under no further liability in respect of such accident under this section.

SCHEDULE

Policy No......................	Issued at

Title of Contract
Name(s) and address(es) of Insured
  (a) Principal
  (b) Contractor(s)
Location of Contract Site

Description of the Insured Items

Section I – Property Insured	Sum Insured
1. Contract Works (Permanent and Temporary Works, including all Materials to be incorporated therein)	
(a) Contract Price	1. (a)
(b) Materials or Items supplied by the Principal	(b)
2. Construction Plant and Equipment	2.
3. Construction Machinery according to attached list	3.
	Total
Limit of indemnity under this Policy in respect of Clearance of Debris	

Deductibles:

Amounts to be borne by the Insured in respect of each and every occurrence for loss of or damage to contract works and/or construction plant and equipment (Schedule Section I items 1 and 2) arising out of

(a) earthquake, storm, hurricane, cyclone, flood, inundation, subsidence, landslide, collapse	the first
(b) any other cause, except fire, lightning and explosion	the first

Amount to be borne by the Insured in respect of each and every occurrence for loss of or damage to construction machinery (Schedule Section I item 3) arising out of

(a) earthquake, storm, hurricane, cyclone, flood, inundation, subsidence, landslide, collapse	the first
(b) any other cause, except fire, lightning and explosion	the first

Section II – Third Party Liability	
1. Limit of indemnity in respect of any one accident or series of accidents arising out of one event	
(a) for bodily injury for any one person	1. (a)
(b) for property damage	(b)
2. Total limit of indemnity under this Policy	2.

Period of Insurance (subject to the Provision
concerning the Period of Cover)

(a) Construction Period	From:	To:
(b) Maintenance Period	From:	To:

Premium
(subject to adjustment in accordance with the Provisions
and Conditions of the Policy)

In Witness whereof the Undersigned being duly authorised by the Insurers and on behalf of the Insurers has/have hereunto set his/her hand(s)
this... day of.. 20...............

Appendix C

SPECIMEN

Policy No.....................

NOTIFICATION OF LOSS OR DAMAGE FOR CONTRACTORS' ALL RISKS INSURANCE

Claim No.:

Name and Address of Insured:

Address of Site:

Name of supervising Engineer:

Name of his Deputy:

Nearest railway station:

Advisable Route of Approach to Site
from Railway Station or otherwise:

Questions	Answers
1. When did the loss or damage occur? (State date and hour)	
2. What was the cause of the damage and how did it happen?	
3. Give details of the damage to (a) works (b) construction machinery (please attach sketches, photos, etc.)	

	Questions	Answers
4.	Is Public Liability involved?	
5.	How will the damage be repaired (give details)	
6.	What are the estimated costs for the repair of the damage to (a) works? (b) construction machinery?	
7.	Is anyone responsible for the damage? Is there any possibility of recovery?	
8.	Remarks	

The Undersigned Insured declares to have answered the above questions conscientiously and truthfully.

Dated at .... this ....day of ....20................

Applicant's Signature ....

Appendix D

SPECIMEN

EMPLOYER'S LIABILITY INSURANCE

WHEREAS the Insured named in the Schedule hereto has made a Proposal which shall be the basis of this Contract and shall be deemed to be incorporated herein and has paid or agreed to pay the premium stated in the Schedule this Policy witnesses that the Insurance hereinafter contained has been effected for the Period of Insurance stated in the Schedule with 'The Honourable Insurance Company Ltd.'*, hereinafter called 'the Company'.

NOW this policy witnesses that subject to the terms and conditions hereinafter contained or endorsed hereon the Company will indemnify the Insured against liability at law for damages and claimant's costs and expenses in respect of death bodily injury or disease caused during any Period of Insurance to any Employee arising out of and in the course of employment by the Insured in the Business within the Geographical Limits provided that in respect of death bodily injury or disease sustained or contracted outside Great Britain Northern Ireland the Isle of Man the Island of Jersey the Island of Alderney or the Island of Guernsey the action for damages is brought against the Insured in a court of law in such territories. The indemnity provided by this policy is deemed to be in accordance with the provisions of any law relating to compulsory insurance of liability to employees in the Republic of Ireland Great Britain Northern Ireland the Isle of Man the Island of Jersey the Island of Alderney or the Island of Guernsey. The Insured shall repay to the Company all sums paid by the Company which the Company would not have been liable to pay but for the provisions of such law.

It is further understood and agreed that where any contract or agreement entered into by the Insured with any public authority company firm or person (hereinafter called 'the principal') so requires the Company shall
- (a) indemnify the Insured against liability arising in connection with and assumed by the Insured by virtue of such contract or agreement
  or
- (b) indemnify the Principal in like manner to the Insured in respect of the Principal's liability arising from the performance of such contract or agreement

but only as far as concerns liability as defined in this Policy to an Employee of the Insured provided always that the conduct and control of all claims is vested in the Company.

The Company will also pay
- (a) all other costs and expenses
- (b) the Solicitor's fee for representation of the Insured at
  - (i) any coroner's inquest or inquiry in respect of any death
  - (ii) proceedings in the court of summary jurisdiction arising out of any alleged breach of statutory duty resulting in bodily injury death or disease which may be the subject of indemnity under this Policy

incurred with the Company's written consent.

For and on behalf of

*The name of the Company is a fictitious one and is used only for the sake of clarity.

EXCEPTION

The Company shall not be liable in respect of any legal liability of whatsoever nature directly or indirectly caused by or contributed to by or arising from

(a) ionising radiations or contamination by radioactivity from any nuclear fuel or from any nuclear waste from the combustion of nuclear fuel

(b) the radioactive toxic explosive or other hazardous properties of any explosive nuclear assembly or nuclear component thereof where such liability is

(i) the liability of any Principal

(ii) assumed by the Insured under agreement and would not have attached in the absence of such agreement.

EXTENSION

This Policy subject to its terms and conditions indemnifies

(a) any Employee being a member of the Insured's first-aid or medical organisation (other than a qualified medical practitioner) against liability at law for damages and claimant's costs and expenses in respect of medical or surgical treatment given to any other Employee in connection with any bodily injury death or disease sustained by such other Employee and arising out of and in the course of employment in the Business

(b) (i) the insured in respect of bodily injury or disease sustained by Employees when engaged in private duties (including duty as chauffeur) for Directors Partners and Senior Officials of the Insured

(ii) in like manner to the Insured any Director Partner or Senior Official of the Insured in respect of the employment on private business duties (including duty as chauffeur) of any Employee by such person

provided always that the Insured shall have arranged with such person for the conduct and control of all claims to be vested in the Company and that such person shall as though he were the Insured observe fulfil and be subject to the terms and conditions of this Policy in so far as they can apply.

INTERPRETATIONS

For the purposes of this Policy

1. 'Proposal' shall mean any signed proposal form and declaration and any information supplied by or on behalf of the Insured in addition thereto or in substitution therefore

2. 'Employee' shall mean any

(a) person under a contract of service or apprenticeship with the Insured

(b) labour master and persons supplied by him

(c) person employed by labour only sub-contractors

(d) self employed person

(e) person hired from any public authority company firm or individual

While working for the Insured in connection with the Business

3. 'Business' shall include the provision and management of canteens social sport and welfare organisations for the benefit of the Insured's employees and first aid fire and ambulance services.

SPECIMEN*

ENDORSEMENT 'XYZ'

This Policy does not indemnify the Insured in respect of any claim arising in connection with:-

(a)   the making of sewers or other excavations exceeding in any part a depth of ten feet from the surface
(b)   the use of explosives
(c)   quarrying tunnelling demolition water diversion pile driving dam construction or work within or behind dams
(d)   the construction alterations or repair of buildings involving the use at any time of mechanically-driven lifts cranes or hoists other than those designed for a maximum load not exceeding two tons
(e)   the construction alteration or repair of towers steeples chimneys shafts viaducts bridges docks or wells.

*This endorsement is a specimen document and may not appear in some policies. If it does, the Conditions stipulated therein must be checked against the work being insured and changed if found contradictory.

APPENDIX D

CONDITIONS

1. The Policy and Schedule shall be read together and any word or expression to which a specific meaning has been given shall bear such meaning wherever it may occur.

2. The Insured shall give written notice to the company as soon as possible after receiving information of any claim or loss or any occurrence for which there may be liability under this Policy with full particulars thereof. Every letter claim writ summons and process shall be forwarded to the Company on receipt. No admission offer promise payment or indemnity shall be made or given by or on behalf of the Insured without the written consent of the Company which shall be entitled to take over and conduct in the name of the Insured the defence or settlement of any claim or to prosecute in the name of the Insured for its own benefit any claim and shall have full discretion in the conduct of any proceedings and in the settlement of any claim and the Insured shall give all such assistance as the Company may require.

3. If at any time any claim arises under this Policy the Insured is or would but for the existence of this Policy be entitled to indemnity under any other policy or policies the Company shall not be liable except in respect of any excess beyond the amount which would have been payable under such other policy or policies had this insurance not been effected.

4. The first premium and all renewal premiums that may be accepted are to be regulated by the amount of wages salaries and other earnings paid by the Insured to employees during each Period of Insurance. The name of every employee together with the amount of wages salary and other earnings shall be properly recorded and the Insured shall at all times allow the Company to inspect such records and shall supply the company with a correct amount of all such wages salaries and other earnings paid during any Period of Insurance within one month from the expiry date of such Period of Insurance. If the amount so paid shall differ from the amount on which premium has been paid the difference in premium shall be met by a further proportionate payment to the Company or by a refund by the Company as the case may be.

5. The Company may cancel this Policy by sending seven days notice to the Insured at the Insured's last known address. The Insured shall thereupon become entitled to a proportionate return of premium.

6. The due observance and fulfilment of the terms and conditions of this Policy by the Insured insofar as they relate to anything to be done or complied with by the Insured shall be a condition precedent to any liability of the Company to make any payment under the Policy.

7. The Insured shall take all reasonable precautions to prevent accidents and disease.

THE POLICY DOES NOT COVER:
Any liability, loss or damage directly or indirectly occasioned by or happening through or in consequence of:
(a) war, invasion, act of foreign enemy, hostilities (whether war be declared or not) civil war
(b) Rebellion, revolution, insurrection, military or usurped power or confiscation or nationalisation or requisition or destruction of or damage to property by or under the order of any government or Public or Local Authority.

ARBITRATION CLAUSE

All the differences arising out of this Policy shall be referred to the decision of an Arbitrator to be appointed in writing by the parties in difference or if they cannot agree upon a single Arbitrator to the decision of two Arbitrators, one to be appointed in writing by each of the parties within one calendar month, after having been required in writing so to do by either of the parties, or, in case the Arbitrators do not agree, of an Umpire appointed in writing by the Arbitrators before entering upon the reference. The Umpire shall sit with the Arbitrators and preside at their meeting, and the making of an Award shall be a condition precedent to any right of action against the Insurers. If the Insurers shall disclaim liability to the Insured for any claim hereunder and such claim shall not within twelve calendar months from the date of such disclaimer have been referred to arbitration under the provisions herein contained, then the claim shall for all purposes be deemed to have been abandoned and shall not thereafter be recoverable hereunder.

SCHEDULE

SPECIMEN

INSURED:	
ADDRESS:	
THE BUSINESS/OCCUPATION:	
THE PREMISES:	

PERIOD OF INSURANCE:-   From: To:
(Both days inclusive) and any subsequent period for which the Insured shall pay and the Company shall agree to accept renewal.

REFERENCE NO.		POLICY NO.	
RENEWAL DATE		CLASS	
BROKER'S NAME AND ADDRESS:			
ANNUAL PREMIUM		FIRST PREMIUM	

PROPERTY INSURED	As follows or as detailed in the Specification attached hereto which is declared to be incorporated in and to form part of this Policy.
EXCESS	The Insured shall pay: a) The first £ of each and every claim or b) The first £ of all claims payable in any one period aggregated.
SUM INSURED OR LIMIT OF LIABILITY	£ Total Sum Insured by this Policy being % of the Total Specification Sum Insured or as per Separate Amounts shown for each Section
OTHER INTERESTS	
DELETED AT INCEPTION	
ENDORSEMENTS	'XYZ'
WARRANTIES	
SECTIONS/ EXTENSIONS Included	

Broker Code			Date			Date	
U/W Year			Due Date				
Currency			Admin. Des P.P.L.			Territory Code	

TYPED BY		EXAMINED BY		DATE SIGNED	

Appendix E

SPECIMEN

PROPOSAL

PROFESSIONAL INDEMNITY INSURANCE

Section 1 – General – To be completed by all Proposers

1. Full Title of Firm ....
2. Business/Practice ....
3. Address(es) of Firm ....
   ....
   ....
   ....
4. Date Established ....
5. During past 10 years has the title under (1) above been changed or any merger consolidation or purchase of other business taken place ....
   If 'Yes' give details ....
6.

Name of Each Partner/Principal	Qualifications & Date Qualified	How long as Partner/Principal	Previous Firm(s)

7. Total number of Staff:

Qualified		Non Qualified	
Full time	Part time	Full time	Part time

8. Is Partner/Principal in full time attendance at each office ....
9. Total Indemnity Required   £
10. Amount of Excess Proposer prepared to carry   £
11. State Proportion of the Business/Practice Relating (or which is expected to relate) to Work Overseas and Territories involved ....
12. Does the Business/Practice or any of its Partners/Principals have any Association with or interests in any other Company/Firm or Organisation ....
    If 'Yes' give details ....
    ....
13. Does Proposer contract to undertake any form of Manufacture, Construction, Erection or Supply ..
    If 'Yes' give details ....
    ....
14. Has Proposer any other Professional Indemnity Insurance in force ....
    If 'Yes' state (a) Limit of Indemnity ....
                   (b) Name of Insurer ....
15. Has any proposal for insurance of this nature on behalf of the Proposer, their Predecessors in business, or present Principals, been declined, or any policy cancelled or renewal refused ....

16. Have any claims made against the Proposer or any partner, director, consultant or employee, for neglect error or omission in relation to professional duties . . . .
If 'Yes' give details . . . .
17. Is any partner/principal, consultant or employee, after enquiry, aware of any circumstances which might give rise to a claim against the firm its principals or partners or its predecessors . . . .
18. Are any of the following extensions of cover required:
a) Partners Previous Business . . . .
If 'Yes' state for which partners . . . .
b) Former Partners . . . .
If 'Yes' state for which partners and date from which connection with the business/practice ceased

    c) Loss of Documents . . . .   Limit of indemnity required £ . . . .
    d) Libel & Slander . . . .   Limit of indemnity required £ . . . .
    e) Dishonesty of Employees . . . .
    f) Fidelity . . . .
        i) Limit of indemnity required £ . . . .
        ii) Has the firm any insurance in force already . . . .
       iii) Has the firm sustained any loss as a result of fraud or dishonesty or is it aware of any circumstances of fraud or dishonesty affecting any person covered hereunder . . . .
       iv) Are any members of the staff allowed to handle cash or negotiable documents . . . .
          If so give details . . . .
       v) Does any partner or responsible official carry out regular independent checks of the cash books . . . .
       vi) What system of check is in force and how often is it applied . . . .
      vii) What references are required when engaging staff . . . .

Declaration and Signature
I/We declare that the statements and particulars in this proposal form are true and that I/we have not suppressed or misstated any material facts. I/We agree that this proposal together with any information in connection therewith supplied by or on my/our behalf shall form the basis of any contract of Insurance effected herein. I/We undertake to inform the Company of any material alterations affecting the above statements. Signing this proposal form does not bind the proposer or Company to complete this insurance.

Dated this . . . . day of . . . . 20 . . . . .

Signature of Partner or Principal . . . .

PLEASE COMPLETE THE APPROPRIATE SECTION(S)
Section 2 – Accountants
19. State Gross Fees/Turnover for each of the past years

	Last Year	Previous Year
(a) Ireland and United Kingdom	£	£
(b) U.S.A. & Canada	£	£
(c) All other overseas contracts	£	£

20.

		Percentage of each amount as follows		
(a) Please insert Annual Fees for past year in respect of:	Amount	Ireland	U.K.	Elsewhere
i) Auditing & Preparation of Accounts	£			
ii) Taxation	£			
iii) Examination of Business and Company Accounts	£			
iv) Insolvencies Liquidations and Receiverships	£			
v) Executorships and Trusteeships	£			
vi) Management Consultancy	£			
vii) Other work (please give details)	£			
(b) Please state				
i) Average fee any one client	£			
ii) Maximum fee any one client	£			
(c) Estimated fees for current 12 months	£			

Section 3 – Consulting Engineers/Architects

21. State gross fees received during past 12 months from each of the following works:

	Ireland	U.K.	Elsewhere
i) Consulting Engineering	£	£	£
a) Civil	£	£	£
b) Structural	£	£	£
c) Soil and foundation	£	£	£
d) Mechanical	£	£	£
e) Electrical	£	£	£
f) Heating and ventilating	£	£	£
ii) Architectural	£	£	£
ii) Town Planning/Quantity Surveying	£	£	£
iv) Structural Surveyor Inspection Reports	£	£	£
v) Valuations on existing properties	£	£	£
vi) Any other work (full details please)	£	£	£
. . . .			
TOTAL			

22. a) Please give fees received during past five years

Year	Ireland	U.K.	Elsewhere
20	£	£	£
20	£	£	£
20	£	£	£
20	£	£	£
20	£	£	£
	£	£	£

    b) Estimated fees fo current 12 months

	Ireland	U.K.	Elsewhere
	£	£	£

23. a) Does Practice/Business engage in any of the following tyes of work:

If 'Yes' please insert % of Total Work applicable.

1. Bridges/Flyovers/Tunnels/Dams/Mines
2. Harbours/Jetties/Sea Defences
3. Marine Surveys
4. Bulk Handling Equipment/Hoppers/Silos
5. Other Mechanical Plant/Equipment
6. Fertiliser/Ammonia/Urea Plants
7. Chemicals/Petro Chemicals/Chemical or Oil Refineries
8. Nuclear/Atomic Projects
9. Sewerage/Water Schemes
10. Hospitals/Universities/Schools
11. Factories
12. Housing
13. High rise (i.e. 10 storeys or more per block)
    Please give total number of storeys in highest block
    completed during past 10 years . . . .

b) Approximately what percentage of Total Work is derived
from the following:
i)   Industrialised Systems Building
ii) Restoration Work
iii) Reinforced/Prestressed Concrete
iv) Soil testing/Foundations/Piles/Underpinning
v)   Government Departments
vi) Local Authorities

Total % of Work

24. a) Have you at any time enteed into Contracts under seal . . . .
    b) If 'Yes' please give following details of each such contract carried out during the past 12 years

Starting Date of Construction	Nature of Contract	Total Contract Value	Completion Date	Gross Fees Received

25. Please give details of the three largest new operations where construction is likely to commence in the coming 12 months.

Starting Date	Type of Contract	Total Contract Value	Approx. Completion Date	Professional Services Provided

NOTE: Cover will exclude any work which the firm carries out as members of a consortium. If cover is required please supply a copy of the Consortium Agreement.

Section 4 – Surveyors, Auctioneers, Valuers & Estate Agents
26. Please insert Gross Fees/Turnover for each of past 2 years

	Last year	Previous year
a) Ireland and U.K.	£	£
b) USA & Canada	£	£
c) All other overseas contracts	£	£

27. a) Total Gross
   Annual fees for
   past year in
   respect of:

	Amount	Percentage of each amount as follows		
		Ireland	U.K.	Elsewhere
Quantity Surveying	£			
Surveying & Valuing	£			
Auctioneering	£			
Estate Agency	£			
Property Management	£			
Insurance agency with authority to issue cover notes or certificates	£			
Architectural design and planning work (see question 22)	£			
Project Management	£			
Other Work give details:	£			
b) Estimated fees for current 12 months	£			

28. State maximum certified building value any one contract £. . . . .

29. a) What system is in force to prevent time limits being overlooked in Rent Act or Landlord and
   Tenant Act cases . . . .
   b) How often does partner check to ensure the system is being properly enforced.

417

Appendix F

SPECIMEN

PROFESSIONAL INDEMNITY POLICY

WHEREAS the Insured named in the Schedule has applied to 'The Honourable Insurance Company Ltd.' (hereinafter* called the 'Company' or the 'Insurer') for the insurance contained herein and has paid the premium.

NOW this policy witnesses that subject to the terms and conditions hereinafter contained or endorsed thereon the Company will indemnify the Insured or the Insured's Executors, Administrators and Assignees against Loss as more fully set forth in the Policy and the Schedule during the period of Insurance stated in the said Schedule or during any subsequent period as may be mutually agreed upon between the Insured and the Company.

PROVIDED that the liability of the Company shall not exceed the limits of liability expressed in the said Schedule or such other limits of liability as may be substituted therefore by memorandum hereon or attached hereto signed by or on behalf of the Company.

For and on behalf of

*The name of the Company is a fictitious one, and is used only for the sake of clarity.

WHEREAS the Insured named in the Schedule has/have submitted a written proposal containing particulars and statements which (together with any other information which may have been supplied) it is agreed shall be the basis of this contract and are to be considered as incorporated herein and in consideration of the premium stated in the Schedule.

INSURING CLAUSE

NOW THEREFORE, We, the Company hereby agree to indemnify the Insured up to but not exceeding in the aggregate the sum stated in the Schedule for any sum or sums which the Insured may become legally liable to pay in accordance with the law of any country arising from any claim or claims made against them during the period stated in the Schedule in consequence of any act of neglect, error or omission or for breach of statutory duty as a result of a breach of professional duty in the professional conduct of their business, as stated in the Schedule, by the Insured or any partner or previous partner or any person or party employed or engaged by the Insured including specialist designers or consultants acting on the Insured's behalf and for whom the Insured are responsible.

COSTS AND EXPENSES

FURTHER, it is understood and agreed that the Company will pay in addition to the sum stated in the Schedule, the costs and expenses incurred with the Company's written consent in the defence and/or settlement of any claim. Such written consent not to be unreasonably withheld. However, if a payment in excess of the amount of indemnity available under this insurance has to be made to dispose of a claim made against the Insured the Company's liability in respect of such costs and expenses shall be such proportion of the total costs and expenses incurred as the amount of the indemnity available under this insurance bears to the total amount paid to dispose of the claim.

EXCESS

PROVIDED ALWAYS THAT the Company shall be liable only, in respect of any claim hereunder, for that part of the claim which exceeds the amount stated as 'the Excess' in the Schedule. The expression 'claim' shall mean claim or series of claims arising out of any one act of neglect, error or omission or for breach of statutory duty as a result of a breach of professional duty in the professional conduct of their business, as stated in the Schedule, by the Insured or any partner or previous partners or any person, or party employed or engaged by the Insured including specialist designers or consultants acting on the Insured's behalf and for whom the Insured are responsible.

DEFINITION OF EMPLOYEES

For the purposes of this insurance an Employee shall mean any person past or present:
(a) under a contract of service or apprenticeship with the Insured including any trainee or consultant
(b) director of the Insured
(c) hired to or borrowed by the Insured
(d) undertaking study or work experience.

EXCLUSIONS

1. Any goods or products manufactured, constructed, altered, repaired, serviced, treated, sold, supplied or distributed by the Insured or from any other business or occupation even though the same may be carried on by the Insured in conjunction with their business as stated in the Schedule.

2. Bodily injury, sickness, disease or death sustained by any person arising out of and in the course of his employment by the Insured under a contract of service or apprenticeship with the Insured.

3. The ownership, use, occupation or leasing of property mobile and/or immobile by, to or on behalf of the Insured.

4. Loss or destruction of or damage to any property whatsoever or any loss or expense whatsoever resulting or arising therefrom or any consequential loss or any legal liability of whatsoever nature directly or indirectly caused by or contributed to by or arising from:
   (i) ionising radiations or contamination by radioactivity from any nuclear fuel or from any nuclear waste from the combustion of nuclear fuel.
   (ii) the radioactive, toxic, explosive or other hazardous properties of any explosive nuclear assembly or nuclear component thereof.

5. Any consequences of war, invasion, act of foreign enemy, hostilities (whether war be declared or not), civil war, rebellion, revolution, insurrection or military or usurped power. This exclusion shall not apply in respect of liability or loss caused by explosive devices provided that no state of war exists in the country where the contract is undertaken.

CONDITIONS

1. (a) Notice shall be given by the Insured in writing, as soon as practicable, to the Company of any claim or intimation to the Insured of possible claim made against the Insured which may give rise to a claim under this Policy.

(b) If during the currency hereof the Insured shall become aware of any occurrence which may subsequently give rise to a claim against them under this policy and shall during the currency hereof give written notice to the Company of such occurrence any claim which may subsequently be made against the Insured arising out of that occurrence shall be deemed for the purpose of this insurance to have been made during the currency hereof.

(c) The Insured shall not admit liability for or settle or make or promise any payment in respect of any claim which may be the subject of indemnity hereunder or incur any costs or expenses in connection therewith without the written consent of the Company who if they so wish shall be entitled to take over and conduct in the name of the Insured the defence and/or settlement of any such claim for which purpose the Insured shall give all such information and assistance as the Company may reasonably require.

2. **AGREEMENT TO PAY CLAIM** Companies agree to pay claims which may arise under this insurance without requiring the Insured to dispute any claim unless a Senior Counsel or Lawyer of comparable standing in the territory concerned advise that the same could be contested with a reasonable prospect of success by the Insured and the Insured consents to such claim being contested but such consent is not to be unreasonably withheld. In the event of any dispute arising between the Insured and the Company as to what constitutes an unreasonable refusal to contest a claim at Law, the President of the Association of Consulting Engineers shall nominate a Referee to decide this point (only) and the decision of such Referee shall be binding on both parties.

3. **OTHER INSURANCE** If at the time any claim arises under this insurance and the Insured is or would but for the existence of this insurance be entitled to indemnity under any other policy or policies, the Company shall not be liable except in respect of any excess beyond the amount which would have been payable under such other policy or policies had this insurance not been affected.

4. **TIMBER DISEASE** It is hereby understood and agreed that this policy shall apply to any claim arising or resulting from or in connec-tion with timber disease of any description, woodworm, beetle infestation or any other vermin or insect or any consequential loss or damage arising therefrom provided that:

(i) all reports shall be in writing and shall have been prepared by (a) a Fellow or Professional Associate of the Royal Institute of Chartered Surveyors or (b) a qualified Architect or (c) a person with not less than five years experience of structural surveying, and

(ii) the Insured or a qualified representative of the Insured shall have made a detailed inspection of the building and have fully reported on the condition of the timber and drawn attention to the existence of any defect observed and also of the possibility of such defect becoming more extensive. Further the report must include the following clause in respect of all timber or woodwork not surveyed. 'We have not inspected woodwork or other parts of the structure which are covered, unexposed or inaccessible and we are therefore unable to report that any such part of the property is free from defect'.

5. **SUBROGATION** If any payment is made under this Insurance in respect of a claim here-under Companies are thereupon subrogated to all the Insured's rights of recovery thereto however the Company shall not exercise any such rights against any Employee or former Employee of the Insured unless the claim has been brought about or contributed to by any dishonest fraudulent criminal or malicious act of the Employee, it is understood that Companies shall at all times retain all the Insured's rights of recovery against any person or party who is not an Employee of the Insured or former Employee of the Insured.

6. **FRAUDULENT CLAIMS** If the Insured shall make any claim knowing the same to be false or fraudulent as regards amount or otherwise this insurance shall become void and all claims thereunder shall be forfeited.

7. **DISCLAIMER OF LIABILITY BY COMPAN-IES** In the event of Companies at any time being entitled to avoid this Insurance ab initio by reason of any materially inaccurate or misleading information given to the Company in the Proposal Form or at any time during the negotiations leading to the inception of this Insurance or as a result of failure to disclose material facts before the conclusion of the Insurance or for any other reason at law, the Company may at their election instead of avoiding this Insurance ab initio give notice to the Insured that they regard this Insurance as of full force and effect so that there shall be excluded from the indemnity afforded here-

under any claim which has arisen or which may arise and which is related to circumstances which ought to have been disclosed to the Company in the Proposal Form or which arises out of materially inaccurate or misleading information given to the Company. This Insurance shall then continue as if the same had specifically endorsed ab initio the particular claim or possible claim referred to in the said notice.

8. **ARBITRATION** All differences arising out of this policy shall be referred to the decision of a single Arbitrator to be appointed in writing by the parties in difference or dispute, if the parties can agree on the identity of the single Arbitrator. If the parties cannot agree upon a single Arbitrator, the difference and disputes shall be referred to the decision of two Arbitrators, one to be appointed in writing by each of the parties, within one calendar month after having been required in writing so to do by either of the parties, or, in case the Arbitrators do not agree, of an umpire to have been appointed in writing by the Arbitrators before entering upon the reference. If the Arbitrators cannot agree upon the umpire within one calendar month from the date of finalisation of their appointments, then the Chairman for the time being of the Irish Branch of the Chartered Institute of Arbitrators shall on the application in writing of either Arbitrator appoint an umpire within one calendar month of the application. The umpire shall sit with the Arbitrators and preside at their meeting. The costs of the reference and of the Award shall be at the discretion of the Arbitrator, Arbitrators or umpire making the Award. The Award shall be a condition precedent to any liability of the Company or any right of action against the Company in respect of any claim. If the Company shall disclaim liability to the Insured for any claim hereunder and such claim shall not within twelve calendar months from the date of such disclaimer have been referred to Arbitration under the provisions hereunder contained then the claim shall for all purposes be deemed to have been abandoned and shall not thereafter be recoverable hereunder.

POLICY EXTENSIONS

1. The insurance provide by this policy is extended to indemnify the Insured within the terms of this Policy for any claims made against the Insured in respect of work undertaken by any firm company or individual with whom the Insured is acting jointly.

2. **Dishonesty:** This policy is extended to indemnify the Insured for their legal liability arising from any claim made against them during the period specified in the Schedule by reason of any dishonest, fraudulent, criminal or malicious act or omission of any Employee of the Insured provided always that no indemnity shall be afforded hereby to any person committing or condoning such dishonest, fraudulent, criminal or malicious act or omission and the sums payable under this policy shall only be for the balance of liability in excess of the amounts recoverable from the dishonest or fraudulent person or persons.

3. **Libel and Slander:** Notwithstanding anything contained herein to the contrary this Policy is extended to indemnify the Insured against liability at law for damages and claimants costs and expenses in respect of claims made against the Insured and notified to the Company during any period of Insurance for Libel or Slander committed in good faith by reason of words written or spoken by the Insured or by any Employee in the course of business. The excess shall not apply to this extension.

4. **Infringement of Copyright:**
   (a) This policy is extended to cover costs incurred in prosecuting any claims for an injunction and/or for damages for Infringement of Copyright vested in the Insured notified to the Company during the period of insurance stated in the Schedule provided always that the Company shall not be required to incur any or any further obligation to meet such costs under this section where the Insured's course of action is not one which would be under the circumstances reasonable to pursue. In the event of any dispute arising between the Company and the Insured as to the reasonableness of pursuing any such course of action the opinion of a Senior Counsel or Lawyer of comparable standing in the territory concerned (to be mutually agreed upon by the Company and the Insured) and the Company shall be obtained and his decision shall be binding.
   (b) Further it is understood and agreed that this insurance is extended to indemnify the Insured arising out of infringement of copyright, design, patent and/or innocent breach of confidential information of trade secrets.

The self-insured excess applicable to this extension alone is £........ in respect of each and every claim.

5. **Fees Recovery Extension:** The Company will pay all costs incurred by the Insured in connection with legal proceedings taken by the Insured for the recovery of professional fees in accordance with the appointment subject to the following conditions.

(a) the Insured must advise the Company immediately of their intention to institute such proceedings.

(b) no claims shall attach unless the Company have been advised by their legal advisers that such action could be pursued with the probability of success.

The self-insured excess applicable to this extension alone is £......... in respect of each and every claim.

6. **Loss of Documents Extension:** Notwithstanding anything contained herein to the contrary this Policy is extended to indemnify the Insured against:

(a) any legal liability that they may incur in consequence of any documents (as defined herein) either the property of or entrusted to the Insured or in the custody of any person to or with whom such documents have been entrusted lodged or deposited having been discovered during the period specified in the Schedule to be damaged destroyed lost or mislaid and which after diligent search cannot be found.

(b) the costs and expenses of whatsoever nature incurred by the Insured in replacing or restoring such documents.

Definition – the term 'documents' shall mean deeds, wills, agreements, maps, plans, records, books, letters, certificates, computer systems records, forms and documents of whatever nature whether written printed or reproduced by any other method (other than bearer coupons bank notes currency notes and negotiable instruments).

The self-insured excess applicable to this extension is £....... in respect of each and every claim.

7. **Legal Defence Extension:** This Policy is extended to cover costs charges and expenses (which are not otherwise covered by this Policy) of legal representation of the Insured at any proceedings before any duly constituted court or tribunal of enquiry or otherwise having the like power to compel attendances of witnesses (but not any hearing before any domestic or disciplinary body of any Institute or Association) at which the Insured in the opinion of the Company shall be represented by reason of any conduct on their part which might be relevant to and which might give rise to or has given rise to a claim under the Insuring Clause hereof or by reasons of any prejudice which might be occasioned to the Insured's professional reputation it is agreed and understood that:

(a) this indemnity will only extend to circumstances notified to the Company during the period of insurance stated in the Schedule and

(b) the Company shall not be liable to pay any penalty fine or award of costs made against the Insured and

(c) no costs charges and expenses of any kind other than those incurred with the written consent of the Company shall be payable hereunder and

(d) the Company shall be entitled if they so desire to nominate a Solicitor and if appropriate a Barrister.

8. **Environmental Impairment Extension:** In so far as the same may arise out of the negligent act, error or omission of the Insured or for breach of statutory duty as a result of a breach of professional duty in the professional conduct of their business by the Insured as defined herein this Policy will indemnify the Insured in respect of Compensation which the Insured shall become legally liable to pay for personal injury, property damage or impairment or diminution of or other interference with any other right or amenity protected by law caused or contributed to by the omission, discharge, dispersal, disposal, seepage, release or escape of any liquid, solid gas or the generation of smells, vibrations, lights, electricity, radiation, changes in temperature or any other sensory phenomena which directly or indirectly cause irritation contamination and/or pollution of the environment and for the cost of litigation or for expenses incurred in moving, nullifying or clearing up harmful substances.

9. **Arbitration Extension:** If the Underwriter shall disclaim liability to the Insured in respect of any claims made upon the latter, the Insured shall be at liberty, without prejudice to his claim for indemnity hereunder, to settle or compromise such claims or submit to any judgement or arbitration award in respect thereof and any such settlement compromise, judgement or arbitration award shall be accepted by the Company as the amount payable by the Company to the Insured subject to the terms of this policy and subject to the liability of the Company being established by arbitration as herein provided.

THE INSURED IS REQUESTED TO READ
THIS POLICY AND, IF INCORRECT,
RETURN IT IMMEDIATELY FOR
ALTERATION

Schedule

SPECIMEN

INSURED:	
ADDRESS:	
THE BUSINESS/OCCUPATION:	
THE PREMISES:	

PERIOD OF INSURANCE:-   From: To:
(Both days inclusive) and any subsequent period for which the Insured shall pay and the Company shall agree to accept renewal.

REFERENCE NO.		POLICY NO.	
RENEWAL DATE		CLASS	

BROKER'S NAME AND ADDRESS:	
ANNUAL PREMIUM	FIRST PREMIUM

SUBJECT OF INSURANCE	As follows or as detailed in the Specification attached hereto which is declared to be incorporated in and to form part of this Policy.
EXCESS	The Insured shall pay: (a) The first £ of each and every claim or (b) The first £ of all claims payable in any one period aggregated.
LIMIT OF LIABILITY	£ Total Sum insured by this Policy being % of the Total Specification Sum Insured or as per Separate Amounts shown for each Section.
OTHER INTERESTS	
DELETED AT INCEPTION	
ENDORSEMENTS	
WARRANTIES	
SECTIONS/ EXTENSIONS Included	

Broker Code		Bord. Date					Bord. U/W Code Class				
U/W Year		Due Date									
Currency		Admin. Des P.P.L.				Territory Code					

TYPED BY		EXAMINED BY	DATE SIGNED	

Appendix G

DO YOU KNOW YOUR CONSTRUCTION INSURANCE INTELLIGENCE QUOTIENT (C.I.I.Q.)

Answer in Page:

1. When was insurance first transacted and where? — 3–4

2. In the design and construction of a project, which branches of the law are you likely to encounter? — 12

3. What is the difference between the word 'hazard' and 'risk'? — 28

4. What is the meaning of Risk Management and how should risks be allocated? — 36

5. Write down ten risks that exist in a construcion contract? — Chapter 3

6. Do you know what risks eventuated in the case of the Hyatt-Regency Hotel in Kanses City, U.S.A., The Emley Moor Tower, U.K.; and the Vaiont Reservoir in Italy? — 111, 67, 121

7. Explain the different between 'Responsibility', 'Liability' and 'Indemnity' in relation to a construction contract. — Chapter 5

8. To whom is responsibility owed in construction? — 145

9. What are the most significant developments in the Law of Torts in the past ten years? — 166

10. For how long are you liable in respect of a negligent act under Contract and under Tort? When does accrual start? — 223

11. What is the meaning of subrogation? — 232

12. What are the features of an insurance contract? — 217, 263

13. What are the characteristics of a construction contract? — 33, 187

14. What are the responsibilities of an insurance broker? — 208

15. What is the difference between defective design and faulty design? — 243

16. How many insurance policies are required under the Standard Form of Contract you use? What are they called? — Figure 8.3

17. What is the difference in a Public Liability policy between joint insurance and the insurance provided under the 'Principal Clause'? — 250

18. What conditions are normally imposed by your Public Liability policy and (or) Professional Indemnity policy regarding settlement of claims made against you? — Chapter 13

19. What is a 'claims made' Professional Indemnity Policy? — 365

20. What exclusions can one find in Professional Indemnity policy and (or) Public Liability policy? — 348–356

21. What is the concept of project insurance, and what is the Difference-in-Conditions Method in construction insurance? — 388

Appendix H

THE INSURANCE CLAUSES OF THE FIDIC CONDITIONS OF CONTRACT FOR WORKS OF CIVIL ENGINEERING CONSTRUCTION, PART I – GENERAL CONDITIONS, 4TH EDITION 1987, REPRINTED 1992

Care of Works

20.1 The Contractor shall take full responsibility for the care of the Works and materials and Plant for incorporation therein from the Commencement Date until the date of issue of the Taking-Over Certificate for the whole of the Works, when the responsibility for the said care shall pass to the Employer. Provided that:

(a) if the Engineer issues a Taking-Over Certificate for any Section or part of the Permanent Works the Contractor shall cease to be liable for the care of that Section or part from the date of issue of the Taking-Over Certificate, when the responsibility for the care of that Section or part shall pass to the Employer, and

(b) the Contractor shall take full responsibility for the care of any outstanding Works and materials and Plant for incorporation therein which he undertakes to finish during the Defects Liability Period until such outstanding Works have been completed pursuant to Clause 49.

Responsibility to Rectify Loss of Damage

20.2 If any loss or damage happens to the Works, or any part thereof, or materials or Plant for incorporation therein, during the period for which the Contractor is responsible for the care thereof, from any cause whatsoever, other than the risks defined in Sub-Clause 20.4, the Contractor shall, at his own cost, rectify such loss or damage so that the Permanent Works conform in every respect with the provisions of the Contract to the satisfaction of the Engineer. The Contractor shall also be liable for any loss or damage to the Works occasioned by him in the course of any operations carried out by him for the purpose of complying with his obligations under Clauses 49 and 50.

Loss or Damage Due to Employer's Risks

20.3 In the event of any such loss or damage happening from any of the risks defined in Sub-Clause 20.4, or in combination with other risks, the Contractor shall, if and to the extent required by the Engineer, rectify the loss or damage and the Engineer shall determine an addition to the Contract Price in accordance with Clause 52 and shall notify the Contract or accordingly, with a copy to the Employer. In the case of a combination of risks causing loss or damage any such determination shall take into account the proportional responsibility of the Contractor and the Employer.

Employer's Risks

20.4 The Employer's risks are:

(a) war, hostilities (whether war be declared or not), invasion, act of foreign enemies,

(b) rebellion, revolution, insurrection, or military or usurped power, or civil war,

(c) ionising radiations, or contamination by radio-activity from any nuclear fuel, or from any nuclear waste from the combustion of nuclear fuel, radioactive toxic or explosive or other hazardous properties of any explosive nuclear assembly or nuclear component thereof,

(d) pressure waves caused by aircraft or other aerial devices travelling at sonic or supersonic speeds,

(e) riot, commotion or disorder, unless solely restricted to employees of the Contractor or of his Subcontractors and arising from the conduct of the Works,

(f) loss or damage due to the use or occupation by the Employer of any Section or part of the Permanent Works, except may be provided for in the Contract,

(g) loss or damage to the extent that it is due to the design of the Works, other than any part of the design provided by the Contractor or for which the Contractor is responsible, and

(h) any operation of the forces of nature against which an experienced contractor could not reasonably have been expected to take precautions.

Insurance of Works and Contractor's Equipment

21.1 The Contractor shall, without limiting his or the Employer's obligations and responsibilities under Clause 20, insure:

(a) the Works, together with materials and Plant for incorporation therein, to the full replacement cost (the term 'cost' in this context shall include profit),

(b) an additional sum of 15 per cent of such replacement cost, or as may be specified in Part II of these Conditions, to cover any additional costs of and incidental to the rectification of loss or damage including professional fees and the cost of demolishing and removing any part of the Works and of removing debris of whatsoever nature, and

(c) the Contractor's Equipment and other things brought onto the Site by the Contractor, for a sum sufficient to provide for their replacement at the Site.

Scope of Cover

21.2 The insurance in paragraphs (a) and (b) of Sub-Clause 21.1 shall be in the joint names of the Contractor and the Employer and shall cover:

(a) the Employer and the Contractor against all loss or damage from whatsoever cause arising, other than as provided in Sub-Clause 21.4, from the start of work at the Site until the date of issue of the relevant Taking-Over Certificate in respect of the Works or any Section or part thereof as the case may be, and

(b) the Contractor for his liability:

(i) during the Defects Liability Period for loss or damage arising from a cause occurring prior to the commencement of the Defects Liability Period, and

(ii) for loss or damage occasioned by the Contractor in the course of any operations carried out by him for the purpose of complying with his obligations under Clauses 49 and 50.

Responsibility for Amounts not Recovered

21.3 Any amounts not insured or not recovered from the insurers shall be borne by the Employer or the Contractor in accordance with their responsibilities under Clause 20.

Exclusions

21.4 There shall be no obligation for the insurances in Sub-Clause 21.1 to include loss or damage caused by:

(a) war, hostilities (whether war be declared or not), invasion, act of foreign enemies,

(b) rebellion, revolution, insurrection, or military or usurped power, or civil war,

(c) ionising radiations, or contamination by radio-activity from any nuclear fuel, or from any nuclear waste from the combustion of nuclear fuel, radio-active toxic explosive or other hazardous properties of any explosive nuclear assembly or nuclear component thereof, or

(d) pressure waves caused by aircraft or other aerial devices travelling at sonic or supersonic speeds.

Damage to Persons and Property

22.1 The Contractor shall, except if and so far as the Contract provides otherwise, indemnify the Employer against all losses and claims in respect of:

(a) death of or injury to any person, or

(b) loss of or damage to any property (other than the Works).

which may arise out of or in consequence of the execution and completion of the Works and the remedying of any defects therein, and against all claims, proceedings, damages, costs, charges and expenses whatsoever in respect thereof or in relation thereto, subject to the exceptions defined in Sub-Clause 22.2.

Exceptions

22.2 The 'exceptions' referred to in Sub-Clause 22.1 are:

(a) the permanent use or occupation of land by the Works, or any part thereof,

(b) the right of the Employer to execute the Works, or any part thereof, on, over, under, in or through any land,

(c) damage to property which is the unavoidable result of the execution and completion of the Works, or the remedying of any defects therein, in accordance with the Contract, and

(d) death of or injury to persons or loss of or damage to property resulting from any act or neglect of the Employer, his agents, servants or other contractors, not being employed by the

Contractor, or in respect of any claims, proceedings, damages, costs, charges and expenses in respect thereof or in relation thereto or, where the injury or damage was contributed to by the Contractor, his servants or agents, such part of the said injury or damage as may be just and equitable having regard to the extent of the responsibility of the Employer, his servants or agents or other contractors for the injury or damage.

Indemnity by Employer
22.3 The Employer shall indemnify the Contractor against all claims, proceedings, damages, costs, charges and expenses in respect of the matters referred to in the exceptions defined in Sub-clause 22.2.

Third Party Insurance (including Employer's Property)
23.1 The Contractor shall, without limiting his or the Employer's obligations and responsibilities under Clause 22, insure, in the joint names of the Contractor and the Employer, against liabilities for death of or injury to any person (other than as provided in Clause 24) or loss of or damage to any property (other than the Works) arising out of the performance of the Contract, other than the exceptions defined in paragraphs (a), (b) and (c) of Sub-Clause 22.2

Minimum Amount of Insurance
23.2 Such insurance shall be for at least the amount stated in the Appendix to Tender.

Cross Liabilities
23.3 The insurance policy shall include a cross liability clause such that the insurance shall apply to the Contractor and to the Employer as separate insureds.

Accident or Injury to Workmen
24.1 The Employer shall not be liable for or in respect of any damages or compensation payable to any workman or other person in the employment of the Contractor or any Subcontractor, other than death or injury resulting from any act or default of the Employer, his agents or servants. The Contractor shall indemnify and keep indemnified the Employer against all such damages and compensation, other than those for which the Employer is liable as aforesaid, and against all claims, proceedings, damages, costs, charges, and expenses whatsoever in respect thereof or in relation thereto.

Insurance Against Accident to Workmen
24.2 The Contractor shall insure against such liability and shall continue such insurance during the whole of the time that any persons are employed by him on the Works. Provided that, in respect of any persons employed by any Subcontractor, the Contractor's obligations to insure as aforesaid under this Sub-Clause shall be satisfied if the Subcontractor shall have insured against the liability in respect of such persons in such manner that the Employer is indemnified under the policy, but the Contractor shall require such Subcontractor to produce to the Employer, when required, such policy of insurance and the receipt for the payment of the current premium.

Evidence and Terms of Insurances
25.1 The Contractor shall provide evidence to the Employer prior to the start of work at the Site that the insurances required under the Contract have been effected and shall, within 84 days of the Commencement Date, provide the insurance policies to the Employer. When providing such evidence and such policies to the Employer, the Contractor shall notify the Engineer of so doing. Such insurance policies shall be consistent with the general terms agreed prior to the issue of the Letter of Acceptance. The Contractor shall effect all insurances for which he is responsible with insurers and in terms approved by the Employer.

Adequacy of Insurances
25.2 The Contractor shall notify the insurers of changes in the nature, extent or programme for the execution of the Works and ensure the adequacy of the insurances at all times in accordance with the terms of the Contract and shall, when required, produce to the Employer the insurance policies in force and the receipts for payment of the current premiums.

Remedy on Contractor's Failure to Insure
25.3 If the Contractor fails to effect and keep in force any of the insurances required under the Contract, or fails to provide the policies to the Employer within the period required by Sub-Clause 25.1, then and in any such case the Employer may effect and keep in force any such insurances and pay any premium as may be necessary for that purpose and from time to time deduct the amount so paid from any monies due or to become due to the Contractor, or recover the same as a debt due from the Contractor.

Compliance with Policy Conditions
25.4 In the event that the Contractor or the Employer fails to comply with conditions imposed by the insurance policies effected pursuant to the Contract, each shall indemnify the other against all losses and claims arising from such failure.

Appendix I

THE INSURANCE CLAUSES OF THE FIDIC CONDITIONS OF CONTRACT FOR ELECTRICAL AND MECHANICAL WORKS, PART I – GENERAL CONDITIONS, 3RD EDITION 1987, REPRINTED 1988

Risk and Responsibility

Allocation of Risk and Responsibility
37.1 The Risks of loss of or damage to physical property and of death and personal injury which arise in consequence of the performance of the Contract shall be allocated between the Employer and the Contractor as follows:
(a) the Employer: the Employer's Risks as specified in Sub-Clause 37.2
(b) the Contractor: the Contractor's Risks as specified in Sub-Clause 37.3

Employer's Risks
37.2 The Employer's Risks are:
(a) war and hostilities (whether war be declared or not), invasion, act of foreign enemies;
(b) rebellion, revolution, insurrection, military or usurped power or civil war insofar as it relates to the country in which the Works are located or countries through which plant must be transported.
(c) ionising radiation or contamination by radioactivity from any nuclear fuel, or from any nuclear waste from the combustion of nuclear fuel, radioactive toxic explosives or other hazardous properties of any explosive nuclear assembly or nuclear components thereof;
(d) pressure waves caused by aircraft travelling at sonic or supersonic speed;
(e) riot, commotion or disorder, unless solely restricted to the employees of the Contractor or of his Subcontractors;
(f) use or occupation of the Works or any part thereof by the Employer;
(g) fault, error, defect or omission in the design of any part of the Works by the Engineer, Employer or those for whom the Employer is responsible;
(h) the use or occupation of the Site by the Works or as any part thereof, or for the purposes of the Contract; or interference, whether temporary or permanent with any right of way, light, air or water or with any easement, wayleaves or right of a similar nature which is the inevitable result of the construction of the Works in accordance with the Contract;
(i) the right of the Employer to construct the Works or any part thereof on, over, under, in or through any land;
(j) damage (other than that resulting from the Contractor's method of construction) which is the inevitable result of the construction of the Works in accordance with the Contract;
(k) the act, neglect or omission or breach of contract or of statutory duty of the Engineer, the Employer or other contractors engaged by the Employer or of their respective employees or agents;
and all risks which an experienced contractor could not have foreseen or, if foreseeable, against which measures to prevent loss, damage or injury from occurring could not reasonably have been taken by such contractor.

Contractor's Risks
37.3 The Contractor's Risks are all risks other than those identified as the Employer's Risks.

© FIDIC 1987

428

Care of the Works and Passing of Risk

Contractor's Responsibility for the Care of the Works

38.1 The Contractor shall be responsible for the care of the Works or any Section thereof from the Commencement Date until the Risk Transfer Date applicable thereto under Sub-Clause 38.2

    The Contractor shall also be responsible for the care of any part of the Works upon which any outstanding work is being performed by the Contractor during the Defects Liability Period until completion of such outstanding work.

Risk Transfer Date

38.2 The Risk Transfer Date in relation to the Works or a Section thereof is the earliest of either:
- (a) the date of issue of the Taking-Over Certificate, or
- (b) the date when the Engineer is deemed to have issued the Taking-Over Certificate or the Works are deemed to have been taken over in accordance with Clause 29, or
- (c) the date of expiry of the notice of termination when the Contract is terminated by the Employer or the Contractor in accordance with these Conditions.

Passing of Risk of Loss of or Damage to the Works

39.1 The risk of loss of or damage to the Works or any Section thereof shall pass from the Contractor to the Employer on the Risk Transfer Date applicable thereto.

Loss or Damage Before Risk Transfer Date

39.2 Loss of or damage to the Works or any Section thereof occurring before the Risk Transfer Date shall:
- (a) to the extent caused by any of the Contractor's Risks, be made good forthwith by the Contractor at his own cost, and
- (b) to the extent caused by any of the Employer's Risks, be made good by the Contractor at the Employer's expense if so required by the Engineer within 28 days after the occurrence of the loss or damage. The price for making good such loss and damage shall be in all circumstances reasonable and shall be agreed by the Employer and the Contractor, or in the absence of agreement, shall be fixed by arbitration under Clause 50.

Loss or Damage After Risk Transfer Date

39.3 After the Risk Transfer Date, the Contractor's liability in respect of loss of or damage to any part of the Works shall, except in the case of Gross Misconduct, be limited:
- (a) to the fulfilment of the Contractor's obligations under Clause 30 in respect of defects therein, and
- (b) to making good forthwith loss or damage caused by the Contractor during the Defects Liability Period.

Damage to Property and Injury to Persons

Contractor's Liability

40.1 Except as provided under Sub-Clause 41.1, the Contractor shall be liable for and shall indemnify the Employer against all losses, expenses and claims in respect of any loss or damage to physical property (other than the Works), death or personal injury occurring before the issue of the last Defects Liability Certificate to the extent caused by:
- (a) defective design, material or workmanship of the Contractor, or
- (b) negligence or breach of statutory duty of the Contractor, his Subcontractors or their respective employees and agents.

Employer's Liability

40.2 The Employer shall be liable for and shall indemnify the Contractor against all losses, expenses or claims in respect of loss of or damage to any physical property or of death or personal injury whenever occurring, to the extent caused by any of the Employer's Risks.

Accidents

41.1 The Contractor shall be liable for and shall indemnify the Employer against all losses, expenses or claims arising in connection with the death of or injury to any person employed by the Contractor or his Subcontractors for the purposes of the Works, unless caused by any acts or defaults of the Engineer, the Employer or other contractors engaged by the Employer or by their respective employees or agents. In the latter cases the Employer shall be liable for and shall indemnify the Contractor against all losses, expenses and claims arising in connection therewith.

Limitations of Liability

Liability for Indirect or Consequential Damage
42.1 Neither party shall be liable to the other for any loss of profit, loss of use, loss of production, loss of contracts or for any other indirect or consequential damage that may be suffered by the other, except:
(a) as expressly provided in Clause 27, and
(b) those provisions of these Conditions whereby the Contractor is expressly entitled to receive profit.

Maximum Liability
42.2 The liability of the Contractor to the Employer under these Conditions shall in no case exceed the sum stated in the Preamble or, if no such sum is stated, the Contract Price.

Liability after Expiration of Defects Liability Period
42.3 The Contractor shall have no liability to the Employer for any loss of or damage to the Employer's physical property which occurs after the expiration of the Defects Liability Period unless caused by Gross Misconduct of the Contractor.

Exclusive Remedies
42.4 The Employer and the Contractor intend that their respective rights, obligations and liabilities as provided for in these Conditions shall alone govern their rights under the Contract and in relation to the Works.
Accordingly, the remedies provided under the contract in respect of or in consequence of:
(a) any breach of contract, or
(b) any negligent act or omission, or
(c) death or personal injury, or
(d) loss or damage to any property
are, save in the case of Gross Misconduct, to be to the exclusion of any other remedy that either may have against the other under the law governing the Contract or otherwise.

Mitigation of Loss or Damage
42.5 In all cases the party claiming a breach of Contract or a right to be indemnified in accordance with the Contract shall be obliged to take all reasonable measures to mitigate the loss or damage which has occurred or may occur.

Foreseen Damages
42.6 Where either the Employer or the Contractor is liable in damages to the other these shall not exceed the damage which the party in default could reasonably have foreseen at the date of the Contract.

Insurance

The Works
43.1 The Contractor shall insure the Works in the joint names of the Contractor and the Employer to their full replacement value with deductible limits not exceeding those stated in the Preamble.
(a) from the Commencement Date until the Risk Transfer Date against any loss or damage caused by any of the Contractor's Risks and any other risks specified in the Preamble, and
(b) during the Defects Liability period against loss or damage which is caused either:
(i) by the Contractor in completing any outstanding work or complying with his obligations under Clause 30, or
(ii) by any of the Contractor's Risks which occurred prior to the Risk Transfer Date.

Contractor's Equipment
43.2 The Contractor shall insure Contractor's Equipment for its full replacement value whilst in transit to the Site, from commencement of loading until completion of unloading at the Site and while on the Site against all loss or damage caused by any of the Contractor's Risks.

Third Party Liability
43.3 The Contractor shall insure against liability to third parties for any death or personal injury and loss of or damage to any physical property arising out of the performance of the Contract and occurring before the issue of the last Defects Liability Certificate.
Such insurance shall be effected before the Contractor begins any work on the Site. The insurance shall be for not less than the amount specified in the Preamble.

Employees

43.4 The Contractor shall insure and maintain insurance against his liability under Sub-Clause 41.1.

General Requirements of Insurance Policies

43.5 The Contractor shall:

(a) whenever required by the Employer produce the policies or certificates of any insurance which he is required to effect under the Contract together with receipts for the premiums,

(b) effect all insurances for which he is responsible with an insurer and in terms approved by the Employer, and

(c) make no material alterations to the terms of any insurance without the Employer's approval. If an insurer makes any material alteration to the terms the Contractor shall forthwith notify the Employer, and

(d) in all respects comply with any conditions stipulated in the insurance policies which he is required to place under the Contract.

Permitted Exclusions from Insurance Policies

43.6 The insurance cover effected by the Contractor may exclude any of the following:

(a) the cost of making good any part of the Works which is defective or otherwise does not comply with the Contract provided that it does not exclude the cost of making good any loss or damage to any other part of the Works attributable to such defect or non-compliance,

(b) indirect or consequential loss or damage including any reductions in the Contract Price for delay,

(c) wear and tear, shortages and theft,

(d) risks relating to vehicles for which third party or other insurance is required by law.

Remedies on the Contractor's Failure to Insure

43.7 If the Contractor fails to produce evidence of insurance cover as stated in Sub-clause 43.5. (a) then the Employer may effect and keep in force such insurance. Premiums paid by the Employer for this purpose shall be deducted from the Contract Price.

Amounts not Recovered

43.8 Any amounts not recovered from the insurers shall be borne by the Employer or Contractor in accordance with their responsibilities under Clause 37.

Force Majeure

Definition of Force Majeure

44.1 Force Majeure means any circumstances beyond the control of the parties, including but not limited to:

(a) war and other hostilities, (whether war be declared or not), invasion, act of foreign enemies, mobilisation, requisition or embargo;

(b) ionising radiation or contamination by radioactivity from any nuclear fuel or from any nuclear waste from the combustion of nuclear fuel, radioactive toxic explosives, or other hazardous properties of any explosive nuclear assembly or nuclear components thereof;

(c) rebellion, revolution, insurrection, military or usurped power and civil war;

(d) riot, commotion or disorder, except where solely restricted to employees of the Contractor.

Effect of Force Majeure

44.2 Neither party shall be considered to be in default or in breach of his obligations under the Contract to the extent that performance of such obligations is prevented by any circumstances of Force Majeure which arise after the date of the Letter of Acceptance or the date when the Contract becomes effective, whichever is the earlier.

Notice of Occurrence

44.3 If either party considers that any circumstances of Force Majeure have occurred which may affect performance of his obligations he shall promptly notify the other party and the Engineer thereof.

Performance to Continue

44.4 Upon the occurrence of any circumstances of Force Majeure the Contractor shall endeavour to continue to perform his obligations under the Contract so far as reasonably practicable. The Contractor shall notify the Engineer of the steps he proposes to take including any reasonable alternative means for performance which is not prevented by Force Majeure. The Contractor shall not take any such steps unless directed so to do by the Engineer.

Additional Costs caused by Force Majeure

44.5 If the Contractor incurs additional costs in complying with the Engineer's directions under Sub-clause 44.4, the amount thereof shall be certified by the Engineer and added to the Contract Price.

Damage Caused by Force Majeure

44.6 If in consequence of Force Majeure the Works shall suffer loss or damage the Contractor shall be entitled to have the value of the work done, without regard to the loss or damage that has occurred, included in a Certificate of Payment.

Termination in Consequence of Force Majeure

44.7 If circumstances of Force Majeure have occurred and shall continue for a period of 182 days then, notwithstanding that the Contractor may by reason thereof have been granted an extension of Time for Completion of the Works, either party shall be entitled to serve upon the other 28 days' notice to terminate the Contract. If at the expiry of the period of 28 days Force Majeure shall still continue the Contract shall terminate.

Payment on Termination for Force Majeure

44.8 If the Contract is terminated under Sub-clause 44.7 the contractor shall be paid the value of the work done.

The Contractor shall also be entitled to receive:

(a) the amounts payable in respect of any preliminary items so far as the work or service comprised therein has been carried out and a proper proportion of any such item in which the work or service comprised has only been partially carried out,

(b) the cost of materials or goods ordered for the Works or for use in connection with the Works which have been delivered to the Contractor or of which the Contractor is legally liable to accept delivery. Such materials or goods shall become the property of and be at the risk of the Employer when paid for by the Employer and the Contractor shall place the same at the Employer's disposal,

(c) the amount of any other expenditure which in the circumstances was reasonably incurred by the Contractor in the expectation of completing the whole of the Works,

(d) the reasonable cost of removal of Contractor's Equipment from the Site and the return thereof to the Contractor's works in his country or to any other destination at no greater cost, and

(e) the reasonable cost of repatriation of the Contractor's staff and workmen employed wholly in connection with the Works at the date of such termination.

Release from Performance

44.9 If circumstances of Force Majeure occur and in consequence thereof under the law governing the Contract the parties are released from further performance of the Contract, the sum payable by the Employer to the Contractor shall be the same as that which would have been payable under Sub-clause 44.8 if the Contract had been terminated under Sub-clause 44.7.

Force Majeure Affecting Engineer's Duties

44.10 The provisions of Clause 44 shall also apply in circumstances where the Engineer is prevented from performing any of his duties under the Contract by reason of Force Majeure.

Appendix J

THE INSURANCE CLAUSES OF THE ICE CONDITIONS OF CONTRACT, MEASUREMENT VERSION, 7TH EDITION, 1999

Care of the Works

20 (1) (a) The Contractor shall save as in paragraph (b) hereof and subject to sub-clause (2) of this Clause take full responsibility for the care of the Works and materials plant and equipment for incorporation therein from the Works Commencement Date until the date of issue of a Certificate of Substantial Completion for the whole of the Works when the responsibility for the said care shall pass to the Employer.

(b) If the Engineer issues a Certificate of Substantial Completion for any Section or part of the Permanent Works the Contractor shall cease to be responsible for the care of that Section or part from the date of issue of that Certificate of Substantial Completion when the responsibility for the care of that Section or part shall pass to the Employer.

(c) The Contractor shall take full responsibility for the care of any work and materials plant and equipment for incorporation therein which he undertakes during the Defects Correction Period until such work has been completed.

Excepted Risks

(2) The Excepted Risks for which the Contractor is not liable are loss or damage to the extent that it is due to

(a) the use or occupation by the Employer his agents servants or other contractors (not being employed by the Contractor) of any part of the Permanent Works

(b) any fault defect error or omission in the design of the Works (other than a design provided by the Contractor pursuant to his obligations under the Contract)

(c) riot war invasion act of foreign enemies or hostilities (whether war be declared or not)

(d) civil war rebellion revolution insurrection or military or usurped power

(e) ionising radiations or contamination by radioactivity from any nuclear fuel or from any nuclear waste from the combustion of nuclear fuel radioactive toxic explosive or other hazardous properties of any explosive nuclear assembly or nuclear component thereof and

(f) pressure waves caused by aircraft or other aerial devices travelling at sonic or supersonic speeds.

Rectification of loss or damage

(3) (a) In the event of any loss or damage to

(i)   the Works or any Section or part thereof or

(ii) materials plant or equipment for incorporation therein

while the Contractor is responsible for the care thereof (except as provided in sub-clause (2) of this Clause) the Contractor shall at his own cost rectify such loss or damage so that the Permanent Works conform in every respect with the provisions of the Contract and the Engineer's instructions. The Contractor shall also be liable for any loss or damage to the Works occasioned by him in the course of any operations carried out by him for the purpose of complying with his obligations under Clauses 49 and 50.

(b) Should any such loss or damage arise from any of the Excepted Risks defined in sub-clause (2) of this Clause the Contractor shall if and to the extent required by the Engineer rectify the loss or damage at the expense of the Employer.

(c) In the event of loss or damage arising from an Excepted Risk and a risk for which the Contractor is responsible under sub-clause (1)(a) of this Clause then the Engineer shall when determining the expense to be borne by the Employer under the Contract apportion the cost of rectification into that part caused by the Excepted Risk and that part which is the responsibility of the Contractor.

Insurance of Works etc.

21 (1) The Contractor shall without limiting his or the Employer's obligations and responsibilities under Clause 20 insure in the joint names of the Contractor and the Employer the Works together with materials plant and equipment for incorporation therein to the full replacement cost plus an additional 10% to cover any additional costs that may arise incidental to the rectification of any loss or damage including professional fees cost of demolition and removal of debris.

Extent of cover

(2) (a) The insurance required under sub-clause (1) of this Clause shall cover the Employer and the Contractor against all loss or damage from whatsoever cause arising other than the Excepted Risks defined in Clause 20(2) from the Works Commencement Date until the date of issue of the relevant Certificate of Substantial Completion.

(b) The insurance shall extend to cover any loss or damage arising during the Defects Correction Period from a cause occurring prior to the issue of any Certificate of Substantial Completion and any loss or damage occasioned by the Contractor in the course of any operation carried out by him for the purpose of complying with his obligations under Clauses 49, 50 and 51.

(c) Nothing in this Clause shall render the Contractor liable to insure against the necessity for the repair or reconstruction of any work constructed with materials or workmanship not in accordance with the requirements of the Contract unless the Bill of Quantities provides a special item for this insurance.

(d) Any amounts not insured or not recovered from insurers whether as excesses carried under the policy or otherwise shall be borne by the Contractor or the Employer in accordance with their respective responsibilities under Clause 20.

Damage to persons and property

22 (1) The Contractor shall except if and so far as the Contract provides otherwise and subject to the exceptions set out in sub-clause (2) of this Clause indemnify and keep indemnified the Employer against all losses and claims in respect of

(a) death of or injury to any person or

(b) loss of or damage to any property (other than the Works)

which may arise out of or in consequence of the construction of the Works and the remedying of any defects therein and against all claims demands proceedings damages costs charges and expenses whatsoever in respect thereof or in relation thereto.

Exceptions

(2) The exceptions referred to in sub-clause (1) of this Clause which are the responsibility of the Employer are

(a) damage to crops being on the Site (save in so far as possession has not been given to the Contractor)

(b) the use or occupation of land provided by the Employer for the purposes of the Contract (including consequent losses of crops) or interference whether temporary or permanent with any right of way light air or water or other easement or quasi-easement which are the unavoidable result of the construction of the Works in accordance with the Contract

(c) the right of the Employer to construct the Works or any part thereof on over under in or through any land

(d) damage which is the unavoidable result of the construction of the Works in accordance with the Contract and

(e) death of or injury to persons or loss of or damage to property resulting from any act neglect or breach of statutory duty done or committed by the Employer his agents servants or other contractors (not being employed by the Contractor) or for or in respect of any claims demands proceedings damages costs charges and expenses in respect thereof or in relation thereto.

Indemnity by Employer

(3) The Employer shall subject to sub-clause (4) of this Clause indemnify the Contractor against all claims demands proceedings damages costs charges and expenses in respect of the matters referred to in the exceptions defined in sub-clause (2) of this Clause.

Shared responsibility

(4) (a) The Contractor's liability to indemnify the Employer under sub-clause (1) of this Clause shall be reduced in proportion to the extent that the act or neglect of the Employer his agents servants or other contractors (not being employed by the Contractor) may have contributed to the said death injury loss or damage.

(b) The Employer's liability to indemnify the Contractor under sub-clause (3) of this Clause in respect of matters referred to in sub-clause (2)(e) of this Clause shall be reduced in proportion to the extent that the act or neglect of the Contractor or his sub-contractors servants or agents may have contributed to the said death injury loss or damage.

Third party insurance

23 (1) The Contractor shall without limiting his or the Employer's obligations and responsibilities under Clause 22 insure in the joint names of the Contractor and the Employer against liabilities for death of or injury to any person (other than any operative or other person in the employment of the Contractor or any of his sub-contractors) or loss of or damage to any property (other than the Works) arising out of the performance of the Contract other than those liabilities arising out of the exceptions defined in Clause 22(2)(a) (b) (c) and (d).

Cross liability clause

(2) The insurance policy shall include a cross liability clause such that the insurance shall apply to the Contractor and to the Employer as separate insured.

Amount of insurance

(3) Such insurance shall be for at least the amount stated in the Appendix to the Form of Tender.

Accident or injury to operatives etc.

24 The Employer shall not be liable for or in respect of any damages or compensation payable at law in respect or in consequence of any accident or injury to any operative or other person in the employment of the Contractor or any of his sub-contractors save and except to the extent that such accident or injury results from or is contributed to by any act or default of the Employer his agents or servants and the Contractor shall indemnify and keep indemnified the Employer against all such damages and compensation (save and except as aforesaid) and against all claims demands proceedings costs charges and expenses whatsoever in respect thereof or in relation thereto.

Evidence and terms of insurance

25 (1) The Contractor shall provide satisfactory evidence to the Employer prior to the Works Commencement Date that the insurances required under the Contract have been effected and shall if so required produce the insurance policies for inspection. The terms of all such insurances shall be subject to the approval of the Employer (which approval shall not unreasonably be withheld). The Contractor shall upon request produce to the Employer receipts for the payment of current insurance premiums.

Excesses

(2) Any excesses on the policies of insurance effected under Clauses 21 and 23 shall be no greater than those stated in the Appendix to the Form of Tender.

Remedy on Contractor's failure to insure

(3) If the Contractor shall fail upon request to produce to the Employer satisfactory evidence that there is in force any of the insurances required under the Contract then the Employer may effect and keep in force any such insurance and pay such premium or premiums as may be necessary for that purpose and from time to time deduct the amount so paid from any monies due or which may become due to the Contractor or recover the same as a debt due from the Contractor.

Compliance with policy conditions

(4) Both the Employer and the Contractor shall comply with all conditions laid down in the insurance policies. Should the Contractor or the Employer fail to comply with any condition imposed by the insurance policies effected pursuant to the Contract each shall indemnify the other against all losses and claims arising from such failure.

THE INSURANCE CLAUSES OF THE ICE CONDITIONS OF CONTRACT, DESIGN AND CONSTRUCT, 2ND EDITION, 2001

Care of the Works

20 (1) (a) The Contractor shall save as in paragraph (b) hereof and subject to sub-clause (2) of this Clause take full responsibility for the care of the Works and for materials plant and equipment for incorporation therein from Commencement Date until the date of issue of a Certificate of Substantial Completion for the whole of the Works when the responsibility for the said care shall pass to the Employer.

(b) If the Employer's Representative issues a Certificate of Substantial Completion for any Section or part of the Permanent Works the Contractor shall cease to be responsible for the care of that Section or part from the date of issue of that Certificate of Substantial Completion when the responsibility for the care of that Section or part shall pass to the Employer. Provided always that the Contractor shall remain responsible for any damage to such completed work caused by or as a result of his other activities on the Site.

(c) The Contractor shall take full responsibility for the care of any outstanding work and materials plant and equipment for incorporation therein which he undertakes to finish during the Defects Correction Period until such outstanding work has been completed.

Expected risks

(2) The Expected Risks for which the Contractor is not liable are loss and damage to the extent that it is due to

(a) the use or occupation by the Employer his agents servants or other contractors (not being employed by the Contractor) of any part of the Permanent Works

(b) any fault error or omission in the design of the Works for which the Contractor is not responsible under the Contract

(c) riot war invasion of foreign enemies or hostilities (whether war be declared or not)

(d) civil war rebellion revolution insurrection or military or usurped power

(e) ionising radiations or contamination by radioactivity from any nuclear fuel or from any nuclear waste from the combustion of nuclear fuel radioactive toxic explosive or other hazardous properties of any explosive nuclear assembly or nuclear component thereof and

(f) pressure waves caused by aircraft or other aerial devices travelling at sonic or supersonic speeds.

Rectification of loss or damage

(3) (a) In the event of any loss or damage to

(i) the Works or any Section or part thereof or

(ii) materials plant or equipment for incorporation therein

while the Contractor is responsible for the care thereof (except as provided in sub-clause (2) of this Clause) the Contractor shall at his own cost rectify such loss or damage so that the Permanent Works conform in every respect with the provisions of the Contract. The Contractor shall also be liable for any loss or damage to the Works occasioned by him in the course of any operations carried out by him for the purpose of complying with his obligations under Clauses 49 and 50.

(b) Should any loss or damage arise from any of the Excepted Risks defined in sub-clause (2) of this Clause the Contractor shall if and to the extent required by the Employer's Representative rectify the loss or damage at the expense of the Employer.

(c) In the event of loss or damage arising from an Excepted Risk and a risk for which the Contractor is responsible under sub-clause (1)(a) of this Clause then the cost of rectification shall be apportioned accordingly.

Insurance of Permanent and Temporary Works etc

21 (1) The Contractor shall without limiting his or the Employer's obligations and responsibilities under Clause 20 insure in the joint names of the Contractor and the Employer the Permanent and Temporary Works together with materials plant and equipment for incorporation therein to the full replacement cost plus an additional 10% to cover any additional costs that may arise incidental to the rectification of any loss or damage including professional fees cost of demolition and removal of debris.

Extent of cover

(2) (a) The insurance required under sub-clause (1) of this Clause shall cover the Employer and the Contractor against all loss or damage from whatsoever cause arising other than the Excepted Risks defined in Clause 20(2) from the Commencement Date until the date of issue of the relevant Certificate of Substantial Completion.

(b) The insurance shall extend to cover any loss or damage arising during the Defects Correction Period from a cause occurring prior to the issue of any Certificate of Substantial Completion and any loss or damage occasioned by the Contractor in the course of any operation carried out by him for the purpose of complying with his obligations under Clauses 49 and 50.

(c) Nothing in this Clause shall render the Contractor liable to insure against the necessity for the repair or reconstruction of any work constructed with materials or workmanship not in accordance with the requirements of the Contract unless the Contract otherwise requires.

(d) Any amounts not insured or not recovered from insurers whether as excesses carried out under the policy or otherwise shall be borne by the Contractor or the Employer in accordance with their respective responsibilities under Clause 20.

Damage to persons or property

22 (1) The Contractor shall except if and so far as the Contract provides otherwise and subject to the exceptions set out in sub-clause (2) of this Clause indemnify and keep indemnified the Employer against all losses and claims in respect of

(a) death of or injury to any person or

(b) loss of or damage to any property (other than the Works)

which may arise out of or in consequence of the design and construction of the Works and the remedying of any defects therein and against all claims demands proceedings damages costs charges and expenses whatsoever in respect thereof or in relation thereto.

Exceptions

(2) The exceptions referred to in sub-clause (1) of this Clause which are the responsibility of the Employer are

(a) damage to crops being on the Site (save in so far as possession has not been given to the Contractor)

(b the use or occupation of land provided by the Employer for the purposes of the Contract (including consequent losses of crops) or interference whether temporary or permanent with any right of way light air or water or other easement or quasi-easement which is the unavoidable result of the construction of the Works in accordance with the Contract

(c) the right of the Employer to construct the Works or any part thereof on over under in or through any land

(d) damage which is the unavoidable result of the construction of the Works in accordance with the Employer's Requirements including any design for which the Contractor is not responsible under the Contract

(e) death of or injury to persons or loss of or damage to property resulting from any act neglect or breach of statutory duty done or committed by the Employer his agents servants or other contractors (not being employed by the Contractor) or for or in respect of any claims demands proceedings damages costs charges and expenses in respect thereof or in relation thereto.

Indemnity by Employer

(3) The Employer shall subject to sub-clause (4) of this Clause indemnify the Contractor against all claims demands proceedings damages costs charges and expenses in respect of the matters referred to in the exceptions defined in sub-clause (2) of this Clause.

Shared responsibility

(4) (a) The Contractor's liability to indemnify the Employer under sub-clause (1) of this Clause shall be reduced in proportion to the extent that the act or neglect of the Employer his agents' servants or other contractors (not being employed by the Contractor) may have contributed to the said death injury loss or damage.

(b) The Employer's liability to indemnify the Contractor under sub-clause (3) of this Clause in respect of matters referred to in sub-clause (2)(e) of this Clause shall be reduced in proportion to the extent that the act or neglect of the Contractor to his sub-contractors servants or agents may have contributed to the said death injury loss or damage.

Third party insurance

23 (1) The Contractor shall without limiting his or the Employer's obligations and responsibilities under Clause 22 insure in the joint names of the Contractor and the Employer against liabilities for death of or injury to any person (other than any operative or other person in the employment of the Contractor or any of his sub-contractors) or loss of or damage to any property (other than the Works) arising out of the performance of the Contract other than those liabilities arising out of the exceptions defined in Clause 22 (2)(a)(b)(c) and (d).

(2) The insurance policy shall include a cross liability clause such that the insurance shall apply to the Contractor and to the Employer as separate insured.

(3) Such insurance shall be for at least the amount stated in the Appendix to the Form of Tender.

Accident or injury to operatives etc

24 The Employer shall not be liable for or in respect of any damages or compensation payable at law in respect or in consequence of any accident or injury to any operative or other person in the employment of the Contractor or any of his sub-contractors save and except and to the extent that such accident or injury results from or is contributed to by any act or default of the Employer his agents or servants and the Contractor shall indemnify and keep indemnified the Employer against all such damages and compensation (save and except as aforesaid) and against all claims demands proceedings costs charges and expenses whatsoever in respect thereof or in relation thereto.

Evidence and terms of insurance

25 (1) The Contractor shall provide satisfactory evidence to the Employer prior to the Commencement Date that the insurances required under the Contract have been effected and shall if so required produce the insurance policies for inspection. The terms of all such insurances shall be subject to the approval of the Employer (which approval shall not unreasonably be withheld). The Contractor shall upon request produce to the Employer receipts for the payment of current insurance premiums.

Excesses

(2) Any excesses on the policies of insurance effected under Clauses 21 and 23 shall be as stated by the Contractor in the Appendix to the Form of Tender.

Remedy on Contractor's failure to insure

(3) If the Contractor fails upon request to produce to the Employer satisfactory evidence that there is in force any of the insurances required under the Contract then the Employer may effect and keep in force any such insurance and pay such premium or premiums as may be necessary for that purpose and from time to time deduct the amount so paid from any monies due or which may become due to the Contractor or recover the same as a debt due from the Contractor.

Compliance with policy conditions

(4) Both the Employer and the Contractor shall comply with all conditions laid down in the insurance policies. Should the Contractor or the Employer fail to comply with any condition imposed by the insurance policies effected pursuant to the Contract each shall indemnify the other against all losses and claims arising from such failure.

Appendix L

THE INSURANCE CLAUSES OF THE FIDIC CONDITIONS OF CONTRACT FOR CONSTRUCTION, 1ST EDITION, 1999

Risk and Responsibility

Indemnities

17.1 The Contractor shall indemnify and hold harmless the Employer, the Employer's Personnel, and their respective agents, against and from all claims, damages, losses and expenses (including legal fees and expenses) in respect of:

(a) bodily injury, sickness, disease or death, of any person whatsoever arising out of or in the course of or by reason of the Contractor's design (if any), the execution and completion of the Works and the remedying of any defects, unless attributable to any negligence, wilful act or breach of the Contract by the Employer, the Employer's Personnel, or any of their respective agents, and

(b) damage to or loss of any property, real or personal (other than the Works), to the extent that such damage or loss:

   (i) arises out of or in the course of or by reason of the Contractor's design (if any), the execution and completion of the Works and the remedying of any defects, and

   (ii) is attributable to any negligence, wilful act or breach of the Contract by the Contractor, the Contractor's Personnel, their respective agents, or anyone directly or indirectly employed by any of them.

The Employer shall indemnify and hold harmless the Contractor, the Contractor's Personnel, and their respective agents, against and from all claims, damages, losses and expenses (including legal fees and expenses) in respect of (1) bodily injury, sickness, disease or death, which is attributable to any negligence, wilful act or breach of the Contract by the Employer, the Employer's Personnel, or any of their respective agents, and (2) the matters for which liability may be excluded from insurance cover, as described in sub-paragraphs (d)(i), (ii) and (iii) of Sub-Clause 18.3 [Insurance Against Injury to Persons and Damage to Property].

Contractor's Care of the Works

17.2 The Contractor shall take full responsibility for the care of the Works and Goods from the Commencement Date until the Taking-Over Certificate is issued (or is deemed to be issued under Sub-Clause 10.1 [Taking Over of the Works and Sections]) for the Works, when responsibility for the care of the Works shall pass to the Employer. If a Taking-Over Certificate is issued (or is so deemed to be issued) for any Section or part of the Works, responsibility for the care of the Section or part shall then pass to the Employer.

After responsibility has accordingly passed to the Employer, the Contractor shall take responsibility for the care of any work which is outstanding on the date stated in a Taking-Over Certificate, until this outstanding work has been completed.

If any loss or damage happens to the Works, Goods or Contractor's Documents during the period when the Contractor is responsible for their care, from any cause not issued in Sub-Clause 17.3 [Employer's Risks], the Contractor shall rectify the loss or damage at the Contractor's risk and cost, so that the Works, Goods and Contractor's Documents conform with the Contract.

The Contractor shall be liable for any loss or damage caused by any actions performed by the Contractor after a Taking-Over Certificate has been issued. The Contractor shall also be liable for any loss or damage which occurs after a Taking-Over Certificate has been issued and which arose from a previous event for which the Contractor was liable.

APPENDIX L

Employer's Risks

17.3   The risks referred to in Sub-Clause 17.4 below are:

(a) war, hostilities (whether war be declared or not), invasion, act of foreign enemies,

(b) rebellion, terrorism, revolution, insurrection, military or usurped power, or civil war, within the Country,

(c) not, commotion or disorder within the Country by persons other than the Contractor's Personnel and other employees of the Contractor and Subcontractors,

(d) munitions of war, explosive materials, ionising radiation or contamination by radio activity, within the Country, except as may be attributable to the Contractor's use of such munitions, explosives, radiation or radio-activity,

(e) pressure waves caused by aircraft or other aerial devices travelling at sonic or supersonic speeds,

(f) use or occupation by the Employer of any part of the Permanent Works, except as may be specified in the Contract,

(g) design of any part of the Works by the Employer's Personnel or by others for whom the Employer is responsible, and

(h) any operation of the forces of nature which is Unforeseeable or against which an experienced contractor could not reasonably have been expected to have taken adequate preventative precautions.

Consequences of Employer's Risks

17.4   If and to the extent that any of the risks listed in Sub-Clause 17.3 above results in loss or damage to the Works, Goods or Contractor's Documents, the Contractor shall promptly give notice to the Engineer and shall rectify this loss or damage to the extent required by the Engineer.

If the Contractor suffers delay and/or incurs Cost from rectifying this loss or damage, the Contractor shall give a further notice to the Engineer and shall be entitled subject to Sub-Clause 20.1 [Contractor's Claims] to:

(a) an extension of time for any such delay, if completion is or will be delayed, under Sub-Clause 8.4 [Extension of Time for Completion], and

(b) payment of any such Cost, which shall be included in the Contract price. In the case of sub-paragraphs (f) and (g) of Sub-Clause 17.3 [Employer's Risks], reasonable profit on the Cost shall also be included.

After receiving this further notice, the Engineer shall proceed in accordance with Sub-Clause 3.5 [Determinations] to agree or determine these matters.

Intellectual and Industrial Property Rights

17.5   In this Sub-Clause, 'infringement' means an infringement (or alleged infringement) of any patent, registered design, copyright, trade mark, trade name, trade secret or other intellectual or industrial property right relating to the Works; and 'claim' means a claim (or proceedings pursuing a claim) alleging an infringement.

Whenever a Party does not give notice to the other Party of any claim within 28 days of receiving the claim, the first Party shall be deemed to have waived any right to indemnity under this Sub-Clause.

The Employer shall indemnify and hold the Contractor harmless against and from any claim alleging an infringement which is or was:

(a) an unavoidable result of the Contractor's compliance with the Contract, or

(b) a result of any Works being used by the Employer:

(i) for a purpose other than that indicated by, or reasonably to be inferred from, the Contract, or

(ii) in conjunction with any thing not supplied by the Contractor, unless such use was disclosed to the Contractor prior to the Base Date or is stated in the Contract.

The Contractor shall indemnify and hold the Employer harmless against and from any other claim which arises out of or in relation to (i) the manufacture, use, sale or import of any Goods, or (ii) any design for which the Contractor is responsible.

If a Party is entitled to be indemnified under this Sub-Clause, the indemnifying Party may (at its cost) conduct negotiations for the settlement of the claim, and any litigation or arbitration which may arise from it. The other Party shall, at the request and cost of the indemnifying Party, assist in contesting the claim. This other Party (and its Personnel) shall not make any admission which might be prejudicial to the indemnifying Party, unless the indemnifying Party failed to take over the conduct of any negotiations, litigation or arbitration upon being requested to do so by such other Party.

Limitation of Liability

17.6 Neither Party shall be liable to the other Party for loss of use of any Works, loss of profit, loss of any contract or for any indirect or consequential loss or damage which may be suffered by the other Party in connection with the Contract, other than under Sub-Clause 16.4 [Payment on Termination] and Sub-Clause 17.1 [Indemnities].

The total liability of the Contractor to the Employer, under or in connection with the Contract other than under Sub-Clause 4.19 [Electricity, Water and Gas], Sub-Clause 4.20 [Employer's Equipment and Free-Issue Material], Sub-Clause 17.1 [Indemnities] and Sub-Clause 17.5 [Intellectual and Industrial Property Rights], shall not exceed the sum stated in the Particular Conditions or (if a sum is not so stated) the Accepted Contract Amount.

The Sub-Clause shall not limit liability in any case of fraud, deliberate default or reckless misconduct by the defaulting Party.

Insurance

General Requirements for Insurances

18.1 In this Clause, 'insuring Party' means, for each type of insurance, the Party responsible for effecting and maintaining the insurance specified in the relevant Sub-Clause.

Wherever the Contractor is the insuring Party, each insurance shall be effected with insurers and in terms approved by the Employer. These terms shall be consistent with any terms agreed by both Parties before the date of the Letter of Acceptance. This agreement of terms shall take precedence over the provisions of this Clause.

Wherever the Employer is the insuring Party, each insurance shall be effected with insurers and in terms consistent with the details annexed to the Particular Conditions.

If a policy is required to indemnify joint insured, the cover shall apply separately to each insured as though a separate policy had been issued for each of the joint insured. If a policy indemnifies additional joint insured, namely in addition to the insured specified in this Clause, (i) the Contractor shall act under the policy on behalf of these additional joint insured except that the Employer shall act for Employer's Personnel, (ii) additional joint insured shall not be entitled to receive payments directly from the insurer or to have any other direct dealings with the insurer, and (iii) the insuring Party shall require all additional joint insured to comply with the conditions stipulated in the policy.

Each policy insuring against loss or damage shall provide for payments to be made in the currencies required to rectify the loss or damage. payments received from insurers shall be used for the rectification of the loss or damage.

The relevant insuring Party shall, within the respective periods stated in the Appendix to Tender (calculated from the Commencement Date), submit to the other Party:

(a) evidence that the insurances described in this Clause have been effected, and

(b) copies of the policies for the insurances described in Sub-Clause 18.2 [insurance for Works and Contractor's Equipment] and Sub-Clause 18.3 [Insurance against Injury to Persons and Damage to Property].

When each premium is paid, the insuring Party shall submit evidence of payment to the other Party. Whenever evidence or policies are submitted, the insuring Party shall also give notice to the Engineer.

Each Party shall comply with the conditions stipulated in each of the insurance policies. The insuring Party shall keep the insurers informed of any relevant changes to the execution of the Works and ensure that insurance is maintained in accordance with this Clause.

Neither Party shall make any material alteration to the terms of any insurance without the prior approval of the other Party. If an insurer makes (or attempts to make) any alteration, the Party first notified by the insurer shall promptly give notice to the other Party.

If the insuring Party fails to effect and keep in force any of the insurances it is required to effect and maintain under the Contract, or fails to provide satisfactory evidence and copies of policies in accordance with this Sub-Clause, the other Party may (at its option and without prejudice to any other right or remedy) effect insurance for the relevant coverage and pay the premiums due. The insuring Party shall pay the amount of these premiums to the other Party, and the Contract Price shall be adjusted accordingly.

Nothing in this Clause limits the obligations, liabilities or responsibilities of the Contractor or the Employer, under the other terms of the Contract or otherwise. Any amounts not insured or not recovered from the insurers shall be borne by the Contractor and/or the Employer in accordance with these obligations, liabilities or responsibilities. However, if the insuring Party fails to effect and keep in force an insurance which is available and which it is required to effect

441

and maintain under the Contract, and the other Party neither approves the omission nor effects insurance for the coverage relevant to this default, any moneys which should have been recoverable under this insurance shall be paid by the insuring Party.

Payments by one Party to the other Party shall be subject to Sub-Clause 2.5 [Employer's Claims] or Sub-Clause 20.1 [Contractor's Claims], as applicable.

Insurance for Works and Contractor's Equipment

18.2   The insuring Party shall insure the Works, Plant, Materials and Contractor's Documents for not less than the full reinstatement cost including the costs of demolition, removal of debris and professional fees and profit. This insurance shall be effective from the date by which the evidence is to be submitted under sub-paragraph (a) of Sub-Clause 18.1 [General Requirements for Insurances], until the date of issue of the Taking-Over Certificate for the Works.

The insuring Party shall maintain this insurance to provide cover until the date of issue of the Performance Certificate, for loss or damage for which the Contractor is liable arising from a cause occurring prior to the issue of the Taking-Over Certificate, and for loss or damage caused by the Contractor in the course of any other operations (including those under Clause 11 [Defects Liability]).

The insuring Party shall insure the Contractor's Equipment for not less than the full replacement value, including delivery to Site. For each item of Contractor's Equipment, the insurance shall be effective while it is being transported to the Site and until it is no longer required as Contractor's Equipment.

Unless otherwise stated in the Particular Conditions, insurances under this Sub-Clause:

(a) shall be effected and maintained by the Contractor as insuring Party,

(b) shall be in the joint names of the Parties, who shall be jointly entitled to receive payments from the insurers, payments being held or allocated between the Parties for the sole purpose of rectifying the loss or damage,

(c) shall cover all loss and damage from any cause not listed in Sub-Clause 17.3 [Employer's Risks],

(d) shall also cover loss or damage to a part of the Works which is attributable to the use or occupation by the Employer of another part of the Works, and loss or damage from the risks listed in sub-paragraphs (c), (g) and (h) of Sub-Clause 17.3 [Employer's Risks], excluding (in each case) risks which are not insurable at commercially reasonable terms, with deductibles per occurrence of not more than the amount stated in the Appendix to Tender (if an amount is not so stated, this sub-paragraph (d) shall not apply), and

(e) may however exclude loss of, damage to, and reinstatement of:

    (i)   a part of the Works which is in a defective condition due to a defect in its design, materials or workmanship (but cover shall include any other parts which are lost or damaged as a direct result of this defective condition and not as described in sub-paragraph (ii) below),

    (ii) a part of the Works which is lost or damaged in order to reinstate any other part of the Works if this other part is in a defective condition due to a defect in its design, materials or workmanship,

    (iii) a part of the Works which has been taken over by the Employer, except to the extent that the Contractor is liable for the loss or damage, and

    (iv) Goods while they are not in the Country, subject to Sub-Clause 14.5 [Plant and Materials Intended for the Works].

If, more than one year after the Base Date, the cover described in sub-paragraph (d) above ceases to be available to a commercially reasonable terms, the Contractor shall (as insuring Party) give notice to the Employer, with supporting particulars. The Employer shall then (i) be entitled subject to Sub-Clause 2.5 [Employer's Claims] to payment of an amount equivalent to such commercially reasonable terms as the Contractor should have expected to have paid for such cover, and (ii) be deemed, unless he obtains the cover at commercially reasonable terms, to have approved the omission under Sub-Clause 18.1 [General Requirements for Insurances].

Insurance against Injury to Persons and Damage to Property

18.3   The insuring Party shall insure against each Party's liability for any loss, damage, death or bodily injury which may occur to any physical property (except things insured under Sub-Clause 18.2 [Insurance for Works and Contractor's Equipment]) or to any person (except persons insured under Sub-Clause 18.4 [Insurance for Contractor's Personnel]), which may arise out of the Contractor's Personnel]), which may arise out the Contractor's performance of the Contract and occurring before the issue of the Performance Certificate.

This insurance shall be for a limit per occurrence of not less than the amount stated in the Appendix to Tender, with no limit on the number of occurrences. If an amount is not stated in the Appendix to Tender, this Sub-Clause shall not apply.

Unless otherwise stated in the Particular Conditions, the insurances specified in this Sub-Clause:

(a) shall be effected and maintained by the Contractor as insuring Party,

(b) shall be in the joint names of the Parties,

(c) shall be extended to cover liability for all loss and damage to the Employer's property (except things insured under Sub-Clause 18.2) arising out of the Contractor's performance of the Contract, and

(d) may however exclude liability to the extent that it arises from:

    (i) the Employer's right to have the Permanent Works executed on, over, under, in or through any land, and to occupy this land for the Permanent Works,

    (ii) damage which is an unavoidable result of the Contractor's obligations to execute the Works and remedy any defects, and

    (iii) a cause listed in Sub-Clause 17.3 [Employer's Risks], except to the extent that cover is available at commercially reasonable terms.

Insurance for Contractor's Personnel

18.4   The Contractor shall effect and maintain insurance against liability for claims, damages, losses and expenses (including legal fees and expenses) arising from injury, sickness, disease or death of any person employed by the Contractor or any other of the Contractor's Personnel.

    The Employer and the Engineer shall also be indemnified under the policy of insurance, except that this insurance may exclude losses and claims to the extent that they arise from any act or neglect of the Employer or of the Employer's Personnel.

    The insurance shall be maintained in full force and effect during the whole time that these personnel are assisting in the execution of the Works. For a Subcontractor's employees, the insurance may be effected by the Subcontractor, but the Contractor shall be responsible for compliance with this Clause.

Force Majeure

Definition of force Majeure

19.1   In this Clause, 'Force Majeure' means an exceptional event or circumstance:

(a) which is beyond a Party's control,

(b) which such Party could not reasonably have provided against before entering into the Contract,

(c) which, having arisen, such Party could not reasonably have avoided or overcome, and

(d) which is not substantially attributable to the other Party.

Force Majeure may include, but is not limited to, exceptional events or circumstances of the kind listed below, so long as conditions (a) to (d) above are satisfied:

    (i) war, hostilities (whether war be declared or not), invasion, act of foreign enemies,

    (ii) rebellion, terrorism, revolution, insurrection, military or usurped power, or civil war,

    (iii) riot, commotion, disorder, strike or lockout by persons other than the Contractor's Personnel and other employees of the Contractor and Sub-contractors,

    (iv) munitions of war, explosive materials, ionising radiation or contamination by radio activity, except as may be attributable to the Contractor's use of such munitions, explosives, radiation or radio activity, and

    (v) natural catastrophes such as eathquake, hurricane, typhoon or volcanic activity.

Notice of Force Majeure

19.2   If a Party is or will be prevented from performing any of its obligations under the Contract by Force Majeure, then it shall give notice to the other Party of the event or circumstances constituting the Force Majeure and shall specify the obligations, the performance of which is or will be prevented. The notice shall be given within 14 days after the Party became aware, (or should have become aware), of the relevant event or circumstance constituting Force Majeure.

    The Party shall, having giving notice, be excused performance fo such obligations for so long as such Force Majeure prevents it from performing them.

    Notwithstanding any other provision of this Clause, Force Majeure shall not apply to obligations of either Party to make payments to the other Party under the Contract.

Duty to Minimise Delay

19.3   Each Party shall at all times use all reasonable endeavours to minimise any delay in the performance of the Contract as a result of Force Majeure.

    A Party shall given notice to the other Party when it ceases to be affected by the Force Majeure.

Consequences of Force Majeure

19.4   If the Contractor is prevented from performing any of his obligations under the Contract by Force Majeure of which notice has been given under Sub-Clause 19.2 [Notice of Force Majeure], and suffers delay and/or incurs Cost by reason of such Force Majeure, the Contractor shall be entitled subject to Sub-Clause 20.1 [Contractor's Claims] to:

(a) an extension of time for any such delay, if completion is or will be delayed, under Sub-Clause 8.4 [Extension of Time for Completion], and

(b) if the event or circumstance is of the kind described in sub-paragraphs (i) to (iv) of Sub-Clause 19.1 [Definition of Force Majeure] and, in the case of sub-paragraphs (ii) to (iv), occurs in the Country, payment of any such Cost.

After receiving this notice, the Engineer shall proceed in accordance with Sub-Clause 3.5 [Determinations] to agree to determine these matters.

Force Majeure Affecting Subcontractor

19.5   If any Subcontractor is entitled under any contract or agreement relating to the Works to relief from force majeure on terms additional to or broader than those specified in this Clause, such additional or broader force majeure events or circumstances shall not excuse the Contractor's non-performance or entitle him to relief under this Clause.

Optional Termination, Payment and Release

19.6   If the execution of substantially all the Works in progress is prevented for a continuous period of 84 days by reason of Force Majeure of which notice has been given under Sub-Clause 19.2 [Notice of Force Majeure], or for multiple periods which total more than 140 days due to the same notified Force Majeure, then either Party may give to the other Party a notice of termination of the Contract. In this event, the termination shall take effect 7 days after the notice is given, and the Contractor shall proceed in accordance with Sub-Clause 16.3 [Cessation of Work and Removal of Contractor's Equipment].

Upon such termination, the Engineer shall determine the value of the work done and issue a Payment Certificate which shall include:

(a) the amounts payable for any work carried out for which a price is stated in the Contract;

(b) the Cost of Plant and Materials ordered for the Works which have been delivered to the Contractor, or of which the Contractor is liable to accept delivery; this Plant and Materials shall become the property of (and be at the risk of) the Employer when paid for by the Employer, and the Contractor shall place the same at the Employer's disposal;

(c) any other Cost or liability which in the circumstances was reasonably incurred by the Contractor in the expectation of completing the Works;

(d) the Cost of removal of Temporary Works and Contractor's Equipment from the Site and the return of these items to the Contractor's works in his country (or to any other destination at no greater cost); and

(e) the Cost of repatriation of the Contractor's staff and labour employed wholly in connection with the Works at the date of termination.

Release from Performance under the Law

19.7   Notwithstanding any other provision of this Clause, if any event or circumstance outside the control of the Parties (including, but not limited to, Force Majeure) arises which makes it impossible or unlawful for either or both Parties to fulfil its or their contractual obligations or which, under the law governing the Contract, entitles the Parties to be released from further performance of the Contract, then upon notice by either Party to the other Party of such event or circumstance

(a) the Parties shall be discharged from further performance, without prejudice to the rights of either Party in respect of any previous breach of the Contract, and

(b) the sum payable by the Employer to the Contractor shall be the same as would have been payable under Sub-Clause 19.6 [Optional Termination, Payment and Release] if the Contract had been terminated under Sub-Clause 19.6.

Appendix M

THE INSURANCE CLAUSES OF THE IEI CONDITIONS OF CONTRACT, 3RD EDITION, 1980, REPRINTED IN 1990

Care of the Works

20 (1) The Contractor shall take full responsibility for the care of the Works from the date of the commencement thereof until 14 days after the Engineer shall have issued a Certificate of Completion for the whole of the Works pursuant to Clause 48. Provided that if the Engineers shall issue a Certificate of Completion in respect of any Section or part of the Permanent Works before he shall issue a Certificate of Completion in respect of the whole of the Works the Contractor shall cease to be responsible for the care of that Section or part of the Permanent Works 14 days after the Engineer shall have issued the Certificate of Completion in respect of that Section or part and the responsibility for the care thereof shall thereupon pass to the Employer. Provided further that the Contractor shall take full responsibility for the care of any outstanding work which he shall have undertaken to finish during the Period of Maintenance until such outstanding work is complete.

Responsibility for Reinstatement

(2) In case any damage loss or injury from any cause whatsoever including the negligence or default of the Employer his servants or agents (save and except the Excepted Risks as defined in sub-clause (3) of this Clause) shall happen to the Works or any part thereof while the Contractor shall be responsible for the care thereof the Contractor shall at his own cost repair and make good the same so that at completion the Permanent Works shall be in good order and condition and in conformity in every respect with the requirements of the Contract and the Engineer's instructions. To the extent that any such damage loss or injury arises from any of the Excepted Risks the Contractor shall if required by the Engineer repair and make good the same as aforesaid at the expense of the Employer. The Contractor shall also be liable for any damage to the Works occasioned by him in the course of any operations carried out by him for the purpose of completing any outstanding work or of complying with his obligations under Clauses 49 and 50.

Excepted Risks

(3) The 'Expected Risk' are riot war invasion act of foreign enemies hostilities (whether war be declared or not) civil war rebellion insurrection or military or usurped power ionising radiations or contamination by radioactivity from any nuclear fuel or from any nuclear waste from the combustion of nuclear fuel radioactive toxic explosive or other hazardous properties of any explosive nuclear assembly or nuclear component thereof pressure waves caused by aircraft or other aerial devices travelling at sonic or supersonic speeds or use by occupation by the Employer his agents servants or other contractors (not being employed by the Contractor) of any part of the Permanent Works or a cause due to fault defect error omission in the design of the Works (other than a design provided by the Contractor pursuant to his obligations under the Contract) except and to the extent that the damage loss or injury arising from an 'Excepted Risk' is attributable to fault defect error or omission on the part of the Contractor.

Insurance of Works, etc

21 (1) Without limiting his obligations and responsibilities under Clause 20 the Contractor shall insure in the joint names of the Employer and the Contractor:-
(a) the Permanent Works and the Temporary Works (including for the purpose of this Clause any unfixed materials or other things whether on the Site or otherwise allocated to the Contract in the Contractor's statements for incorporation therein) to their full value with excess limits which shall not exceed the relevant figures stated in the Appendix to the Form of Tender;

(b) the Constructional Plant (including for the purpose of this Clause any such plant whether on the Site or otherwise allocated to the Contract in the Contractor's statements) to its full value with excess limits which shall not exceed the relevant figures stated in the Appendix to the Form of Tender;

against all loss or damage from whatever cause arising (other than the Exclusions as defined in sub-clause (2) of this Clause) for which he is responsible under the terms of the Contract and in such manner that the Employer and Contractor are insured for the period stipulated in Clause 20 (1) and are also insured for loss or damage arising during the Period of Maintenance from such cause occurring prior to the commencement of the Period of Maintenance and for any loss or damage occasioned by the Contractor in the course of any operation carried out by him for the purpose of complying with his obligations under Clauses 49 and 50.

Insurance Exclusions

(2) Provided that without limiting his obligations and responsibilities as aforesaid nothing contained in this Clause shall render the Contractor liable to insure:

(a) against the necessity for the repair or reconstruction of any work constructed with materials and workmanship not in accordance with the requirements of the Contract unless the Bill of Quantities shall provide a special item for this insurance in respect of specific elements.

(b) in respect of the 'Excepted Risks' as defined in Clause 20 (3).

(c) in respect of consequential loss including penalties for delay and non-completion wear and tear shortages and pilferages.

(d) in respect of mechanically propelled vehicles to which the Road Traffic Acts apply.

Liability and Indemnity for Damage to Persons or Property

22 (1) The Contractor shall be liable for and shall indemnify and keep indemnified the Employer against all losses and claims for injuries or damage (save as otherwise provided in paragraph (b) hereof) to any person or property whatsoever (other than the Works for which insurance is required under Clause 21 but including surface or other damage to land being the Site suffered by any persons in beneficial occupation of such land) which may arise out of or inconsequence of the construction and maintenance of the Works and against all claims demands proceedings damages costs charges and expenses whatsoever in respect thereof or in relation thereto.

Provided always that:-

(a) the Contractor's liability to indemnify the Employer as aforesaid shall be limited to the sum stated in the Appendix to the Form of Tender and (in accordance with the operation of sub-clause (2) of this Clause) shall be reduced proportionately to the extent that the act or neglect of the Employer his servants or agents may have contributed to the said loss injury or damage;

(b) nothing herein contained shall be deemed to render the Contractor liable for or in respect of or to indemnify the Employer against any compensation or damage for or with respect to:-

(i) damage to crops being on the Site (save in so far as possession has not been given to the Contractor);

(ii) the use or occupation of land (which has been provided by the Employer) by the Works or any part thereof or for the purpose of constructing completing and maintaining the Works (including consequent losses of crops) or interference whether temporary or permanent with any right of way light air or water or other easement or quasi-easement which are the unavoidable result of the construction of the Works in accordance with the Contract;

(iii) the right of the Employer to construct the Works or any part thereof on over under in any land;

(iv) damage (other than that resulting from the Contractor's method of construction) which is the unavoidable result of the construction of the Works in accordance with the Contract;

(v) injuries or damage to persons or property resulting from any act or neglect or breach of statutory duty done or committed by the Engineer or the Employer his agents servants or other contractors (not being employed by the Contractor) or for or in respect of any claims demands proceedings damages costs charges and expenses in respect thereof or in relation thereto;

(vi) the 'Excepted Risks' as set out in Clause 20 (3);

(vii) mechanically propelled vehicles to which the Road Traffic Acts apply.

Indemnity by Employer

(2) The Employer will save harmless and indemnify the Contractor from and against all claims demands proceedings damages costs charges and expenses in respect of the matters referred to in the proviso to sub-clause (1) of this Clause provided always that the Employer's liability to

indemnify the Contractor (in accordance with sub-clause (1)(a) of this Clause) shall be reduced proportionately to the extent that the act or neglect of the Contractor or his sub-contractor's servants or agents may have contributed to the said injury or damage.

Public Liability Insurance

23 (1) The Contractor shall take out before commencing the Works and maintain a Public Liability Insurance Policy (but without limiting his obligations and responsibilities under Clause 22) in the joint names of the Employer and the Contractor to insure against any damage loss or injury which may occur to any property or to any person by or arising out of the execution of the Works or in the carrying out of the Contract otherwise than due to the matters referred to in proviso (b) to sub-clause 22(1). The Contractor shall insure during the Period of Maintenance for any loss or damage occasioned by him for the purpose of complying with his obligations under Clauses 49 and 50.

Amount and Terms of Insurance

(2) Such insurance shall be for an amount which is at least equal to that stated in the Appendix to the Form of Tender and with excess limits which shall not exceed those stated in the said Appendix.

(3) The Public Liability Insurance Policy shall contain a cross liability clause such that all losses and claims for injuries or damage to any person or property whatsoever (other than the Works for which insurance is required under sub-clause 21 (1)) are covered as if the Employer and the Contractor are separate insureds.

Accident or Injury to Workmen

24 (1) The Employer shall not be liable for or in respect of any damages or compensation payable at law in respect or in consequence of any accident or injury to any workman or other person in the employment of the Contractor or any sub-contractor or to any person employed by the Employer whose services may for the time being be loaned or made available to the Contractor or his sub-contractors save and except to the extent that such accident or injury results from or is contributed to by any act or default of the Employer his agents or servants and the Contractor shall indemnify and keep indemnified the Employer against all such damages and compensation (save and except as aforesaid) and against all claims demands proceedings costs charges and expenses whatsoever in respect thereof or in relation thereto.

Employer's Liability Insurance

(2) The Contractor shall take out before commencing the Works and maintain an Employer's Liability Insurance Policy to insure against his liability under sub-clause (1) of this Clause with excess limits not exceeding those stated in the Appendix to the Form of Tender and shall continue such insurance during the whole of the time that any persons are employed by him on the Works. Provided always that in respect of any persons employed by any sub-contractor the Contractor's obligation to insure as aforesaid under this sub-clause shall be satisfied if the sub-contractor shall have insured against the liability in respect of such persons in such manner that the Employer is indemnified under the policy but the Contractor shall require such sub-contractor to produce to the Employer when required such policy of insurance and the receipt for payment of the current premium.

Production of Insurance Policies

25 (1) The Contractor shall before commencing the Works and whenever required produce to the Employer for inspection any policy or policies of insurance required by Clauses 21, 23 and 24 or any other insurance which he may be required to effect and keep in force together with the receipts in respect of premiums paid under such policy or policies.

Remedy on Contractor's Failure to Insure

(2) If the Contractor shall fail upon request to produce to the Employer satisfactory evidence that there is in force the insurance referred to in Clauses 21, 23 and 24 or any other insurance which he may be required to effect under the terms of the Contract then and in any such case the Employer may effect and keep in force deduct the amount so paid by the Employer as aforesaid from any monies due or which may become due to the Contractor or recover the same as a debt due from the Contractor.

Terms of Insurance

(3) All insurances required under Clauses 21, 23 and 24 shall be effected with Insurers and in terms approved by the Employer (which approval shall not be unreasonably withheld). During the currency of the Contract any material alteration to such insurance made at the Contractor's

request shall be immediately notified by the Contractor to the Employer and shall be subject to the approval of the Employer which approval shall not be unreasonably withheld. In the case of any material alteration made by the Insurer the Contractor shall immediately provide written evidence to the Employer of such alteration. Such alteration shall not release the Contractor in any way from his obligations under the Contract.

WAR CLAUSE

Works to Continue for 28 Days on Outbreak of War
65 (1) If during the currency of the Contract there shall be an outbreak of war (where war is declared or not) in which the State shall be engaged on a scale involving general mobilisation of the armed forces of the State the Contract shall for a period of 28 days reckoned from midnight on the date that the order for general mobilisation is given continue so far as is physically possible to execute the Works in accordance with the Contract.

Effect of Completion Within 28 Days
(2) If at any time before the expiration of the said period of 28 days the Works shall have been completed or completed so far as to be usable all provisions of the Contract shall continue to have full force and effect save that:-
   (a) the Contractor shall in lieu of fulfilling his obligations under Clauses 49 and 50 be entitled at his option to allow against the sum due to him under the provisions hereof the cost (calculated at the prices ruling at the beginning of the said period of 28 days) as certified by the Engineer at the expiration of the Period of Maintenance of repair rectification and making good any work for the repair rectification or making good of which the Contractor would have been liable under the said Clauses had they continued to be applicable;
   (b) the Employer shall not be entitled at the expiration of the Period of Maintenance to withhold payment under Clause 60 (5)(c) of the second half of the retention money or any part thereof except such sum as may be allowable by the Contractor under the provisions of the last preceding paragraph which sum may (without prejudice to any other mode of recovery thereof) be deducted by the Employer from such second half.

Right of Employer to Determine Contract
(3) If the Works shall not have been completed as aforesaid the Employer shall be entitled to determine the Contract (with the exception of this Clause and Clauses 66 and 68) by giving notice in writing to the Contractor at any time after the aforesaid period of 28 days has expired and upon such notice being given the Contract shall (except as above mentioned) forthwith determine but without prejudice to the claims of either party in respect of any antecedent breach thereof.

Removal of Plant on Determination
(4) If the Contract shall be determined under the provisions of the last preceding sub-clause the Contractor shall with all reasonable despatch remove from the Site all his Constructional Plant and shall give facilities to his sub-contractors to remove similarly all Constructional Plant belonging to them and in the event of any failure so to do the Employer shall have the like powers as are contained in Clause 53 (8) in regard to failure to remove Constructional Plant on completion of the Works but subject to the same condition as is contained in Clause 53 (9).

Payment on Determination
(5) If the Contract shall be determined as aforesaid the Contractor shall be paid by the Employer (insofar as such amounts or items shall not have been already covered by payment on account made to the Contractor) for all work executed prior to the date of determination at the rates and prices provided in the Contract and in addition:-
   (a) the amounts payable in respect of any preliminary items so far as the work or service comprised therein has been carried out or performed and a proper proportion as certified by the Engineer of any such items the work or service comprised in which has been partially carried out or performed;
   (b) the cost of materials or goods reasonably ordered for the Works which shall have been delivered to the Contractor or of which the Contractor is legally liable to accept delivery (such materials or goods becoming the property of the Employer upon such payment being made by him);
   (c) a sum to be certified by the Engineer being the amount of any expenditure reasonably incurred by the Contractor in the expectation of completing the whole of the Works in so far as such expenditure shall not have been covered by the payments in this sub-clause before mentioned;
   (d) any additional sum payable under sub-clause (6)(b)(c) and (d) of this Clause;
   (e) the reasonable cost of removal under sub-clause (4) of this Clause.

Provisions to Apply as from Outbreak of War

(6) Whether the Contract shall be determined under the provisions of sub-clause (3) of this Clause or not the following provisions shall apply or be deemed to have applied as from the date of the said outbreak of war notwithstanding anything expressed in or implied by the other terms of the Contract viz:-

(a) The Contractor shall be under no liability whatsoever whether by way of indemnity or otherwise for or in respect of damage to the Works or to property (other than property of the Contract or property hired by him for the purposes of executing the Works) whether of the Employer or of third parties or for or in respect of injury or loss of life to persons which is the consequence whether direct or indirect of war hostilities (whether war has been declared or not) invasion act of the State's enemies civil war rebellion revolution insurrection military or usurped power and the Employer shall indemnify the Contractor against all such liabilities and against all claims demands proceedings damages costs charges and expenses whatsoever arising thereout or in connection therewith.

(b) If the Works shall sustain destruction or any damage by reason of any of the causes mentioned in the last preceding paragraph the Contractor shall nevertheless be entitled to payment for any part of the Works so destroyed or damaged and the Contractor shall be entitled to be paid by the Employer the cost of making good any such destruction or damage so far as may be required by the Engineer or as may be necessary for the completion of the Works on a cost basis plus such profit as the Engineer may certify to be reasonable.

(c) The terms of the Contract Price Fluctuations Clause shall continue to apply but if subsequent to the outbreak of war the plant index figures therein shall cease to be published then the last published plant index figure shall be used thereafter as the basis for calculating the variation in plant costs for the remaining certificates of the Contract.

(d) Damage or injury caused by the explosion whenever occurring of any mine bomb shell grenade or other projectile missile or munition of war and whether occurring before or after the cessation of hostilities shall be deemed to be the consequence of any of the events mentioned in sub-clause (6)(a) of this Clause.

Appendix N

THE INSURANCE CLAUSES OF THE IEI CONDITIONS OF CONTRACT, 4TH EDITION 1995, REPRINTED IN 1998

Care of the Works

20 (1) (a) The Contractor shall save as in paragraph (b) hereof and subject to sub-clause (2) of this Clause take full responsibility for the care of the Works together with materials and Plant for incorporation therein from the Works Commencement Date until the date of issue of a Certificate of Substantial Completion for the whole of the Works when the responsibility for the said care shall pass to the Employer.

(b) If the Engineer issues a Certificate of Substantial Completion for any Section or part of the Permanent Works the Contractor shall cease to be responsible for the care of that Section or part from the date of issue of such Certificate of Substantial Completion when the responsibility for the care of the Section or part shall pass to the Employer.

(c) The Contractor shall take full responsibility for the care of any outstanding work together with materials and Plant for incorporation therein which he undertakes to finish during the Defects Correction Period until such outstanding work has been completed.

Employer's risks

(2) The Employer's risks for which the Contractor is not liable are loss or damage to the extent that it is due to

(a) the use or occupation by the Employer his agents servants or other contractors (not being employed by the Contractor) of any part of the Permanent Works except as may be provided for in an amendment to this Clause.

(b) any fault defect error or omission in the design of the Works (other than a design provided by the Contractor pursuant to his obligations under the Contract)

(c) riot war invasion act of foreign enemies or hostilities (whether war be declared or not)

(d) civil war rebellion revolution insurrection or military or usurped power

(e) ionising radiations or contamination by radioactivity from any nuclear fuel or from any nuclear waste from the combustion of nuclear fuel radioactive toxic explosive or other hazardous properties of any explosive nuclear assembly or nuclear component thereof and

(f) pressure waves caused by aircraft or other aerial devices travelling at sonic or supersonic speeds.

Rectification of loss or damage

(3) (a) In the event of any loss or damage to

(i) the Works or any Section or part thereof or

(ii) materials or Plant for incorporation therein

while the Contractor is responsible for the care thereof (except as provide in sub-clause (2) of this Clause) the Contractor shall at his own rectify such loss or damage so that the Permanent Works confirm in every respect with the provisions of Contract and the Engineer's instructions. The Contractor shall also be liable for any loss or damage to the Works occasioned by him in the course of any operations carried out by him for the purpose of complying with his obligations under Clauses 49 and 50.

(b) Should any such loss or damage arise from any of the Employer's Risks defined in sub-clause (2) of This Clause the Contractor shall if and to the extent required by the Engineer rectify the loss or damage at the cost of the Employer.

(c) In the event of loss or damage arising from a combination of an Employer's Risk and a risk for which the Contractor is responsible under sub-clause (1)(a) of this Clause then the Engineer shall when determining the cost to be borne by the Employer under the Contract apportion the cost of rectification into that part caused by the Employer's Risk and that part which is the responsibility of the Contractor.

Insurance of Works etc.

21 (1) The Contractor shall without limiting his or the Employer's obligations and responsibilities under Clause 20 insure in the joint names of the Contractor and the Employer the Works together with materials and Plant for incorporation therein and Contractor's Equipment (including for the purpose of this Clause any unfixed materials or Plant or other things whether on the Site or otherwise allocated to the Contract in the Contractor's statements for incorporation therein) to the full replacement cost plus an additional 10% to cover any additional costs that may arise incidental to the rectification of any loss or damage including professional fees cost of demolition and removal of debris.

Extent of cover

(2) (a) The insurance required under sub-clause (1) of this clause shall cover the Employer and the Contractor against all loss or damage from whatsoever cause arising other than from
    (i) any fault defect error or omission in the design of the Works for which the Contractor is responsible under the Contract or
    (ii) the Employer's Risks as defined in Sub-clause 20(2)

from the Works Commencement Date until the date of issue of the relevant Certificate of Substantial Completion.

(b) The insurance shall extend to cover any loss or damage arising during the Defects Correction Period from a cause occurring prior to the issue of any Certificate of Substantial Completion and any loss or damage occasioned by the Contractor in the course of any operation carried out by him for the purpose of complying with his obligations under Clauses 49 and 50.

(c) Nothing in this Clause shall render the Contractor liable to insure against the necessity for the repair or reconstruction of any work constructed with materials Plant or workmanship not in accordance with the requirements of the Contract unless the Bill of Quantities shall provide a special item for this insurance.

(d) Any amounts not insured or not recovered from insurers whether as excesses carried under the policy or otherwise shall be borne by the Contractor or the Employer in accordance with their respective responsibilities under Clause 20.

Damage to persons and property

22 (1) The Contractor shall except if and in so far as the Contract provides otherwise and subject to the exceptions set out in sub-clause (2) of the Clause indemnify and keep indemnified the Employer against all losses and claims in respect of
    (a) death of or injury to any person or
    (b) loss or damage to any property (other than the Works) which may arise out of or in consequence of the execution of the Works and the remedying of any defects therein and against all claims demands proceedings damages costs charges and expenses whatsoever in respect thereof or in relation thereto.

Exceptions

(2) The exceptions referred to in sub-clause (1) of this Clause which are the responsibility of the Employer are
    (a) damage to crops being on the Site (save in so far as possession has not been given to the Contractor)
    (b) the use or occupation of land (provided by the Employer) by the Works or any part thereof or for the purpose of the construction and completion of the Works (including consequent losses of crops) or interference whether temporary or permanent with any right of way light air or water or other easement or quasi-easement which are the unavoidable result of the construction of the Works in accordance with the Contract
    (c) the right of the Employer to construct the Works or any part thereof on over under in or through any land
    (d) damage (other than that resulting from the Contractor's method of construction) which is the unavoidable result of the construction of the Works in accordance with the Contract and
    (e) death of or injury to persons or loss of or damage to property resulting from any act neglect or breach of statutory duty done or committed by the Employer his agents servants or other contractors (not being employed by the Contractor) or for or in respect of any claims demands proceedings damages costs charges and expenses in respect thereof or in relation thereto.

Indemnity by Employer

(3) The Employer shall subject to sub-clause (4) of this Clause indemnify the Contractor against all claims demands proceedings damages costs charges and expenses in respect of the matters referred to in the exceptions defined in sub-clause (2) of this Clause.

Shared responsibility

(4) (a) the Contractor's liability to indemnify the Employer under sub-clause (1) of this Clause shall be reduced in proportion to the extent that the act or neglect of the Employer his agents servants or other contractors (not being employed by the Contractor) may have contributed to the said death injury loss or damage.

(b) the Employer's liability to indemnify the Contractor under sub-clause (3) of this Clause in respect of matters referred to in sub-clause (2)(e) of this Clause shall be reduced in proportion to the extent that the act or neglect of the Contractor or his sub-contractors servants or agents may have contributed to the said death injury loss or damage.

Third party insurance

23 (1) The Contractor shall without limiting his or the Employer's obligations and responsibilities under Clause 22 insure in the joint names of the Contractor and the Employer against liabilities for death of or injury to any person (other than any operative or other person in the employment of the Contractor or any of his sub-contractors) or loss or damage to any property (other than the Works) arising out of the construction and completion of the Contract other than the exceptions defined in Sub-clauses 22(2)(a)(b)(c) and (d).

Cross liability clause

(2) The insurance policy shall include a cross liability clause such that the insurance shall apply to the Contractor and to the Employer as separate insureds.

Amount of insurance

(3) Such insurance shall be for at least the amount stated in the Appendix to the Form of Tender.

Accident or injury to workpeople

24 (1) The Employer shall not be liable for or in respect of any damages or compensation payable at law in respect or in consequence of any accident or injury to any operative or other person in the employment of the Contractor or any of his sub-contractors or to any person employed by the Employed whose services may for the time being be formally seconded in writing to the Contractor or his sub-contractors save and except to the extent that such accident or injury results from or is contributed to by any act or default of the Employer his agents or servants and the Contractor shall indemnify and keep indemnified the Employer against all such damages and compensation (save and except as aforesaid) and against all claims demands proceedings costs charges and expenses whatsoever in respect thereof or in relation thereto.

Employer's liability insurance

(2) The Contractor shall take out before commencing the Works and maintain an Employer's liability insurance policy to insure against his liability under sub-clause (1) of this Clause with excess limits not exceeding those stated in the Appendix to the Form of Tender and shall continue such insurance during the whole of the time that any persons are employed by him on the Works. Provided always that in respect of any persons employed by any sub-contractor the Contractor's obligation to insure as aforesaid under his sub-clause shall be satisfied if the sub-contractor shall have insured against the liability in respect of such persons in such manner that the Employer is indemnified under the policy but the Contractor shall require such sub-contractor to produce to the Employer when required such policy of insurance and the receipt for payment of the current premium.

Evidence and terms of insurance

25 (1) The Contractor shall provide satisfactory evidence to the Employer prior to the Works Commencement Date that the insurances required under the Contract have been effected and shall if so required produce the insurance policies for inspection. The terms of all such insurances shall be subject to the approval of the Employer (which approval shall not unreasonably be withheld). The Contractor shall upon request produce to the Employer receipts for the payment of current insurance premiums.

Excesses

(2) Any excesses on the insurance policies effected under Clauses 21, 23 and 24 shall be as stated in the Appendix to the Form of Tender.

Remedy on Contractor's failure to insure

(3) If the Contractor shall fail upon request to produce to the Employer satisfactory evidence that there is in force any of the insurance policies required under the Contract then and in any such

case the Employer may effect and keep in force any such insurance and pay such premium or premiums as may be necessary for that purpose and from time to time deduct the amount so paid from any monies due or which may become due to the Contractor or recover the same as a debt due from the Contractor.

WAR CLAUSE

Works to continue for 28 days on outbreak of War

65   (1) If during the currency of the Contract there shall be an outbreak of war (whether war is declared or not) in which the State shall be engaged on a scale involving general mobilisation of the armed forces of the State the Contractor shall for a period of 28 days reckoned from midnight on the date that the order for general mobilisation is given continue so far as is physically possible to execute the Works in accordance with the Contract.

Effect of substantial completion within 28 days

(2) If at any time before the expiration of the said period of 28 days the Works shall have been substantially completed or substantially completed so far as to be usable all provisions of the Contract shall continue to have full force and effect save that

(a) the Contractor shall in lieu of fulfilling his obligations under Clauses 49 and 50 be entitled at his option to allow against the sum due to him under the provisions hereof the cost (calculated at the prices ruling at the beginning of the said period of 28 days) as certified by the Engineer at the expiration of the Defects Correction Period of repair rectification and making good of which the Contractor would have been liable under the said Clauses had they continued to be applicable

(b) the Employer shall not be entitled at the expiry of the Defects Correction Period to withhold payment under sub-clause 60 (6)(c) of the second half of the retention money or any part thereof except such sum as may be allowable by the Contractor under the provisions of the last preceding paragraph which sum may (without prejudice to any other mode of recovery thereof) be deducted by the Employer from such second half.

Right of Employer to determine Contract

(3) If the Works shall not have been substantially completed as aforesaid the Employer shall be entitled to determine the Contract (with the exception of this Clause and Clauses 66 and 68) by giving notice in writing to the Contractor at any time after the aforesaid period of 28 days has expired and upon such notice being given the Contract shall (except as above mentioned) forthwith determine but without prejudice to the claims of either party in respect of any antecedent breach thereof.

Removal of Contractor's Equipment on determination

(4) If the Contract shall be determined under the provision of the last preceding sub-clause the Contractor shall will all reasonable despatch remove from Site all his Contractor's Equipment and shall give facilities to his sub-contractors to remove similarly all Contractor's Equipment belonging to them and in the event of any failure so to do the Employer shall have the like powers as are contained in sub-clauses 53 (3) in regard to failure to remove Contractor's Equipment on completion of the Works but subject to the same condition as is contained in sub-clause 53 (2).

Payment on determination

(5) If the Contract shall be determined as aforesaid the Contractor shall be paid by the Employer (insofar as such amounts or items shall not have been already covered by payment on account made to the Contractor) for all work executed prior to the date of determination at the rates and prices provided in the Contract and in addition

(a) the amounts payable in respect of any preliminary items so far as the work or service comprised therein has been carried out or performed and a proper proportion as certified by the Engineer of any such items the work or service comprised in which has been partially carried out or performed

(b) the cost of materials good or Plant reasonably ordered for the Works which have been delivered to the Contractor or of which the Contractor is legally liable to accept delivery (such materials good or Plant becoming the property of the Employer upon such payment being made by him)

(c) a sum to be certified by the Engineer being the amount of any expenditure reasonably incurred by the Contractor in the expectation of completing the whole of the Works in so far as such expenditure shall not have been covered by the payments in this sub-clause before mentioned

(d) any additional sum payable under sub-clauses (6)(b)(c) and (d) of this Clause and

(e) the reasonable cost of removal under sub-clause (4) of this Clause.

Provisions to apply as from outbreak of war

(6) Whether the Contract shall be determined under the provisions of sub-clause (3) of this Clause or not the following provisions shall apply or be deemed to have applied as from the date of the said outbreak of war notwithstanding anything expressed in or implied by the other terms of the Contract viz

    (a) The Contractor shall be under no liability whatsoever by way of indemnity or otherwise for or in respect of damage to the Works or to property (other than property of the Contractor or property hired by him for the purposes of executing the Works) whether of the Employer or of third parties or for or in respect of injury or loss of life to persons which is the consequence whether direct or indirect of war hostilities (whether war has been declared or not) invasion act of the State's enemies civil war rebellion revolution insurrection military or usurped power and the Employer shall indemnify the Contractor against all such liabilities and against all claims demands proceedings damages costs charges and expenses whatsoever arising thereout or in connection therewith.

    (b) If the Works shall sustain destruction or any damage by reason of any of the causes mentioned in the last preceding paragraph the Contractor shall nevertheless be entitled to payment for any part of the Works so destroyed or damaged and the Contractor shall be entitled to be paid by the Employer the cost of making good any such destruction or damage so far as may be required by the Engineer or as may be necessary for the completion of the Works on a cost basis plus such profit as the Engineer may certify to be reasonable.

    (c) The terms of the Contract Price Fluctuations Clause shall continue to apply but if subsequent to the outbreak of war the plant index figures therein shall cease to be published then the last published plant index figure shall be used thereafter as the basis for calculating the variation in plant costs for the remaining certificates of the contract.

    (d) Damage or injury caused by the explosion whenever occurring of any mine bomb shell grenade or other projectile missile or munition of war and whether occurring before or after the cessation of hostilities shall be deemed to be the consequence of any of the events mentioned in sub-clause (6)(a) of this Clause.

Appendix O

LIABILITY AND INSURANCE CLAUSES OF THE CLIENT/CONSULTANT MODEL SERVICES AGREEMENT

(The White Book) 3rd Edition, 1998

LIABILITY AND INSURANCE

16. **Liability between the Parties**

   16.1 **Liability of the Consultant**
   The Consultant shall only be liable to pay compensation to the Client arising out of or in connection with the Agreement if a breach of Clause 5 (i) is established against him.

   16.2 **Liability of the Client**
   The Client shall be liable to the Consultant if a breach of his duty to the Consultant is established against the Client.

   16.3 **Compensation**
   If it is considered that either party is liable to the other, compensation shall be payable only on the following terms:

   (i) Such compensation shall be limited to the amount of reasonably foreseeable loss and damage suffered as a result of such breach, but not otherwise.

   (ii) In any event, the amount of such compensation will be limited to the amounts specified in Clause 18.1

   (iii) If either party is considered to be liable jointly with third parties to the other, the proportion of compensation payable by him shall be limited to that proportion of liability which is attributable to his breach.

17. **Duration of Liability**

   Neither the Client nor the Consultant shall be considered liable for any loss or damage resulting from any occurrence unless a claim is formally made on him before the expiry of the relevant period stated in the Particular Conditions, or such earlier date as may be prescribed by law.

18. **Limit of Compensation and Indemnity**

   18.1 **Limit of Compensation**
   The maximum amount of compensation payable by either party to the other in respect of liability under Clause 16 is limited to the amount stated in the Part II. This limit is without prejudice to any Agreed Compensation specified under Clause 31(ii) or otherwise imposed by the Agreement.

   Each party agrees to waive all claims against the other in so far as the aggregate of compensation which might otherwise be payable exceeds the maximum amount payable.

   If either party makes a claim for compensation against the other party and this is not established the claimant shall entirely reimburse the other for his costs incurred as a result of the claim.

18.2 Indemnity

So far as the applicable law permits, the Client shall indemnify the Consultant against the adverse effects of all claims including such claims by third parties which arise out of or in connection with the Agreement:

   (i) except insofar as they are covered by the insurances arranged under the terms of Clause 19.

   (ii) made after the expiry of the period of liability referred to in Clause 17.

18.3 Exceptions

Clauses 18.1 and 18.2 do not apply to claims arising:

   (i) from deliberate default or reckless misconduct.

   (ii) otherwise than in connection with the performance of obligations under the Agreement.

19. Insurance for Liability and Indemnity

The Client can request in writing that the Consultant

   (i) insures against his liability under Clause 16.1,

   (ii) increases his insurance against liability under Clause 16.1 over that for which he was insured at the date of the Client's first invitation to him for a proposal for the Services,

   (iii) insures against public/third party liability,

   (iv) increases his insurance against public/third party liability over that for which he was insured at the date of the Client's first invitation to him for a proposal for the Services,

   (v) effects other insurances.

If so requested, the Consultant shall make all reasonable efforts to effect such insurance or increase in insurance with an insurer and on terms acceptable to the Client.

The cost of such insurance or increase in insurance shall be at the expense of the Client.

20. Insurance for Client's Property

Unless otherwise requested by the Client in writing the Consultant shall make all reasonable efforts to insure on terms acceptable to the Client:

   (i) against loss or damage to the property of the Client supplied or paid for under Clause 6.

   (ii) against liabilities arising out of the use of such property.

The cost of such insurance shall be at the expense of the Client.

LIST OF CASES

A.M.F. International Ltd. v. Magnet Bowling & G.P. Trentham Ltd., [1968] 1 W.L.R. 1028 –
254, 257

Allied Properties v. Blume (1972), 25 CA, 3d, 848 – 154

Aluminium Products (Qld) Pty. Ltd. v. David Hill and Others (1980), 3 BCLRS 103 – 175

Anns and Others v. London Borough of Merton [1977] 2 All ER 492 – 176

Appleby v. Myers (1867) L.R. 1 C.P. 615 – 230

Applegate v. Moss and Archer v. Moss (1971), 1 All ER 747, U.K. – 101, 173

B.L. Holdings Limited v. Robert J. Wood & Partners (1979) 12 BLR 1 – 151

*Balcomb and Another v. Wards Construction (Medway) Ltd. and Others, and Pettybridge
and Another v. Wards Construction (Medway) Ltd. and Others* (1981) 259, E.G.765 – 60

Batty and Another v. Metropolitan Property Realisation Ltd. and Others 1978, 2 All E.R.
445 – 60, 178

Bolam v. Friern Hospital Management Committee [1957] 2 All ER 118 – 66, 151

Bond Air Services v. Hill (1955), 2QB 417 – 368

Bottomley v. Bannister and Otto v. Bolton & Norris [1936] 2 K.B. 46 – 170, 172

Bowen v. Paramount Builders [1977] NZLR 394 – 56

Brickfield Properties v. Newborough [1971] 3 All E.R. 328 – 152

Brickfield Properties v. Newton [1971] 1 WLR 862 – 155

Cagne v. Bertran (1954), 43 Cal. 2d 481, 275 p. 2d15 – 153

Cambridge Water Co. Ltd. v. Eastern Counties Leather plc. [1994], 1 All ER – 150

Candler v. Crane, Christmas, [1951] 1 All ER 426 – 170

Cartledge and Others v. Jobling & Sons Ltd. [1963] AC 758 – 177

*Chariot Inns Limited v. Assicurazioni Generali S.P.A. and Coyle Hamilton Hamilton Phillips
Ltd.* (Supreme Court 1981) – 185, 210

Charon v. Singer Sewing Machines (1968) 112 S.J. 536 – 230

City of Brantford v. Kemp & Wallace-Carruthers & Associates Ltd. (1960), 23 DLR – 59, 150

City of Brantford v. Kemp and Wallace-Carruthers & Associates Ltd. (1960), 23 DLR (2d),
640 (Canada) – 150

Claude R. Ogden and Co. Pty. Ltd. v. Reliance Fire Sprinkler Co. Pty. Ltd. [1975] 1 Lloyd's
Rep. 52. P. 73, per J. MacFarlan – 212

Clay v. A. J. Crump & Sons Ltd. [1964] 3 All ER 687 – 170

Commonwealth Construction Co. Ltd. v. Imperial Oil Ltd. [1976] 69 DLR (3d) 558 – 182

Croudace Construction Ltd. v. Cawoods Concrete Products Ltd. (1978), 8 BLR 20 – 350

Department of the Environment v. Thomas Bates & Sons Ltd. [1991] 1 A.C. 499 – 177

Dickson & Co. v. Devitt (1917), 86 L.J.K.B. 313 – 212

District of Surrey v. Church, 1977, Canada, 76 DLR – 59, 175

Donoghue v. Stevenson [1932] AC562 – 6, 168, 177

Dorset Yacht Co. v. Home Office (1970) – 170

Dutton v. Bognor Regis Urban District Council [1972] 1 Q. B. 373 – 170, 173

Eames London Estates Ltd. and Others v. North Herts District Council & Others (1981), 259 E.G. – 59, 162

East Ham Corporation v. Bernard Sunley & Sons Ltd. [1966] 3 All ER 619 – 161

East Suffolk Rivers Catchment Board v. Kent (1941) A.C. 74 – 172

Eckersley, T. E. and Others v. Binnie & Partners & Others, Court of Appeal, (1988) Con LR1 – 155

English Industrial Estates Corp. v. G. Wimpey, & Co. Ltd. (1973) – 232

Equitable Debenture Assets Corporation Ltd. v. William Moss and Others (1984), 2 CLD–12–01 – 165

Esso Petroleum Co. Ltd. v. Mardon [1976] 1 Q.B. 801 – 173

Fenton v. Thorley & Company Limited [1903] A.C. 443 – 192

Finlay v. Murtagh [1979] I.R. 249 – 174

Gallagher v. N. McDowell Ltd. [1961] N.I. 26 – 170

Geer v. Bennett (1970), Florida, USA, G.F.I.P. Vol. V, No. 1 – 164

Glasgow Corporation v. Muir (1943) – 67

Gold v. Patman and Fotheringham [1958] 2 All ER 497 – 198

Governors of the Peabody Donation Fund v. Sir Lindsay Parkinson & Co. Ltd. & Others (1984), 3 All ER, 529 – 110

Gravely v. The Providence Partnership (1977), Federal Appellate Court, USA, G.F.I.P. Vol. VIII No. 1 – 154

Greaves (Contractors) Limited v. Baynham Meikle & Partners [1975] 1 WLR 1095 – 154, 155, 156

Hatcher v. Black and Others (1954) – 164

Hawkins, George v. Chrysler (UK) Limited and Burne Associates (1986) 38 BLR 36 – 156

Hedley Byrne & Co. Ltd. v. Heller & Partners Ltd. [1964] 2 All ER 575 – 152, 170, 171, 174, 178, 371

Home Office v. Dorset Yacht Co. Ltd. [1970] 2 All ER 294 – 170

Independent Broadcasting Authority v. EMI Electronics and BICC Construction, (1980) 14 BLR 1 – 61, 67, 156

Investors in Industry Commercial Properties v. South Bedfordshire District Council & Others [1986] 1 All ER 787 – 111

Junior Books Ltd. v. Veitchi Co. Ltd. [1982] 3 W.L.R. 477 – 178

Kitchens of Sara Lee Inc. v. A. L. Jackson Co. et al. (1972), Lake County, Illinois, USA, G.F.I.P. Vol. III No. 2 – 152, 153

Kruger, W. C. and Associates v. Robert D. Krause Engineering Co. and Albuquerque Testing Laboratory, Inc. – 56

Leicester Guardians v. Trollope (1911), LXXV JP 197 – 160

Lewis v. Anchorage Asphalt Paving Co. (1975), Alaska Supreme Court, USA, G.F.I.P., Vol. VII No. 6 – 164

Lloyd v. Grace, Smith & Co. [1912] AC 716 – 160

London County Council v. Cattermoles (Garages) Ltd. [1953] 2 All ER 582 – 160

MacPherson v. Buick Motor Co. (1916) – 168

Makin (F & F) Ltd. v. London & North Eastern Ry. Co. [1943] 1 K.B. 467, C.A. per Lord Greene, M.R. P. 474 – 193

Manufacturers' Mutual Insurance v. The Queensland Government Railway & Another (1968), Q.W.N. 12 – 243

McCollum (R.D.) Ltd. v. Economical Mutual Insurance Co. [1962] O.R. 850 per Lancheville, – 193

McNealy v. The Pennine Insurance Company, [1978] 2 Lloyd's Rep. 18 – 209

Medjuck & Budovitch Ltd v. Adi Ltd 33 NBR 2nd 271 (80 Apr. 271, paragraph 110) – 155

Mills v. Smith [1963] 2 All ER 1078, per J.P. Paul, 1079 – 194

Ministry of Housing and Local Government v. Sharp [1970] 1 All ER 1009 – 170, 171

Moneypenny v. Hartland, 1826, 2 C. & P. 378 – 59

The Moorcock (1889) LR 14 PD64 – 150

Morash v. Lockhart and Ritchie Ltd. (1979), 95 D.L.R. (3d) 647 – 213

Moresk v. Hicks (1966) 2 Lloyd's Rep. 338, 116 – 152

Morgan, Brian v. Park Developments Ltd., Judgment delivered on 2nd February, 1983 – 177

Morren v. Swinton and Pendlebury Borough Council [1965] 1 WLR 576 – 160

Mowbray v. Merryweather [1895] 2 Q.B. 640 – 254

Munro Brice & Co. v. War Risks Association (1918), 2 KB 78 – 368

Murphy v. Brentwood District Council, [1991] 1 A.C. 398 – 177

Nelson Lumber Co. Ltd. v. Koch (1980), III DLR (Canada) – 152

Nelson v. Union Wire Rope Corporation (1964), 199 N.E. Rep. (2d) 769 – 170, 171

Norta Wallpapers (Ireland) v. Sisk and Sons (Dublin) Limited (1978) IR 114 – 156

Oldschool v. Gleeson (1976), 4 BLR 103 – 161

Osman v. J. Ralph Moss Ltd. [1979] 1 Lloyd's Rep. 313 – 208–210

Paris v. Stepney Borough Council (1951) – 67

Pentagon Construction (1969) Co. Ltd. v. United States Fidelity and Guarantee Co. (1978), 1 Lloyd's Reports, 93 – 351

Petrofina (U.K.) and Others v. Magnaload Ltd. and Others, [1983] 3 All ER 35 – 182

Photo Production Ltd. v. Securicor Transport Ltd., [1980] 1 All ER 556 – 238, 240

Pirelli General Cable Works Ltd. v. Oscar Faber and Partners [1983] 1 All ER 65 – 176, 177

(1) *QV Ltd (formerly Holbeach Marsh Co-operative Ltd.); (2) QV Foods Ltd. (formerly QV Ltd.) v. (1) Fredrick F. Smith (Trading as Fredrick F. Smith Associates); (2) D. A. Green & Sons Ltd. (Defendants) and Eternit UK Ltd. (Formerly Eternit TAC Ltd) (Third party)*, (1998) QBD Official Referees' Business – 66

Raineri v. Miles [1981] A.C. 1050, 1086 – 188, 326

Redhead v. Midland Railway Co. (1869) – 67

Rylands v. Fletcher (1868) L.R., 3 H.L., 330 – 149, 198

Samuels v. Davis [1943] 1 KB 526 – 157

Saunders v. Broadstairs Local Board (1980), Reported on Hudson's Building Contracts, 4th Edition, Vol. 2 – 160

Sealand of the Pacific- v. Robert C. McHaffie Ltd. 1974, 51 DLR, Canada – 69

Shanklin Pier Ltd. v. Detel Products Ltd. (1951), 2 All ER 471 – 155

Smith v. South Wales Switchgear Co. Ltd. [1978] 1 All ER 18, and *Photo Products Ltd. v. Securicor Transport Ltd.* [1980] 1 All ER 556 – 181, 238

Sparham – Souter and Others v. Town and Country Developments (Essex) Ltd. [1976] 2 All ER 65 – 175–177

Strong and Pearl v. Allison and Co. Ltd. (1926), 25 L 1.L. Rep. 504 – 212

Sutcliffe v. Chippendale and Edmondson, 1971, 18 BLR, 149, U.K. – 103

Sutcliffe v. Thackrah (1974), 1 All ER 859 – 163

Trim Joint District School Board of Management v. Kelly, [1914] A.C. 667, H.L. – 192

Turner v. Garland and Christopher (1853), Hudson's Building Contract, 4th Edition, Vol. 2, page 1 – 67

Ultramares Corporation v. Touche (1931), 255 N.E. 170, per C.J. Cardozo – 178, 358

Victoria University of Manchester v. Hugh Wilson and Lewis Womersely and Others (1984) CILL 126 – 165

Viking Grain Storage Limited v. T. H. White Installations Limited and Another (1985) 33 BLR 103 – 158

Vlado Vulic v. Bohdam Bilinsky and Others (1982) NSW Supreme Court No. 17700/78 – 175

Voli v. Inglewood Shire Council [1963] 10, CLR 74 – 173

Vonesck v. Hirsch and Stevens Inc. (1974), Wisconsin Supreme Court, USA, G.F.I.P., Vol. V. No. 1 – 164

Whitehouse v. Jordan [1981] 1 WLR 246; 125 SJ 167; [1981] 1 All ER 267 – 154

Wilcox v. Norberg and Wiggins Insurance Agencies Ltd. [1979] 1.W.W.R. 414 – 212

Williamson and Vellmer Engineering v. Sequoia Insurance Company – 105, 369

Winterbottom v. Wright (1842) – 168

Xerox Corporation v. Turner Construction Company, et. al. (1973), G.F.I.P. Vol. IV No. 4 – 154

Young & Marten v. McManus Childs [1969] 1 AC 454; [1968] 3 WLR 630 – 155

INDEX

Note: Acts of UK Parliament will be found grouped together under the heading Acts of Parliament

Accidents 27, 30–32, 36, 64, 73, 85, 87,
        93, 97, 100, 104, 130, 144, 167,
        191–194, 209, 211, 220, 260,
        261, 283, 284, 295, 305, 338,
        342, 343, 346–348, 355, 357,
        367, 372, 387, 389, 402, 404,
        405, 411, 427, 429, 435
    definition 31, 192, 194
    kind of 31, 32
    insurance 5, 179, 167, 191
    workmen 192, 194, 283, 427
Accountant 170, 174, 220, 263, 414
Accrual date 175, 178, 424
Acid rain 64
Acts of Parliament
    Insurance Act, 1601 4
    Limitation Acts 177
    London Government Act, 1963 110
    Marine Insurance Act, 1601 52
    National Insurance (Industrial Injuries)
        Act, 1965 194
    Peak Forest Canal Act, 1794 193
    Public Health Act 172
    Unfair Contract Terms Act, 1977 241
    Workman's Compensation Acts 192,
        194
Advance payment guarantees 206
American Consulting Engineers' Council
        363, 365, 375
American Institute of Architects 56, 133,
        164
Anglo-Saxon Law 4
Anteriority cover 369, 372
Arabic 4, 18, 28
Arbitration 32, 113, 199, 250, 316, 327,
        331, 353, 366, 404, 411, 420,
        421, 422, 429, 440
Architects 17, 56, 59, 103, 110, 111, 116,
        117, 133, 151–154, 160,
        163–165, 170–175, 358, 360,
        363, 372, 376, 412, 418
Armando Salles de Oliveira dam 128
Association of Consulting Engineers 222,
        376, 420
Association of Consulting Engineers of
        Ireland 6
Association of Soil and Foundation
        Engineers 376
Australia 124, 125, 166, 173, 174
Authors quoted
    Abrahamson Max W. 132, 257, 290,
        380
    Alderson, Judge 170
    Atkin, Lord 169, 170, 212
    Bowen, L. J. 150
    Buckmaster, Lord 169
    Bunni, Nael G. 19, 24, 39, 63, 133,
        222, 224
    Carroll, Miss Justice M. 179
    Channell, J. 160
    Coke, Lord 3
    Cornes, D. L. 163
    Denning, Lord 7, 148, 155–157, 164,
        168, 170, 172, 174, 176, 177
    Dilhorne, Lord 240
    Duncan-Wallace, Ian. N. 37, 164, 198,
        200, 230, 250, 254, 318
    Du Parcq, Lord Justice 157
    Fay, Judge Edgar 162
    Gardam, David 11
    Genecki, Paul L. 359, 374, 375
    Grove, Jesse B. 12, 37, 130, 131, 132,
        137
    Hart, Chris 180
    Harris, Sir William 11
    Heywood, John 60
    Iselin, D. C. 19
    James, Williams 52

Jung, Carl Gustav 26
Keane, David 174
Kenny, Judge 174
Le Corbusier 17
Lindley, Lord 192
Lloyd, Professor Dennis 166
Lloyd, J. 182
Lloyd, Dennis 166
Lloyd, Humphrey 37, 131
Loomis, Carol J. 180
Loreburn, Lord 192
MacMillan, Lord 169
McHaffie, Robert C. 69
McNair, Mr Justice 66, 151
Mercier, Claude Y. 179, 363, 373
Miller, Peter O. 18, 19, 20, 21, 23, 63
Molineaux, Charles 12
Paull, J. 102
Pope, Alexander 23
Purcell, Philip F. 183
Reid, Lord 163
Sachs, L. J 172
Scarman, Lord 157
Tyas, J. G. M. 14
Uff, John 79
Walker, Professor David M. 142, 149,
    290, 303
Wilberforce, Lord 241
Wildavsky, A. 52
Williams, James 52
Windeyer, J. 244
Average clause 253, 343, 345

Babylon/Babylonians 1, 3, 68, 169, 190
Belgium 202
Bhopal 64,65
Bid bonds 205
Big Thompson Canyon 40
Bill of Quantities 245, 246, 249, 267, 285,
    434, 444, 451
Bodily injury 133, 197, 296, 312, 314,
    318, 322, 332, 334, 340, 347,
    356–358, 367, 404, 408, 439
Bonds and guarantees 205
Bonham's case 3
'Bottomry' 3, 4
Breach of contract 14, 60, 144, 165,
    173–175, 241, 252, 271, 291,
    382, 428
  fundamental 102
Breach of statutory duty 12, 145, 239,
    256, 281, 290, 361, 408, 419,
    429, 434, 437, 446, 451
Bridges & Highways Corps 16
British Insurance Association 199
Brokers 105, 127, 207, 208, 210–214,

    220, 305, 354, 360, 363, 372,
    375, 391, 395
  duties 210
Building Research Station 84

Caconde Dam 128
Camden School 83
Canada 55, 59, 69, 106, 125, 150, 152,
    175, 179, 180, 182, 193, 212,
    373, 390
Causation 195, 201, 381
Cedent 214
Certificate of Completion 232
  Cert of Substantial Comp. 232, 234,
    248
Certification 88, 116, 355, 360
Champerty 65
Civil Engineering Contractors' Association
    7
Civil engineering 7, 8, 10, 14–17, 24, 33,
    50, 70, 87, 92, 139, 145, 188,
    189, 220, 222, 280, 318, 340,
    343, 348, 354, 339, 350, 363,
    387
  definition 16
  origins 16, 17
Civil Law 5, 145, 326, 327
Civil war 228, 229, 233, 235, 242, 247,
    272, 281, 290, 291, 301, 315,
    323, 329, 349
Claims 65, 79, 110, 178, 179, 199, 223,
    228, 229, 248, 252, 255, 256,
    258, 259, 262–264, 269, 270,
    291, 295, 296, 280, 283, 284,
    291, 292, 293, 297, 300, 314,
    315, 320–324, 330, 332–336,
    347, 348, 353–358, 361,
    364–377, 381, 384, 385, 391,
    392, 394–396
Clauses of insurance 9, 15–17, 270, 279,
    280, 287, 329, 339
  FIDIC's Green Form 299–305, 339, 340
    clause 6 301, 302–304
    clause 13 300, 303–305
    clause 14 300, 301, 303–305
  FIDIC's New Suite of Major Contracts
    16, 24, 114, 117, 128, 133, 137,
    226, 231, 308, 339
    clause 17 137, 304, 308, 310–314,
    317, 323–328
    clause 18 308, 311–326, 441
    clause 19 188, 189, 223, 308, 312,
    324–328, 443
  differences 310
  redrafting 329
  FIDIC Old Red Book

clause 20 189, 228–230, 232, 445
clause 21 245, 247, 445
clause 22 253, 255, 445
clause 23 257, 447
clause 24 258, 447
clause 25 261, 270, 447
re-drafting 279
re-drafting priorities 279
ICE Form 226
clause 20 234
clause 21 248
clause 22 254, 256
clause 23 259
clause 24 259
clause 25 262
Clauses of standard forms of contract 9
15, 221, 270
clause 2 188, 290, 302, 304, 306, 320
clause 8 188, 230, 300, 324
clause 13 188, 275, 300, 303, 326
clause 40 188
clause 48 231, 232, 249, 269, 282,
283, 290
clause 49 228, 231, 233, 249, 275,
285
clause 50 285
clause 59 380
Code of Ethics 21, 146
Co-Insurance 207, 214, 367
Collapse 17, 23, 29, 30, 55, 67, 80, 81,
83, 86, 87, 91, 94–96, 111, 113,
119, 121. 122, 127, 142, 198,
199, 210, 382
Common Law 2–6, 13, 13, 37, 66, 67,
133, 144, 145, 148, 149, 151,
155, 166, 172, 173, 174, 175,
179, 212, 326, 346, 355
Commonwealth Construction Ltd. 182
Commotion 17, 106, 281, 291, 301, 315,
232, 329,
Company suretyship 206
Competition, price 19–23, 117
Completion, certificate of 445
Composite policies 340–343, 349
Computers 18, 19, 76
Concorde project 61
Conditions of Engagement of the
Professionals in construction
221–224, 257, 326, 339, 455
Consequential loss 119, 285, 308, 317,
332, 349, 382, 383
Construction 28, 29, 31–34, 37–44,
48–52, 130, 131, 133, 137,
182–207, 217–220, 279–281, 284,
287, 288, 293–295, 309, 311,
318, 332, 338, 339, 340–348,

350–360, 363, 366, 367, 384,
385, 389–393
computers and 18
contract 6–12, 16, 28, 56, 70, 108,
109, 130, 133, 137, 138, 146,
153, 155, 164, 185, 207, 279,
288, 309, 326, 327, 350, 358,
360, 386
inherent characteristics 187
cost over-run 38, 61
costs 19, 20, 24, 44, 106, 114, 117,
188, 191, 309, 345
defective 190
defects in 2, 33, 177, 296
hazards in 39, 40
injuries and 40
insurance 1, 130, 184, 186, 187, 192,
215, 217, 219, 326, 349,388, 394,
396
see construction insurance
law and 12, 148, 150, 186, 386, 388
projects 33, 53, 97, 107, 114, 117,
128, 130, 141, 156, 187,
210,218, 219, 347, 351, 364,
386, 387, 396
inherent characteristics 33, 187
inherent risks 36, 42, 48
uniqueness 50, 190
risks in 32, 33, 34, 46, 52, 53, 132,
133, 134, 139, 190, 289
society and 17, 145
see also following entry
Construction insurance 1, 2, 5, 184, 186,
187, 192, 195, 215, 217, 219,
326, 349, 388, 394, 396
interaction with law/insurance 7, 12,
195, 386
policies required 196, 261, 318, 329
alternative methods 379, 387, 388
composite policies 340–343, 349
conditions, compliance with 353, 382
damage to persons and property 195,
198, 253, 288, 255, 256
exclusions 203, 219, 229, 247, 263,
270, 271, 381
gaps and overlaps 16, 204, 220, 225,
227, 318, 380, 387, 388
gross misconduct 430
history 218
in joint names 343, 360
law and 7
loss must occur 382
policies 204, 205
documentation 218
features 263
exclusions from 203, 219, 229, 247,

263, 271, 341, 348, 349, 356,
  359, 381, 383
  required 196, 261, 318
third party 229, 257, 258, 259 260,
  312, 357, 392, 394
understanding of, lacking 24, 25
workmen, accidents to 257
Works 243, 248, 270, 340
wrap-up 394
see also Clauses of insurance; Insurance;
  Liabilities; Risk; Risk Listing
Construction plant, insurance of 284
Consulting Engineers' Advancement Society
  Inc. 377, 378
Contract 14, 16, 44, 138, 149, 150, 164,
  184, 187, 221
  Breach of, fundamental 240, 241
  characteristics 187, 302
  construction, *see* Construction Contract
  contractor to complete 187, 188, 230
  definition 7
  function of 11, 12
  developments in 150, 164, 386
  implied terms 150, 152, 158, 165
  insurance 105, 106, 148, 185, 195, 197,
    219
  law of 150, 164, 174, 249, 318
  developments in 150, 164, 386
  model insurance clauses 279, 287, 329,
    338
  revision 141
  risk sharing and 142
  standard forms 9, 11, 15, 17, 24, 44,
    128, 133, 137, 141, 142, 134,
    135, 188, 221, 222, 226, 242,
    298, 306, 317, 318, 339, 387
  vast sums 167, 189, 197
  see also Clauses of insurance
Contractors 65, 87, 88, 91, 96, 104, 110,
  111, 116, 132, 133, 182, 185,
  204, 279, 281
  duties 164
  failure to insure 229, 262, 263, 286,
    301, 427, 431, 435, 438, 447, 452
  indemnity 224, 240, 256, 260, 269,
    271, 283, 292
  insurance required 339
  owner's interference 146
  risks 260, 281, 282, 283, 284,287, 290,
    292
  see also following entry
Contractors' All Risk Insurance 15, 197,
  243, 287, 297, 340, 341, 343,
  344, 345, 347, 348, 350, 351,
  352, 356, 379, 380, 383, 385,
  387, 390

claims settlement 354
conditions 197
definitions 342–343, 349
endorsements 341, 354
exclusions 197, 341, 348
memoranda 241, 354
operative clause 341–344, 356
perils insured 355
policy wording 341
recital clause 341, 342
schedule 342–346, 356
signature clause 354, 341
specimen documents 400
Contractors' Public Liability 356, 380,
  390
Contribution 5, 183, 184, 186, 254, 280,
  360
Contributory responsibility 52
Coyle Hamilton Phillips Limited 185,
  210
Cross liability clause 229, 237, 259, 285,
  356
Custom 42, 44, 160, 193, 361, 379

Dams 40, 73, 128
Date of accrual 175, 178, 424
Decennial insurance 201–203, 338
Decisions, reasoned 26
Defensive design 19
Dentists 363
Design 1, 6, 12, 6, 17, 18, 19, 20, 48, 43,
  55, 56, 59
  defective 29, 38, 67, 75, 85, 86, 94,
    133, 195, 251, 338, 352, 355,
    384, 392, 393
  defensive 19
  duty of care in 151, 161, 163
  faulty 90, 243, 244, 245, 384
  see also following entry
Design professionals 18, 38, 47, 150, 151,
  204, 372–379
  insurance 358, 359, 362
  owners interference 152
  risks and 204
  supervision and 161, 378
  see also Professional Indemnity
    insurance,
Difference-in-Conditions insurance 295,
  338, 388, 389, 380
Diyala bridge 81
Duty of care 14, 111, 151, 161, 163, 169,
  170, 171, 178, 202

Economic loss 27, 168, 171, 172, 173,
  178, 360
Egypt 1, 2

Electrical and mechanical contracts 294, 295, 280
Emley Moor 67, 242
Employees 32, 40, 42, 99, 106, 107, 119, 120, 139, 144, 147, 160, 178, 197, 239, 240, 242, 260, 261, 276, 281, 291, 340, 344, 349, 356, 357, 359, 362, 365, 371, 404, 409, 411, 414, 425, 428, 429, 431, 440, 443, 419
  indemnity by 365
  liability 408
  see also following entry
Employers' Liability Insurance 197, 340, 356, 357
Endorsement 186, 219, 357, 362, 440, 410
Engineering 19, 33, 50, 56, 63, 79, 87, 88, 92, 96, 101, 103, 104, 105, 111, 117, 127, 131, 137, 145, 159, 164, 180, 188, 189, 198, 200, 220, 222, 224, 227, 230, 232, 243, 244, 245, 249, 253, 254, 257, 265, 280, 295, 296, 280, 298, 299, 318, 339, 340, 343, 348, 354, 363, 369, 374, 380, 384, 387, 415, 425, 457, 460
  nature of 17
  science and 17
Engineers 55, 58, 87, 103–108, 110, 116–118, 122, 145, 149, 152, 159, 161–164, 166, 201, 202, 220, 222, 298, 307, 360, 363, 365, 376, 377, 378, 401, 415, 420, 445
  professional indemnity insurance 257
  role of during construction 161
Environmental impairment 371
Equity 12, 15, 42
Erection All Risks Insurance 15, 76, 197, 287, 295, 340, 387
Euclides da Cunha 128
Evidence, Law of 6
Excess 9, 200, 219, 253, 237, 271, 341, 343, 344, 347, 348, 355, 361, 362, 366, 357, 370, 374, 376, 377, 388, 389, 391, 393, 411, 412, 413, 419, 420, 422, 423, 445, 446, 447, 452, 467
Exclusion of liability clause 102
Expressed terms 150

Federation of Civil Engineering Contractors 8
FIDIC 9–11, 16, 18–20, 24, 63, 108, 109, 114, 128, 133, 141, 149, 151, 159, 184, 189, 202, 221, 222, 224, 226, 227, 230, 231, 238, 249, 279, 280, 287, 288, 298, 299, 303, 304, 306, 308, 310, 313, 318, 327, 328, 329, 330, 331, 332, 339, 342, 345, 350, 352
  Policy Statement 108, 161
  Standing Committee on Professional Liability 288, 352, 376
  Supervision 108
  See also following entry and Clauses of contract
FIDIC Forms of Contract 9, 133,137, 339
  The New Red Book 16, 58, 189, 222, 226, 298, 299, 304, 305, 306, 308, 313, 314, 318, 319, 323, 329, 338, 339, 379, 380, 387, 425, 428, 439
  The New Suite of Contract 1, 16, 24, 114, 128, 133, 137, 318, 326, 327, 328
  The Old Red Book 114, 141, 188, 189, 299, 304, 309, 310, 326, 329, 331, 339
  The Old Yellow Book 226, 279, 280, 329, 339
  The White Book 329, 339
  Orange Book 141, 159
  Silver Book 16, 141, 226
  see also Clauses of contract
Fiemme valley 39
Fire 5, 17, 31, 38, 64, 84, 90, 92, 93, 95, 100, 108, 116, 120, 122, 125, 129, 130, 133, 182, 183, 191, 203, 210–213, 220, 230, 235, 240, 297, 355, 360
Fire Extended Cover 17, 340
Fire insurance 5, 7, 211, 213
Fitness for purpose 127, 128, 153–159, 180, 307, 310, 324
Floods 123, 124, 125
Force Majeure 254, 300, 302, 323, 325, 326, 327, 328, 338, 431, 432, 444, 445
France 5, 17, 120, 390
  building defects analysis 63
  Civil Code 5
  Torts, law of 166

General Accident Insurance Company Ltd. 211
General Conditions of Contract (Ireland) 16, 107, 165, 239, 240, 278, 341, 344, 356
Graminha 128
Greeks 4
Gross misconduct 283, 284, 292, 293, 429, 430

Guarantees and bonds 205
Guerrero, Mexico 77

Hammurabi 1–3, 6, 21, 169
  Code 1, 169
    Commercial Risk 2
    Construction 1, 6
Harbour Bay Condominium 53
Hazard 44, 61, 74, 75, 77, 79, 81, 92,
      164, 303, 347, 366, 395, 399, 424
  analysis 34, 36, 123, 396
  awareness 27
  classification 27
  construction 39, 190
  definition 27, 28, 30, 124, 303
  natural 27, 33, 40, 89 123, 124
  risk and 28, 38, 396
Health & Safety 6, 31, 32, 40, 87, 146,
      177, 288
  EU Directives 6, 34
  Executive in the UK 99, 100
  Regulations 33, 34, 100
  Statistics 31, 32, 40
Holland see Netherlands
Hong Kong 130, 131
Hostilities 242, 228, 229, 233, 235, 242,
      247, 281, 290, 291, 329, 349,
      400, 411, 419, 425, 426, 428,
      431, 433, 436, 440, 443, 445,
      449, 450, 454
Hyatt-Regency Hotel 39, 112

ICE Conditions of Contract 8, 11, 222,
      224, 225, 231, 232, 242, 250,
      252, 278, 345, 380, 433, 436
  Insurance and 232, 242, 252, 380, 433
  see also Clauses of insurance
IGRA document for Design and Supervision
      of Construction of Works 221
IGRA document for Pre-Investment Studies
      222, 339
Imperial Oil Limited 182
Implied terms 150, 152, 158, 165
Indemnity 25, 81, 105, 130, 137, 142,
      158, 169, 180, 208, 209, 210,
      217, 223, 253, 254, 255, 280,
      283, 303, 318, 331, 335, 338,
      358, 359, 360, 361, 379, 383,
      386, 389, 390, 391, 393, 398,
      403, 404, 408, 411, 413, 414,
      424, 427, 434, 437, 440, 441,
      446, 451, 454, 457
  clauses 224, 225, 239, 240, 235, 254,
      257, 260, 263, 279, 280, 283,
      290, 293, 318
  construction insurance 2, 5

contract of 355
  definition 145
  law and 180
  limit of 220, 224, 263, 340, 341, 343,
      344, 346, 347, 356, 361, 362,
      383, 385, 388, 389, 391, 393,
      399, 404, 405, 414
  through
    insurance contract 184
    law 181
Instincts 26
Institution of Civil Engineers 8, 16, 19, 23,
      24, 145, 222
Institution of Engineers of Ireland 9, 16
Insurance
  agent 210
  brokers 207, 305, 295, 291
    duties 208
  clauses see Clauses of insurance
  contract 105, 106, 148, 185, 195, 197,
      264, 344, 346, 353, 354, 381,
      382
  causation 195, 201, 381
  consequential loss 285, 382, 383
  decennial see Decennial insurance
  definition 4
  departmentation 220
  disclosure of relevant information 382
  documentation 218
  evidence of 261
  forms of 209, 214, 390
  fortuity 381
  general principles 186, 264, 275, 355,
      381
  history 218
  indemnity contract, as 184
  joint see joint insurance 237, 250, 259,
      270, 391
  law and 7, 24, 180
  legal rules 184–186
  market 25, 199, 200, 294, 317, 318,
      371, 375, 388, 391, 393
  non-negligence see Non-negligence
      insurance
  no profit 185
  period 217, 382
  premiums 262, 435, 438, 452
  project 220, 294, 338, 352, 388–396
    advantages 392
    disadvantages 393
  religion and 4
  requirements 199, 209, 229, 237, 261,
      286, 338, 387
  role of 25, 158
  see also Construction insurance;
      Indemnity; Joint insurance; CAR;

Liability; Professional; Public;
  Responsibilities; and Risks
Insurrection 242, 272, 228, 229, 233, 235,
  247, 271, 290, 291, 301, 315,
  323, 329, 249
International Federation of Consulting
  Engineers, 9, 20, 43, 298, 376
Interruption insurance 389
Inundation 33, 355
Iraq 81
Ireland 19, 156, 170, 174, 177, 188, 206,
  210, 226, 354, 390
Irpina 77
Islamic Law 4, 5, 28, 166, 213
Ismaning Institute 74
Itaipu Hydroelectric plant 42, 390
Italy 18, 63, 77, 121, 124, 190

James Bay Project 390
Japan 77, 84, 124, 125
Joint Contracts Committee 11
  Tribunal 17
Joint insurance 210, 237, 243, 250, 259,
  270, 391, 424
Justinian, Emperor 5

Kansas City Hyatt-Regency Hotel
  walkways 111
Kleinwort-Benson 83
Kuwait 95

Law
  construction and 12
  history 5
  insurance and 7
  Islamic see Islamic law
  see also Civil Law; Common Law;
    Equity
Law Reform Committee 176, 459
Lawyers 18, 23, 109, 113, 386
Leicester University, reading room 83
Liability 24, 48, 56, 66, 130, 176, 178,
  184, 186, 195, 199, 200–202,
  213, 220–225, 232, 236–244,
  252, 254, 257, 261, 280, 285,
  292, 301, 344, 346, 349, 353,
  355, 360–364, 372, 376, 380,
  383, 384, 387, 389, 420, 422
  absolute 198
  allocation 281
  cross 272, 237, 258, 259, 269, 273,
    284, 292, 294, 300, 340, 359,
    369, 379, 387, 390
  definition 144
  employers' 357
  joint and several 60, 175

law's development and 138
  contract law 150, 164
  future 179, 180
  tort 166
  legal 181, 264, 340, 359, 362, 372,
    419, 421
  levels of 149
  limitation 223, 257, 308, 312, 317,
    331, 430, 441
  period of limitation 6, 176, 201
  professional 56, 180, 202, 224, 289,
    352, 365, 375
  responsibility and 138, 195, 224, 230,
    279, 329
  retrospective 364, 365
  simple 24
  site staff 160
  skill & care 151, 153–159, 175, 188,
    212, 244, 326
  strict 24, 180, 188, 198, 356
  torts and 66, 213
  vicarious 359
    definition 359
Liability insurance 70, 179, 180, 213, 296,
  356, 357
Life insurance/assurance 220, 338
Lloyd's Underwriters 207, 213
Logic 139, 143
Longarone 122
Loomis, Carol J. 180
Loreburn, Lord 192
Loss of Life 28, 29, 30, 265, 266, 279,
  281
Loss of Profits insurance 389
Loss prevention 288, 355, 358, 377, 378,
  393

Machinery insurance 5, 422
Maintenance bonds 206
Manufacturers' Risk insurance 251
Marine Cum Erection Polity 295
Marine Insurance 4, 5, 7, 52, 77, 196
Military or usurped power, 228, 229, 233,
  235, 272, 290, 301, 315, 323,
  400, 411, 419, 425, 426, 428,
  431, 433, 436, 441, 444, 447,
  451, 452, 547
Mirani, Australia 243
Mitigation 12, 30, 34, 36, 39, 51, 70, 130,
  141, 142, 203, 288, 313, 353,
  386, 388, 430
Moorcock, The 150
Mohammed, Prophet 5, 213
Munich Reinsurance Company 15, 186,
  220, 372
Mutual insurance companies 213

Napoleon Bonaparte 5
National Insurance Schemes 396
National Program Administrator Inc. 373
National Society of Professional Engineers
    56, 164
Natural hazards 27, 33, 123, 134, 125
Natural Justice, Rules of 15
Negligence 6, 7, 14, 24, 109, 149, 153,
    164, 168, 179, 349, 355, 356,
    397
  contributory 14
  definition 66
  Law of 6, 8
    contract 7
    extension of 148, 167
    tort 6, 14
Neighbour, definition 169
Netherlands 390
New Zealand 2, 36, 37, 175, 377, 378
New Zealand Architects' Co-operative
    Society Ltd. 378
Non-negligence insurance 198, 200
North Sea Oil Fields 61
Nuclear reaction 61

Operation 14, 15, 179, 205, 248
Operative clause 219, 342–344, 356,
    368
Outstanding work 231, 236, 238, 249

Parana River 42
Patent, infringement 316
Performance bonds 205
Period of Limitation 6
Personal Injury 133, 137, 138, 270, 422,
    428, 429, 430
Physicians 362
Pioneer River 243
Posteriority Cover 370
Pre-Investment Studies 222
Principal clause 343, 424
Probability 28, 30, 31, 48, 52, 53, 61, 78,
    347, 365, 369, 422
  assessment of risk of occurrence 28, 30,
    31, 34, 36, 39, 40, 43, 48, 50,
    53,.77, 81, 123, 130, 253, 303,
    381
  theories of 34, 50, 52, 53
Products Liability Insurance 296
Professional Indemnity insurance 197, 305,
    338, 356, 358–378
  breach of warranty 372
  claims 361, 364, 365, 367–377
    causes of 358
    settlement 373
  conditions 360, 361, 368, 369

cover, cancellation of 372
dishonesty 370
documents, loss of 370
environmental impairment 371
exceptions/exclusions 367
excess 376, 377, 378
fees, legal costs in claiming 371
group schemes 374, 376, 378
indemnity, limit of 361, 362, 364, 365,
    366, 370, 378
infidelity 370
insured 370–378
insuring clause 361, 368, 369
liability exposure 358
libel and slander 370
memoranda 361, 369
partners 369, 414
  previous business 369
  retired 370
policy wording 375
premium 376, 377
QC clause 367, 369
schedule 361, 362
signature clause 361, 362
specimen 368, 369
subrogation 371
Professional services, price competition and,
    19
Project Insurance 220, 225, 294, 338, 352,
    338–394, 379, 389, 424
Property, definition,
  Insurance 7, 183, 196, 340, 344
Proprietary insurance companies 213
Public Liability Insurance 197, 200, 224,
    293, 296, 339, 340, 356, 367,
    375, 376, 389, 447

Quality 18–23, 26, 85, 86, 88, 95, 110,
    114, 116, 117, 126, 130, 147,
    153, 156, 158, 165, 184, 186,
    190, 203, 208, 395

Radioactivity 203, 235, 409, 419, 428,
    431, 435, 436, 450
Rebellion 235, 247, 291, 315, 232, 349,
    419, 454
Reinsurance 15, 39, 186, 213–217, 220,
    372
Reinsurance Treaty 214
Responsibility
  allocation 107, 109
  contractor, of 146
  definition 117, 144
  design, duty of care 151, 161
  fitness for purpose 180
  for care of the works 228–230, 233,

282, 290, 300, 314, 330, 349,
425, 433,445, 450
liability and 139, 148
nature 159
others in the project, towards 146
owner 146, 147, 161, 141,142, 145
professional team 146
site staff selection 160
society towards 145, 146
supervision 161
towards
employees and staff 147
own self 147
parties 146
society 145, 146
Retention money bonds 206
Retrocession 214, 215, 217
Retrospective liability 346, 365
RIAI Form of Contract 8, 17, 188
RIBA Standard Form of Contract, 8, 17,
133, 188, 198, 199, 318
Rio Pardo 128
Riot 17, 242, 237, 228, 235, 234, 281,
290, 291, 301, 315, 323, 329,
384, 400, 425, 428, 431, 433,
455, 450
Risk
allocation 37, 38, 48, 47, 130–132
rules of 37, 38, 131–132
criteria 45, 131, 288
of risks of injury or loss, 133, 134,
137, 138, 280
of risks of economic loss or time loss,
133, 135, 136–138
analysis 37
assessment 34, 36, 88
constraints 49
methods 51
calculation 29, 31
classification 41, 45
by chronology 43
feasibility stage 43, 53
design stage 43, 53, 61
construction stage, 53, 70
post-construction stage 53, 114
contractor's 133–138, 159, 237, 292
cultivation 33
defective materials, insurance 70, 205,
251, 252, 284, 350, 352, 402
definition 28
design 237, 242, 329, 338, 352, 385,
393
employers' 133–138, 159, 189, 228,
230, 233, 237, 242, 280, 287, 292
evaluation 36
excepted 242

foreseeable 29, 195
government covered 382
hazard and 30, 52
insurability 25, 194, 405
limitation 194
international scale 195
in major projects 48
management 36
definition 36
steps in 36
normal 137, 281, 288
political 78, 107, 126, 127, 128
origin of word 28
quantification 36
sharing 25, 38, 132, 141, 150, 376
special 137, 225, 236–238, 242, 265,
269, 272–277, 280, 290, 304,
327, 348, 389
definitions 291
uninsurable 44, 133, 195, 203, 205,
293, 381
Risks listing
acts of God 70, 79, 121
acts of Man 78, 80, 97, 98
arson 107, 114, 119
burglary 102, 107
checking, inadequate 68
choice of site 56
collapse 80, 83, 845, 86, 87, 88, 91,
94–97, 111, 113, 119, 121, 122,
128
communication, and lack of 68, 102,
117
conflict 99, 110
contractor, choice of 70
corrosion 95
cyclone 76, 77, 89
dangerous substances 64, 84
defective design 85, 86, 94
defective temporary works and their
design 94, 96
defective workmanship and material 85,
86
design, inappropriate 64
dispute resolution 113
duration of construction, extended 81
earthquake 77, 122–126
equipment, inadequate 69
explosion and fire 92
extended duration of construction, 81
fatigue 119
finance, adequacy 60
fire 92, 119
fitness for purpose 127, 128
flood 72, 124, 125
foreseeable problems 69

fraud and infidelity 101
geology 79
ground movement 87, 91
haste 68
human error 97, 125
hurricane, tornado and whirlwind 74, 124
illegal activities 113
impact 101, 106, 116, 126
incompetence 107
inefficiency and delays 108
innovation 66, 67, 83
inspection 60, 109, 110, 116, 117
insurers' conditions, failure to comply with 105
knowledge, lack of 68, 86
local residents', acceptability by 81
maintenance 70, 90, 114, 122, 127, 129
malicious acts 107
man-made 80, 81, 114, 126
mechanical and electrical breakdown 88
    inadequate performance 69
natural hazards 123, 124, 125
negligence and lack of care 66, 100
nominated subcontractor, choice of 70
nuclear reaction 61
oscillation 93
owners' brief to professional team 55
owners choice of professional team 53
political 78, 107, 126, 127, 128
problems, failure to account for 69
programming 102
project, operational faults 90, 128
rainfall, excessive 70
removal of support 84
resistance to fire 114
riot and civil commotion 106
safety 114, 116, 118
safety precautions, lack of 70
serviceability 118
site choice 56
site management, inadequate 90
site supervision, inadequate 108
soil investigation, adequacy 57
stability of government 78
state of the art, codes and technical knowledge 66
storms 123, 124, 125
strike 107
subsidence 56, 75, 91
support, removal of 84
surveys and inspections, adequacy 60
technical complexity 83
temperature extremes 75
topography 79, 121
theft and burglary 102

tornado 124
underground obstructions, 80
unforeseen physical obstruction 58
untested and unproven techniques 69
variation from contract documents 111
volcano 124
vibration 93, 118
war 61
wear & tear 114, 128
wind and storms 73, 124
Risk Transfer Date 429, 430
Roman 4, 166
Royal Institute of the Architects of Ireland 8, 19, 174

Sabotage 90, 107
Sao Paulo 84
Saudi Arabia 202
Scientist, function 63
Serviceability 114, 118, 372
Settlement 56, 59, 118, 119, 153, 172, 185, 200, 210, 248, 263, 264, 276, 316, 353, 354, 355, 364, 369, 373, 375, 382, 384, 391, 392, 403, 404, 411, 424, 440
Simcoe and Erie General Insurance Company 373
Singapore 119
Site staff, selection of 160
Smeatonian Society 18
Société Minoteries Tunisiennes 118
Society of Civil Engineers, American 110
South Wales Switchgear Co. Ltd. 181, 329, 459
Standard Forms of Contract 8, 9, 11, 15, 17, 24, 79, 128, 137, 141, 142, 188, 221, 226, 235, 242, 298, 306, 317, 318, 339, 387
Building, JCT 17, 133, 188
clauses of insurance, see Clauses of insurance
FIDIC 9, 16, 108, 133, 137, 141, 222, 298, 339
Green Book 226, 298, 299–304, 340
New Red Book 16, 57, 226, 298, 305, 312, 340
New Yellow Book 16, 226, 298, 305, 340
Old Red Book 9, 16, 58, 114, 133, 188, 224, 279, 303
Old Yellow Book 9, 226, 279, 303
Silver Book 16, 226, 298, 305, 308, 340
White Book 221–224, 326, 339, 455
ICE 8, 9, 11, 12, 16, 44, 57, 132, 133,

136, 159, 163, 188, 221, 222, 224, 226, 227, 230, 231, 232, 236, 237, 241, 242
  Civil engineering 8
  Design/Build and Design/Construct 133, 158, 159, 176, 190, 219, 394, 397, 436
  Risks under 137
Standards 18, 66, 78, 169, 190, 198, 243, 244, 349
  B.S. 4778: 1991 27, 28, 36, 303
  As/NZS 3951: 1995 28, 29
Statistics 25
  Fatal injuries 31, 32, 40–42
  Health & Safety 31, 32
  Major injuries 32, 40–42
  Other injuries 32, 40, 41
Statutory duty, breach 12, 239, 256, 281, 290, 361, 408, 419, 422, 429, 434, 437, 446, 451
Strike 107
Subjection 144, 303
Subrogation 250, 371, 391–393, 420, 424
Subsidence 56, 75, 91, 198, 199, 355, 405
  Reasons for 75
Sum insured 199, 203, 207, 217, 219, 263, 271, 293, 354, 383, 384, 391, 402, 403, 405, 415, 423
Sun Alliance Insurance Group 211
Supervision 108, 110, 126, 152, 161, 164, 222, 339, 351
Surgeon 157, 362
Surveyors 363, 416
Sydney Opera House 61

Tangshan, China 77
Tendering 8, 252
Theft 102, 355, 401, 431
Third Party Liability 284, 292, 312, 383, 387, 404, 430, 456
Torts 15, 166
  accrual date 178

definition 15
economic loss 168
law of 14, 166, 168, 176, 179, 356
  developments in 166, 180
  legal principles 166, 180
Toxic waste disposal 362
Transit insurance 294–296, 338, 344, 430

U.M.B. Chrysler 239
Unavoidable result of the execution of the Works 228, 255–257, 281, 288, 302, 322, 332, 333, 426, 434, 437, 443, 444, 451
Unavoidable result of compliance with the Contract 311, 316, 331, 440
Under-insurance 253, 343
Union Carbide 64
United States 5, 26, 31, 74, 114, 120, 164, 153, 351, 359, 374, 377, 390, 460
Utmost good faith 185, 208, 218, 264, 362, 382

Vaiont Reservoir 121, 424
Venezuela Civil Code 202
Vicarious liability 144, 357
Victor O. Schinnerer & Co. Inc. 374, *see also* footnotes

War 61, 142, 195, 200, 203, 212, 269, 272, 281, 290, 295, 301, 315, 323, 329, 330, 338, 341, 349, 400, 411
Westerberg, Normal 19
Wichita Falls 74
Workmanship 2, 38, 70, 78, 85–88, 90, 111, 114, 133, 142, 147, 155, 158, 178, 195, 205, 248, 251, 252, 264, 283–285, 293, 295, 321, 335, 338, 350–352, 355, 338, 350–352, 355, 382, 402, 429, 434, 437, 442, 446, 451
  definition of faulty workmanship 245
  comparison with faulty design 245
Wrap-up Insurance 388, 391–396

Made in the USA
Middletown, DE
25 September 2021

49087711R00276